AF577327

ADVANCED ENGINEERING CHEMISTRY

Second Edition

ADVANCED ENGINEERING CHEMISTRY

Second Edition

M. SENAPATI

INFINITY SCIENCE PRESS LLC
Hingham, Massachusetts
New Delhi, India

Publisher: David F. Pallai

INFINITY SCIENCE PRESS LLC
11 Leavitt Street
Hingham, MA 02043
Tel. 877-266-5796 (toll free)
Fax 781-740-1677
info@infinitysciencepress.com
www.infinitysciencepress.com

This book is printed on acid-free paper.

M. Senapati *Advanced Engineering Chemistry, Second Edition.*
ISBN: 978-0-9778582-9-3 0-9778582-4-3

Library of Congress Cataloguing-in-Publication Information
Senapati, M. (Manas)
Advanced engineering chemistry / M. SENAPATI. – 2nd ed.
p. cm.
ISBN 978-0-9778582-9-3 (hardcover : alk. paper)
1. Chemical engineering. I. Title.
TP155.S36 2007
660–dc22

2007002210

7 8 9 5 4 3 2 1

Our titles are available for adoption, license, or bulk purchase by institutions, corporations, etc. For additional information, please contact the Customer Service Dept. at 877-266-5796 (toll free).

Faith is the surest guide
in the darkest days

— The Mother

Contents

PART 1

PART 1

Chapter 1

STRUCTURE AND BONDING

1.1 INTRODUCTION

John Dalton's concept of indivisibility of the atom is discarded by the discovery of various fundamental particles.

An *atom* consists of three fundamental particles:

Electron: A negatively charged particle with almost negligible mass, which was discovered by J. J. Thomson from the studies carried out on cathode rays in the discharge tube. The name electron was proposed by Stoney.

$$\text{Charge } (q) = -\,1.602 \times 10^{-19} \text{ coulomb}$$

$$\text{Mass } (m) = 9.1 \times 10^{-28} \text{ g}$$

Proton: A positively charged particle with mass nearly 1840 times heavier than the mass of an electron. The proton was discovered in the anode ray experiment. The magnitude of charge is the same as that of an electron.

$$\text{Charge } (q) = +\,1.602 \times 10^{-19} \text{ coulomb}$$

$$\text{Mass } (m) = 1.66 \times 10^{-24} \text{ g}$$

Neutron: A neutral particle with no charge, which was discovered by James Chadwick (1932). Its mass is nearly the same as that of a proton.

$$\text{Charge } (q) = 0$$

$$\text{Mass } (m) = 1.66 \times 10^{-24} \text{ g}$$

The atom is the smallest unit of an element that retains the characteristic properties of the element.

Thomson's Model of the Atom

After the discovery of protons and electrons, Thomson suggested his model of the atom. According to Thomson, an atom consists of a sphere of positive electricity in which electrons are embedded like the plums in a pudding. This model could not explain the experimental facts and was discarded.

1.2 RUTHERFORD MODEL OF THE ATOM

Discovery of the Nucleus

Rutherford (1911) bombarded a thin sheet of gold (10^{-4} mm) with high-speed α-particles. α-particles (He^{2+}) are obtained from a radioactive substance. A thin lead plate with a hole serves to form a beam of α-particles. A circular screen coated with zinc sulphide was placed on the opposite side of the foil.

After carrying out a series of experiments, Rutherford interpreted his observations as follows:

- Most α-particles passed through the foil without any deflection.
- A few α-particles were deflected through large angles.
- A very few particles (one in about 10,000) retraced their own path.

Rutherford (1913) concluded the following:

- The atom is spherical and mostly hollow.
- Most of the space in an atom is empty.
- Each atom consists of a small, heavy positively charged part at its center known as the *nucleus*.
- Electrons are distributed in the extranuclear part of the atom.
- The extranuclear electrons are not stationary. They are moving around the nucleus at a high speed in different paths called *orbits*.

According to Rutherford's planetary model, the atom is just like a solar system where electrons are moving in the extranuclear part around the nucleus like planets around the sun. The electrons are thus also known as planetary electrons. The centrifugal force from the rotation of the electrons balances the force of attraction.

Drawbacks of the Rutherford Model

This model does not obey the Maxwell law of electrodynamics. An electron while moving around the nucleus should continuously lose energy and would ultimately fall into the nucleus (Fig. 1.1). We know electrons never fall into the nucleus.

The model also could not explain the emission of various spectral lines during the emission of the hydrogen spectrum.

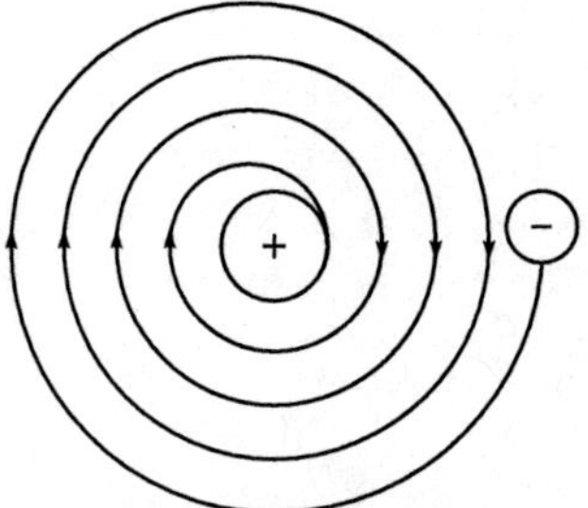

Fig. 1.1 Gradual decrease in the radius of orbit.

1.3 FAILURE OF CLASSICAL MECHANICS

Classical mechanics, the laws of motion, was introduced by Isaac Newton. According to classical mechanics, as the energy is radiated by the electron, the orbital radius must also change, and ultimately the electron should fall into the nucleus. The energy should therefore be radiated in a continuum, and the observed line spectra indicating discrete energy changes is not expected. The classical theory therefore cannot explain the formation of line spectra. Classical mechanics failed when applied to very small particles and the transfer of very small quantities of energy. Classical mechanics is essentially based on deterministic laws, which means that the position and velocity of a particle are known at any time. Thus, a particle in the classical sense has always been associated with the dynamic property of the exact location, exact velocity, and exact path of movement. A series of experimental observations relating to energy distribution in black-body radiations, photoelectric effect, linespectra of atoms, and so on have shown that the laws of classical mechanics fail to account for the observed behaviour of very small particles.

1.4 BLACK-BODY RADIATION

A black-body absorbs completely all radiations falling upon it and emits radiations of all wavelengths. A cavity with highly insulating walls and a small orifice that is heated to a constant high temperature would serve as a black-body. Radiations of all wavelengths would be emitted from the orifice, and they may be treated as black-body radiations. For experimental purposes, a black-body can be very well approximated, as noted previously, by a small hole in the wall of an enclosure. Nearly all the radiation entering the hole will be absorbed by successive reflections inside the enclosure, and thus the hole almost exactly fulfil's the definition of a black-body. The radiation emerging from the hole will also be very nearly equal to that of a black-body, dependent only on the temperature of the enclosure and independent of the nature of the interior matter. Such black-body radiation was investigated in a complete series of experiments by O. Lummer and E. Pringsheim from 1897 to 1899. The typical curves from a black-body at different temperatures are shown in Fig. 1.2. The ordinate is the intensity E_λ, which is defined so that $E_\lambda \, d\lambda$ gives the rate of emission of energy per unit area of surface, in the wavelength range λ to $\lambda + d\lambda$ (Fig. 1.3).

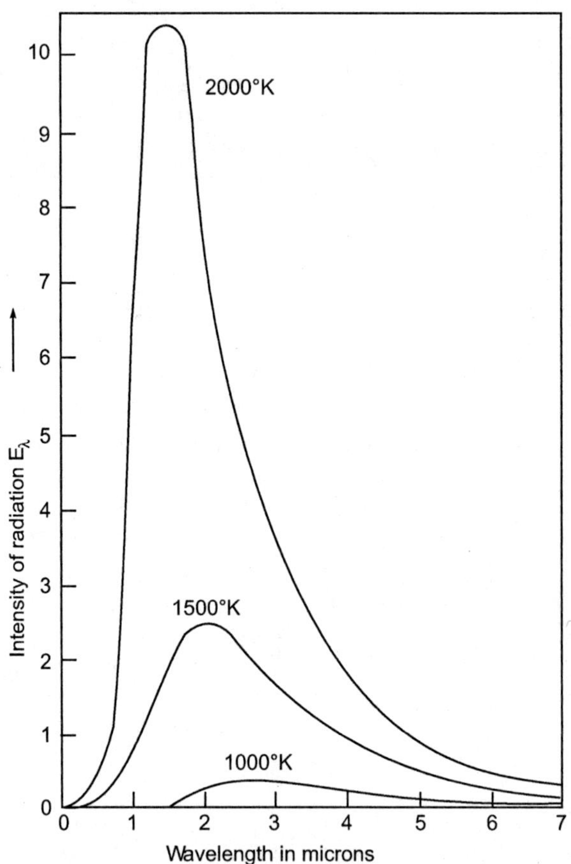

Fig. 1.2 Emission of radiation from a black-body at different temperatures. The area under the curve between two different wavelengths gives the energy in calories emitted by a cm^2 surface of a black-body in the specified range of wavelengths.

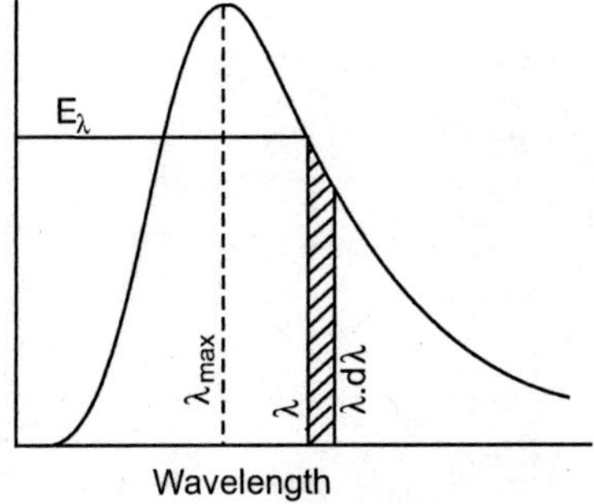

Fig. 1.3 Definition of the intensity of radiation E_λ. The shaded area (E_λ $d\lambda$) gives the rate of emission of energy in the wavelength range λ to λ + $d\lambda$ from the unit area of surface.

The total intensity E, that is the total energy radiated per unit area of surface per unit time is

$$E = \int_0^\infty E_\lambda \, d\lambda$$

The British scientists Rayleigh and Jeans developed a unified law of radiation that is accepted as the correct result of the laws of classical mechanics. An expression for the Rayleigh-Jeans law is given by

$$E_\lambda = 2\pi ckT / \lambda^4$$

where, E_λ = Intensity of radiation

c = Speed of light

T = Absolute temperature

k = Boltzmann's constant

λ = Wavelength

The Rayleigh-Jeans law is successful at long wavelengths but fails at short wavelengths. The equation predicts that oscillators of very short wavelength are strongly excited even at room temperature. This absurd result, which implies that a large amount of energy is radiated in the high frequency region of the electromagnetic spectrum is called the *ultraviolet catastrophe.* According to classical mechanics, objects should glow in the dark; there should be no darkness.

1.5 THE PHOTOELECTRIC EFFECT

In an evacuated vessel, if suitable light is allowed to fall on a clean metal surface, electrons are emitted from the surface. This phenomenon is called the *photoelectric effect.* Usually a radiation in the ultraviolet region and also in some cases in the visible region produces such an effect. The experimental characteristics of the photoelectric effect are as follows:

1. No electrons are ejected, regardless of the intensity of the radiation, unless its frequency exceeds a threshold value characteristic of the metal.
2. The kinetic energy of the ejected electrons increases linearly with the frequency of the incident radiation but is independent of the intensity of the radiation.
3. Even at low intensities, electrons are ejected immediately if the frequency is above the threshold.

In 1905, Einstein explained the photoelectric effect on the basis of Planck's concept of quantum energy. Einstein suggested that the emission or absorption of radiant energy takes place in quanta. Each *quantum*, which is like a particle travelling with the velocity of light, gives up its entire energy to the electron on the surface of a metal. A portion of this energy called the work function (W) is utilized in liberating the electron from other forces of attraction. The balance of the energy gives the electron its kinetic energy $\left(\frac{1}{2}mv^2\right)$. Hence we have

$$h\upsilon = \mathrm{W} + \frac{1}{2}mv^2$$

where, υ is the frequency of the incident radiation. This is the Einstein equation for photoelectric emission.

Laws of the Photoelectric Effect

- The rate of emission of photoelectrons from a metal surface is directly proportional to the intensity of incident light.
- Maximum kinetic energy of photoelectrons is directly proportional to the frequency of incident light.
- Maximum kinetic energy of photoelectrons is independent of the intensity of incident radiation.
- There is no time lag between the incidence of light and the emission of photoelectrons.
- For emission of photoelectrons, the frequency of incident light must be equal to or greater than the threshold frequency.

1.6 ATOMIC AND MOLECULAR SPECTRA

The most compelling evidence for the quantization of energy comes from the observation of the frequencies of radiation emitted and absorbed by the molecules and atoms.

The prominent feature of any molecular and atomic spectra is the emission or absorption of radiation at a series of discrete frequencies. This can be understood if the energy of the molecules or atoms is also confined to discrete values because their energy can be emitted or absorbed only in discrete amounts.

For instance, if the energy of an atom decreases by ΔE, the energy is carried away as radiation of frequency $\upsilon = \frac{\Delta E}{h}$, and a line appears in the spectrum.

1.7 ELECTROMAGNETIC RADIATIONS

Clark Maxwell (1864) observed that an alternating current of high frequency is capable of radiating continuous energy in the form of waves. He termed these waves as *electromagnetic waves* or *electromagnetic radiations*. They are so named because they have both electric and magnetic properties. The radiations—visible, ultraviolet, infrared, X-rays, radiowaves,—are called electromagnetic radiations. They travel through space in wave motion. The electromagnetic radiations have the following important characteristics.

- The electromagnetic waves consist of electric and magnetic fields oscillating perpendicular to each other, and both are perpendicular to the direction of the radiation propagation.
- These waves do not require any medium for propagation.
- All electromagnetic waves travel with the same speed (*i.e.*, 3.0×10^8 m sec^{-1} or speed of light).

All waves have the five following characteristics by which they can be recognized:

Wavelength (λ): The distance between two consecutive crests or troughs of the wave as shown in Fig. 1.4. It is denoted by the Greek letter lambda (λ).

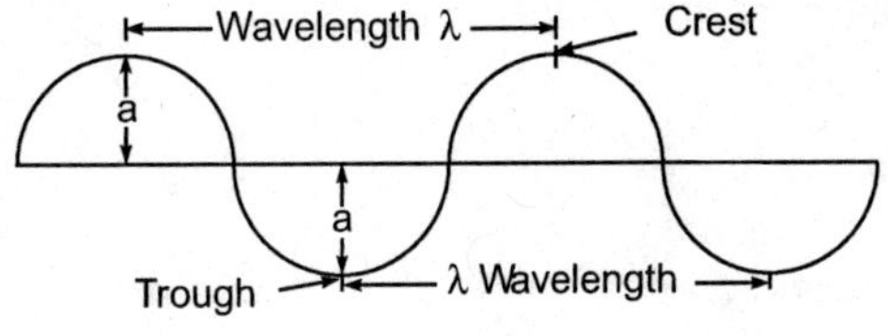

Fig. 1.4

Unit: Angstrom (Å)

$$1\ \text{Å} = 10^{-10}\ \text{meter}$$

Frequency (υ): The number of times a wave passes through a fixed point per unit time is called frequency. It is denoted by the Greek letter nu (υ).

Unit: Cycles per second or Hertz (Hz)

Wave number ($\bar{\upsilon}$)**:** The number of wavelengths per centimeter; the reciprocal of wavelength denoted by $\bar{\upsilon}$ (nu bar).

$$\bar{\upsilon} = \frac{1}{\lambda}$$

Unit: cm^{-1}, $meter^{-1}$

Amplitude (a): The depth of trough and height of crest. This is the measure of intensity of radiation and is denoted by the letter '*a*'.

Velocity (c): The distance travelled by a wave in one second. It is denoted by the letter '*c*'.

$$c = \upsilon \times \lambda$$
$$c = 3 \times 10^8 \text{ m/s}$$

The Electromagnetic Spectrum

The electromagnetic spectrum is defined as the arrangement of various types of electromagnetic radiations in terms of increasing (or decreasing) wavelengths (or frequency). The complete range of electromagnetic waves is called the *electromagnetic spectrum.* The spectrum obtained by white light is called the *continuous spectrum.* The visible light is the form of electromagnetic radiation that lies in the wavelength range of 3800 Å–7600 Å. The region of 3800 Å the wavelength corresponds to the violet color, and the region of 7600 Å wavelength corresponds to the red color. If the wavelength is less than 3800 Å, the radiation is called ultraviolet light, and if the wavelength is greater than 3800 Å, the radiation is called infrared light. The wavelengths of various waves increase in the following order:

Cosmic rays < γ-rays < X-rays < UV rays < Visible rays < IR rays < Microwaves < Radiowaves

Planck's Quantum Theory

Electromagnetic wave theory could not explain the phenomena of photoelectric effect and black-body radiations. Max Planck (1900) put forward a theory known as the quantum theory of radiation to explain these phenomena. The main features of Planck's quantum theory are the following:

- The radiant energy is not emitted or absorbed continuously but discontinuously in the form of small packets of energy called *quantum*. The quantum energy of light is termed as *photon.*
- The energy associated with each quantum is directly proportional to the frequency of the radiation:

$$E \propto \upsilon$$

or $$E = h\upsilon, \text{ where } h \text{ is called Planck's constant}$$

$$h = 6.62 \times 10^{-34} \text{ Joules sec.}$$

$$= 6.62 \times 10^{-24} \text{ erg sec.}$$

- The total amount of energy emitted or absorbed by a body will be some whole number multiple of quantum by an integer 'n'

$$E = nh\upsilon$$

where, $$n = 1, 2, 3, ..., \text{etc.}$$

Again we know that

$$\upsilon = \frac{c}{\lambda}$$

$$\therefore \qquad E = \frac{hc}{\lambda}$$

This equation represents that a wave of higher frequency or lower wavelength will be more energetic and vice versa, for example violet light of high frequency has more energy while the red light of low frequency is associated with less energy.

1.8 THE BOHR MODEL OF THE ATOM

Neil Bohr (1913) proposed his quantum mechanical structure of the atom with the help of Planck's quantum theory of radiation to overcome the drawbacks of Rutherford's model of the atom. The important postulates are as follows:

- An atom consists of a heavy, positively charged nucleus. The electrons revolve around the nucleus in definite circular orbits called stationary states.
- Each of the fixed circular orbits is associated with a certain amount of energy; these energy amounts are called energy levels. The energy associated with different energy levels increases with the increase in distance from the nucleus. The energy levels are named as K, L, M, N, ..., or numberd as 1, 2, 3, 4, ... from the nucleus.
- The angular momentum of an electron in closed shell is quantized as an integral multiple of $\frac{h}{2\pi}$

$$\text{Angular momentum} = \text{Moment of inertia} \times \text{Angular velocity}$$

$$= mr^2 \times \frac{v}{r} = mvr$$

$$mvr = n\frac{h}{2\pi}$$

where $n = 1, 2, 3, \ldots$

h = Planck's constant

m = Mass of electron

v = Velocity of electron

r = Radius of orbit

- The energy is emitted or absorbed discontinuously in the form of quantas or small packets when electrons jump from one level to other. When an electron jumps from a higher energy state to a lower energy state, energy is emitted. When an electron jumps from a lower energy state to a higher energy state, energy is absorbed.

$$E_2 - E_1 = h\upsilon$$

E_2 = Energy of higher energy level

E_1 = Energy of lower energy level

- The lowest energy level is called the ground state and is where the electron is most stable.

Hydrogen Spectrum

Although the hydrogen atom has only one electron, its spectrum consists of several lines. A sample of hydrogen gas contains an infinite number of molecules (1 mole = 6.023×10^{23} molecules). Let a sample of hydrogen gas is exposed to electromagnetic radiation of whole range. The electron of each hydrogen atom is excited to a higher energy level. Excited electrons have the tendency to come down to the ground state with the release of energy. When the radiation of released energy is passed through the prism or spectrometer, a series of lines are observed. The hydrogen spectrum is therefore a line spectrum. Balmer was the first scientist to observe this series in the visible region, so the series is called the *Balmer series.* He discovered a relationship between the wave number and the position of lines which is

$$\bar{\upsilon} = R_H\left[\frac{1}{2^2} - \frac{1}{n^2}\right] \tag{1.1}$$

where R_H is the Rydberg constant equal to 109737 cm^{-1}, and n is an integer with values of 3, 4, 5, 6, and so on.

The nature of this relationship forced Balmer and Ritz to realize that other series of spectral lines might exist. Ritz (1908) gave a rule, known as the Ritz combination principle, which was applicable to the spectrum of hydrogen. According to this principle, the wave number in any line in a series can be represented as a difference of two square terms, one of which is constant, and the other varies throughout the series. Thus,

$$\bar{\upsilon} = R_H \left[\frac{1}{n_1^2} - \frac{1}{n_2^2} \right] \tag{1.2}$$

On the basis of this equation, four other series have later been discovered in the hydrogen spectrum (Fig. 1.5). For all the series, n_1 = any lower shell, and n_2 = any higher shell.

Following are the series of the hydrogen spectrum:

Lyman series: This series of lines falls in the ultraviolet region of the electromagnetic spectrum. For the Lyman series, $n_1 = 1$ and $n_2 = 2, 3, 4, \ldots$.

Balmer series: This series falls in the visible region of the electromagnetic spectrum. The different lines in this series are due to the transition of electrons from various higher energy levels in various hydrogen atoms to the second shell. For the Balmer series, $n_1 = 2$ and $n_2 = 3, 4, 5, 6, \ldots$.

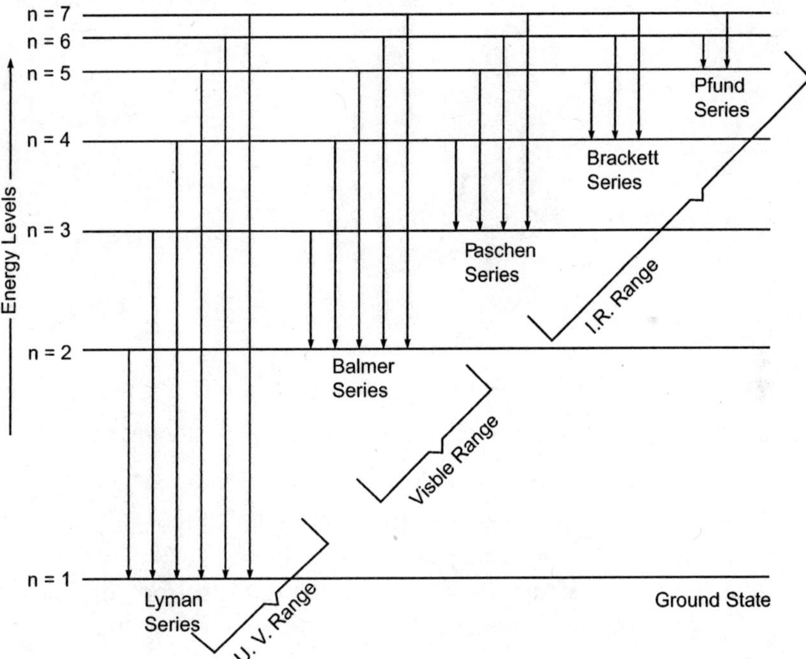

Fig. 1.5 Energy level diagram of the hydrogen spectrum.

Paschen series: This series appears in the near infrared region of the electromagnetic spectrum. The different lines in this series are due to the transitions of electrons from various higher energy levels in various hydrogen atoms to the third shell. For the Paschen series, $n_1 = 3$ and $n_2 = 4, 5, 6, \ldots$.

Brackett series: The lines of this series appear in the far infrared region. The different lines in this series appear due to the transitions of electrons from various higher energy levels in various hydrogen atoms to the fourth shell. For the Brackett series, $n_1 = 4$ and $n_2 = 5, 6, 7, \ldots$.

Pfund series: The lines appear in the far infrared region of the higher wavelength. The different lines in this series are due to the transitions of electrons from various higher energy levels in various hydrogen atoms to the fifth shell. For the Pfund series, $n_1 = 5$ and $n_2 = 6, 7, 8, \ldots$.

Advantages of Bohr's Model

- Bohr's theory provides more comprehensive treatment regarding extranuclear configuration.
- It accounts for the spectra of hydrogen line species such as singly ionized helium (He^+), doubly ionized lithium (Li^{2+}), and triply ionized beryllium (Be^{3+}), which have only one electron.
- The radius of various orbits in Bohr's model of the atom have been determined on the basis of Bohr's postulates. The expression for the radius of the nth orbit is

$$r_n = \frac{n^2h^2}{4\pi^2 m\, Z e^2 k}$$

where, r_n = Radius of the nth orbit

h = Planck's constant

m = Mass of electron (9.1×10^{-31}kg)

Z = Atomic number of the element

e = Charge on the electron (1.6×10^{-19} C)

k = Coulomb's law constant (9×10^9 Nm^2/c^2)

The greater value of 'n' corresponds to the bigger radius.

- Bohr's theory explains why the electron is not merging with the nucleus and the atom does not collapse.
- It helps in calculating the energy of an electron in a particular orbit of hydrogen. The expression for energy of an electron in the nth orbit of hydrogen is

$$E_n = -\frac{2k^2\pi^2 me^4 Z^2}{n^2h^2}$$

where, k = Coulomb's law constant (9×10^9 Nm^2/c^2)

e = Charge on electron (1.6×10^{-19} C)

Z = Atomic number of element (1 for hydrogen)

m = Mass of the electron (9.1×10^{-31} kg)

h = Planck's constant (6.62×10^{-34} J sec^{-1})

Substituting the constant values,

$$E_n = -\frac{21.79 \times 10^{-19}}{n^2} \text{ J/atom} = -\frac{1312}{n^2} \text{ KJ/mole}$$

Hence, for each value of 'n', there exists different energy levels for the electron.

Drawbacks of Bohr's Model

Following are the drawbacks of Bohr's Model:

- Bohr's model fails to explain the spectra of multi-electron atoms.
- According to Bohr, the circular orbits of electrons are planar; however, modern research shows that electrons move around the nucleus in three dimensional space.
- The model fails to explain the cause of chemical combination and shapes of molecules.
- Heisenberg's uncertainty principle does not satisfy Bohr's theory. According to Heisenberg, the position and momentum of a subatomic particle such as an electron cannot be determined simultaneously.
- The model cannot satisfy wave mechanics, which suggests that the electrons have both wave and particle nature.
- The model fails to explain the splitting of spectral lines in a magnetic field (Zeeman effect).
- The model fails to explain the splitting of spectral lines in an electrical field (Stark effect).

1.9 THE SOMMERFELD MODEL OF THE ATOM

Sommerfeld extended Bohr's model of the atom and suggested that the electronic paths around the nucleus are elliptic with different eccentricities of which Bohr's circular orbit is a special case.

1.10 WAVE NATURE OF THE PARTICLE

Electrons behave both as particles and waves. This type of dual behavior is also exhibited by light and even atoms. For example, in a discharge tube, electrons behave as particles, whereas in electron diffraction experiments, they act as electromagnetic waves. Phenomena such as photoelectric effect, reflection, and refraction of light can be explained on the basis of the particle nature of light, whereas the common phenomena of interference and diffraction can be explained by the wave theory of light. Thus, light (like electrons) is considered to have a dual nature as a particle and as a wave. This joint particle and wave character of matter and radiation is called *wave-particle duality*. Louis de Broglie suggested that dual nature (wave and particle) should not be confined to radiations alone but should also be extended to matter. Bohr's theory treats the electron exclusively as a particle. According to de Broglie, however, every material particle such as an electron should have not only the particle properties but also the wave properties. He proposed a relation between momentum and wavelength of a particle in motion.

$$\lambda = \frac{h}{P} = \frac{h}{mv}$$

where, h = Planck's constant

P = Momentum

λ = Wavelength

m = Mass of electron

v = Velocity of moving electron.

This is known as de Broglie's equation, which is derived as follows:

$$E = h\upsilon \qquad \text{(Planck's relation for energy of radiation)}$$

$$E = mc^2 \qquad \text{(Einstein's relation for motion of particles)}$$

Combining the two equations,

$$h\upsilon = mc^2 \qquad [c = \text{Velocity of light}]$$

or, $$\frac{hc}{\lambda} = mc^2$$

or, $$\frac{h}{\lambda} = mc$$

or, $$\lambda = \frac{h}{mc}$$

Now, momentum $$P = \frac{h}{\lambda}$$

$\therefore$ $$\lambda = \frac{h}{P} = \frac{h}{mv}$$

The velocity 'v' (a particle property) is thus connected to wavelength 'λ' (a wave property) by Planck's constant 'h'. The validity of this hypothesis was demonstrated by Davisson and Germer (1927). They reported the diffraction of a beam of electrons by a nickel crystal, and the values of wavelengths calculated from the angles of maximum electron intensity were also in agreement with the de Broglie relation. Again, G.P. Thomson (1928) showed that a beam of electrons was diffracted when passed through a thin gold foil. These experiments conclude that particles have wave-like properties. de Broglie's concept is significant only for small particles (such as an electron) and insignificant for large particles (such as a baseball). A particle with a high linear momentum has a short wavelength. Macroscopic particles have such high momenta even when they are moving slowly that their wavelengths are undetectably small, and the wave-like properties cannot be observed.

1.11 HEISENBERG'S UNCERTAINTY PRINCIPLE

Heisenberg (1926) proposed a principle, known as the *uncertainty principle,* which is the result of the principle of quantum mechanics. The principle states that "It is impossible to know exactly both the position and the momentum of an electron or any other microscopic particle simultaneously with accuracy".

A mathematical expression of the uncertainty relation was derived by Heisenberg from the general laws of quantum mechanics. For the pair of variables, position and momentum, the relation is

$$\Delta x \, . \, \Delta p \geq \frac{h}{2\pi}$$

where, Δx is the uncertainty in position, and Δp is the uncertainty in momentum. The product of these two uncertainties cannot be less than the Planck constant h.

To measure the momentum of a particle with absolute accuracy, Δx must become zero.

So, $$\Delta p = \frac{h}{\Delta x} = \frac{h}{0} = \infty$$

This shows that the momentum of the particle being measured becomes completely indeterminate. Conversely, if the momentum is known exactly ($\Delta p = 0$), then the position must be completely uncertain ($\Delta x = \infty$).

The uncertainty principle has no significance for macroscopic particles, where classical mechanics holds good.

The implications of the uncertainty principle are listed here:

- Heisenberg's uncertainty principle invalidates the Bohr-Sommerfeld theory because in this theory, electrons are assigned to definite orbits with precise velocities.
- Wave-particle duality is an essence of nature.

- The classical mechanics become inapplicable in the dynamics of microparticles. Wave mechanics becomes the fruitful theory in this domain. The Heisenberg uncertainty principle may also be stated in terms of energy and time:

$$\Delta E \, . \, \Delta t \geq \frac{h}{2\pi}$$

where, ΔE is the uncertainty in an energy measurement, and Δt is the uncertainty in the time at which the measurement is made. Energy can be precisely known, *i.e.*, ΔE can be zero, only if Δt is infinite. Thus, a system can be said to have an accurately defined energy only if it remains in that state for an infinite time.

1.12 THE QUANTUM MECHANICAL MODEL OF THE ATOM

Erwin Schrödinger developed a new model of the atom. He incorporated the idea of quantization and the conclusion of the de Broglie principle and the Heisenberg principle in his model. In this model, the behavior of the electron in an atom is described by the mathematical equation known as the Schrödinger wave equation:

$$\frac{\partial^2 \psi}{\partial x^2} + \frac{\partial^2 \psi}{\partial y^2} + \frac{\partial^2 \psi}{\partial z^2} + \frac{8\pi^2 m}{h^2} (E - V) \psi = 0$$

where x, y, and z are the three space coordinates.

m = Mass of electron

h = Planck's constant

E = Total energy

V = Potential energy

ψ = Wave function (amplitude of the wave)

The preceding equation can also be expressed as

$$\overline{V}^2 \psi + \frac{8\pi^2 m}{h^2} (E - V) \psi = 0$$

Here, $\overline{V}^2$ is known as the *Laplacian operator.*

Interpretation of the Wave Function

The interpretation of the wave function in terms of the location of the particle is based on a suggestion made by Max Born. The only dependent variable in the Schrödinger equation is the wave amplitude or wave function ψ, and it depends on the coordinates of the particle. The particular values of ψ are called *eigen functions* or *characteristic wave functions*, whereas the values of energy that correspond to the eigen functions are called *eigen values* of the system. The eigen function for an electron is called an *atomic orbital*. The value of ψ^2 at any point is taken to represent the probability of finding the electron at that point at a given instant. The greater the value of ψ^2, the greater is the probability of finding the electron at that point at a given instant. Accordingly, the electrons are regarded as a diffused cloud rather than a small individual.

1.13 QUANTUM NUMBERS

Quantum numbers are the integers used to designate that an electron is present in an atom. Quantum numbers may be defined as a set of numbers that display complete information about the size, shape, and orientation of the orbital. The quantum numbers follow:

Principal quantum number (n): The principal energy level, shell, or orbit introduced by Bohr. It is denoted by 'n' and has integral values. The number of electrons associated with a shell is equal to $2n^2$.

Principal Quantum Number	*Name of Shell*	*Number of Electrons ($2n^2$)*
$n = 1$	K	2
$n = 2$	L	8
$n = 3$	M	18
$n = 4$	N	32

Principal quantum number refers to the average distance of the electron from the nucleus. It gives energy.

Azimuthal quantum number (l): Introduced by Sommerfeld, this number is also called angular momentum quantum number, is denoted by 'l', and has all integral values from 0 to $(n - 1)$. It gives the name of the subshell or orbital associated with the main shell. The value of 'l' refers to the shape of the subshell. The maximum number of electrons that can be accommodated in each subshell is $(4l + 2)$

Principal Quantum Number	*Azimuthal Quantum Number*	*Name of Orbital*	*Number of Electrons*
$n = 1$	$l = 0$	s	2
$n = 2$	$l = 0$ $l = 1$	s p	2 6

Principal Quantum Number	*Azimuthal Quantum Number*	*Name of Orbital*	*Number of Electrons*
$n = 3$	$l = 0$ $l = 1$ $l = 2$	s p d	2 6 10
$n = 4$	$l = 0$ $l = 1$ $l = 2$ $l = 3$	s p d f	2 6 10 14

Magnetic quantum number (m): Zeeman introduced this number which describes the behavior of the electron in a magnetic field. This is due to the fact that a revolving electron possesses angular momentum that results in the development of a small magnetic field. Total value of 'm' for a given value of 'n' is n^2 and for a given value of 'l' is $(2l + 1)$. The values of 'm' signify the possible number of orientations of a subshell and range between $(-l, ... 0 .... +l)$. This number explains the splitting of spectral lines in the magnetic field (Zeeman effect).

Name of Subshell	*Magnetic Quantum Number*	*Orientations of Orbitals*
$l = 0, s$ $l = 1, p$ $l = 2, d$ $l = 3, f$	0 $-1, 0, +1$ $-2, -1, 0, +1, +2$ $-3, -2, -1, 0, +1, +2, +3$	s p_x, p_y, p_z $d_{xy}, d_{xz}, d_{yz}, d_{x^2-y^2}, d_{z^2}$ Seven

Spin quantum number (s): Schmidt introduced the number which represents the direction of spin of the electron around its own axis. The electron can spin clockwise or anticlockwise. Therefore, the spin quantum number can have only two values:

$$s = +\frac{1}{2} \qquad \text{(clockwise spin)}$$

$$s = -\frac{1}{2} \qquad \text{(anticlockwise spin)}$$

The two spins of the electrons in an orbital are usually represented by arrows pointing in the opposite direction: ↑ and ↓, respectively. The first electron entering in an orbital always has an clockwise spin (↑), and the second electron in an orbital has an anticlockwise spin (↓).

1.14 SHAPES OF ATOMIC ORBITALS

Orbit

An *orbit* is a well defined circular path around the nucleus in which the electron revolves. The maximum number of electrons in an orbit is $2n^2$, where 'n' is the number of the orbit (Fig. 1.6 (*b*)).

Orbital

An *orbital* is a three-dimensional region of space around the nucleus in which the probability of finding the electron is maximum (90–95%). The maximum number of electrons in an orbital is two (Fig. 1.6 (*a*)).

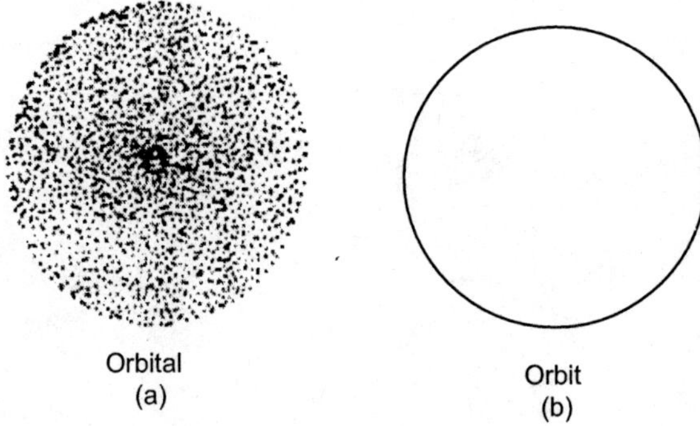

Fig. 1.6 Orbital and orbit.

- **s-orbital (l = 0, m = 0):** Because the *s* subshell has only one orientation, the probability of finding the electron is the same in all directions from the nucleus. Thus, *s*-orbital is nondirectional and is spherically symmetrical in shape (wave function is independent of angle). The greater the value of 'n', the greater the volume of *s*-orbital. Thus, a 2*s* orbital is larger than a 1*s* orbital (Fig. 1.7). A 2*s* orbital consists of inner- and outer spheres. The region in between the spheres where electron density is zero is called a *node*, a *nodal plane*, or *nodal surface*. The number of nodal planes in the *s*-orbital of any energy level 'n' = $n - 1$.

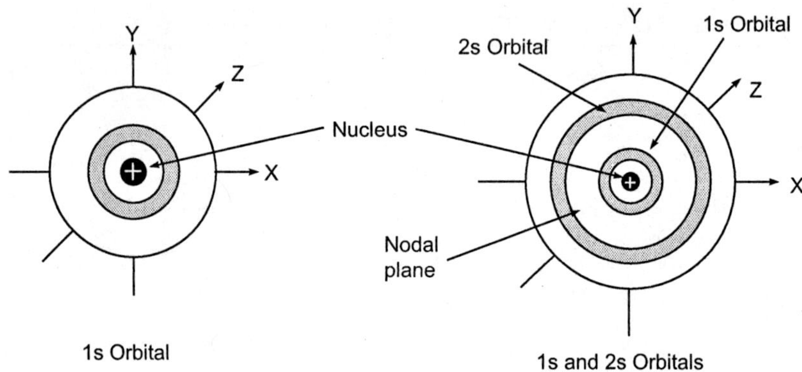

Fig. 1.7 Shape of s-orbitals.

- **p-orbital (l = 1, m = – 1, 0, + 1):** The p-subshell has three orientations directed along the x, y, and z axes, hence the p-subshell has three orbitals: p_x, p_y, and p_z. A p-orbital has a dumbbell shape and is directional. The probability of finding the electrons is equal in both the lobes. The spatial distribution of these three $2p$ orbitals are shown in Fig. 1.8.

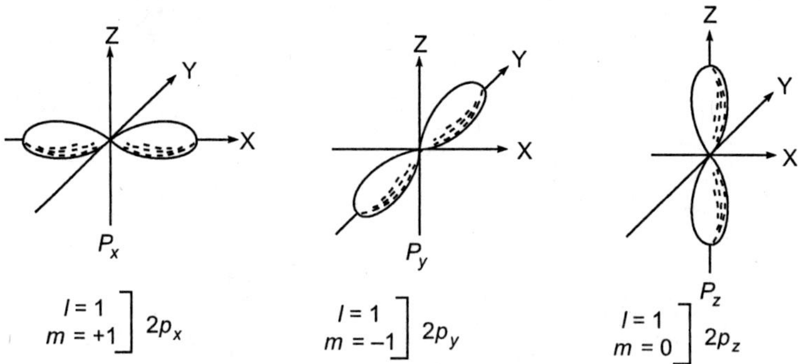

Fig. 1.8 Shape of 2p orbitals.

- **d-orbital (l = 2, m = – 2, – 1, 0, + 1, + 2):** The d-subshell has five orientations and thus five orbitals: d_{xy}, d_{yz}, d_{xz}, $d_{x^2-y^2}$, and d_{z^2}. The d_{xy}, d_{yz}, and d_{xz} orbitals lie in the xy, yz, and xz planes, respectively. The lobes are 45° with the axes and lie in between the axes. The $d_{x^2-y^2}$ has four lobes of electron density along the x- and y axes. The d_{z^2} orbital is symmetrical around the z-axis and has a ring in the xy plane. Except the d_{z^2} orbital, all other d-orbitals are double dumbbell in shape (Fig. 1.9)

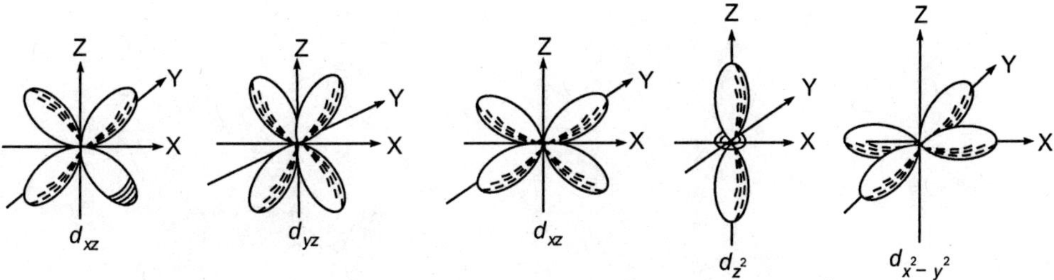

Fig. 1.9 Shape of d-orbitals.

- **f-orbital (l = 3, m = – 3, – 2, – 1, 0, + 1, + 2, + 3):** The *f*-subshell has seven orientations and thus seven orbitals designated as, $f_{x^3}, f_{y^3}, f_{z^3}, f_{x^2-y^2}, f_{y^2-z^2}, f_{z^2-x^2}$, and f_{xyz}.

1.15 THE MOLECULAR ORBITAL THEORY

In the molecular orbital theory proposed by Hund and Mulliken, it is accepted that electrons should not be regarded as belonging to particular bonds but should be treated as spreading throughout the entire molecule. According to the molecular orbital theory, a covalent bond is formed when two half-filled orbitals of two atoms come nearer and then overlap each other to form a new bigger orbital known as the *Molecular Orbital* (MO). The MO surrounds the atomic nuclei of the two atoms, and each of the electrons that are now paired in the molecular orbital is electrostatically attracted by both the nuclei. The energy of the MO is less than the sum of the energies of the two atomic orbitals, and consequently the resulting molecule is more stable than the two separate atoms. Like atomic orbitals, the molecular orbitals also obey the Paulis exclusion principle, and thus one MO can accommodate only two electrons with antiparallel spins, for examples, the formation of the hydrogen molecule.

Let ψ_A (1s) be the wave function of hydrogen atom A and ψ_B (1s) be the wave function of hydrogen atom B. The MO wave function of the hydrogen molecule is expressed as a combination of two atomic orbital wave functions as

$$\psi_{M.O.} = \psi_A (1s) \pm \psi_B (1s)$$

The expression is applicable to the diatomic molecule, and this type of approach for obtaining the MO of a molecule is known as *Linear Combination of Atomic Orbitals* (LCAO). According to the LCAO method, the linear combination of atomic orbitals can take place either by adding (constructive) or subtracting (destructive) the wave functions of the atomic orbitals involved. Two atomic orbitals overlap each other to result in two molecular orbitals, namely the bonding and antibonding orbitals (Fig. 1.10)

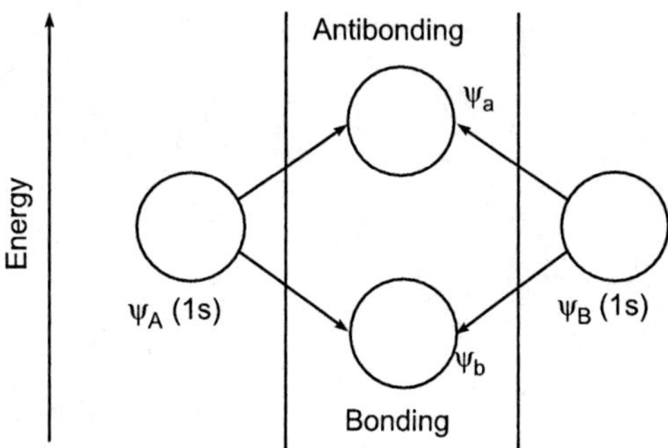

Fig. 1.10 Formation of molecular orbitals from atomic orbitals.

The wave functions for these molecular orbitals are given by the following equations:

$$\psi_b = \psi_A(1s) + \psi_B(1s) \tag{1.3}$$

$$\psi_a = \psi_A(1s) - \psi_B(1s) \tag{1.4}$$

ψ_b and ψ_a represent the wave functions for bonding and antibonding molecular orbitals, respectively. ψ_b^2 and ψ_a^2 give the probabilities of finding electrons in the two molecular orbitals formed according to the preceding equations.

Therefore,

$$\psi_b^2 = (\psi_A + \psi_B)^2 = \psi_A^2 + \psi_B^2 + 2.\psi_A.\psi_B$$

ψ_b^2 is greater than $\psi_A^2 + \psi_B^2$, that is the probability of finding electrons in the molecular orbital is greater than that of the parent atomic orbitals. Hence, the MO ψ_b has a lower energy than that of the parent atomic orbitals, which leads to the formation of a stable molecule and is therefore termed a bonding MO. Similarly,

$$\psi_a^2 = (\psi_A - \psi_B)^2 = \psi_A^2 + \psi_B^2 - 2.\psi_A.\psi_B$$

ψ_a^2 is less than $\psi_A^2 + \psi_B^2$, that is the probability of finding electrons in the molecular orbital obtained by the LCAO approach in accordance with Equation 1.4 is less than that of the parent atomic orbitals. Hence, the MO ψ_a has an energy higher than that of the parent atomic orbitals and is therefore referred to as *antibonding MO,* which cannot form a chemical bond. In the bonding orbital, the electron density is concentrated between the two nuclei, whereas in the antibonding MO, the electron density moves away from the region between the two nuclei. Antibonding orbitals are marked with an asterisk (*).

The probability of bonding molecular orbital $\left(\psi_b^2\right)$ formation is greater than that of antibonding molecular orbital $\left(\psi_a^2\right)$ formation. It is observed as follows,

$$
\begin{aligned}
\psi_b^2 &= (\psi_A + \psi_B)^2 \\
&= \psi_A^2 + \psi_B^2 + 2 . \psi_A \cdot \psi_B \\
&= \left(\psi_A^2 + \psi_B^2 - 2 . \psi_A \cdot \psi_B\right) + 4 . \psi_A \cdot \psi_B \\
&= (\psi_A - \psi_B)^2 + 4 . \psi_A \cdot \psi_B = \psi_a^2 + 4\, \psi_A \cdot \psi_B
\end{aligned}
$$

1.16 SHAPES OF MOLECULAR ORBITALS

When two atomic orbitals overlap along the internuclear axis, the resulting molecular orbital is called a sigma (σ) molecular orbital. On the other hand, when two atomic orbitals overlap sideways, the resulting molecular orbital is called a *pi* (π) molecular orbital. The combination of atomic orbitals to form molecular orbitals are illustrated here:

- **Sigma (σ) and sigma star (σ*) molecular orbitals:** Overlapping of the two atomic orbitals taking place along their internuclear axes results in MOs known as sigma (σ) orbitals, the bond formed is known as a σ bond. The MO is called a sigma orbital because it resembles an *s*-orbital when viewed along the axis and also because it has zero orbital angular momentum around the internuclear axis. If the atoms A and B are like atoms, σ and σ* orbitals are symmetric about the bond axis or the internuclear axis. It is evident that a σ bonding MO has no nodal plane containing the molecular axis. The MO formed by the combination of the orbitals of the two hydrogen atoms is shown in Fig. 1.11.

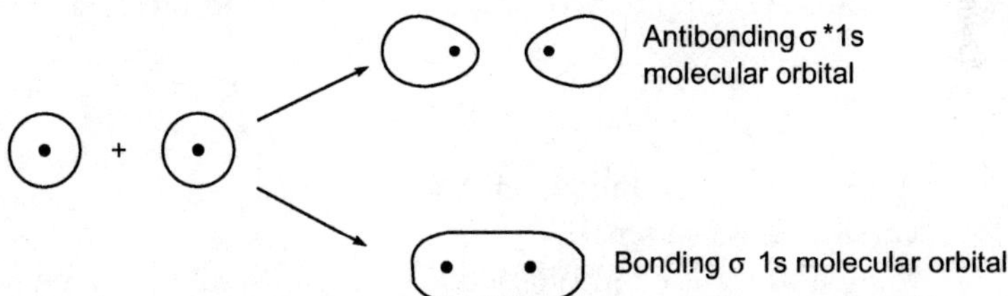

Fig. 1.11 Molecular orbital formation by combination of 1s atomic orbitals.

- **Pi (π) and Pi star (π*) molecular orbitals:** Overlapping of two atomic orbitals taking place sideways, that is in their parallel axes, results in MOs known as Pi (π) orbitals; the bond formed is known as a π bond. A π orbital looks like a *p*-orbital when viewed along the axis of the molecule and has one unit of orbital angular momentum around the internuclear axis. The overlap may be constructive or destructive and results in a bonding or an antibonding π orbital. The MOs arising from the combination of the two atomic p_y orbitals as well as the two atomic p_z orbitals are all unsymmetric about the internuclear axis. The formation of the MOs by combination of *p*-orbitals is shown in Fig. 1.12.

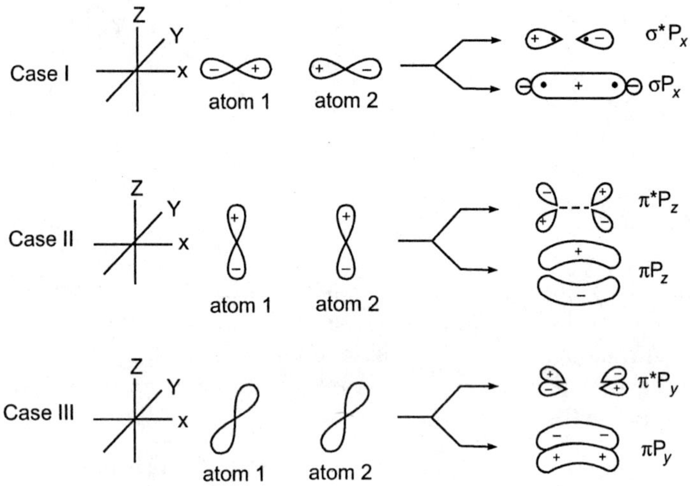

Fig. 1.12 Formation of MOs by combination of the p atomic orbital.

1.17 THE OVERLAP INTEGRAL

The extent to which two atomic orbitals on different atoms overlap is measured by the overlap integral S:

$$S = \int \psi_A \psi_B \, d\tau$$

$d\tau$ is the small volume of space in terms of infinitesimal distance along the three axes *dx*, *dy*, and *dz*.

$$dx\, dy\, dz = d\tau$$

- If the atomic orbital ψ_A on A is small and the orbital ψ_B on B is large or vice versa, then S is small.
- If ψ_A and ψ_B are simultaneously large, then S may be large.
- If the two normalized atomic orbitals are identical, then S = 1.

1.18 SUMMARY OF THE FUNDAMENTAL IDEAS OF THE MOLECULAR ORBITAL THEORY

Hund-Mulliken proposed the molecular orbital theory. According to this theory:

- Orbitals with the same energy level of each bonded atom involved in molecule formation lose their identity and merge together to give rise to an equivalent number of new molecular orbitals. Half of the MOs so formed are bonding and, half are antibonding. The bonding MO are represented by σ, π, δ, and so on; antibonding MO are represented by σ*, π*, δ*, and so on.

- A bonding MO has lower energy and greater stability than its parent atomic orbitals; whereas an antibonding MO has higher energy and lower stability than its parent atomic orbitals.
- MO formation is always accompanied by loss of energy.
- Two *s*-orbitals can overlap end to end to form one sigma bonding MO (σ) and one sigma antibonding MO (σ*).
- Two *p*-orbitals can overlap sideways to form one pi bonding MO (π) and one pi antibonding MO (π*).
- The energies of σ-type bonding orbitals are lower than those of π-type orbitals.
- A σ2*p* bonding orbital lies at a lower energy than a π2*p* bonding orbital. This sequence is correct, when the energy difference between 2*s* and 2*p* is large as in O, F, and Ne. Hence, the energies of MOs formed in O_2, F_2, and Ne_2 increases as follows (Fig. 1.13):

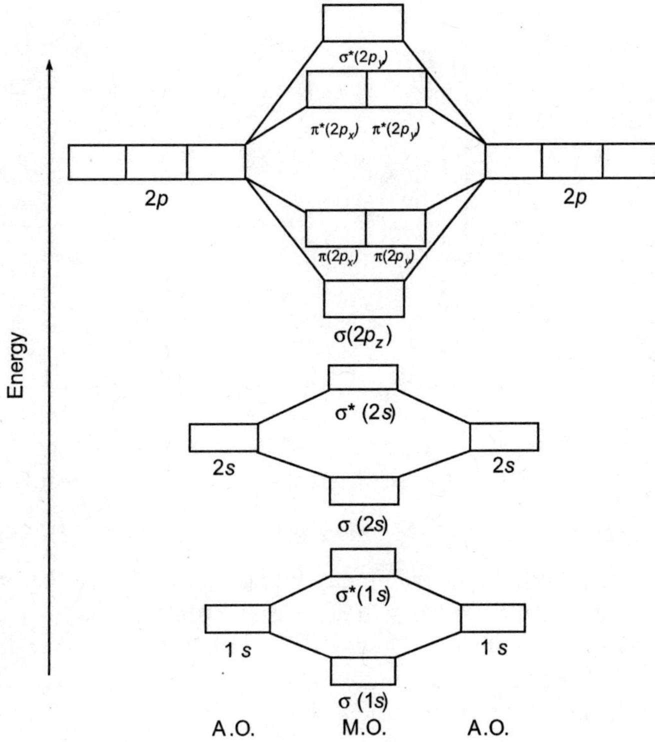

Fig. 1.13 MO energy level diagram for O_2 to Ne_2.

- **For O_2 to Ne_2:** $\sigma 1s < \sigma^*1s < \sigma 2s < \sigma^*2s < \sigma 2p_z < \pi 2p_x = \pi 2p_y < \pi^*2p_x = \pi^*2p_y < \sigma^*2p_z$. When the energy difference between 2*s* and 2*p* is smaller as in Li through N, the reversal energies of σ2*p* and π2*p* takes place. The energies of MOs of homonuclear diatomic species of elements from Li_2 through N_2 actually increase as follows (Fig. 1.14).

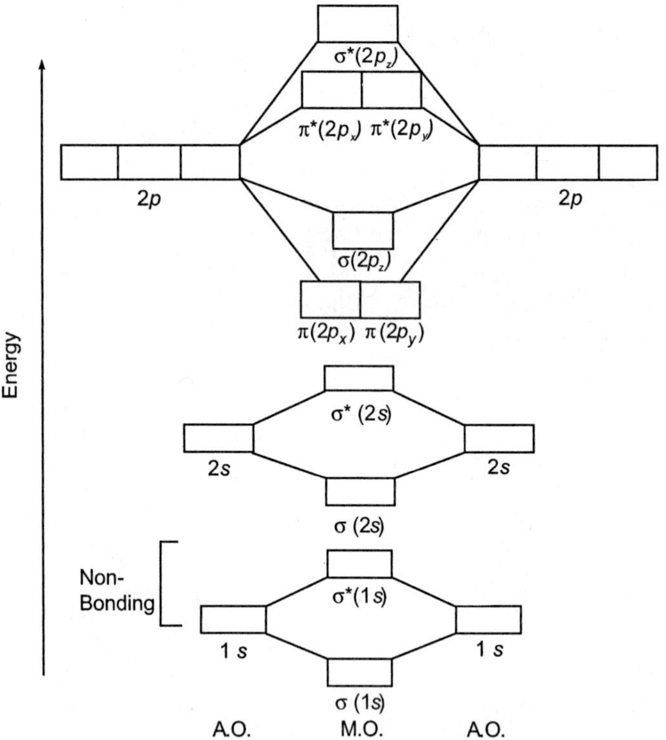

Fig. 1.14 MO energy level diagram for Li_2 to N_2.

- **For H_2 to N_2:** $\sigma 1s < \sigma^*1s < \sigma 2s < \sigma^*2s < \pi 2p_x = \pi 2p_y < \sigma 2p_z < \pi^*2p_x = \pi^*2p_y < \sigma^*2p_z$. The number of covalent bond in a molecule is given by *bond order.*

  Bond order $= \frac{1}{2}$ (Number of bonding electrons – Number of antibonding electrons) Bond order may be integer or fractional but must be positive. If bond order is zero or negative, the molecule is unstable and does not exist. The higher the bond order, the greater the stability of the molecule. The higher the bond order, the higher is the dissociation energy. Bond order is inversely proportional to bond length.
- Molecules or ions with one or more unpaired electrons in any one or more MOs are paramagnetic in nature, whereas those with no unpaired electron in an MO are diamagnetic.

1.19 BONDING OF HOMONUCLEAR DIATOMIC MOLECULES

H_2 molecule

Configuration of H: $1s^1$

$$\text{MO configuration} = \sigma (1s^2)\, \sigma^* (1s^0)$$

$$\text{Bond order} = \frac{2-0}{2} = 1$$

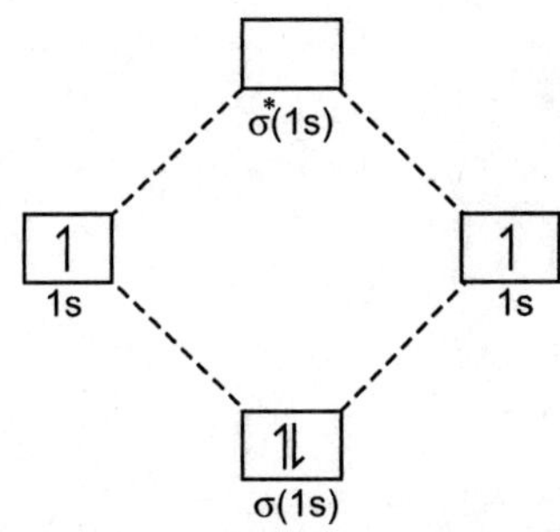

Magnetic property: Diamagnetic, because all electrons are paired.

Li_2 molecule

Configuration of Li = $1s^2\, 2s^1$

MO configuration: $\sigma (1s^2)\, \sigma^* (1s^2)\, \sigma (2s^2)$

$$\text{Bond order} = \frac{4-2}{2} = 1$$

Magnetic property: Diamagnetic, because all electrons are paired.

He_2 molecule

Configuration of He: $1s^2$

MO configuration: $\sigma (1s^2)\, \sigma^* (1s^2)$

$$\text{Bond order} = \frac{2-2}{2} = 0$$

Molecule does not exist.

H_2^+ molecule

Configuration of H: $1s^1$

MO configuration: $\sigma (1s^1)\, \sigma^* (1s^0)$

$$\text{Bond order} = \frac{1-0}{2} = \frac{1}{2}$$

Magnetic property: Paramagnetic, because one electron is unpaired.

O_2 molecule

Configuration of O: $1s^2\, 2s^2\, 2p^4$

MO configuration: $\sigma\ (1s^2)\ \sigma^*\ (1s^2)\ \sigma\ (2s^2)\ \sigma^*\ (2s^2)$

$\sigma\ (2p_z^2)\ [\pi\ (2p_x^2)\ \pi\ (2p_y^2)]\ [\pi^*\ (2p_x^1)\ \pi^*\ (2p_y^1)]$

$$\text{Bond order} = \frac{8-4}{2} = 2$$

The MO diagram is shown in Fig. 1.15.

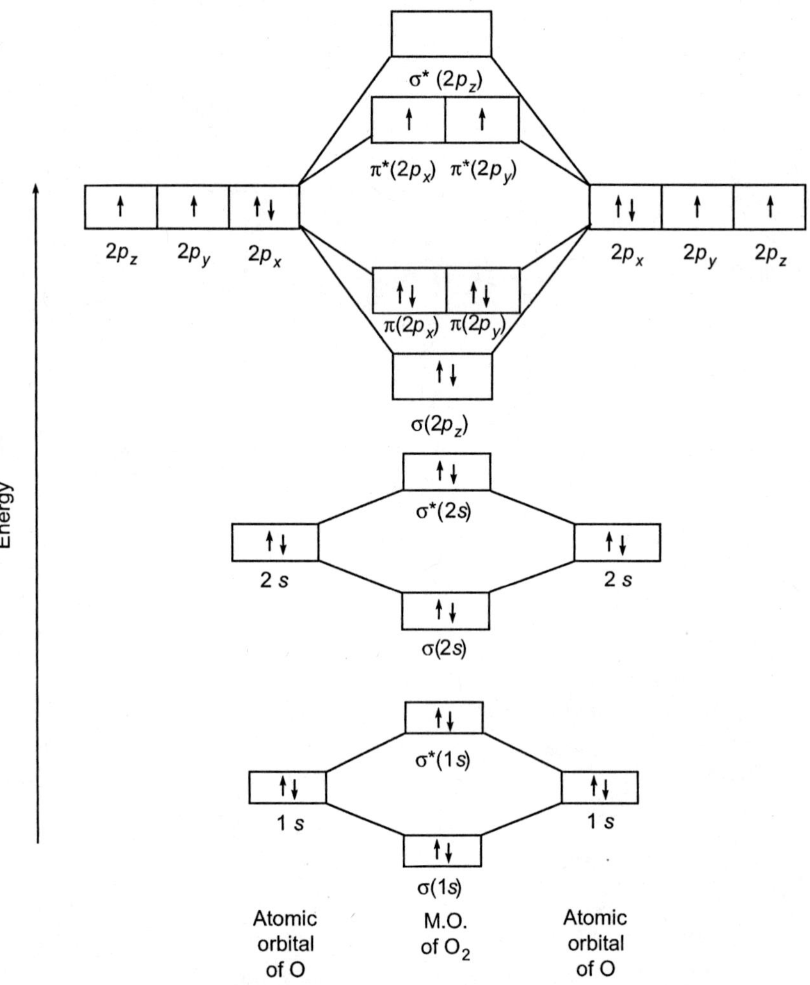

Fig. 1.15 MO diagram of the oxygen molecule (O_2).

Magnetic property: Paramagnetic, because two electrons are unpaired.

N_2 molecule

Configuration of N: $1s^2\ 2s^2\ 2p^3$

MO configuration: $\sigma\ (1s^2)\ \sigma^*\ (1s^2)\ \sigma\ (2s^2)\ \sigma^*\ (2s^2)\ [\pi\ (2p_x^2)\ \pi\ (2p_y^2)]\ \sigma\ (2p_z^2)$

$$\text{Bond order} = \frac{10-4}{2} = 3$$

The MO diagram is shown in Fig. 1.16.

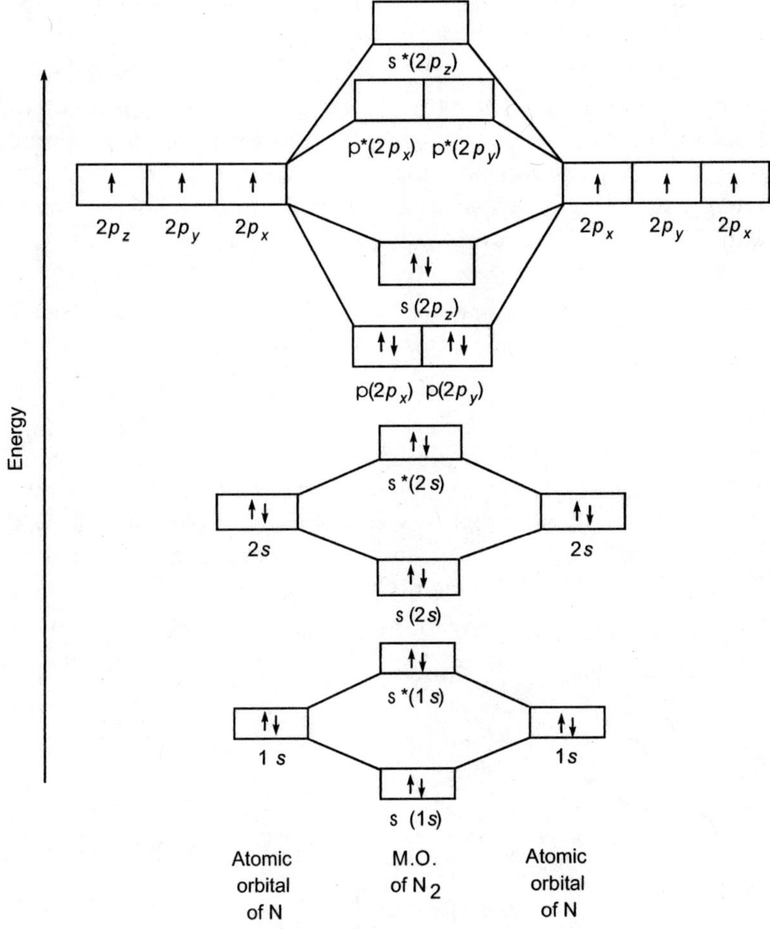

Fig. 1.16 MO diagram of the nitrogen (N_2).

Magnetic property: Diamagnetic, because all electrons are paired.

1.20 BONDING OF HETERONUCLEAR DIATOMIC MOLECULES

A heteronuclear diatomic molecule is formed from atoms of two different elements, such as CO and HCl. The electron distribution in the covalent bond between the atoms is not symmetrical because it is energetically favorable for the electron pair to be found closer to one atom than the other. This results in a *polar bond,* a covalent bond in which the electron pair is shared unequally by the two atoms. The bond in HF, for example, is polar, with the electron pair closer to the F atom. The accumulation of the electron pair near the F atom results in that atom having a net negative charge, called a partial negative charge, and is denoted by δ^-. There is a matching partial positive charge, δ^+, on the H atom. A polar bond consists of two electrons in an orbital of the form.

$\psi = C_A\, \psi_A + C_B\, \psi_B$ with unequal coefficients. The relative proportions of atomic orbital ψ (A) and ψ (B) are C_A^2 and C_B^2, respectively. In the case of a homopolar covalent bond, $C_A^2 = C_B^2$ and for pure ionic bond either $C_A = 0$ or $C_B = 0$. During the formation of a bonding MO, the atomic orbital with lower energy makes the larger contribution; whereas during the formation of an antibonding MO, the dominant component comes from an atomic orbital with higher energy.

HF molecule: The general form of the molecular orbitals of HF is

$$\psi = C_H \psi_H + C_F \psi_F$$

where ψ_H is an H (1*s*) orbital and ψ_F is an F (2*p*) orbital. The H (1*s*) orbital lies 13.6 eV below the zero of energy, and the F (2*p*) orbital lies at 18.6 eV. Hence, the bonding σ orbital in HF is mainly F (2*p*), and the antibonding σ orbital is mainly H (1*s*) orbital in character. The two electrons in the bonding orbital are most likely to be found in the F (2*p*) orbital. So, there is a partial negative charge on the F atom and a partial positive charge on the H atom. The MO electronic structure of HF is

$$\sigma\,(1s^2)\ \sigma\,(2s^2)\ \sigma\,(2p_z^{\,2})\ [\pi\,(2p_x^{\,2})\ \pi\,(2p_y^{\,2})]$$

The MO diagram is shown in Fig. 1.17.

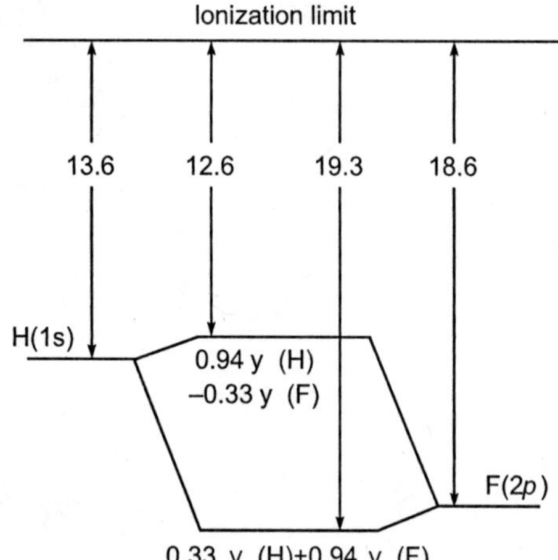

Fig. 1.17 The AO energy levels of H and F and MO formed by their combination. The bonding MO has the F-atom character predominantly (about 88%); whereas the antibonding MO has the H-atom character predominantly (about 88%).

The MO energy level diagram is distorted (Fig. 1.17). The greater the difference in the atomic number and the electronegativity of the atoms, the greater the distortion.

NO molecule:

N $(1s^2\ 2s^2\ 2p^3)$ + O $(1s^2\ 2s^2\ 2p^4)$ → NO, $\sigma\ (1s^2)$

$\sigma^*\ (1s^2)\ \sigma\ (2s^2)\ \sigma^*\ (2s^2)\ \sigma\ (2p_z^2)\ [\pi\ (2p_x^2)\ \pi\ (2p_y^2]$

$[\pi^*\ (2p_x^1)\ \pi^*\ (2p_y^0)]$ or $[\pi^*\ (2p_x^0)\ \pi^*\ (2p_y^1)]$

$$\text{Bond order} = \frac{1}{2}\ (10 - 5) = 2.5$$

The MO diagram is shown in Fig. 1.18.

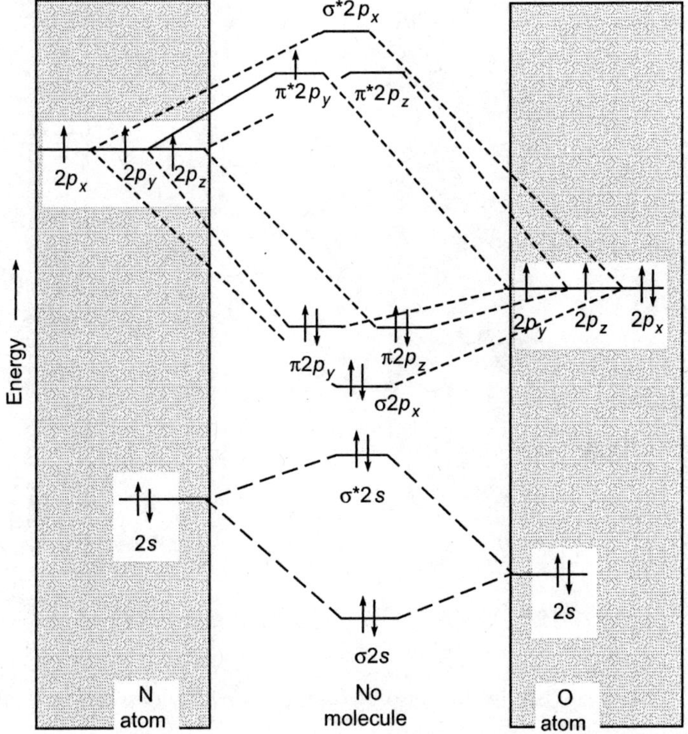

Fig. 1.18 The AO energy levels of N and O and the MO formed by their combination NO.

The molecule is paramagnetic, because it contains an unpaired electron in the $\pi^*\ (2p_x)$ or $\pi^*\ (2p_y)$ orbital. Because there is a significant difference of about 250 KJ mol^{-1} in the energy of the AOs involved, the combination gives less effective MOs than in O_2 or N_2. Therefore, the bond between N and O is weaker than might be expected.

CO molecule:

C ($1s^2$ $2s^2$ $2p^2$) + O ($1s^2$ $2s^2$ $2p^4$) → CO, σ ($1s^2$) σ* ($1s^2$) σ ($2s^2$) σ* ($2s^2$) [π ($2p_x^2$) π ($2p_y^2$)] σ ($2p_z^2$)

$$\text{Bond order} = \frac{1}{2}(10 - 4) = 3$$

The molecule exists with a triple bond (see Fig. 1.19).

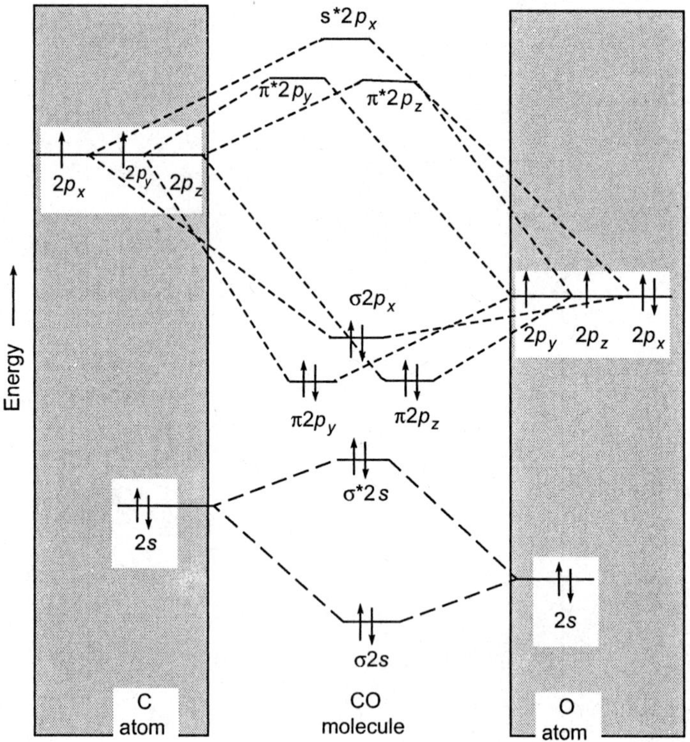

Fig. 1.19 The AO energy levels of C and O and the MO formed by their combination CO.

The molecule is diamagnetic, because all electrons are paired.

CN^- molecule: Six electrons of C and eight electrons of N^- occupy the MOs as in the homonuclear diatomic molecule.

σ ($1s^2$) σ* ($1s^2$) σ ($2s^2$) σ* ($2s^2$) [π ($2p_x^2$) π ($2p_y^2$)] σ ($2p_z^2$)

$$\text{Bond order} = \frac{1}{2}(10 - 4) = 3$$

CN^- ions exists with a triple bond [: C ≡ N :]$^-$.

1.21 MOLECULAR ORBITALS FOR POLYATOMIC SYSTEMS

The MOs of polyatomic molecules are built in the same manner as in diatomic molecules, with the only difference being that we use more AOs to construct them. As for diatomic molecules, polyatomic molecular orbitals spread over the entire molecule. A MO has the general form:

$$\psi = \sum_i c_i \, \psi_i$$

where ψ_i is an atomic orbital, and the sum extends over all the valence orbitals of all the atoms in the molecule. The principal difference between diatomic and polyatomic molecules lies in the greater range of shapes that are possible. A diatomic molecule is necessarily linear, but a triatomic molecule, for example, may be either linear or angular with a characteristic bond angle.

1.22 METALLIC BOND

The bonding that holds the metal atoms firmly together by force of attraction between metal ions and the mobile electrons is called *metallic bonding*. X-ray analysis of metal crystal has revealed that each atom in a metal crystal is surrounded by 8 or 12 other metal atoms. In metal atoms, the valence electrons are few, so it is not possible for a metal atom to form 8 to 12 covalent bonds with neighboring atoms. There are three major theories applied for bonding in metals:

Free electron theory: The theory proposed by Drude and Lorentz states that metals having 1, 2, or 3 electrons in the outermost shells that are electropositive lose their electron easily because of low ionization energy values and form free electrons and the remaining portion of the atom with a kernel carrying a positive charge. The free electrons are mobile in nature. Therefore, the metal crystal is represented by an arrangement of positively charged kernels in a sea of mobile electrons that are shared by each kernel to give metallic bonds (Fig. 1.20). This theory cannot account for the discrepancy between the observed and predicted specific heats of metals. It also fails to explain why some solids are conductors, some are insulators, and some are semiconductors.

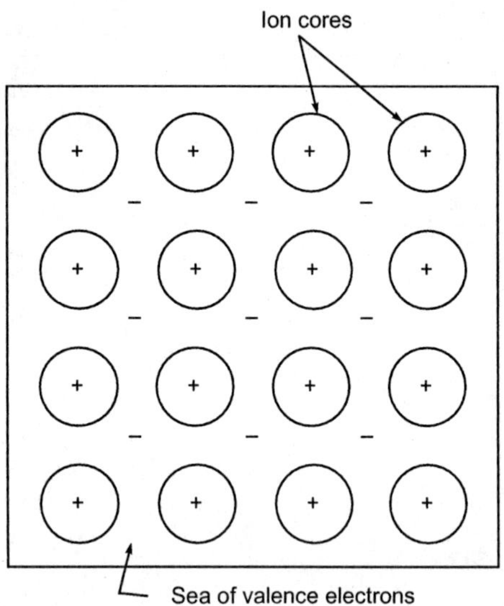

Fig. 1.20 Schematic illustration of metallic bonding.

Valence bond theory: According to this theory, the metallic bond is essentially covalent in origin, and the metallic structure involves resonance of electron pair bonds between each atom and its neighbors. A potassium atom with $4s^1$, that is one electron in its outermost shell, may be shared with one of its neighbors to form a normal two-electron bond.

$$\begin{array}{cccccc} K & K-K & K & K-K & K-K & K-K^+ \\ | & & | & & & \\ K & K-K & K & K-K & K-K & K^+-K \end{array}$$

These are some of the resonance structures of K that account for large resonance energy and large cohesive force.

Molecular orbital theory: According to this theory, the metallic bond results from the delocalization of the free electrons orbital over the atoms of a metal structure.

Consider the construction of a crystal of a sodium metal by adding Na atoms one at a time, forming first Na_2 then Na_3, Na_4, and Na_N (where N is a very large number of the order of 10^{23} in 1 cm^3 of a solid).

In the diatomic Na_2 molecule, each Na atom has the electronic configuration [Ne] $3s^1$ with a single $3s$ valence electron. Two $3s$ atomic orbitals, one from each Na atom, overlap to form two MOs.

The MO diagram is shown here:

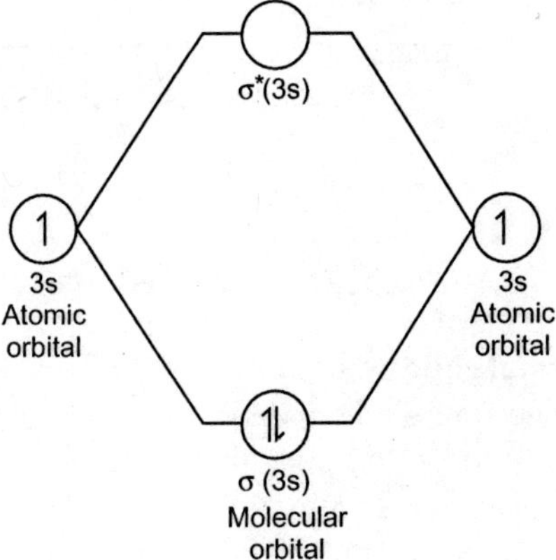

The bonding MO σ (3*s*) is occupied with two valence electrons, whereas the antibonding MO σ* (3*s*) is vacant.

1.23 PROPERTIES OF METALLIC BOND

Following are the properties of metallic bond:

Nondirectional bond: Metallic bond is nondirectional. Because of the nondirectional arrangement of the positive kernels in space, the metals are crystalline.

Weak bonds: Metallic bond is weak. The valence electrons in a metallic bond are attracted simultaneously by a large number of kernels. Thus, the net binding energy between an electron and a kernel is very small, which creates a weak bond. This bond is even weaker than a covalent bond because in a covalent bond, the shared pair of electrons are comparatively more strongly attracted by two nuclei.

High conductivity of metals: The presence of the mobile valence electrons causes high conductivity. In an electric field, the valence electrons move readily and conduct electricity throughout the metal from one end to the other (Fig. 1.21). Electricity simply directs the movement of electrons. As the temperature of a metal is raised, the kinetic energy of positive kernels and consequently, their vibration, is increased. This evidently interferes with the directed flow of electrons through the metal crystal. As a result of this, the resistance of a metal increases and its conductivity decreases with the rise in temperature.

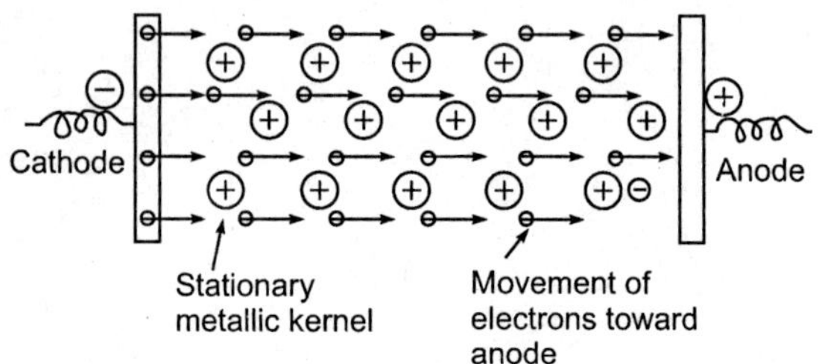

Fig. 1.21 Conduction of electricity by metals.

High thermal conductivity of metals: This is also due to the presence of mobile valence electrons. When one part of a metal is heated, the mobile valence electrons at the heated part acquire a large amount of kinetic energy. Because energized electrons in a metallic crystal are free to move, these electrons move rapidly to the colder part of the metal. During the process, some of their thermal energy is transferred to these parts. So, the temperature of colder parts is also raised and heat gets conducted throughout the metal (Fig. 1.22).

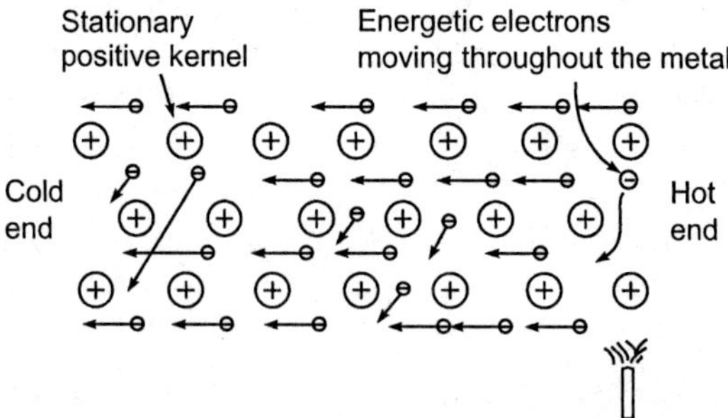

Fig. 1.22 Thermal conductivity of metals.

Bright metallic luster: This is also due to the presence of highly mobile valence electrons in metallic crystals. Metals have a low IP, and their valence electrons are mobile. When light (E = $h\upsilon$) falls on a metal surface, the mobile electrons start vibrating with frequency (υ). These vibrating electrons emit electromagnetic radiations in the form of light. This light gets reflected and the metal surface shows lustre.

An exception: Is that spongy or finely divided metals, such as platinum, appear grey or black. The reason is that such metals have irregular surfaces that trap the light because of a large number of short-distance reflections.

Softness, malleability, and ductility: Metals are malleable (can be beaten into sheets) and ductile (can be drawn into wires). This property is due to the nondirectional nature of the metallic bond. Under the influence of an external stress, the position of one group of kernels can be changed relative to the neighboring groups of kernels, without changing their relative positions with respect to the surrounding sea of electrons (Fig. 1.23). The ease with which the positive kernels in a metallic crystal can be moved past the nearest neighbors explains why metals are malleable. This also explains why metals such as sodium and potassium are soft and can even be cut with a knife. This delocalized electron model can also explain the fact that only a small force is required to bend a straight copper or aluminium wire, but for straightening out a bent wire, much more force is required. Not only that, even after straightening, a small kink always remain around the spot. The formation of a sharp bend involves separating metallic kernels from their adjacent electrons as well as from their nearest kernel neighbors, thereby the previous pattern of kernel and electrons is disturbed and a somewhat different new pattern is set up in the crystal with new planes and edges. It is difficult to restore the original pattern because the new edges and planes formed, during bending, do not ordinarily fit together exactly to restore the original pattern.

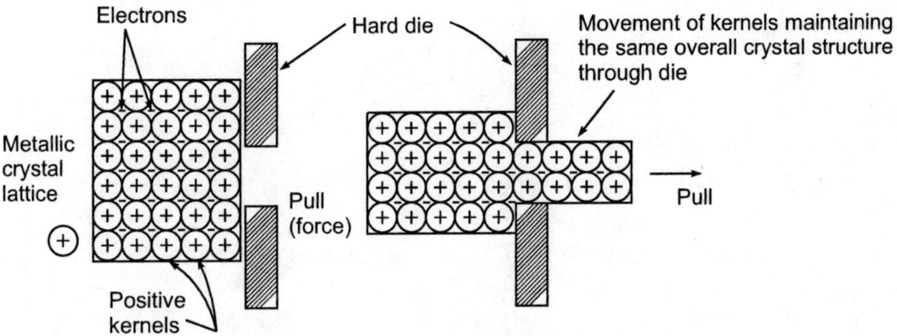

Fig. 1.23 Relative displacement of metal kernels when external stress is applied. Explanation of malleability and ductility.

High tensile-strength or tenacity: Metals possess high tenacity, that is they can resist stretching force without breaking. A strong electrostatic force of attraction between the positively charged kernels and the sea of electrons is responsible for the high tenacity of metals. On the other hand, crystalline substances with covalent bonds do not possess higher density. This is due to the absence of any electrostatic force of attraction in their crystals, because no oppositely charged units are present in their lattice.

Elasticity: Metals show elasticity, that is, they regain their original form after the applied stress is removed. The elasticity of metals is also due to the ease with which positive kernels in metallic lattice can slide from one site to another.

Hardness and rigidity: Except alkali metals, all metals are very hard. Hardness of metal is directly proportional to the number of valence electrons but inversely proportional to the size of kernel. Due to larger kernel size and only one valence electron, the alkali metals form weak metallic bonds and are thus soft.

SOLVED NUMERICAL PROBLEMS

1. *Calculate the frequency of the radiation in Hertz of wavelength 3500 Å, given velocity of light c = 3 × 10¹⁰ cm/sec.*

Sol. $$\upsilon = \frac{c}{\lambda}$$

Given, $$c = 3 \times 10^{10} \text{ and } \lambda = 3500 \times 10^{-8} \text{ cm}$$

∴ $$\upsilon = \frac{3 \times 10^{10}}{3500 \times 10^{-8}} = 8.57 \times 10^{14} \text{ Hertz}$$

2. *Calculate the wave number of the lines having frequency of 5×10^{16} cycles/sec.*

Sol. Given, $$c = 3.0 \times 10^{8} \text{ m/sec.}$$

$$\upsilon = 5 \times 10^{16} \text{ cycles/sec.}$$

Now, wave number

$$\bar{\upsilon} = \frac{1}{\lambda} = \frac{\upsilon}{c} = \frac{5 \times 10^{16}}{3 \times 10^{8}} = 1.66 \times 10^{8} \text{ cycles . m}^{-1}$$

3. *Calculate the wavelength, frequency, and wave number of a light wave whose time period is 2.4×10^{-10} sec.*

Sol. The time period 'T' is the time to complete one cycle.

$$T = \frac{1}{\upsilon}$$

or, $$\upsilon = \frac{1}{T} = \frac{1}{2.4 \times 10^{-10} \text{ sec}} = 4.16 \times 10^{9} \text{ sec}^{-1}$$

$$\lambda = \frac{c}{\upsilon} = \frac{3 \times 10^{8} \text{ m/sec}}{4.16 \times 10^{9} \text{ sec}^{-1}} = 7.2 \times 10^{-2} \text{ m}$$

$$\bar{\upsilon} = \frac{1}{\lambda} = \frac{1}{7.2 \times 10^{-2} \text{ m}} = 13.88 \text{ m}^{-1}$$

4. *An electron is moving with a kinetic energy of 4.55×10^{-25} J. Calculate de Broglie wavelength for it.*

Sol. $$\text{K.E.} = \frac{1}{2} m V^2 = 4.55 \times 10^{-25} \text{ J}$$

$$V^2 = \frac{2 \times 4.55 \times 10^{-25}}{9.1 \times 10^{-31}} \qquad [\because m = 9.1 \times 10^{-31} \text{ kg}]$$

$$V^2 = 10^{6}$$

$$V = 10^3 \text{ m/sec}$$

Because de Broglie wavelength,

$$\lambda = \frac{h}{mV}$$

$$= \frac{6.626 \times 10^{-34}}{9.1 \times 10^{-31} \times 10^3}$$

$$= 7.28 \times 10^{-7} \text{ m}$$

5. *Calculate the wavelength of an iron ball of mass 10 mg moving with a velocity of 10^5 cm per sec.*

Sol. $$\lambda = \frac{h}{mV} = \frac{6.626 \times 10^{-27}}{0.01 \times 10^5} = 6.626 \times 10^{-30} \text{ cm/sec}$$

$$\left(m = 10 \text{ mg} = \frac{10}{1000} \text{ g} = 0.01 \text{ g}\right)$$

6. *Calculate the momentum of a particle that has a wavelength of 2 Å. ($h = 6.6 \times 10^{-34}$ kg m^2 sec^{-1})*

Sol. We know that,

$$\lambda = \frac{h}{P}$$

Given, $$\lambda = 2 \text{ Å} = 2 \times 10^{-10} \text{ m}, h = 6.6 \times 10^{-34} \text{ kg m}^2 \text{ s}^{-1}.$$

$\therefore$ $$P = \frac{6.6 \times 10^{-34} \text{ kg m}^2 \text{ s}^{-1}}{2 \times 10^{-10} \text{ m}} = 3.3 \times 10^{-24} \text{ kg m s}^{-1}$$

7. *Calculate the uncertainty in the velocity of an electron, if the uncertainty in its position is 1.0×10^{-10} m. ($h = 6.6 \times 10^{-34}$ kg m^2 s^{-1}, $m = 9.1 \times 10^{-31}$ kg).*

Sol. $$\Delta V \geq \frac{h}{4\pi \, m \, \Delta x}$$

$\therefore$ $$\Delta V \geq \frac{6.6 \times 10^{-34} \text{ kg m}^2 \text{ s}^{-1}}{4 \times 3.143 \times 9.1 \times 10^{-31} \text{ kg} \times 10^{-10} \text{ m}}$$

$$\geq 5.768 \times 10^5 \text{ m s}^{-1}.$$

8. *A dust particle has mass equal to 10^{-11} g and velocity 10^{-4} cm sec^{-1}. The error in measurement of velocity is 0.1%. Calculate the uncertainty in its position.*

Sol. Velocity, $$V = 10^{-4} \text{ cm s}^{-1}$$

$$\Delta V = \frac{0.1 \times 10^{-4}}{100} = 1 \times 10^{-7} \text{ cm s}^{-1}$$

$$\Delta V \times \Delta x \geq \frac{h}{4\pi m}$$

$$\therefore \qquad \Delta x \geq \frac{6.625 \times 10^{-34}}{4 \times 3.143 \times 10^{-11} \times 10^{-7}} \geq 0.527 \times 10^{9} \text{ cm.}$$

9. *Calculate the energy per photon for radiation of wavelength 650 pm.*

Sol.

$$E = h\upsilon = \frac{hc}{\lambda}$$

$$= \frac{6.625 \times 10^{-34} \text{ Js} \times 3 \times 10^{8} \text{ m s}^{-1}}{650 \times 10^{-12} \text{ m}} = 3.058 \times 10^{-16} \text{ J}$$

10. *A particle is moving with a velocity of 6.5 × 10^7 m s^{-1}, and the wavelength associated with the particle is 5 × 10^{-11} m. Find the momentum of the particle.*

Sol.

$$\lambda = \frac{h}{mV} = \frac{h}{P}$$

$$\text{Momentum (P)} = \frac{h}{\lambda} = \frac{6.625 \times 10^{-34} \text{ kg m}^2 \text{ s}^{-1}}{5 \times 10^{-11} \text{ m}}$$

$$= 1.25 \times 10^{-23} \text{ kg m s}^{-1}$$

11. *Calculate the maximum velocity of an electron emitted from the surface of sodium metal if the photons of wavelength 200 nm strike it. The work function for the metal is 6.95 × 10^{-12} ergs. The mass of the electron is 9.11 × 10^{-31} kg.*

Sol. Energy of photon

$$= \frac{hc}{\lambda} = \frac{6.626 \times 10^{-27} \times 3 \times 10^{10}}{200 \times 10^{-7}}$$

$$= 0.09939 \times 10^{-10} \text{ erg}$$

$$= 9.939 \times 10^{-12} \text{ erg}$$

Kinetic energy of photo electron

$$= \text{Energy of photon} - \text{Work function}$$

$$= 9.939 \times 10^{-12} - 6.95 \times 10^{-12}$$

$$= 2.989 \times 10^{-12} \text{ erg}$$

Thus,

$$\frac{1}{2} mV^2 = 2.989 \times 10^{-12}$$

$$V^2 = \frac{2 \times 2.989 \times 10^{-12}}{9.1 \times 10^{-28}} = 0.6569 \times 10^{16}$$

$$V = 0.8105 \times 10^{8} \text{ cm/sec}$$

$$= 0.8105 \times 10^6 \text{ m/sec}$$
$$= 8.105 \times 10^5 \text{ m/sec}$$

12. *Show that the wavelength of a 150 gm rubber ball travelling at a velocity of 50 m/sec. is short enough to be observed.*

Sol.
$$\lambda = \frac{h}{mV} = \frac{6.626 \times 10^{-31}}{150 \times 10^{-3} \times 50} = 8.834 \times 10^{-30} \text{ m}$$

Its wavelength is so short that it does not fall in visible range, so we cannot observe it.

13. *Calculate the uncertainty in velocity of a base ball (mass = 0.1 kg), if the uncertainty in its position is 100 pm.*

Sol. Here $m = 0.1$ kg; $\Delta x = 100$ pm $= 100 \times 10^{-12}$ m $= 10^{-10}$ m

Now, $\Delta x \,.\, m\,(\Delta V) \geq \frac{h}{4\pi}$

$$\Delta V \geq \frac{h}{4\pi\, m \,.\, \Delta x}$$
$$\geq \frac{6.6 \times 10^{-34} \text{ kg m}^2 \text{ s}^{-1}}{4 \times 3.14 \times 0.1 \text{ kg} \times 10^{-10} \text{ m}}$$
$$\geq 5.25 \times 10^{-24} \text{ m s}^{-1}$$

14. *Write down the MO configurations for O_2 and O_2^-, compare their bond lengths, and predict their magnetic property.*

Sol. MO configuration of

O_2: $\sigma (1s^2)\ \sigma^* (1s^2)\ \sigma (2s^2)\ \sigma^* (2s^2)\ \sigma (2p_z^2)\ [\pi (2p_x^2) = \pi (2p_y^2)]\ [\pi^* (2p_x^1) = \pi^* (2p_y^1)]$

O_2^-: $\sigma (1s^2)\ \sigma^* (1s^2)\ \sigma (2s^2)\ \sigma^* (2s^2)\ \sigma (2p_z^2)\ [(\pi (2p_x^2) = \pi (2p_y^2)]\ [\pi^* (2p_x^2)\ \pi^* (2p_y^1)]$

$$\text{B.O. of } O_2 = \frac{1}{2}(10 - 6) = 2$$

$$\text{B.O. of } O_2^- = \frac{1}{2}(10 - 7) = 1.5$$

$\therefore$ Bond length of $O_2^- > O_2$

Both are paramagnetic, because they contain unpaired electrons.

15. *What is the bond order of N_2^+?*

Sol. MO configuration of N_2^+ is

$$\sigma (1s^2)\ \sigma^* (1s^2)\ \sigma (2s^2)\ \sigma^* (2s^2)\ [\pi (2p_x^2)\ \pi (2p_y^2)]\ \sigma (2p_z^1)$$

$\therefore$
$$\text{B.O.} = \frac{1}{2}[N_b - N_a]$$

$$= \frac{1}{2}(9 - 4) = 2.5$$

16. *Which of the following pairs is isoelectronic?*
(1) *N_2 and CO* (2) *O_2^- and N_2* (3) *Li_2 and Be_2.*
Sol. (1) Number of valency electrons in

$$N_2 = 2 \times 5 = 10$$

$$CO = 4 + 6 = 10$$

17. *Arrange O_2, O_2^+, O_2^-, and O_2^{2-} in increasing order of stability.*

Sol. O_2^{2+} (B.O. = 1.0) < O_2^- (B.O. = 1.5) < O_2 (B.O. = 2.0) < O_2^+ (B.O. = 2.5)

18. *Write the molecular orbital of N_2^-.*
Sol. Number of electrons in N_2^- = 7 + 8 = 15
MO configuration of N_2^- is
$\sigma(1s^2)\,\sigma^*(1s^2)\,\sigma(2s^2)\,\sigma^*(2s^2)\,[\pi(2p_x^2) = \pi(2p_y^2)]\,\sigma(2p_z^2)\,[\pi^*(2p_x^1) = \pi^*(2p_y^0)]$
OR $[\pi^*(2p_x^0) = \pi^*(2p_y^1)]$

EXERCISES

1. What is Rutherford's nuclear model of the atom? Mention the drawbacks of Rutherford's model of the atom.
2. State Heisenberg's uncertainty principle.
3. Write de Broglie's equation. How are wave and particle nature of moving electrons experimentally demonstrated?
4. Write notes on:
   (*a*) Wave-particle duality
   (*b*) Uncertainty principle
5. What are the quantum numbers? Discuss the types and what they signify.
6. What is Planck's quantum theory?
7. Describe the failures of classical mechanics.
8. (*a*) Define the linear combination of atomic orbitals.
   (*b*) What is the difference between ψ and ψ^2?
9. (*a*) Discuss briefly the molecular orbital theory.
   (*b*) Write the molecular orbital configuration of a CO molecule.
10. Draw the molecular orbital energy diagram of O_2, explain its magnetic character, and find its bond order.
11. What does bonding and antibonding molecular orbitals mean?
12. What is metallic bonding? Explain the properties of metals on the basis of the electron sea model.

13. Discuss the theories of metallic bonding.
14. Explain the paramagnetic character of oxygen by using molecular orbital theory.
15. (*a*) What is Schrödinger's wave equation? Explain the significance of ψ.
    (*b*) Explain why O_2 is paramagnetic while F_2 is diamagnetic, and compare the bond length of N_2 and N_2^+.

MULTIPLE CHOICE QUESTIONS

1. The Rutherford scattering experiment is related to the size of the
   (*a*) nucleus (*b*) atom (*c*) electron (*d*) neutron
2. The dual nature of radiations was proposed by
   (*a*) Max-Planck (*b*) de Broglie (*c*) Einstein (*d*) none of these
3. Classical mechanics was introduced by
   (*a*) Einstein (*b*) Newton (*c*) Thomson (*d*) none of these
4. The value of Planck's constant is
   (*a*) 6.6×10^{-32} g m^2 s^{-1} (*b*) 6.6×10^{-34} kg m^2 s^{-1}
   (*c*) 6.6×10^{-33} kg m s^{-1} (*d*) 6.6×10^{-31} kg m s^{-1}
5. The Azimuthal quantum number determines the
   (*a*) size (*b*) spin
   (*c*) orientation (*d*) angular momentum of orbitals
6. The magnetic quantum number of an atom determines the
   (*a*) size (*b*) spin
   (*c*) angular momentum (*d*) orientation of the orbital in space
7. Heisenberg's uncertainty principle is expressed as
   (*a*) $\lambda = \dfrac{h}{m\text{V}}$ (*b*) $\Delta x \,.\, \Delta p \geq \dfrac{h}{4\pi}$
   (*c*) $mvr = \dfrac{nh}{2\pi}$ (*d*) $\text{E} = mc^2$
8. Bohr's model of the atom is based upon
   (*a*) Planck's quantum theory (*b*) Classical mechanics
   (*c*) Heisenberg's uncertainty principle (*d*) Schrödinger's wave equation
9. Any *p*-orbital can accommodate up to
   (*a*) four electrons (*b*) two electrons with parallel spin
   (*c*) six electrons (*d*) two electrons with opposite spin
10. The uncertainty principle was proposed by
    (*a*) de Broglie (*b*) Max-Planck (*c*) Heisenberg (*d*) Sommerfeld
11. The de Broglie wavelength is given by

(*a*) $\lambda = \frac{h}{m}$ (*b*) $\lambda = \frac{h}{mV}$ (*c*) $\lambda = \frac{V}{m}$ (*d*) $\lambda = \frac{h}{mV^2}$

12. Which of the following orbitals have a dumbbell shape?
(*a*) *s* (*b*) *p* (*c*) *d* (*d*) *f*

13. The bond order of NO^+ is
(*a*) 3 (*b*) 2.5 (*c*) 3.5 (*d*) 4

14. The ion that is isoelectronic with CO is
(*a*) CN^- (*b*) O_2^+ (*c*) O_2^- (*d*) N_2^+

15. The molecule is unstable and does not exist if the bond order is
(*a*) zero (*b*) negative (*c*) fractional (*d*) positive
(*e*) both (*a*) and (*b*).

16. The total number of orbitals possible for the quantum number *n* is
(*a*) n (*b*) n^2 (*c*) $2n$ (*d*) $2n^2$

17. The highest number of unpaired electrons are present in
(*a*) Fe^0 (*b*) Fe^{4+} (*c*) Fe^{2+} (*d*) Fe^{3+}

18. Who modified Bohr's theory by introducing elliptical orbits for the electron path?
(*a*) Hund (*b*) Thomson (*c*) Rutherford (*d*) Sommerfeld

19. $E = h\upsilon$ was developed by
(*a*) Einstein (*b*) Schrödinger (*c*) Max-Planck (*d*) Newton

20. The Lorentz explanation is for
(*a*) ionic bonds (*b*) covalent bonds
(*c*) metallic bonds (*d*) hydrogen bonds

21. Which of the following MOs has the lowest energy?
(*a*) $\sigma 2p_z$ (*b*) $\sigma^* 2p_z$ (*c*) $\pi^* 2p_x$ (*d*) $\pi^* 2p_y$

22. Which of the following is paramagnetic?
(*a*) O_2 (*b*) N_2 (*c*) O_2^{2-} (*d*) H_2

23. During the change of NO^+ to NO, the electron is added to the
(*a*) σ-orbital (*b*) π-orbital (*c*) σ*-orbital (*d*) π*-orbital

24. Which of the species is diamagnetic?
(*a*) O_2^+ (*b*) O_2 (*c*) O_2^- (*d*) O_2^{2-}

25. Among the following diatomic molecules, the shortest bond length is associated with
(*a*) O_2 (*b*) N_2 (*c*) C_2 (*d*) F_2.

Chapter 2

Spectroscopy and Photochemistry

2.1 INTRODUCTION

The instrumental methods of analysis are now being extensively used for the qualitative analysis and identification of organic compounds. The most useful of these methods are the diffraction method and the spectroscopic method. We will study about spectroscopic methods in this chapter.

2.2 DISPERSION OF WHITE LIGHT AND THE VISIBLE SPECTRUM

The apparatus for producing a spectrum of white light is shown in Fig. 2.1. When a beam of light from a source F (a tungsten filament lamp) is selected by the narrow slit S and allowed to fall on a prism P by a collimating lens L_1, a band of seven colors can be focused on screen T with the help of an object lens L_2. The phenomenon in which white light is separated into various colored components or wavelengths is called *dispersion*. The colored components range from red through orange, yellow, green, blue, and indigo to violet. These colors diffuse gradually into each other. The band of colors is known as a *spectrum*. (Note: spectra is the plural of spectrum.)

The complete optical system for producing and viewing the spectrum is known as an *optical spectroscope*. A simple spectroscope consists of the following parts (Fig. 2.1).

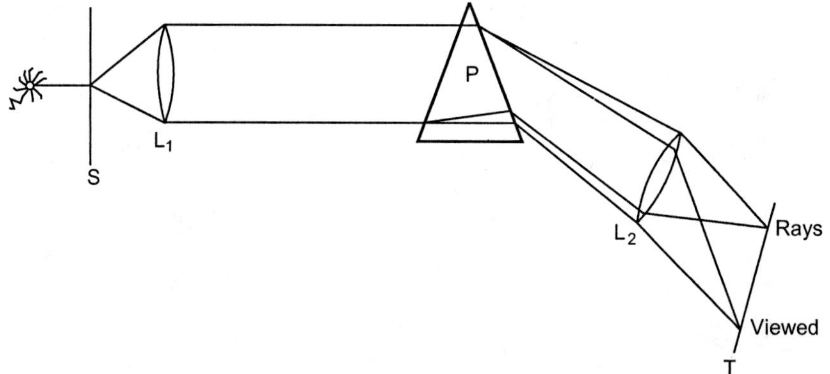

Fig. 2.1 The optical spectroscopy.

Source (F): Generally nonvolatile incandescent solids radiate white light, hence a tungsten filament lamp is often used. Sunlight is also essentially white.

Slit (S): Limits the width of the beam from the source. The slit is used to help narrow the final images of various wavelengths. The slit possesses parallel and sharp edges.

Collimating lens (L_1): Bends the rays of light and makes them parallel before they enter the prism.

Prism (P): This is the essential part of spectroscope used as a dispersing medium. As the prism disperses the light, various components of different wavelengths are sent off in different directions.

Object lens (L_2): This brings the parallel rays of each component to focus and forms a sharp image of the slit in each component.

Detector (T): The images of the slit in various colored components can be viewed on the screen. Some spectroscopes use the magnifying eye piece to see the images. In such cases, the object lens and eye piece form a small telescope.

Various categories of spectroscopes are available. If a device is attached to a spectroscope to measure the wavelength of light, then the instrument is called a *spectrometer*. A *spectrophotometer* is an instrument that uses a photoelectric cell to provide a quantitative measure of light intensity at each wavelength. *Spectroscopy* is the study of the absorption and emission of all forms of electromagnetic radiations by matter.

2.3 ELECTROMAGNETIC RADIATIONS

In 1690, Huygens put forward the *wave theory of light*, in which light travels in the form of waves. All waves (sound waves, water waves, etc.) have the properties of reflection, diffraction, and interference. Generally, the energy travels from the source in all directions like the radii of a circle, hence the name radiation.

Wavelength

The distance travelled by a wave during a complete cycle is called its *wavelength*, which is represented as λ. It is expressed in Angstroms (Å), microns (μ), or millimicrons (mμ).

$$1 \text{ micron} = 10^{-4} \text{ cm}$$

$$1 \text{ Angstrom} = \frac{1}{10} \text{ m}\mu = 10^{-8} \text{ cm}$$

Frequency

The number of wavelengths passing through any point per second is known as the *frequency* of the wave. It is denoted by υ and expressed in the number of cycles per second.

$$E = h\upsilon, \text{ where } h = 6.624 \times 10^{-27} \text{ erg. sec}$$

Velocity

The distance travelled by a wave in unit time is called its *velocity*.

$$c = \upsilon\lambda \tag{2.1}$$

c is the velocity of the wave. It is expressed in cm per sec.

Wave Number

The number of wavelengths in one centimeter is known as the *wave number* or *kayser*. It is denoted by $\bar{\upsilon}$ and is given by

$$\bar{\upsilon} = \frac{1}{\lambda} \tag{2.2}$$

Putting the value of λ from Equation (2.1):

$$\bar{\upsilon} = \frac{\upsilon}{c} \tag{2.3}$$

It is expressed in cm^{-1}.

The electromagnetic radiations (visible light, X-rays, ultraviolet rays, infrared rays, cosmic rays, and radio waves) are transmitted at the same speed, that is 2.99 $\times 10^{10}$ cm/sec, but they differ in frequency and in wavelength. An ordered arrangement of radiations according to wavelength is known as an *electromagnetic spectrum*.

2.4 EMISSION AND ABSORPTION SPECTRA

Classification of spectra is as follows:

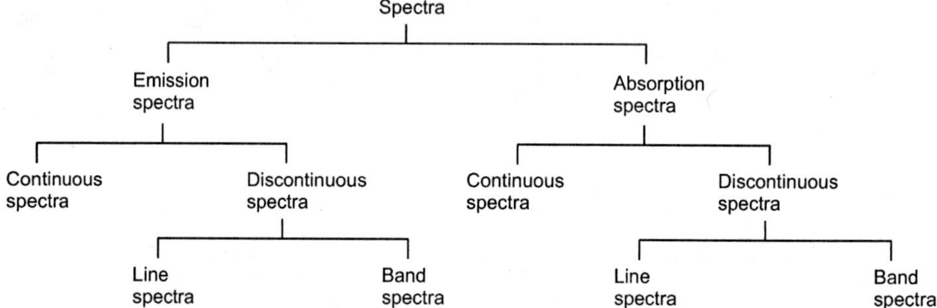

Emission spectra: When the radiations emitted by a luminous source of light are examined directly by a spectroscope, the spectra produced are known as *emission spectra,* for example, visible spectrum obtained from a filament lamp or the rays of the sun.

Absorption spectra: When the light from a source is passed through an absorbing medium, and the resulting radiations are examined by a spectroscope, the spectra obtained are known as *absorption spectra.* Dark lines or bands appear on a bright background in these spectra because certain radiations are absorbed by the absorbing medium under investigation.

Both emission and absorption spectra are again subdivided into two classes on the basis of their appearance:

Continuous spectra: In the visible spectrum, colors diffuse into each other gradually. Thus, the change in color appears to be continuous so such spectra are called *continuous emission spectra*. These are obtained from incandescent solids, hot furnaces, hot iron, molten glass, and so on. If a piece of pure red glass is put in the path of visible light, then the red glass absorbs every radiation except the red one, and the spectrum is called a *continuous absorption spectrum* of red glass.

Discontinuous spectra: These are not continuous in their appearance.

The spectra are again subdivided into:

Line spectra: Line spectra consist of lines of definite wavelengths and are given by gases and vapors of atoms. Hence, they are also called *atomic spectra.* Line spectra of particular atoms, for examples, Na-vapor, appear as bright lines on the dark background in the emission spectra and as dark lines of the same wavelengths on the bright background in the absorption spectra. These lines can be grouped into various series. The sources of line spectra are vapors in flame and metallic arc. In atomic spectroscopy, only electronic transitions are involved.

Band spectra: Band spectra consist of a group of bands that are sharply defined at one end and shaded off at the other end. When a substance with a band spectrum of emission is heated to such a high temperature that its molecules break into atoms, the band spectrum disappears. This indicates that band spectra are produced by molecules; hence they are also called *molecular spectra.* They appear both in emission spectra and absorption spectra. The sources of band spectra are carbon arc cored with metallic chlorides, bromides or fluorides, and a vacuum tube containing gases such as CO_2, N_2, and so on.

2.5 SPECTROMETER

The instrument used for the production, observation, and recording of spectra is called a spectrometer, photometer, spectrograph, and spectrophotometer depending on the various modifications. The basic principle underlying these instruments can be understood with the help of the spectrometer used for the study of absorption spectra. A flow-sheet representation of the absorption spectrometer is shown in Fig. 2.2.

The different units are discussed here:

Source: The function of the source is to provide incident light of the sufficient intensity and the required region of wavelengths. A glass-enclosed tungsten filament is usually used for visible, near-infrared and near-ultraviolet regions.

In ultraviolet region hydrogen (or deuterium), a discharge lamp is used, whereas in an infrared region, an electrically heated rod of rare earth oxides at temperature range 1500—2000°C is used. The beam from the source is focused on the sample by means of lenses and mirrors.

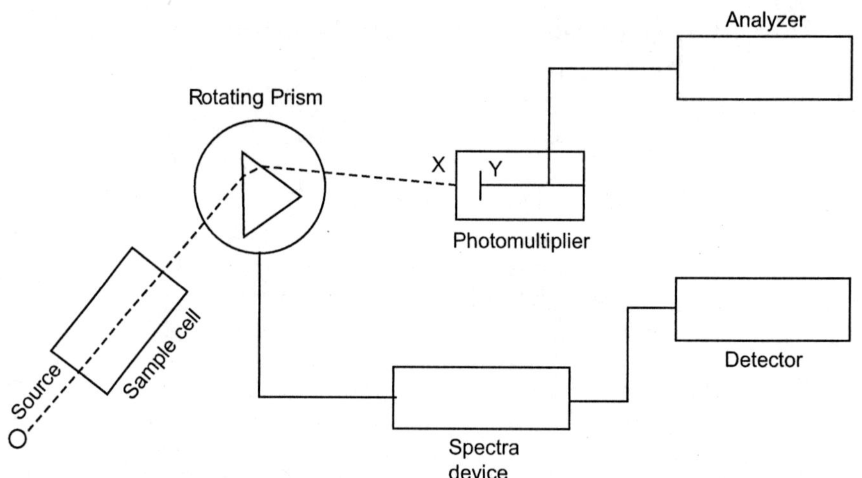

Fig. 2.2 The basic arrangement of instrument units in an absorption spectrometer.

Sample cell: The sample cell is used to hold the compound under investigation. The cell material does not absorb the radiations of the source. This is glass for visible light, quartz for ultraviolet, and alkali halides for infrared radiations. The sample in the cell absorbs characteristic wavelengths and passes them to the analyzer.

Analyzer: This consists of either a rotating prism or a grating. The material a prism is made of is the same as that used for sample cells in various regions of the electromagnetic spectrum. The various wavelengths are separated by the prism and fall on the detector due to the rotating prism; this process is known as *scanning*. It is synchronized with the recorder.

Detector: The detector converts radiant energy into electrical energy. There are two types of detectors:

- **Thermal detectors:** Thermal detectors use the heating effect of the current and are used for middle- and infrared regions, for example, thermocouples, thermopiles, and bolometers.
- **Photodetectors:** Photodetectors use the photoelectric effect and are used in visible-, ultraviolet-, and near-infrared regions. Photomultiplier is extremely sensitive and responds fast. The beam from the analyzer is focused at the entrance slit X of the photomultiplier. When the beam falls on the light-sensitive surface Y of the photomultiplier, electrons are emitted that cause the formation of a small current. This current varies according to the intensity of the beam entering at X.

Recorder: The current is amplified by the amplifier and causes the pen recorder to move on a chart paper to show the percentage absorption at various frequencies or wavelengths. This is known as the absorption spectrum of the sample.

The basic principle for the instrumentation is similar for both the emission and the absorption spectra. The only difference is that the sample itself acts as a source. It is excited with thermal or electrical energy to emit radiations. The emitted radiations are then analyzed in a similar manner as described for the absorption spectra. Usually, a photographic plate is used instead of the detector and the recorder.

2.6 BAND OR MOLECULAR SPECTRA

Band spectra significantly contribute to the determination of molecular structure. The complex structure of band spectra consists of a greater number of closely packed lines that appear as bands.

Classification

Band spectra are classified either according to the molecular energies involved in the production of spectra or according to the region of electromagnetic spectrum in which they are found. Following are the types of band spectra obtained on the basis of these two criteria:

Rotational spectra: These are produced due to the transitions in the rotational energy of the molecule only. Because they are seen in the microwave region of the electromagnetic spectrum, their study is called *microwave spectroscopy.*

Vibrational rotational spectra: These are produced due to simultaneous transitions in the vibrational and rotational energies of the molecule. Their study is called *infrared spectroscopy,* because they are seen in the infrared region of the electromagnetic spectrum.

Electronic spectra: These are produced by the simultaneous transitions in the electronic, vibrational, and rotational energies of the molecule. Their study is called *ultraviolet* and *visible spectroscopy* because they are seen in the ultraviolet and visible region of the electromagnetic spectrum.

2.7 ULTRAVIOLET AND VISIBLE SPECTROSCOPY

Ultraviolet and visible spectroscopy is the study of electronic spectra that are found in the wavelength region 100—8000 Å of the electromagnetic spectrum. The transitions in the energy levels of the outer shell electrons are responsible for the production of spectra in this region. The absorption of light energy (UV energy) involves the promotion of electrons in σ, π, and n-orbitals from the ground state to higher energy states. According to the molecular orbital theory, when a molecule is excited by the absorption of energy (UV or visible light), its electrons are promoted from a bonding to an antibonding orbital. The modes of electronic energy changes that occur when radiations in the UV and the visible region are absorbed by an organic compound are:

- Transition between bonding orbitals and antibonding orbitals, for example:

  $\sigma \rightarrow \sigma^*$ or $\pi \rightarrow \pi^*$ (Fig 2.3). This excited state closely resembles the polar character.

$$C=C \xrightarrow{h\upsilon} C^+-C^-$$

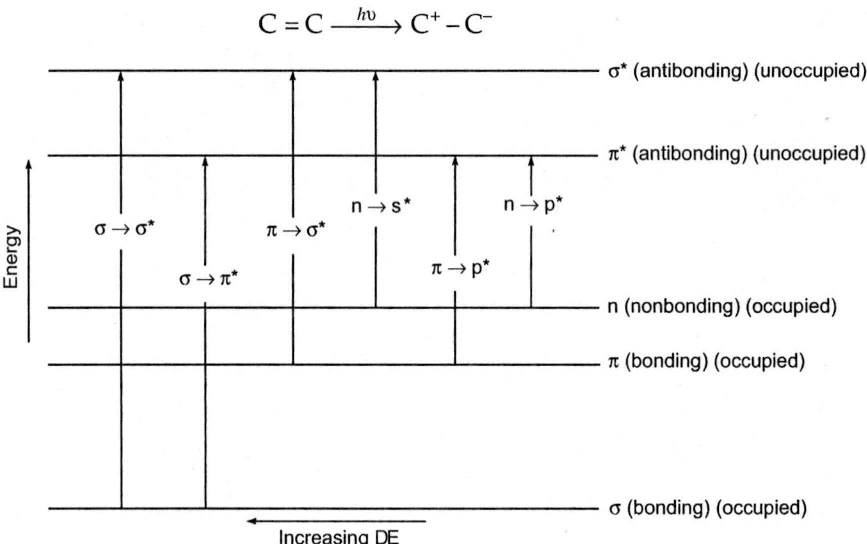

Fig. 2.3 Relative energies required for the various types of electron transitions.

- Promotion of a nonbonding electron (an unshared pair of electrons) into an antibonding sigma or antibonding π, that is:

$$n \rightarrow \sigma^* \text{ or } n \rightarrow \pi^* \text{ (Fig. 2.3)}$$

$$\sigma \rightarrow \sigma^* > n \rightarrow \sigma^* > \pi \rightarrow \pi^* > n \rightarrow \pi^*$$

Since, $E = h\upsilon = \dfrac{hc}{\lambda}$

So only $\pi \rightarrow \pi^*$ and $n \rightarrow \pi^*$ are seen in the near-UV region.

Important Terms

Chromophore: A covalently unsaturated group responsible for the absorption of selective wavelengths are called chromophores. For example:
Simple chromophores, C = C, C ≡ N, C = O, —N = N—, and so on undergo $\pi \rightarrow \pi^*$ transitions in the short wavelength regions of UV radiations.

Auxochrome: A saturated group containing nonbonding electrons that when attached to the chromophore alters the absorption position and intensity, for example:

$$-\ddot{\underset{..}{O}}H, -\ddot{N}H_2-, \ddot{\underset{..}{Cl}}:$$

Bathochromic shift (Red shift): Shift of the absorption position to a longer wavelength region by substitution or the solvent effect. The $n \rightarrow \pi^*$ transition for carbonyl compounds has a bathochromic shift when the polarity of solvent is decreased.

Hypsochromic shift (Blue shift): Shift of the absorption position to a lower wavelength region by substitution or the solvent effect. In case of aniline absorption, maxima occurs at 280 mμ because the unshared pair or nitrogen atom is in conjugation with the π-bond of the benzene ring. In an acidic solution, a blue shift occurs and absorption shifts to 200 mμ. It is because of formation of anilinium ion in acidic solution in which conjugation is removed.

Hyperchromic effect: Increase in the absorption intensity.

Hypochromic effect: Decrease in the absorption intensity.

2.8 INSTRUMENTATION

The general outline of the instrument has already been given in Fig. 2.2. The most suitable source of light is the hydrogen discharge lamp for the ultraviolet region and the tungsten filament lamp for the visible region. The light from the source is dispersed into various wavelengths by a prism or grating. The prism material is quartz for UV and glass for the visible region. The UV spectra of compounds are generally determined either in vapor phase or solution. The compound under investigation should be dissolved in the solvent that does not absorb in that region. Ethyl alcohol (95%) is commonly used for that purpose. The dilute solution of the compound is placed in a sample cell made of quartz for the ultraviolet region and glass for the visible region. Visible and ultraviolet spectra are generally recorded as absorption spectra.

Following are the applications of ultraviolet and visible spectroscopy:

Identification and assignment of structure: The absorption spectrum of an unknown compound is compared with those of a number of known compounds. If the absorption spectrum of the unknown compound matches with a particularly known compound, then the structures of the two will also be similar. This technique is known as *fingerprinting*.

Control of purification: Ultraviolet and visible spectroscopy is used to control the purification of a compound. If a compound is transparent in the UV and visible regions, then its purification is continued until it gives a minimum value of absorbance.

Quantitative analysis: Ultraviolet and visible spectrophotometric methods have been used to determine the unknown concentration of a compound in a solution. The essential condition to develop such a method is that Beer's law, $A = \varepsilon cb$, must be obeyed in the range of concentration to be used. The absorbance A is measured directly from the spectrophotometer at various concentrations C of the solution. A graph is then plotted between A and C as shown in Fig. 2.4. Absorbance A of the solution of unknown concentration is then determined by the spectrophotometer, and its concentration C is then determined from the graph. The extinction coefficient (ε) can also be determined from the slope of the graph.

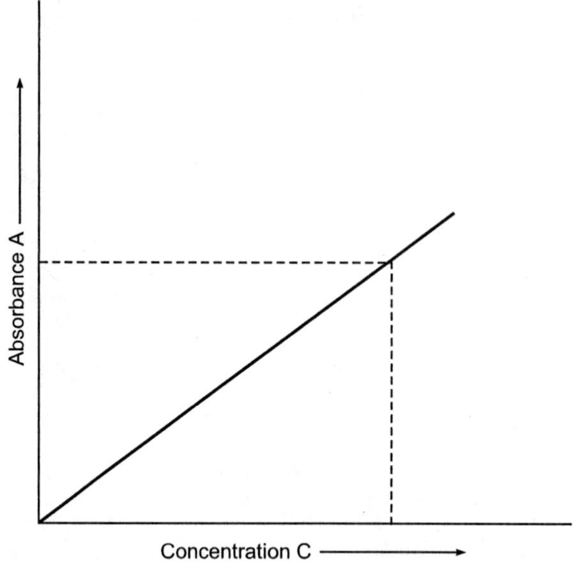

Fig. 2.4 Graph between concentration and absorbance.

2.9 INFRARED SPECTROSCOPY

Infrared radiation refers broadly to that part of the electromagnetic spectrum between the visible and microwave regions of greatest practical use to the organic chemist, which is the limited portion between 4000 cm^{-1} and 666 cm^{-1} (2.5–15.0 μm). Recently, there has been interest in the near-infrared region (0.7–2.5 μm) and the far-infrared region (14.3–50 μm). From the theoretical discussion that follows, it is clear that even a very simple molecule can give an extremely complex spectrum. It is unlikely that any two compounds except optical enantiomorphs give the same infrared spectrum. The organic chemist takes advantage of this. A peak-by-peak correlation is excellent evidence for identity. The infrared can be used to study gases, liquids, and solids. The infrared approach has been the much more prolific method so far in determining the moments of inertia and internuclear distances of simple molecules.

2.10 THEORY

Infrared radiations of frequencies less than about 100 cm^{-1} are absorbed and converted by an organic molecule into energy of molecular rotation. This absorption is quantized; thus, a molecular rotation spectrum consists of discrete lines. Infrared radiation in the range from 1–100 μm is absorbed and converted by an organic molecule into energy of molecular vibration. This absorption is also quantized, but vibrational spectra appear as bands rather than lines because a single vibrational change is accompanied by a number of rotational energy changes. It is with these vibrational-rotational bands particularly those occurring between 4000 cm^{-1} and 666 cm^{-1} (2.5–15 μm). The frequency or wavelength of absorption depends on the relative masses of the atoms, the force constants of bonds, and the geometry of atoms.

Band positions in infrared spectra are presented either as wavelengths or as wave numbers. The micron ($\mu = 10^{-6}$ m) has long been used as a unit for wavelength λ in infrared spectrometry, but recently, the term micrometer ($\mu m = 10^{-6}$ meter) is used. The wave number unit (cm^{-1}) is currently used. The units are related as follows:

$$cm^{-1} = \frac{1}{\mu m} \times 10^4$$

Band intensities are expressed either as transmittance (T) or absorbance (A). *Transmittance* is the ratio of radiant power transmitted by a sample to the radiant power incident on the sample.

$$A = \log_{10}\left(\frac{1}{T}\right)$$

Molecular Vibration

The atoms in a molecule are never stationary, whatever the temperature. In fact, even in a solid near absolute zero temperature, the atoms are in constant oscillation about an equilibrium position. The amplitude of the oscillation of the atom is only of the order 10^{-9} to 10^{-10} cm, whereas their vibrational frequencies are of the order of infrared radiation. The normal modes of vibrations in a molecule are categorized as follows:

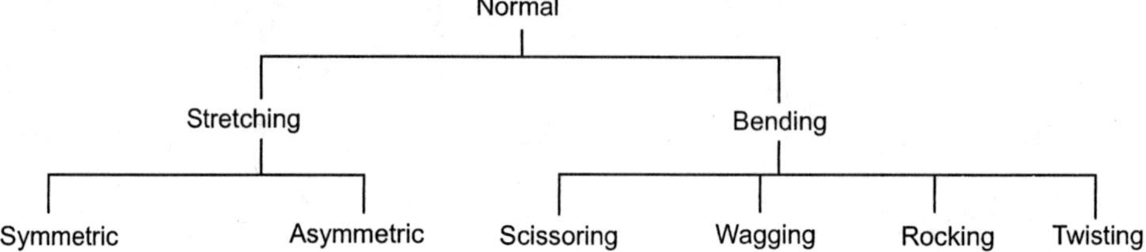

Stretching or valence bond vibrations: A stretching vibration is a rhythmical movement along the bond axis such that the interionic distance is increasing or decreasing. Two types of stretching modes for an AX_2 group, for example, CH_2 in a hydrocarbon molecule, are shown in the preceding figure.

Bending or deformation vibrations: When a three-atom system is a part of a big molecule, bending vibrations may occur. Those vibrations in which the atoms or the group as a whole oscillates perpendicular to the bond axis are called bending or deformation vibrations. Four important types of bending vibrations are as follows:

- *Rocking vibrations* are those in which the vibrating group swings back and forth in the plane of the molecule as shown in Fig. 2.5 (*a*).

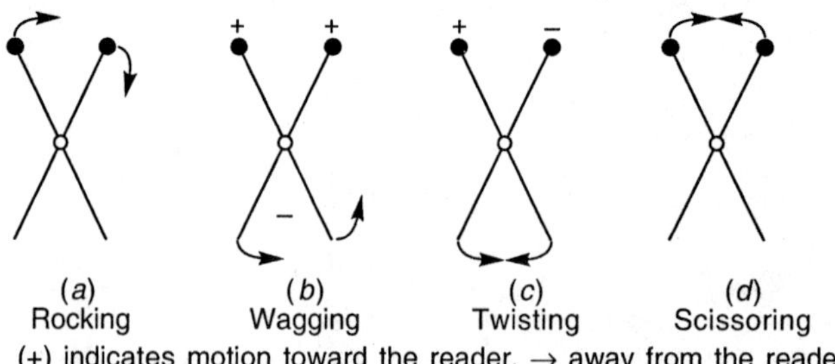

Fig. 2.5 Bending vibrations.

- *Wagging vibrations* are those in which the vibrating group swings back and forth out of the plane of molecule as shown in Fig. 2.5 (*b*).
- *Twisting vibrations* are those in which the vibrating group rotates about the chemical bond that is attached to the rest of the molecule as shown in Fig. 2.5 (*c*).
- *Scissoring vibrations* are those in which two atoms attached to a central atom move away from and toward each other as shown in Fig. 2.5 (*d*).

Number of Fundamental Vibrations

A molecule has as many degrees of freedom as the total degrees of freedom of its individual atoms. Each atom has three degrees of freedom corresponding to cartesian coordinates (X, Y, Z). If there are N uncombined atoms free to move in three dimensions, then the system would have 3N translational degrees of freedom. However, if these atoms are contained in one molecule, there are still 3N degrees of freedom of which three degrees are for the translation of the center of gravity of the molecule. For a nonlinear molecule, three of the degrees of freedom describe rotation and three describe translation. Then,

$$\text{translational} + \text{vibrational} + \text{rotational degrees of freedom} = 3N$$

that is the vibrational degrees of freedom

or $$\text{fundamental vibrations} = 3N - 6.$$

These values are the number of fundamental vibrational frequencies of the molecules, that is the number of different normal modes of vibration. For a linear molecular such as Cl_2, CO_2 or C_2H_2, if the molecule is placed along the Z axis, then it has only finite moments of inertia about the other two axes and only two degrees of rotational freedom. Thus the number of vibrational degrees of freedom would become 3N – 5. Fundamental vibrations involve no change in the center of gravity of the molecule.

2.11 INSTRUMENTATION

The modern double-beam infrared spectrophotometer consists of five principal sections: Source, sampling area, photometer, grating (monochromator), and detector. The experimental arrangement is shown on the next page.

The light source is Nernst glower, which is a molded rod that contains a mixture of the oxides of rare earths heated to 1500°C. Prism or grating is used as the monochromator. Cell containers and prisms are made of metal halides, such as NaCl. Thermocouple or bolometer is used as the detector. The sample may be in the form of a gas, solid, liquid, or solution. The solvents used in general are CCl_4, CS_2, and $CHCl_3$, which have less absorption in the infrared region.

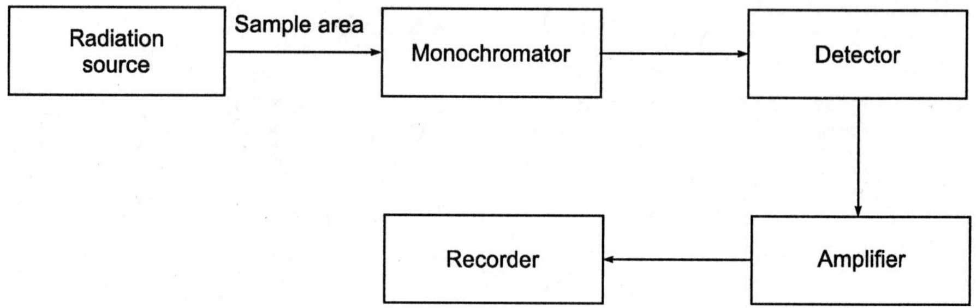

2.12 APPLICATION OF INFRARED SPECTROSCOPY

Structure elucidation and identification of a compound: Different functional groups in a polyatomic molecule absorb at characteristic frequencies. This phenomenon has been helpful in elucidating the structure of organic compounds. For example, if the spectrum of a compound shows a strong absorption at 1600–1800 cm^{-1}, then a carbonyl group should be present. On the other hand, if the spectrum contains no absorption in the region 1600–1800 cm^{-1}, then no carbonyl group is present in the compound. When the presence of different functional groups in the compound is established, then the structure can be elucidated with the help of other available data. The most useful function of infrared spectroscopy is the fingerprint technique, that is the identification of an unknown sample by matching its spectrum with that of a known compound. Fig. 2.6 shows the appearance of most common groups on an infrared spectrum.

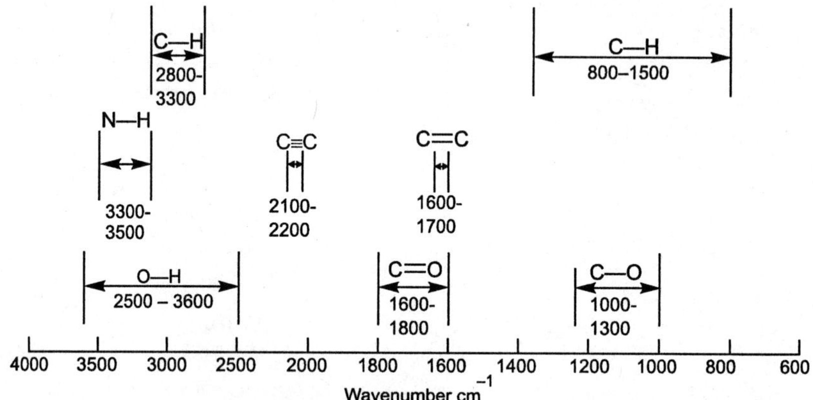

Fig. 2.6 The diagram shows where some of the most common groups appear on an infrared spectrum.

Determination of purity: We know that each particular compound has a characteristic infrared spectrum. In the presence of impurities, extra absorption bands will appear in the spectrum, and some parts of the spectrum will not be so sharp. Thus, the infrared spectrum can be used to determine the purity of a compound.

Study of reaction kinetics: In a chemical reaction, some bonds break and produce new bonds. Accordingly, certain bonds in the infrared spectrum disappear, and new bonds will appear in the course of time. Thus, the progress of organic reactions can be studied.

Determination of the force constant of a bond: Chemists are especially interested to find the force constant of a bond because it is a measure of the strength of the bond. We know that the frequency of absorption υ is related to force constant K by the relation:

$$\upsilon = \frac{1}{2\pi c}\sqrt{\frac{K}{\mu}}$$

or

$$K = 4\pi^2 c^2 \upsilon^2 \mu$$

$\bar{\upsilon}$ is measured from infrared spectroscopy, and K can be calculated for a bond. Table 2.1 shows the values of force constants and the bond energy for some diatomic molecules that absorb in the infrared region. The table clearly shows that the value of K determines the relative strengths of bonds.

TABLE 2.1 Force Constants of Some Bonds

Bond	*Infrared λ_{max} (μ)*	*Force Constant (dynes / cm) × 10^3*	*Bond Energy (kcal/mole)*
H–F	3.44	9.7	135
H–Cl	3.47	4.8	102
H–Br	3.77	4.1	87
H–I	4.48	3.2	71

To ascertain hydrogen bonding in a molecule: Generally, it is not possible to distinguish between intramolecular and intermolecular hydrogen bonding. This can be ascertained by taking a series of IR spectra of the compound at different dilutions. As the dilution is increased, the absorption band diminishes due to intermolecular hydrogen bonding; whereas that due to intramolecular hydrogen bonding remains unchanged.

Provides valuable information about dipole moments, bond lengths, and so on.

Distinguishes positional isomers of a compound.

Examples of IR spectra:

- Benzene (C_6H_6) spectrum is shown in Fig. 2.7, which is fairly complex because absorption due to the benzene molecule as a whole occurs in addition to those characteristics of particular bonds. Various peaks in the spectrum of certain vibration in the molecule have been marked as well as the major peak.

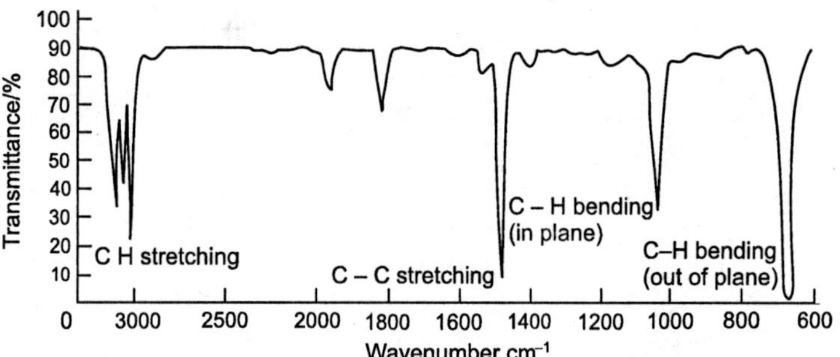

Fig. 2.7 IR spectrum of benzene (in liquid film).

IR spectra are normally plots of percentage transmittance and not absorbance (the two are related inversely). In addition, the wavenumber ($\overline{\upsilon}$), which is the reciprocal of the wavelength (in cm), is employed, and this quantity is related directly to the energy of the transition (i.e., $[\overline{\upsilon}] \propto E$).

- Ethyl ethanoate ($CH_3CO_2CH_2CH_3$) spectrum is shown in Fig. 2.8 with the three characteristic peaks marked.

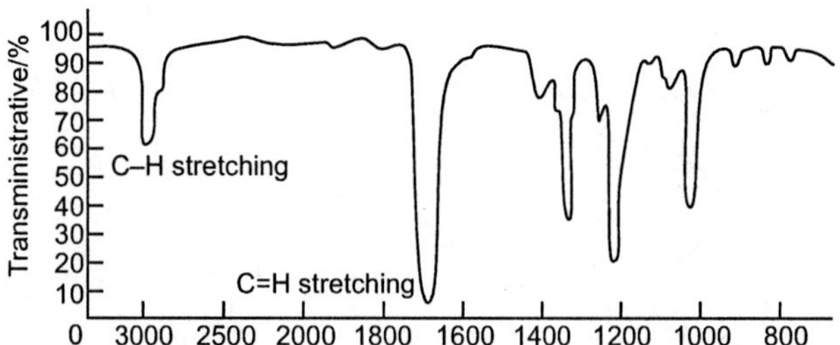

Fig. 2.8 Spectrum of ethyl ethanoate (in liquid film).

- Ethanol (CH_3CH_2OH) spectrum is shown in Fig. 2.9 with the three characteristic peaks.

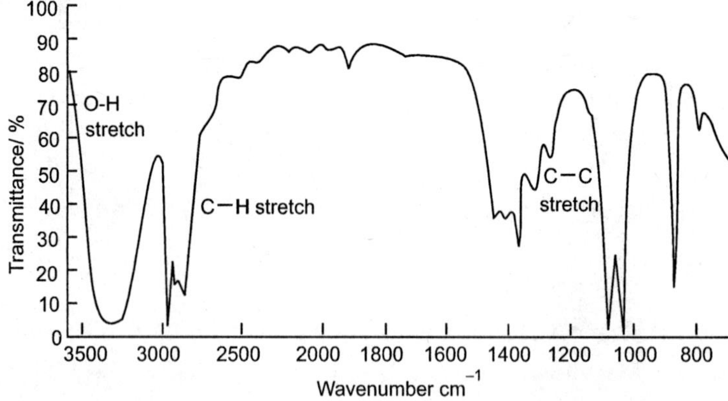

Fig. 2.9 IR spectrum of ethanol.

- Propanone (CH_3COCH_3) spectrum is shown in Fig. 2.10 with the characteristic peaks marked.

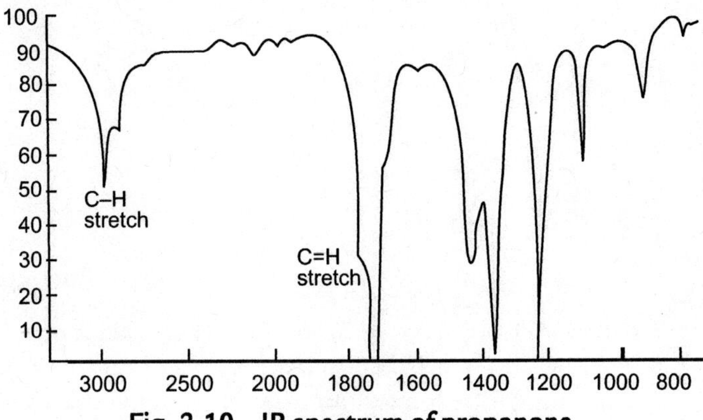

Fig. 2.10 IR spectrum of propanone.

2.13 MICROWAVE SPECTROSCOPY

Rotational spectroscopy involves the transitions between the rotational energy levels of a gaseous molecule having permanent dipole moment on the absorption of radiations falling in the spectral range of 1–100 cm^{-1} (microwave region).

2.14 THEORY

Microwave spectra are due to the result of transitions between the rotational levels of gaseous molecules possessing a permanent dipole moment such as HCl, HBr, CO, NO, H_2O vapor, NH_3, and so on, on the absorption of radiations falling in the microwave region (λ = 3 mm–1.3 m). Homonuclear diatomic molecules such as H_2, Cl_2, N_2, O_2, and so on, and nonpolar polyatomic molecules such as CO_2 do not show microwave spectra. The rotational or microwave spectra are observed only when molecules exist in the gaseous state. The rotational energy levels are not quantized in solid and liquid state. Hence, solid and liquid molecules do not exhibit rotational spectra.

2.15 INSTRUMENTATION

The experimental arrangement of microwave spectrometer is shown schematically, here:

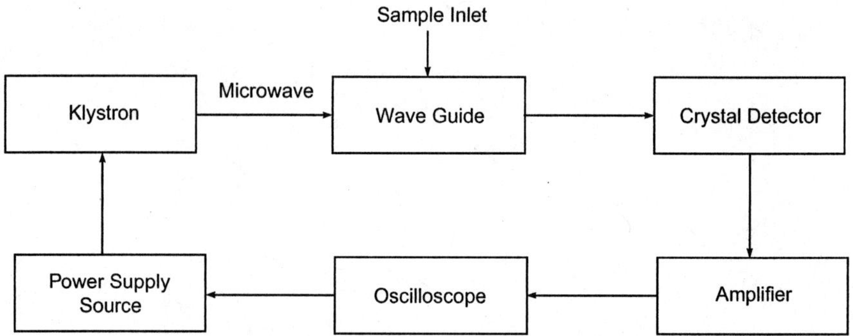

The different units are discussed in the following list:

Klystron: This is a specialized two-cavity vacuum tube. An applied voltage accelerates the electrons, thereby generating a monochromatic microwave radiation. The frequency of radiation generated is dependent on the voltage applied to the tube.

Waveguide: This is a hollow metallic conductor in which electromagnetic waves in the microwave region can be transmitted efficiently from a source (klystron) to crystal detector.

Crystal detector: This utilizes a piezo electric crystal (*e.g.*, quartz, Rochelle salt, barium titanate) to determine the frequency or wavelength of a radiation.

Amplifier and oscilloscope: The vibration of crystal in turn produces an electrical signal, which is amplified by an amplifier and finally displayed as a wave pattern on the oscilloscope screen.

Microwave Spectra of Diatomic Molecules

Consider a diatomic molecule AB of equilibrium bond length r rotating about its axis through center of gravity (C_g). r_1 and r_2 are the distances of the atoms from the center of gravity of the molecule, and m_1 and m_2 are the masses of two atoms A and B, respectively.

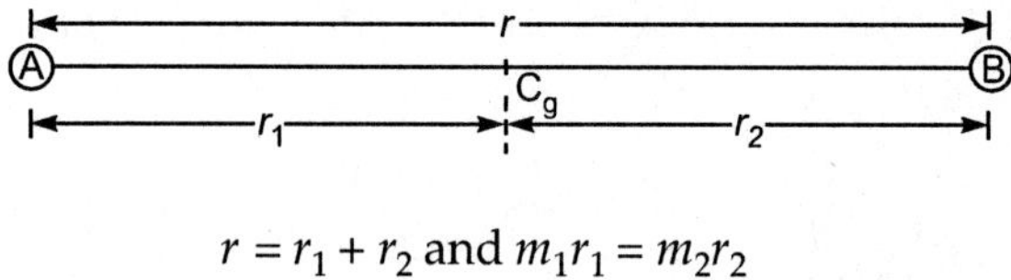

$$r = r_1 + r_2 \text{ and } m_1 r_1 = m_2 r_2$$

Moment of inertia (I) of the molecule

$$I = m_1 r_1^2 + m_2 r_2^2 = \mu r^2$$

where, μ is the reduced mass of the molecules.

The angular momentum (L) of the rotating molecule,

$$L = I\omega \quad (\omega = \text{Angular velocity})$$

The angular momentum is quantized and given by

$$L = \sqrt{J(J+1)} \cdot \left(\frac{h}{2\pi}\right) \qquad (2.4)$$

where, J is the rotational quantum number and has values 0, 1, 2, 3 and so on.

Now, the energy of a rotating molecule is given by

$$E = \frac{1}{2} I\omega^2.$$

The quantized rotational energy levels of a rigid diatomic rotor (a molecule which rotates without undergoing any change in its internuclear distance during rotation) are given by

$$EJ = \frac{1}{2} I\omega^2 = \frac{(I\omega)^2}{2I} = \frac{L^2}{2I} \qquad (2.5)$$

From, Equations (2.4) and (2.5), we get

$$EJ = \frac{\left[J(J+1)h^2\right]}{8\pi^2 I}$$

Usually, E J is reported in cm^{-1}. Now, dividing E J by hc

$$EJ = \frac{\left[hJ(J+1)\right]}{8\pi^2 IC} cm^{-1} = BJ(J+1)$$

where, $$B = \frac{h}{8\pi^2 IC} = \text{Rotational constant.}$$

Selection Rules

- A rotating molecule must possess a permanent dipole moment to interact with the oscillating electric field associated with electromagnetic radiation.
- The rotating molecule can increase or decrease its rotational energy only to either the next higher or next lower energy level when it absorbs or emits a quantum of electromagnetic radiation.

$$\Delta J = \pm 1$$

For absorption of radiation, $\Delta J = +1$, and for emission of radiation $\Delta J = -1$.

From Fig. 2.11, it is clear that transition $J_{0\rightarrow1}$, $J_{2\rightarrow3}$, and $J_{3\rightarrow4}$ occurs at 2B, 4B, 6B, and 8B, respectively. In other words, transitions are equispaced with spacing of 2B.

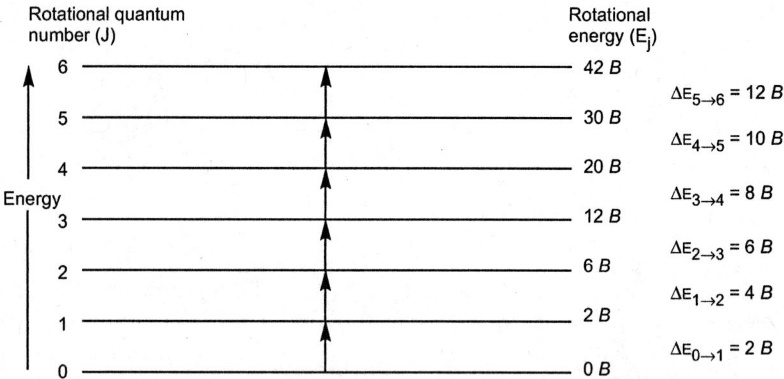

Fig. 2.11 Allowed rotational energies of a diatomic molecule according to selection rule ($\Delta J = \pm 1$).

2.16 APPLICATIONS OF MICROWAVE SPECTROSCOPY

Bond angle and bond distance determination: From the rotational spectrum, which consists of series of equally spaced lines, the value of B can be calculated.

Since,
$$B = \frac{h}{8\pi^2 IC}$$

∴
$$I = \frac{h}{8\pi^2 BC}$$

If we know the atomic masses, reduced mass (μ) can be calculated by using the relation:

$$\mu = \frac{m_1 . m_2}{m_1 + m_2}$$

Since,
$$I = \mu r^2$$

∴
$$r = \sqrt{\frac{I}{\mu}} = \sqrt{\frac{h}{8\pi^2 BC\mu}} = \text{Internuclear distance}$$

Measurement of dipole moment of a molecule: Dipole moments of gaseous polar molecules can be determined by recording the pure rotational spectra in the presence of a strong external field. When an electric field is applied to the sample being studied, the rotational line is split (Stark effect).

Suppose, $\Delta\upsilon$ = shift of rotational frequency caused by an external electric field E,

μ = dipole moment of the polar molecule

then,
$$\Delta\upsilon = (\mu E)^2$$

The value of the dipole moment (μ) can be calculated from the measured value of applied E and measured value of $\Delta\upsilon$.

2.17 PHOTOCHEMISTRY

Photochemistry deals with chemical reactions brought about by the exposure of a system to radiation. When chemical changes occur due to the absorption of photons from the visible and ultraviolet radiations with wavelengths approximately between 10000 Å to 1000 Å, the reactions are called *photochemical reactions*. Photochemistry thus concerns itself with reactions effected with visible and ultraviolet radiations.

2.18 BEER-LAMBERT'S LAW

Beer-Lambert's law governs the absorption of light by the molecules.

When a monochromatic light of intensity I is passed through a solution of concentration, C molar and thickness dx, then the intensity of transmitted light changes by dI because of absorption. Then, probability of absorption of radiation is given by

$$\frac{dI}{I} = -kc\,dx$$

where, k is porportionality constant.

On integrating the preceding expression

$$\int_{I_0}^{I} \frac{dl}{I} = -kc \int_0^I dx$$

$$\ln \frac{I}{I_0} = -kcl$$

or,
$$2.303 \log \frac{I}{I_0} = -kcl$$

or, $$\log \frac{I_0}{I} = \frac{k}{2.303} cl = \varepsilon cl = A$$

where, $\varepsilon = \frac{k}{2.303}$ is called the molar absorptivity coefficient, and $\log \frac{I_0}{I} = A$ is called the absorbance.

$$A = \varepsilon cl$$

The preceding expression is Beer-Lambert's law.

When $\log \left(\frac{I_0}{I}\right)$ is plotted against the concentration of the solution taken in a column of definite thickness, a straight line is obtained. The molar extinction coefficient is ascertained from the slope of the line.

2.19 LAWS OF PHOTOCHEMISTRY

The photochemical reactions are governed by the following two basic principles:

Grotthus-Draper Law: The law states that only those radiations that are absorbed can be effective in producing the chemical change. Although photochemical reactions can occur by the absorption of radiation, all the light absorbed is not effective to bring the chemical change. Some of it may be converted to heat, and some may be remitted as *fluorescence.*

This law is qualitative in nature and does not give any idea of the relation between the light absorbed and the substance undergoing chemical change.

Einstein-Stark Law of Photochemical Equivalence: According to the law, "each molecule taking part in a chemical reaction, when induced by the exposer to light, absorbs one quantum of radiation to cause the reaction."

If the absorbing molecule decomposes or reacts immediately without further reactions, then for every quantum absorbed, one molecule should be involved in the reaction. If υ is the frequency of absorbed radiation, then the energy required by a single atom or molecule in absorbing one quantum depends on υ, and is given by Planck's equation:

$$\Delta E = h\upsilon \text{ per molecule}$$

or, $$\Delta E = Nh\upsilon \text{ where, N = Avogadro's number.}$$

The quantity of energy defined by this equation is called an Einstein.

Einstein, E is the quantity of energy absorbed per mole or one Avogadro's number of photons, that is 6.02×10^{23} photons of reacting substance during the primary photochemical process.

The magnitude of einstein is given by

$$\Delta E = Nh\upsilon = \frac{Nhc}{\lambda}$$

where, c is the velocity of light, and λ is the wavelength of radiation:

$$\Delta E = \frac{6.02 \times 10^{23} \times 6.624 \times 10^{-27} \times 2.9977 \times 10^{10}}{\lambda \times 4.184 \times 10^{10}}$$

$$= 2.859 \times 10^{5} \text{ K cal mole}^{-1}$$

This energy (E) that activates one mole of the reactant, that is the energy corresponding to the Avogadro number of photons, is called *one einstein.* It is evident the energy will be greater the lower the wavelength. So an einstein corresponding to violet or ultraviolet radiations would involve more energy than one in the visible red region.

The efficiency of a photochemical process is often expressed in terms of quantum yield (ϕ), which is defined as the number of molecules reacting per quantum of light absorbed:

$$\phi = \frac{\text{Number of molecules reacting}}{\text{Number of quanta of radiation absorbed}}$$

2.20 TYPICAL PHOTOCHEMICAL REACTIONS

Decomposition of hydrogen iodide: The photochemical decomposition of HI occurs in the radiation of wavelengths in the range 2070–2820 Å, and the quantum yield is 2. The primary photochemical reaction is represented by the equation

$$HI + h\upsilon \rightarrow H + I, \text{ rate} = I_{abs}$$

This is followed by secondary steps:

$$H + HI \rightarrow H_2 + I, \text{ rate} = K_2 [H][HI]$$

and

$$I + I \rightarrow I_2, \text{ rate} = K_3 [I]^2$$

The overall reaction is obtained by adding the three steps:

$$2HI + h\upsilon \rightarrow H_2 + I_2$$

The rate of decomposition of HI:

$$-\frac{d}{dt}[HI] = I_{abs} + K_2 [H][HI] \tag{2.6}$$

Under steady state condition,

$$\frac{d}{dt}[H] = 0$$

That is,
$$\frac{d}{dt}[H] = I_{abs} - K_2 [H][HI] = 0 \tag{2.7}$$

Adding Equations (2.6) and (2.7),

$$-\frac{d}{dt}[HI] = 2I_{abs}$$

Thus, two moles of HI are decomposed by each quantum absorbed. This is confirmed from experimental results. The mechanism is therefore justified.

Hydrogen-bromine reaction: The photochemical reaction between hydrogen and bromine requires a radiation less than 5107 Å, the mechanism being represented as follows:

	Rate constants
$Br_2 + h\upsilon \longrightarrow Br + Br,$	I_{abs}
$Br + H_2 \longrightarrow HBr + H,$	K_2
$H + Br_2 \longrightarrow HBr + Br,$	K_3
$H + HBr \longrightarrow H_2 + Br,$	K_4
$Br + Br \longrightarrow Br_2,$	K_5

The rate of formation of HBr is given by

$$\frac{d}{dt}[HBr] = K_2 [H_2][Br] + K_3 [H][Br_2] - K_4 [HBr][H] \tag{2.8}$$

Under steady state conditions,

$$\frac{d}{dt}[H] = 0, \frac{d}{dt}[Br] = 0$$

We have

$$\frac{d}{dt}[H] = 0 = K_2 [H_2][Br] - K_3 [H][Br_2] - K_4 [HBr][H] \tag{2.9}$$

and,
$$\frac{d}{dt}[Br] = 0 = 2I_{abs} - K_2 [H_2][Br_2] + K_3 [H][Br_2] + K_4 [H][HBr] - K_5 [Br]_2 \tag{2.10}$$

Adding Equations (2.9) and (2.10) :

$$2I_{abs} = K_5 [Br]^2$$

or,
$$[Br] = \sqrt{\frac{2I_{abs}}{K_5}} \tag{2.11}$$

Substitution of this in Equation (2.9) and simplifying:

$$[H] = \frac{K_2\left(\frac{2I_{abs}}{K_5}\right)^{\frac{1}{2}}[H_2]}{K_3[Br_2] + K_4[HBr]} \tag{2.12}$$

Again substituting Equations (2.11) and (2.12) in (2.8):

$$\frac{d}{dt}[HBr] = \frac{2K_2\left(\frac{2I_{abs}}{K_5}\right)^{\frac{1}{2}}[H_2]}{1 + \left[\frac{K_4[HBr]}{K_3[Br_2]}\right]} \tag{2.13}$$

Thus, the rate of the reaction varies as the square root of the intensity of the light.

2.21 FLUORESCENCE AND PHOSPHORESCENCE

Certain molecules after being excited by absorption of visible or ultraviolet radiation emit radiations having frequency less than that of the absorbed radiation. This is known as *fluorescence.* The emission in fluorescence ceases as soon as the light source is removed. Some examples of fluorescent substance are fluorite (CaF_2), uranium, petroleum, organic dyestuff (such as eosin), fluorescin, and so on.

Some molecules after being excited by absorption of visible or ultraviolet radiation emit the radiations slowly and even long after the light source is removed. This is known as *phosphorescence.* Zinc sulphide and alkaline earth sulphides are good examples of such phosphorescent substances.

2.22 CHEMILUMINESCENCE

This phenomenon is the reverse of photochemical reactions. If at ordinary temperatures, light is emitted as a result of chemical reactions, the phenomenon is called *chemiluminescence.*

Some examples are described here:

- Phosphorous oxidizes to produce a glow that is visible in the dark.

$$4P + 5O_2 \longrightarrow 2P_2O_5.$$

- The decay of wood and vegetation over marshy places is accompanied by chemiluminescence.
- The action of atomic hydrogen on mercury causes blue luminescence due to HgH.

2.23 PHOTOSENSITIZATION

Some molecules, that ordinarily do not undergo photochemical reaction by the absorption of light of a particular wavelength, are made to do so in the presence of a foreign substance. These foreign substances, which apparently do not undergo any chemical change but initiate a photochemical change in another substance, are called *sensitizers*; the process is called *photosensitization*. Mercury or cadmium vapors are often used as sensitizers.

The most outstanding example of photosensitization is the photosynthesis of carbohydrates in plants, in which the green coloring matter chlorophyll is the photosensitizer.

$$x\,CO_2 + x\,H_2O + h\upsilon \xrightarrow{\text{Chlorophyll}} (CH_2O)_x + x\,O_2$$

Chlorophyll absorbs visible light in the region 6000–7000 Å, and this absorbed energy is then passed on to the reactants to form the basic molecule of the carbohydrate.

A very interesting use of photosensitization is found in photography.

SOLVED NUMERICAL PROBLEMS

1. *The wave number of the fundamental vibration of* $^{79}Br - {}^{81}Br$ *is 323.2* cm^{-1}. *Calculate the force constant of the bond.*

   (Mass of ^{79}Br *= 78.9183 amu and* ^{81}Br *= 80.9163 amu).*

   Sol. Reduced mass of $^{79}Br - {}^{81}Br$

$$\mu = \frac{78.9183 \text{ g mol}^{-1} \times 80.9163 \text{ g mol}^{-1}}{(78.9183 + 80.9163) \text{ g mol}^{-1} \times 6.023 \times 10^{23} \text{ mol}^{-1}}$$

$$= 6.6333 \times 10^{-23} \text{ g} = 6.6333 \times 10^{-26} \text{ kg}$$

Now,

$$\upsilon = \frac{1}{2\pi c}\sqrt{\frac{k}{\mu}}$$

$\therefore \qquad k = 4\pi^2 c^2 (\bar{\upsilon})^2 \mu$

$\therefore \qquad k = 4\,(3.14)^2\,(3 \times 10^{10}\ \text{cm s}^{-1})^2\,(323.2\ \text{cm}^{-1})^2 \times (6.6333 \times 10^{-26}\ \text{kg})$

$= 246.095\ \text{N m}^{-1}$

2. *Calculate the frequency of the $I = 3 \rightarrow 2$ transition in the pure rotational spectrum of $^{12}C\ ^{16}O$. The equilibrium bond length is 112.81 pm.*

Sol. We know that,

$$\upsilon_{J \rightarrow J+1} = 2B\,(J + 1)\ \text{cm}^{-1}$$

$$\therefore \qquad \upsilon_{2 \rightarrow 3} = 2B\,(2 + 1)\ \text{cm}^{-1} = 6B\ \text{cm}^{-1}$$

Reduced mass of CO,

$$\mu = \frac{m_1 \cdot m_2}{m_1 + m_2} = \frac{12\ \text{g mol}^{-1} \times 16\ \text{g mol}^{-1}}{(12 + 16)\ \text{g mol}^{-1} \times 6.023 \times 10^{23}\ \text{mol}^{-1}}$$

$$= 1.138 \times 10^{-23}\ \text{g} = 1.138 \times 10^{-26}\ \text{kg}$$

Now, $$I = \frac{h}{8\pi^2 BC}$$

or, $$B = \frac{h}{8\pi^2 IC} = \frac{h}{8\pi^2 \mu r^2 C}$$

$$\therefore\ B = \frac{6.625 \times 10^{-34}\ \text{kg m}^2\ \text{s}^{-1}}{8 \times (3.14)^2 \times 1.138 \times 10^{-26}\ \text{kg} \times \left(112.81 \times 10^{-12}\ \text{m}\right)^2 \times 3 \times 10^{10}\ \text{cm s}^{-1}}$$

$$= 1.93\ \text{cm}^{-1}$$

Hence, $\upsilon_{2 \rightarrow 3} = 6 \times 1.93\ \text{cm}^{-1} = 11.58\ \text{cm}^{-1}$.

3. *The internuclear distance of NaCl is 2.36×10^{-10} m. Calculate the reduced mass and the moment of inertia of NaCl, given atomic masses of $Cl = 35 \times 10^{-3}$ kg mol^{-1} and $Na = 23 \times 10^{-3}$ kg mol^{-1}.*

Sol. Reduced mass of NaCl,

$$\mu = \frac{m_1 \cdot m_2}{m_1 + m_2} = \frac{\left(35 \times 10^{-3}\ \text{kg mol}^{-1}\right)\left(23 \times 10^{-3}\ \text{kg mol}^{-1}\right)}{(35 + 23) \times 10^{-3}\ \text{kg mol}^{-1} \times \left(6.023 \times 10^{23}\ \text{mol}^{-1}\right)}$$

$$= 2.304 \times 10^{-26}\ \text{kg}$$

Moment of inertia,

$$I = \mu r^2 = (2.304 \times 10^{-26}\ \text{kg})\,(2.36 \times 10^{-10}\ \text{m})^2$$

$$= 1.283 \times 10^{-45}\ \text{kg m}^2$$

4. *A solution of thickness 3 cm transmits 30% incident light. Calculate the concentration of the solution, given extinction coefficient $\in$ = 4,000 dm^3 mol^{-1} cm^{-1}.*

Sol. Given, $\in = 4{,}000 \text{ dm}^3 \text{ mol}^{-1} \text{ cm}^{-1}$

$$\frac{I}{I_0} = 0.3$$

$$x = 3 \text{ cm.}$$

Absorbance, $$A = \log \frac{I_0}{I} = \log\left(\frac{1}{0.3}\right) = 0.523$$

As, $$A = \in C x$$

$\therefore$ $$C = \frac{A}{\in x} = \frac{0.523}{4{,}000 \text{ dm}^3 \text{ mol}^{-1} \text{ cm}^{-1} \times 3 \text{ cm}}$$

$$= 4.35 \times 10^{-5} \text{ mol dm}^{-3}$$

5. *The fundamental vibration frequency of CO is 2140 cm^{-1}. Calculate the force constant of the molecule, given atomic masses, $^{12}C = 19.9 \times 10^{-27}$ kg and $^{16}O = 26.6 \times 10^{-27}$ kg.*

Sol. The reduced mass of CO = $\mu = \dfrac{m_C \cdot m_O}{m_C + m_O}$

$$= \frac{(19.9 \times 10^{-27} \text{ kg}) \times (26.6 \times 10^{-27} \text{ kg})}{(19.9 + 26.6) \times 10^{-27} \text{ kg}} = 1.14 \times 10^{-26} \text{ kg}$$

Now, $$\upsilon = \frac{1}{2\pi C}\sqrt{\frac{k}{\mu}}$$

$\therefore$ $$k = 4\pi^2 C^2 (\bar{\upsilon})^2 \mu$$

$$= 4\,(3.14)^2\,(3 \times 10^{10} \text{ cm s}^{-1})^2\,(2140 \text{ cm}^{-1})^2\,(1.14 \times 10^{-26} \text{ kg})$$

$$= 1853 \frac{N}{m}$$

6. *(a) Arrange the following in the increasing order of energy:*

(i) IR *(ii) UV-visible* *(iii) Microwave* *(iv) X-ray*

(b) Which of the following will be rotationally active and why?

(i) H_2 *(ii) NO* *(iii) HCl* *(iv) F_2*

Sol. (*a*) Microwave < IR < UV-Visible < X-ray.

(*b*) HCl and NO because they possess permanent dipole moments, so they are rotationally active.

7. *Microwave spectrum of the gaseous HCl molecule exhibits a series of equally spaced lines with interspacing of 20.7 cm*$^{-1}$. *Calculate the internuclear distance of the HCl molecule.*

Sol. $$2B = 20.7 \text{ cm}^{-1} = 0.207 \text{ m}^{-1}$$

or, $$B = 0.1035 \times 10^2 \text{ m}^{-1}$$

Now, $$I = \frac{h}{8\pi^2 BC}$$

$$= \frac{6.625 \times 10^{-34} \text{ kg m}^2 \text{ s}^{-1}}{8\,(3.14)^2 \left(0.1035 \times 10^2 \text{ m}^{-1}\right)\left(3 \times 10^8 \text{ m s}^{-1}\right)}$$

$$= 2.702 \times 10^{-47} \text{ kg m}^2.$$

$$\mu = \frac{m_1 \cdot m_2}{m_1 + m_2}$$

$$= \frac{1 \text{ g mol}^{-1} \times 35.5 \text{ g mol}^{-1}}{(1 + 35.5) \text{ g mol}^{-1} \times 6.023 \times 10^{23} \text{ mol}^{-1}}$$

$$= 1.561 \times 10^{-24} \text{ g} = 1.561 \times 10^{-27} \text{ kg}$$

∴ $$r = \sqrt{\frac{I}{\mu}} = \sqrt{\frac{2.702 \times 10^{-47} \text{ kg m}^2}{1.561 \times 10^{-27} \text{ kg}}} = 1.422 \times 10^{-10} \text{ m}$$

8. *A solution of thickness 2.5 cm transmits 30% of the incident light. What is the concentration of the solution if the molar absorption coefficient is 8000 dm*^{3} *mol*$^{-1}$ *cm*$^{-1}$*?*

Sol. $$l = 2.5 \text{ cm}$$

$$I/I_0 = 30\ \% = 0.3$$

or, $$I_0/I = 3.333$$

Molar absorption coefficient (ε) = 8000 dm^3 mol^{-1} cm^{-1} concentration of the solution (C) = ?

From Lambert-Beer's Law

$$\log I_0/I = \varepsilon\, Cl$$

or, $\log 3.333 = 8000 \text{ dm}^3 \text{ mol}^{-1} \text{ cm}^{-1} \times C \times 2.5 \text{ cm.}$

or, $0.5228 = C \times 20{,}000 \text{ dm}^3 \text{ mol}^{-1}$

or, $C = 2.61 \times 10^{-5} \text{ mol dm}^{-3}$

9. (*a*) *In the visible-UV spectra of CH_3COCH_3 and $CH_2 = CHCOCH_3$, which one will exhibit the higher value of λ_{max} and why?*

(*b*) *Why do the double and triple bonds of $CH_2 = CH_2$ and $CH \equiv CH$ not absorb IR energy?*

Sol. (*a*) In the visible-UV spectra, $CH_2 = CHCOCH_3$ exhibits a higher value of λ_{max} because it has two conjugated chromophores, that is, one double bond (C=C) and a carbonyl group $\left(\begin{matrix} O \\ \| \\ -C- \end{matrix}\right)$.

(*b*) Because of the symmetrical vibrations of C=C double bond and C≡C triple bond, ethylene and acetylene do not absorb IR energy.

10. *Calculate the number of vibrational degree of freedom for the CO_2 molecule.*

Sol. Because CO_2 is a linear molecule

Vibrational degree of freedom = $3N - 5 = 3 \times 3 - 5 = 4$

EXERCISES

1. Explain Lambert-Beer's law and its importance in the spectroscopic methods of structure elucidation.
2. Derive Lambert-Beer's law.
3. Briefly explain vibrational spectroscopy.
4. What are the laws of photochemistry?
5. What is the origin of vibration and electronic spectroscopy? In which spectral ranges do these spectra occur?
6. What is a selection rule?

   Ans. A restriction is the transition between the quantum state of atoms, molecules or metal. Accordingly, the molecule can increase its rotational or vibrational energy only to the next higher or lower energy level when it absorbs or emits a quantum of electromagnetic radiation.
7. Explain the principle and working of a spectrophotometer, giving its diagram as well.
8. Mention the applications and limitations of microwave spectroscopy.
9. (*a*) What is difference between atomic and molecular spectroscopy?

   (*b*) What is an absorption spectra?

10. Write short notes on

(*a*) Klystron (*b*) Vibration spectroscopy

11. What is a microwave?

MULTIPLE CHOICE QUESTIONS

1. Which of the following molecules will not give rotational spectra?

(*a*) N_2 (*b*) CO (*c*) HCl (*d*) HBr

2. The electronic device that generates a microwave is known as a(n)

(*a*) transistor (*b*) klystron (*c*) amplifier (*d*) oscillator

3. Characteristic groups that are responsible for the color of dyestuffs are known as

(*a*) chromosomes (*b*) auxochromes

(*c*) chromospheres (*d*) diazogroups

4. For vibrational spectra, the transition rule is

(*a*) $\Delta \upsilon = \pm 1$ (*b*) $\Delta \upsilon' = \pm 1$ (*c*) $\upsilon = \upsilon'$ (*d*) none of these

5. The electronic transition possible for CH_4 is

(*a*) $\sigma \rightarrow \sigma^*$ (*b*) $\sigma \rightarrow \pi^*$ (*c*) $\pi \rightarrow \pi^*$ (*d*) $n \rightarrow \sigma^*$

6. The bond distance of a gaseous polar molecule in a microwave spectrum is given by

(*a*) $r = \frac{I}{\mu}$ (*b*) $r = \sqrt{\frac{I}{\mu}}$ (*c*) $r = \sqrt{\frac{\mu}{I}}$ (*d*) $r = \frac{\mu}{I}$

7. The value of a force constant in a vibrational spectra is equal to

(*a*) $\pi^2 c^2 (\bar{\upsilon})^2 \mu^2$ (*b*) $\pi^2 c^2 (\bar{\upsilon})^2 \mu$

(*c*) $4\pi^2 c^2 (\bar{\upsilon})^2 \mu$ (*d*) $4\pi^2 c\, \upsilon^2 \mu$

8. The increasing order of a wavelength is

(*a*) X-ray, UV, IR, Microwave (*b*) Microwave, IR, UV, X-ray

(*c*) Microwave, UV, IR, X-ray (*d*) UV, IR, X-ray, Microwave

Chapter 3

Thermodynamic and Chemical Equilibrium

3.1 INTRODUCTION

Thermodynamics, literally meaning a flow of heat, is concerned with the chemical and physical processes that involve the conversion of one form of energy into another form. It is a branch of physical science dealing with the quantitative relation between heat and mechanical energy. Broadly speaking, it deals with the relationship of heat to all other forms of energy, light, kinetic energy, and so on. It aims to determine the efficiency of various types of engines, that is the percentage of heat supplied is converted into mechanical energy. Thermodynamics also helps us determine the criteria for predicting the feasibility of changes, including chemical changes under a given set of conditions, and the extent to which such changes may proceed before attaining equilibrium.

3.2 TERMINOLOGY OF THERMODYNAMICS

Following are important terms in thermodynamics:

Thermodynamic system: Any portion of matter under consideration that is separated from the surroundings by real or imaginary boundaries. The system is called *homogeneous* if it is uniform throughout (*e.g.*, pure solid, gas). On the other hand, the system is called *heterogeneous* if it consists of two or more different phases (*e.g.*, ice in contact with water, alloys).

Surroundings: The rest of the universe around the sysem (system + surrounding = universe). Any real or imaginary layer separating the system and the surroundings is the boundary, through which matter and energy may be exchanged between the two. Ordinarily, surroundings means a water or air bath.

Open system: The system that is capable of exchanging both matter and energy (as heat) with the surroundings, for example, plants and living beings.

Closed system: The system that is capable of exchanging only energy (as heat or work) with the surroundings, for example, a sealed flask containing a gas.

Isolated system: The system that exchanges neither matter nor energy with the surroundings. For such a system, the matter and energy remain constant. There is no such perfectly isolated system, but our universe can be considered as an isolated system because, by definition, it does not have any surroundings.

Extensive property: The property that is independent of the amount of the substance present in the system, for example, total mass, volume, energy, and so on of a system, because they cannot be determined without considering the system as a whole.

Intensive property: The property that is independent of the amount of the substance present in the system, for example, viscosity, surface tension, refractive index, temperature, pressure, density, specific gravity, specific heat, vapor pressure of liquid, and so on.

State of a system: A set of variables, such as pressure, volume, temperature, composition and so on, which describe the system, is known as the state of the system. Thus, when one or more variables undergo change, then the system is said to have undergone a change of state.

Thermodynamic Processes

The thermodynamic processes are as follows:

Isothermal process: A process carried out at constant temperature. When a system undergoes an isothermal process, the system is usually kept in contact with a constant temperature bath (thermostat), and the system maintains its temperature constant by the exchange of heat with the thermostat.

Adiabatic process: A process in which no heat can leave or enter the system (*i.e.*, thermally insulated). Thus for carrying out an adiabatic process, the system is carefully insulated from the surroundings. Note that in an adiabatic process, the temperature of the system may increase or decrease.

Isobaric process: Carried out at constant pressure. For example, a reaction taking place in an open vessel is always at atmospheric pressure, hence such a reaction is isobaric. Volume change always takes place in an isobaric process.

Isochoric process: The process in which the volume of the system is kept constant.

Reversible process: A process is reversible when the energy change in each step of the process can be reversed in direction by merely a small change in a variable (such as temperature, pressure, etc.) acting on the system. Two important requirements for reversibility:

- The change must be performed slowly.
- There must be no friction and no finite temperature differences.

Daniell cell is an example of a reversible process.

Irreversible process: The process in which the system or surroundings are not restored to their initial state at the conclusion of the reverse process. All naturally occurring processes are irreversible. They always tend to proceed in a definite direction, and if left to themselves, they will never proceed in the opposite direction. An irreversible process takes place suddenly or spontaneously without occurring in successive stages in infinitesimal quantities. Some examples of irreversible process are the following:

- Flow of heat from a hotter body to a colder body
- Mixing of two gases
- Expansion of gases from high pressure regions to low pressure regions through orifices

State function: A thermodynamic property that depends only on the state of the system and is independent of the path followed to bring about the change, for example, internal energy (ΔE), enthalpy change (ΔH), free energy change (ΔG), entropy change (ΔS), and so on.

Internal energy: Every chemical system has some internal energy (E), which is a function of the temperature, the chemical nature of the substance, and at times the pressure and volume of the system. It is, therefore, an extensive property. The exact magnitude of the internal energy cannot be determined because the chemical nature includes various indeterminant factors, such as translational, rotational, and vibrational movements of the molecules; the nature of individual atoms; the arrangement and number of electrons; and so on. The change in internal energy accompanying a chemical or physical process is a measurable quantity. This change in internal energy is represented by ΔE. Suppose a system is subjected to a change of pressure and volume only. A and B are the initial and final states of the system, and E_A

and E_B are the internal energies associated with the system in its initial and final states, respectively. The increase in internal energy:

$$\Delta E = E_B - E_A.$$

Internal energy is a state function because it depends only on the initial and final states and not on the path followed between the state, that is no route is specified.

Perfect Differential

It is presumed that z is a single-valued quantity such that its magnitude at any moment or in any given state is determined by the values of x and y at that moment in the given state. If the variables change, then the magnitude of z may change, and this change dz is a perfect differential.

To conclude, dz will be a perfect differential when it is found that any of the following are true:

- z is a single-valued function depending entirely on the instantaneous values of x and y.
- dz between any two specified points or states is independent of the path of transition.
- $\oint dz$ for complete cyclic process is equal to zero.
- $\dfrac{\partial^2 z}{\partial x\, \partial y} = \dfrac{\partial^2 z}{\partial y \,.\, \partial x}$, that is the second differentials of z with respect to x and y carried out in either order become equal to one another.

Thermodynamic functions such as internal energy, heat content, entropy, and so on, depend only on the thermodynamic states, and therefore, their small changes such as dE, dH, dS, and so on are perfect differentials. dq and dW depend on the path of transition, hence they are not perfect differentials.

3.3 THE FIRST LAW OF THERMODYNAMICS

Helmholtz, Clausius, and Kelvin enunciated a generated law of nature, commonly known as the principle of the conservation of energy or *the first law of thermodynamics*. According to this law:

- Energy can neither be created nor destroyed, although it may be changed from one form into another.
- The total energy of an isolated system remains constant, although it may change from one form to another.

The validity of the first law of thermodynamics is based on the fact that it is impossible to construct a perpetual motion machine, that is a machine that can produce energy without an expenditure of energy.

Mathematical Derivation

Suppose a system changes from state A to state B with the energy changes from E_A to E_B. The system, while undergoing change from state A to B, absorbs heat q from the surrounding and also performs some work W. On the other hand, the performance of work by the system tends to lower the energy of the system because performance of work requires expenditure of energy.

Hence, change in internal energy may be given by

$$\Delta E = E_B - E_A = q - W$$

$$\Delta E = q - W \tag{3.1}$$

For an infinitesimally small change, equation (3.1) may be written as

$$dE = dQ - dW \tag{3.2}$$

where dE is the increase in energy, and dq and dW represent small quantities of heat absorbed and external work done by the system, respectively. Equations (3.1) and (3.2) are the mathematical forms of the first law.

Case I. For a cyclic process, returning to the original state, that is, when the initial and final states are identical,

$$\Delta E = 0$$

or,

$$q = W$$

Thus, in such a case, the work done by the gas is equal to the heat absorbed.

Case II. If a system absorbs heat at a constant volume, it does not perform any external work. Thus W = 0 and therefore,

$$\Delta E = q$$

Thus heat absorbed at a constant volume is equal to the increase in internal energy.

Case III. If the system absorbs heat at a constant pressure P, then there will be a volume change. The volume increases from V_A to V_B at constant pressure P. Then work done W in expending against the pressure P is

$$W = P(V_B - V_A) = P\Delta V$$

Thus equation (1) becomes $\Delta E = q - P\Delta V$.

Limitation of the First Law of Thermodynamics

This law only states the exact equivalence of various forms of energy involved but provides no information regarding the feasibility of the process. Moreover, this law does not say whether a gas can diffuse from low pressure to high pressure or whether water itself can flow uphill, and so on.

3.4 APPLICATION OF THE FIRST LAW OF THERMODYNAMICS

1. **Enthalpy or Heat content (H):** Suppose a change of a system from state A to state B is brought about at a constant pressure (P). The volume has increased from V_A to V_B. Suppose heat absorbed by the system during this change is q. Therefore, by the first law of thermodynamics,

$$q = (E_B - E_A) + \text{External work done (W)}$$

where, E_A and E_B are the initial and final internal energies of the system. But work done (W) by the system on the surroundings in increasing volume from V_A to V_B

$$= P(V_B - V_A)$$

$$\therefore \quad q = (E_B - E_A) + P(V_B - V_A)$$

$$= (E_B + PV_B) - (E_A + PV_A)$$

$$= H_B - H_A = \Delta H$$

where the quantity (E + PV) is known as *enthalpy* or *heat content* of the system. Thus, enthalpy change (ΔH) is equal to the heat absorbed (q). Because E is a state (or definite) property, and P and V are also definite properties of a system, it follows that H is also a definite property, depending upon the state of the system. Because H_B and H_A are definite properties, it is obvious that ΔH like ΔE, is a definite or state property, depending only on the initial and final states of the system. Thus, the heat content or enthalpy change,

$$\Delta H = (E_B + PV_B) - (E_A + PV_A)$$

$$= (E_B - E_A) + P(V_B - V_A)$$

or $$\Delta H = \Delta E + P\Delta V$$

At constant pressure, $\Delta H > \Delta E$.

At constant volume, $\Delta H = \Delta E$.

2. **Heat capacity (C):** Heat capacity (C) of a substance is the "amount of heat required to raise the temperature of 1 g through 1°C." Molar heat capacity is the amount of heat required to raise the temperature of 1 mole through 1°C.

Molar heat capacity is of two types:

- Molar heat capacity at constant volume (C_V) is measured by keeping volume constant.
  It is given by

$$C_V = \left(\frac{\partial E}{\partial T}\right)_V$$

- Molar heat capacity at constant pressure (C_P) is measured by keeping pressure constant.
  It is given by

$$C_P = \left(\frac{\partial H}{\partial T}\right)_P = \frac{dq}{dT}$$

Relation Between C_P and C_V:

It is known that

$$C_V = \left(\frac{\partial E}{\partial T}\right)_V \text{ and } C_P = \left(\frac{\partial H}{\partial T}\right)_P$$

The difference between C_P and C_V gives

$$C_P - C_V = \left(\frac{\partial H}{\partial T}\right)_P - \left(\frac{\partial E}{\partial T}\right)_V \tag{3.3}$$

We know that

$$H = E + PV \text{ and } E = f(V, T)$$

Differentiating the preceding equation with respect to temperature at constant pressure,

$$\left(\frac{\partial H}{\partial T}\right)_P = \left(\frac{\partial E}{\partial T}\right)_V + P.\frac{\partial V}{\partial T} + V.\frac{\partial P}{\partial T} \tag{3.4}$$

At constant pressure, $\frac{\partial P}{\partial T} = 0$

So,
$$\left(\frac{\partial H}{\partial T}\right)_P = \left(\frac{\partial E}{\partial T}\right)_V + P\left(\frac{\partial V}{\partial T}\right)_P \tag{3.5}$$

For 1 mole of an ideal gas $PV = RT$

or,
$$\frac{\partial V}{\partial T} = \frac{R}{P} \tag{3.6}$$

Now, $$\left(\frac{\partial H}{\partial T}\right)_P = \left(\frac{\partial E}{\partial T}\right)_V + R$$

or, $$C_P = C_V + R$$

or, $$C_P - C_V = R \tag{3.7}$$

Thus, the difference between molar heat capacities at consant pressure and constant volume is molar gas constant (R) for an ideal gas. The value of constant R is 1.9872 calories.

For solids and liquids,

$$\Delta H = \Delta E \text{ and } \Delta (PV) \approx 0$$

So, $$C_P = C_V.$$

3. **Reversible isothermal expansion of an ideal gas:** Consider n moles of an ideal gas contained in a cylinder fitted by a weightless and frictionless piston. The system is isothermally isolated from the surroundings, that is temperature is kept constant. The gas expands reversibly from volume V_1 to V_2 so that pressure decreases from P_1 to P_2.

   Now in an isothermal expansion, $\Delta E = 0$, because temperature is constant.

   Hence, the work of expansion will be done only at cost of the heat absorbed, that is

$$q = W = \int_{V_1}^{V_2} P\,dV \qquad [\because\ q = \Delta E + W = W] \tag{3.8}$$

For n moles of the gas,

$$PV = nRT$$

or, $$P = \frac{nRT}{V}$$

Substituting the value of P in Equation (3.8)

$$W = \int_{V_1}^{V_2} nRT\,.\frac{dV}{V} = n\,RT \ln \frac{V_2}{V_1}$$

$$= n\,RT \ln \frac{P_1}{P_2}$$

because, $$\frac{V_2}{V_1} = \frac{P_1}{P_2} \text{ at constant temperature.}$$

$$W = 2.303\, n\, RT \log \frac{V_2}{V_1}$$

Here, $R = 1.987$ cal K^{-1} mol^{-1} = 8.316 J K^{-1} mol^{-1}.
The value of W so obtained in an isothermal reversible expansion is often expressed as maximum work (W_{max}), because a reversible expansion gives the maximum possible work.

4. **Adiabatic expansion of an ideal gas:** In an adiabatic process, heat is neither evolved nor absorbed by the system.

So, $$q = 0 \therefore \Delta E = q - W = -W \tag{3.9}$$

During expansion, work is done by the system on the surrounding, so W must be positive. In other words, ΔE is negative, that is, there is a decrease in the internal energy of the system, so the temperature falls.

Let P = External pressure

and ΔV = Increase in volume

$\therefore$ External work done by the system = $P\Delta V$

or, $$\Delta E = -W = -P\Delta V \tag{3.10}$$

If ΔT = the fall in temperature, then,

$$\Delta E = C_V \Delta T \tag{3.11}$$

From Equations (3.10) and (3.11), we get

$$C_V \Delta T = -P dV = -\frac{RT}{V} dV = -RT \frac{dV}{V}$$

or, $$C_V \frac{dT}{T} = -R \frac{dV}{V}$$

or, $$C_V\, d(\ln T) = -R\, d(\ln V)$$

On integration, we get

$$C_V \ln \frac{T_2}{T_1} = -R \ln \frac{V_2}{V_1} = R \ln \frac{V_1}{V_2}$$

or, $$\ln \frac{T_2}{T_1} = \frac{R}{C_V} \ln \frac{V_1}{V_2} \tag{3.12}$$

Now, $$C_P - C_V = R$$

Dividing both the sides by C_V

$$\frac{C_P}{C_V} - 1 = \frac{R}{C_V}$$

But, $$\frac{C_P}{C_V} = R$$

$$\therefore \qquad R - 1 = \frac{R}{C_V} \tag{3.13}$$

From Equations (3.12) and (3.13), we get

$$\ln \frac{T_2}{T_1} = (\gamma - 1) \ln \frac{V_1}{V_2} = \ln \left(\frac{V_1}{V_2}\right)^{\gamma - 1}$$

or, $$\frac{T_2}{T_1} = \left(\frac{V_1}{V_2}\right)^{\gamma - 1} \tag{3.14}$$

or, $$T_1 V_1^{\gamma - 1} = T_2 V_2^{\gamma - 1} \tag{3.15}$$

But, $$T_1 = \frac{P_1 V_1}{R} \text{ and } T_2 = \frac{P_2 V_2}{R}$$

$$\therefore \qquad \frac{P_1 V_1 V_1^{\gamma - 1}}{R} = \frac{P_2 V_2 V_2^{\gamma - 1}}{R}$$

or, $$P_1 V_1^{\gamma} = P_2 V_2^{\gamma} \tag{3.16}$$

Now, $$V_1 = \frac{RT_1}{P_1} \text{ and } V_2 = \frac{RT_2}{P_2} \tag{3.17}$$

From Equations (3.16) and (3.17), we get

$$P_1 \left[\frac{RT_1}{P_1}\right]^{\gamma} = P_2 \left[\frac{RT_2}{P_2}\right]^{\gamma}$$

or, $$P_1^{1-\gamma} . T_1^{\gamma} = P_2^{1-\gamma} . T_2^{\gamma}$$

or, $$\left(\frac{T_1}{T_2}\right)^{\gamma} = \left(\frac{P_2}{P_1}\right)^{1-\gamma} \tag{3.18}$$

3.5 THERMOCHEMISTRY

Chemical reactions are accompanied by evolution or absorption of heat. If a reaction is accompanied by the evolution of heat, it is called an *exothermic reaction*, and if it is accompanied by the absorption of heat, it is called an *endothermic reaction*. The science of thermochemistry is that branch of physical chemistry that deals with the thermal changes accompanying chemical and physical transformations. It deals with transformation of chemical energy into heat energy and vice versa. The energy change that occurs in a chemical reaction is largely due to a change in bond energy, that is change in potential energy that results from the breaking of bonds in the reactants and formation of new bonds in the products. Some thermochemical definitions are given here:

Heat of reaction: The amount of heat evolved or absorbed during a chemical reaction is called the *heat of reaction* (ΔH). This is the difference between the heat contents (or enthalpies) of the products and the reactants; when molar quantities of the reactants indicated by the balanced chemical equation, react completely, or mathematically, the heat of reaction,

$$\Delta H = \text{Enthalpy of product} - \text{Enthalpy of reactant}$$

ΔH will be positive when heat content of the products is greater than heat content of reactants. Heat will be absorbed during the reaction and the reaction will be endothermic. ΔH will be negative when heat content of the reactants is greater than heat content of products. Heat will be evolved during the reaction and the reaction will be exothermic.

Heat of combustion: Heat of combustion of a compound or an element is the change in heat content when 1 mole of the substance is completely burnt in excess of oxygen. Because combustion reactions are always exothermic, the heat of combustion is always negative. For example:

$$CH_4\,(g) + 2O_2\,(g) \longrightarrow CO_2\,(g) + 2H_2O\,(l); \Delta H = -890.2\text{ k J}$$

Heat of neutralization: The heat of neutralization of an acid (or base) is the change in heat content (ΔH) when 1 gram equivalent of the acid (or base) in dilute aqueous solution is neutralized by 1 gram equivalent of a dilute aqueous solution of a base (or acid). For example:

$$HCl\,(aq) + KOH\,(aq) \longrightarrow KCl\,(aq) + H_2O; \Delta H = -57\text{ k J}$$

The heat of neutralization (ΔH) of a strong acid by a strong base and vice versa is – 57 k J, no matter what acid or base is employed. This is because the heat of neutalization is in fact the heat of formation of 1 mole of water from 1 g equivalent H^+ ions (of acid) and 1 g equivalent OH^- ions (of base), whereas the other ions of strong acid and base remain unaltered. Thus,

$$\underset{\text{(of acid)}}{H^+} + \underset{\text{(of base)}}{OH^-} \longrightarrow H_2O; \Delta H = -57\text{ k J}$$

Heat of formation: Heat of formation of a compound is the change in heat content or enthalpy (ΔH); when 1 mole of the compound is formed from its elements in their commonly occurring states. For example:

$$C + 2H_2 \longrightarrow CH_4\,(g); \Delta H_f = -74.8\text{ k J}$$

Heat of fusion: Heat of fusion of a substance is the heat change in converting 1 mole of it from the solid to liquid phase at its melting point. This is always a positive quantity because heat is needed to overcome the intermolecular forces between the constituent particles of a solid. For example:

$$H_2O\,(s) \longrightarrow H_2O\,(l);\, \Delta H = 6.0\ \text{k J}$$

Heat of vaporization: Heat of vaporization of a liquid is the heat change in converting 1 mole of the substance from liquid to the gaseous state at its boiling point. For example:

$$H_2O(l) \longrightarrow H_2O\,(g);\, \Delta H = 42.00\ \text{k J}$$

Heat of sublimation: Heat of sublimation of a substance (solid) is the heat change in converting 1 mole of a solid directly into its vapor at a given temperature, below its melting point. For example:

$$I_2\,(s) \longrightarrow I_2\,(g);\ \Delta H = 62.4\ \text{k J}$$

3.6 VARIATION OF THE HEAT OF REACTION WITH TEMPERATURE (KIRCHHOFF'S EQUATION)

It is a well known fact that heat of reaction varies with temperature. Making use of the first law of thermodynamics, Kirchhoff (1858) deduced certain mathematical expressions that illustrate the variation of the heat of reaction with temperature. Let us consider a reaction in which reactants X (initial state) change into products Y (final state) at constant pressure. The process can be denoted as

$$\underset{\text{Reactant}}{X} \longrightarrow \underset{\text{Product}}{Y}$$

H_y and H_x are the heat contents of the products and reactants in the final and initial states, respectively. The heat of reaction ΔH will therefore be

$$\Delta H = H_y - H_x \tag{3.19}$$

Differentiating this equation, w.r.t. T at constant pressure, we have

$$\left[\frac{\partial(\Delta H)}{\partial T}\right]_P = \left[\frac{\partial H_y}{\partial T}\right]_P - \left[\frac{\partial H_x}{\partial T}\right]_P$$

or,

$$\left[\frac{\partial(\Delta H)}{\partial T}\right]_P = C_P(y) - C_P(x) = \Delta C_P \qquad (3.20) \left[\because \left[\frac{\partial H}{\partial T}\right]_P = C_P\right]$$

ΔC_P represents the difference in the heat capacities at constant pressure of the products and the reactants. Equation (3.20) may also be put in the form

$$\frac{d\Delta H}{dT} = \Delta C_P \tag{3.21}$$

When the reaction takes place at constant pressure and at constant volume, then similarly it can be proved that

$$\frac{d\Delta E}{dT} = C_V \tag{3.22}$$

Equations (3.21) and (3.22) are known as *Kirchhoff's equations* and represent the variation of the heat of reaction with temperature at constant pressure and at constant volume, respectively.

Again Equation (3.21) can be put as

$$d\Delta H = \Delta C_P dT \tag{3.23}$$

Integrating both the sides, we have

$$\int_{H_1}^{H_2} d\Delta H = \int_{T_1}^{T_2} \Delta C_P dT$$

or $$(\Delta H_2 - \Delta H_1) = \Delta C_P (T_2 - T_1)$$

or $$\Delta C_P = \frac{\Delta H_2 - \Delta H_1}{T_2 - T_1} \tag{3.24}$$

Similarly, $$\Delta C_V = \frac{\Delta E_2 - \Delta E_1}{T_2 - T_1} \tag{3.25}$$

Equation (3.23) may also be integrated between a fixed temperature T_1 and T_2:

$$\int_{T_1}^{T_2} d\Delta H = \int_{T_1}^{T_2} \Delta C_P dT$$

$$\Delta H_{T_2} - \Delta H_{T_1} = \int_{T_1}^{T_2} \Delta C_P dT$$

or, $$(\Delta H)_{T_2} - (\Delta H)_{T_1} = \int_{T_1}^{T_2} \Delta C_P dT \tag{3.26}$$

Thus, the change in the heat of reaction at constant pressure per degree change of temperature is represented by the difference in the heat capacities of the products and the reactants at constant pressure. Similarly, the heat of reaction at constant volume per degree change of temperature is given by the difference in the heat capacities of the products and the reactants at constant volume.

Kirchhoff's equation can be used to calculate the heat of reaction at a particular temperature provided the heat of reaction at some other temperature and heat capacities of products and reactants are given.

3.7 HESS'S LAW OF CONSTANT HEAT SUMMATION

According to this law, "the heat change in a particular reaction is always constant and is independent of the manner in which the reaction takes place."

In other words, the overall heat change in a chemical reaction carried out either at constant pressure (or at constant volume) is the same, irrespective of whether the reaction proceeds in one or several stages. This law is a direct consequence of the first law of thermodynamics, according to which heat change ΔH (enthalpy change) or ΔE (internal energy change) of a reaction is a state function, that is dependent only on the initial and final states of the system.

Example: Consider the formation of CO_2. The carbon might be burnt directly to CO_2. On the other hand, carbon might be burnt to carbon monoxide and then the latter burnt to carbon dioxide. The two procedures are as follows:

First procedure (single step)

$$C\,(s) + O_2\,(g) \longrightarrow CO_2\,(g);\quad \Delta H = 393.5\text{ kJ}$$

Second procedure (Double step)

$$C\,(s) + \frac{1}{2}O_2\,(g) \longrightarrow CO\,(g);\quad \Delta H_1 = 110.5\text{ kJ}$$

$$CO\,(g) + \frac{1}{2}O_2\,(g) \longrightarrow CO_2\,(g);\quad \Delta H_2 = 283.0\text{ kJ}$$

By adding $C\,(s) + O_2\,(g) \longrightarrow CO_2\,(g);\quad \Delta H = 393.5\text{ kJ}$

Measurement of Enthalpy and Heat Capacity (Thermochemical Calculations)

Problems that involve the calculation of the heat of reaction are known as thermochemical calculations. The following guidelines are used to solve such problems:

1. Write down the required thermochemical equation indicating the required heat change.

   For example:

$$6C + 3H_2 \rightarrow C_6H_6,\quad \Delta H = x\text{ cal}$$

2. Try to get the preceding equation from the given facts, which can be achieved in the following two ways:
   - Add, subtract, multiply and divide the various given thermochemical reactions as per the requirement by an integer so that the required thermochemical equation can be obtained.
   - If the heat of formation of all the reactants and products are given, heat of reaction may be evaluated as

$$\Delta H = \Delta H_{products} - \Delta H_{reactants}$$

= Heat of combustion of products – Heat of combustion of the reactants

= Bond energy of products – Bond energy of reactants

3. Remember the following relationships:
   - $\Delta H = \Delta E + \Delta nRT$

     where, ΔH is the change in enthalpy at constant pressure.

     Δn = No. of moles of gaseous products – No. of moles of gaseous reactants

     R = Gas constant (2 cal/mole)

     T = Absolute temperature.
   - $\dfrac{\Delta H_2 - \Delta H_1}{T_2 - T_1} = \Delta C_P$ (Kirchhoff's equation)

     where, ΔC_P is the difference in heat capacities of the products and reactants at constant pressure.
     ΔH_1 and ΔH_2 are the enthalpy change at constant pressure at temperature T_1 and T_2, respectively.
   - $\dfrac{\Delta E_2 - \Delta E_1}{T_2 - T_1} = \Delta C_V$ (Kirchhoff's equation)

     where, ΔC_V is the difference in the heat capacities of the products and reactants at constant volume.
     ΔE_1 and ΔE_2 are the heats of reactions at constant volume at temperature T_1 and T_2, respectively.

3.8 BOND ENERGY

Bond energy or heat of formation of the bond in a compound is the amount of energy required to break the bonds of that type present in one mole of the compound. The bond energy is sometimes referred to as the enthalpy or the heat of formation of the bond.

The strength of the bond joining the two atoms together depends upon environment. For example, the energy needed to break the O—H bond in water (H—O—H) to give H and OH groups is 118 K cal mol^{-1}; whereas the energy needed to break the O—H bond in the hydroxyl radical (—O—H) to give O and H atoms is 102 K cal mol^{-1}. The difference in the two values can be related to the fact that OH radicals are much more reactive than oxygen atoms in combining with H atoms to form stable molecules. These bond-breaking reactions can be written as

$$H_2O\,(g) \longrightarrow OH\,(g) + H\,(g);\ \Delta H = 118\ \text{K cal mol}^{-1}$$

$$OH\,(g) \longrightarrow O\,(g) + H\,(g);\quad \Delta H = 102\ \text{K cal mol}^{-1}$$

By adding,

$$H_2O\,(g) \longrightarrow O\,(g) + 2H\,(g);\quad \Delta H = 220\ \text{K cal mol}^{-1}$$

Thus, the total dissociation energy of water into 2H and O is 220 K cal mol^{-1}. Hence, the average bond energy of O—H bond in water is 110 K cal mol^{-1}.

In a polyatomic molecule, such as AB_n, the average bond energy is calculated by dividing the heat of formation of the compound AB_n by n. For example, in the case of a methane (CH_4) molecule having four C—H bonds, the mean bond energy is one-fourth of the heat of formation of methane. Thus, the average bond energy of C—H bonds in methane

$$= \frac{\text{Heat of formation}}{4}$$

$$= \frac{396}{4}\ \text{K cal mol}^{-1} = 99.0\ \text{K cal mol}^{-1}$$

Alternatively, the C—H bond energy in methane may be calculated as the average of the total dissociation energy of the C—H bonds in the series of different dissociation species. Thus, in methane,

$$CH_4\,(g) \longrightarrow CH_3\,(g) + H\,(g);\quad \Delta H = 104\ \text{K cal mol}^{-1}$$

$$CH_3\,(g) \longrightarrow CH_2\,(g) + H\,(g);\quad \Delta H = 106\ \text{K cal mol}^{-1}$$

$$CH_2\,(g) \longrightarrow CH\,(g) + H\,(g);\quad \Delta H = 106\ \text{K cal mol}^{-1}$$

$$CH\,(g) \longrightarrow C\,(g) + H\,(g);\quad \Delta H = 82\ \text{K cal mol}^{-1}$$

So the bond dissociation energy $= \frac{1}{4}$ (104 + 106 + 106 + 82) = 99.5 K cal mol^{-1}.

Applications of Bond Energy

Following are the applications of bond energy:

Calculation of heat of reaction: Consider the reaction $C_2H_4\,(g) + H_2\,(g) \longrightarrow C_2H_6\,(g)$ given that the bond dissociation energy of C = C is 146 K cal mol^{-1}

C—C is 83 K cal mol^{-1}
H—H is 102 K cal mol^{-1}
C—H is 99 K cal mol^{-1}.

Now, Heat of reaction = – [Bond energy of product – Bond energy of reactant]
But in the given reaction:

$$\begin{matrix} H & & H \\ | & & | \\ C & = & C \\ | & & | \\ H & & H(g) \end{matrix} + \begin{matrix} H \\ | \\ H(g) \end{matrix} \longrightarrow \begin{matrix} & & H & & H & & \\ & & | & & | & & \\ H & - & C & - & C & - & H \\ & & | & & | & & \\ & & H & & H(g) & & \end{matrix}$$

- The four C—H bonds remain unaffected.
- The bond (C = C) breaks.
- The bond H—H in H_2 breaks up.
- The product has one C—C bond in addition to two C—H bonds.

$$\therefore \qquad \Delta H_{reaction} = -[(\Delta H_{C-C} + 2\Delta H_{C-H}) - (\Delta H_{C=C} + \Delta H_{H-H})]$$

$$= -(83 + 2 \times 99) + (146 + 102) \text{ K cal} = -33 \text{ K cal.}$$

Calculation of heat of formation: Consider the reaction

$$5\,C(g) + 12\,H\,(g) \longrightarrow \begin{matrix} & & H & & H & & H & & H & & H & & \\ & & | & & | & & | & & | & & | & & \\ H & - & C & - & C & - & C & - & C & - & C & - & H \\ & & | & & | & & | & & | & & | & & \\ & & H & & H & & H & & H & & H & & \end{matrix}$$

Because there are 4 C—C and 12 C—H bonds

$$\Delta H_f = [4 \times \text{B.E. for C—C} + 12 \times \text{B.E. for C—H}]$$

$$= 4 \times 79.3 + 12 \times 98.5 \text{ K cal mol}^{-1}$$

$$= 1{,}501.66 \text{ K cal mol}^{-1}.$$

3.9 THE SECOND LAW OF THERMODYNAMICS

The limitations of the first law of thermodynamics are as follows:

- No definite direction of the flow of heat is indicated in the first law.
- Feasibility of a reaction is not known.

The principle that would determine the direction as well as the extent of a chemical change is provided by the second law of thermodynamics. The statements are as follows:

- According to Clausius, it is impossible to convert heat into an equal amount of work without producing other changes in some parts of the system. In Clausius's own words, "it is impossible for a self acting machine unaided by any external agency to transfer heat from a body at a lower to one at a higher temperature." To calculate the maximum extent to which heat can be converted to work, we make use of the Carnot cycle. The direction of a particular spontaneous transformation can be ascertained with the help of the second law of thermodynamics.
- According to Lord Kelvin, heat cannot spontaneously pass from a colder to a hotter body. The mechanical work can be completely converted to heat, but on the contrary, heat cannot be completely converted into work without making other changes either in the system or the surrounding in a cyclic process.
- According to Ostwald, a perpetual motion machine of the second kind is not possible.

3.10 ENTROPY

Clausius (1854) introduced another most important thermodynamic function called entropy to make more clear and thorough understanding of the second law of thermodynamics. Entropy of a sytem is a measure of randomness or disorder of the system and is denoted by the symbol S. It is a state function (that is, the change in entropy is independent of path). It can also be regarded as the measure of unavailable energy. The change in entropy of a system (or a reaction) is given by

$$\Delta S = S_{final} - S_{initial}$$

$$= S_2 - S_1 = \frac{dq}{T}$$

Entropy is expressed in Joule degree^{-1} mol^{-1} (J K^{-1} mol^{-1}).

Entropy of Physical Changes

The entropy change during physical changes can be expressed as follows:

Entropy of fusion: (solid melts at its melting point)

$$\Delta S_{fusion} = S_{liquid} - S_{solid}$$

$$= \frac{\Delta H_{fusion}}{T_m}$$

where ΔS_{fusion} = molar entropy of fusion, S_{liquid} = molar entropy of the liquid, S_{solid} = molar entropy of the solid, T_m = melting point of the solid on the Kelvin scale, and ΔH_{fusion} = molar enthalpy of fusion.

Entropy of vaporization: (liquid evaporates at its boiling point)

$$\Delta S_{vap} = S_{vap} - S_{liquid} = \frac{\Delta H_{vap}}{T_b}$$

where, ΔS_{vap} = molar entropy of vaporization, S_{vap} = molar entropy of the vapor, S_{liquid} = molar entropy of the liquid, T_b = boiling point of the liquid on the Kelvin scale, ΔH_{vap} = molar enthalpy of vaporization.

Entropy of sublimation: (solid evaporates directly to vapor at a particular temperature)

$$\Delta S_{sublimation} = S_{vap} - S_{solid} = \frac{\Delta H_{sublimation}}{T}$$

where, $\Delta S_{sublimation}$ = molar entropy of sublimation, S_{vap} = molar entropy of the vapor, S_{solid} = molar entropy of the solid, T = sublimation temperature on the Kelvin scale, and $\Delta H_{sublimation}$ = molar enthalpy of sublimation.

Entropy of Chemical Changes

The entropy change of a chemical reaction is given by the difference between the sum of the entropies of all the products and the sum of the entropies of all the reactants.

Consider a chemical reaction of the type

$$a\,A + b\,B + \ldots \rightarrow c\,C + d\,D + \ldots$$

The entropy change of the reaction is given by

$$\Delta S = (c\,S_C + d\,S_D + \ldots) - (aS_A + bS_B + \ldots)$$

where S_A, S_B, S_C, and S_D are the entropies of pure substances A, B, C, and D ... taking part in the reaction.

So, in general

$$\Delta S = \sum S_{products} - \sum S_{reactants}$$

Entropy of one mole of a substance in pure state at one atmospheric pressure and at 25°C, is called as standard entropy of that substance and is denoted as S°. When a reaction involves each reactant and each product in its standard state, the entropy change is said to be standard entropy change. This is denoted by ΔS°.

Thus, $$\Delta S^\circ = \sum S^\circ_{products} - \sum S^\circ_{reactants}$$

Entropy of Universe

Whenever a system absorbs heat from its surroundings or gives off heat to the surroundings, both the system and the surroundings undergo a change in entropy. The heat change of the surroundings is the negative of the heat change of the system, that is

$$q_{system} = -q_{surroundings}$$

The sum of the entropy change of the system plus that of the surroundings is called the entropy change of the universe.

$$\Delta S_{universe} = \Delta S_{system} + \Delta S_{surroundings}$$

For a reversible process,

$$\Delta S_{universe} = \frac{q_{system}}{T} + \frac{q_{surroundings}}{T} = 0$$

The entropy change of the universe for a reversible process is zero.

For an irreversible process

$$\Delta S_{universe} = \Delta S_{system} - \frac{q_{irr}}{T} > 0$$

or $$\frac{q_{rev}}{T} - \frac{q_{irr}}{T} > 0$$

For an irreversible (spontaneous) process, there is always a net increase in the entropy of the universe.

Significance of Entropy (Molecular Interpretation)

Entropy has been regarded as "a measure of disorder or randomness of a system." Thus, when a system goes from a more orderly to a less orderly state, there is an increase in its randomness, and hence entropy of the system increases. Conversely, if the change is one in which there is an increase in orderliness, there is a decrease in entropy. For example, when a solid changes to a liquid, an increase in entropy takes place because with the breaking of the orderly arrangement of the molecules in the crystal to the less orderly liquid state, the randomness increases. Conversely, the progress of solidification brings about a more orderly state, and consequently there is a decrease in entropy. The process of vaporization produces an increase in randomness in the distribution of molecules, hence an increase in entropy. This concept of entropy as a measure of orderliness or randomness, has led to the conclusion that "all substances in their normal crystalline state at the absolute zero temperature would be in the condition of maximum orderly arrangement because all motion has essentially ceased at 0 K." In other words, entropy of a substance at 0 K is minimum.

In any system, only a part of the total energy is made available for useful work because some energy is always dissipated (or wasted) isothermally. This unavailable energy of the system is considered equal to the product of intensity factor and a capacity factor. The intensity factor is referred to as the temperature; whereas the capacity factor is entropy. Thus, we can say that entropy signifies an unavailable form of system energy.

3.11 FREE ENERGY

Free energy of the system (denoted by symbol G) is a thermodynamic state function, which is related to enthalpy and entropy as

$$G = H - TS \tag{3.27}$$

Free energy change (ΔG) of a system (or reaction) is a measure of energy available for doing useful work.

Thus, $$\begin{bmatrix}\text{Energy available}\\ \text{as useful work}\end{bmatrix} = \begin{bmatrix}\text{Total energy}\\ \text{available}\end{bmatrix} - \begin{bmatrix}\text{Nonavailable}\\ \text{form of energy}\end{bmatrix}$$

or $$G = H - TS$$

The change in free energy between two states of system

$$\begin{aligned}\Delta G &= G_2 - G_1\\ &= (H_2 - T_2S_2) - (H_1 - T_1S_1)\\ &= (H_2 - H_1) - (T_2S_2 - T_1S_1) = \Delta H - (T_2S_2 - T_1S_1)\end{aligned}$$

Hence, at constant temperature (*i.e.*, $T_1 = T_2$)

$$\Delta G = \Delta H - T\Delta S \tag{3.28}$$

We have, $$\Delta H = \Delta E + P\Delta V \text{ (at constant pressure)}$$

$$\Delta G = \Delta E + P\Delta V - T.\,\Delta S$$
$$\text{(at constant pressure and temperature)} \tag{3.29}$$

Physical Significance of Free Energy

From the law of conservation of energy, the total heat energy supplied to the system:

$$q = \Delta E + W_{\text{expansion}} + W_{\text{nonexpansion}}$$

$$= \Delta E + P\Delta V + W_{nonexpansion}$$

$$= \Delta H + W_{nonexpansion} \quad (3.30)$$

Now in a reversible change at constant temperature T, we have

$$\Delta S = q/T$$

or,
$$q = T.\Delta S \quad (3.31)$$

From Equations (3.27) and (3.28), we get

$$T\Delta S = \Delta H + W_{nonexpansion}$$

or,
$$\Delta H - T\Delta S = -W_{nonexpansion}$$

or,
$$-(\Delta G)_{P,T} = W_{nonexpansion} \quad (3.32)$$

That is, – ΔG is a measure of nonexpansion or useful work obtainable in a reversible reaction at a constant temperature and pressure. In other words, a decrease in free energy during a process is equal to the useful work obtainable from the process.

Standard free energy change (ΔG^o) is defined as "the free energy change of the process at 298 K when the reactants are converted into the products in their standard states." Thus,

$$\Delta G^o = \Sigma G^o{}_{products} - \Sigma G^o{}_{reactants} \quad (3.33)$$

Relation between standard free energy change (ΔG^o) and equilibrium constant (K_{eq})

$$\Delta G^o = -2.303\ RT \log K_{eq} \quad (3.34)$$

where,
R = Gas constant (8.314 $JK^{-1}\ mol^{-1}$)

T = Temperature in Kelvin

Relation between standard free energy change (ΔG^o) and cell potential

$$-\Delta G^o = W_{max} = \text{Electrical work} = nFE^o$$

$$-\Delta G^o = nFE^o \quad (3.35)$$

where,
n = Number of electrons involved in the cell reaction

E^o = Standard emf (potential) of cell

F = Faraday (96,500 C)

3.12 GIBBS-HELMHOLTZ EQUATION

We know, $G = H - TS$ and $H = E + PV$

$\therefore$ $G = E + PV - TS$

Differentiating, we get $dG = dE + PdV + VdP - TdS - SdT$

But $dq = dE + dW = dE + PdV$

$\therefore$ $dG = dq + VdP - TdS - SdT$

For reversible process, $dS = dq/T$ or $T.dS = dq$

$\therefore$ $dG = VdP - SdT$

If pressure remains constant (*i.e.*, $dP = 0$), then

$$dG = -SdT \tag{3.36}$$

Let G_1 = initial free energy of a system at T, and $G_1 + dG_1$ = free energy of the system at $T + dT$, where dT is infinitesimally small, and pressure is constant:

$$\therefore \quad dG_1 = -S_1 dT \tag{3.37}$$

where S_1 is the entropy of the system in initial state. Now consider that the free energy of the system in final state is G_2 at T. $G_2 + dG_2$ is the free energy of the system at $T + dT$ in the final state, then

$$dG_2 = -S_2 dT \tag{3.38}$$

where S_2 is the entropy of the system in the final state.

Subtracting Equation (3.37) from Equation (3.38), we get

$$dG_2 - dG_1 = -(S_2 - S_1)\,dT$$

or,
$$d\Delta G = -\Delta S dT \tag{3.39}$$

$\therefore$ At constant pressure, $\left[\dfrac{\partial(\Delta G)}{\partial T}\right]_P = -\Delta S$

We know
$$\Delta G = \Delta H - T\Delta S \tag{3.40}$$

or
$$-\Delta S = \frac{\Delta G - \Delta H}{T} \tag{3.41}$$

From Equation (3.40) and Equation (3.41), we get, $\dfrac{\Delta G - \Delta H}{T} = \left[\dfrac{\partial(\Delta G)}{\partial T}\right]_P$

or $$\Delta G = \Delta H + T\left[\frac{\partial(\Delta G)}{\partial T}\right]_P$$

This equation is called the Gibbs-Helmholtz equation in terms of the free energy and enthalpy change at constant pressure.

Applications

The Gibbs-Helmholtz equation is applicable to all processes occurring at constant pressure. It is mainly used for the following:

- Calculating the chemical affinity or the tendency of a reaction to proceed.
- Calculating the maximum work obtainable from a gaseous reaction (Vant Hoff reaction isotherm).
- Deriving the variation of an equilibrium constant with temperature (Vant Hoff reaction isochore).
- Deriving the law of mass action thermodynamically.
- Calculating the heat change ΔH for a process at constant pressure.

Helmholtz (or Work) Function

To know quantitatively the maximum work obtainable from the system during the given change, Helmholtz introduced another function of state A, called the *work function* or *Helmholtz function,* which is defined as

$$A = E - TS \tag{3.42}$$

Because E, T, and S are state functions of the system, it is evident that A is also a state function of the system.

For a change from state 1 to 2 at constant temperature,

$$A_2 - A_1 = (E_2 - E_1) - T\,(S_2 - S_1)$$

or, $$\Delta A = \Delta E - T\,.\,\Delta S \tag{3.43}$$

where, ΔA is the increase in the function A, and ΔE and ΔS are the corresponding increase in internal energy and entropy, respectively.

Suppose the preceding change is carried out reversibly at constant temperature, T.

Heat absorbed by the system is q, then

$$\Delta S = \frac{q}{T} \text{ or } T\,.\,\Delta S = q$$

Substituting this value of T. ΔS in Equation (3.43) we get,

$$\Delta A = \Delta E - q$$

According to the first law of thermodynamics,

$$q = \Delta E + W$$

or, $$-W = \Delta E - q \tag{3.44}$$

Comparing Equation (3.43) with Equation (3.44), we get

$$-\Delta A = W$$

Because W is the maximum work obtainable, that is, by reversible process, it is evident that a decrease in the function A (*i.e.*, $-\Delta A$) gives the maximum work obtainable from the system during the given change. Comparing Equations (3.38) and (3.43) we get

$$\Delta G = \Delta A + P\,.\,dV \text{ (At constant P and T)}$$

But, $$\Delta A = -W$$

$\therefore$ $$\Delta G = -W + P\,.\,dV$$

or, $$-\Delta G = W - P\,.\,dV \tag{3.45}$$

or, $$\begin{bmatrix}\text{Decrease in}\\ \text{free energy}\end{bmatrix} = \begin{bmatrix}\text{Maximum work}\\ \text{obtainable}\end{bmatrix} - \begin{bmatrix}\text{Work done by expansion}\\ \text{against constant external}\\ \text{pressure P}\end{bmatrix}$$

$$= \text{Maximum useful (or nonmechanical) work.}$$

Thus, $-\Delta G$ is a measure of maximum net useful (nonmechanical) work obtainable from the system at a constant temperature and pressure.

Consequently,

$$\text{Net or useful work} = -\Delta G = W - P\,.\,dV \tag{3.46}$$

Application of Free Energy to Gases

Change of free energy with temperature and pressure: We know that

$$G = H - TS \tag{3.47}$$

Again, $$H = E + PV \tag{3.48}$$

So, $$G = E + PV - TS \tag{3.49}$$

On differentiation, Equation (3.49) becomes

$$dG = dE + PdV + VdP - TdS - SdT \tag{3.50}$$

The first law equation for an infinitesimal change is written as

$$dq = dE + dW$$

or,
$$dq = dE + PdV \ (dW = PdV) \tag{3.51}$$

For a reversible process,

$$dS = \frac{dq}{T}$$

or,
$$dq = TdS \tag{3.52}$$

Comparing Equations (3.51) and (3.52),

$$TdS = dE + PdV \tag{3.53}$$

By putting the value of TdS from Equation (3.53) in equation (3.50),

we get
$$dG = dE + PdV + VdP - dE - PdV - SdT$$

or,
$$dG = VdP - SdT \tag{3.54}$$

This equation shows that the change of free energy of a system depends on the change of pressure and temperature.
At constant pressure, $dP = 0$

So,
$$dG = -SdT$$

or,
$$\left(\frac{\partial G}{\partial T}\right)_P = -S \tag{3.55}$$

This equation represents the change of free energy with temperature at a constant pressure.
At a constant temperature, $dT = 0$.
So, Equation (3.53) can be represented as

$$dG = VdP$$

or
$$\left(\frac{\partial G}{\partial P}\right)_T = V \tag{3.56}$$

The preceding equation represents the change of free energy with the pressure at a constant temperature.

Change of free energy for an ideal gas: For an isothermal process, because T is constant:

$$dG = VdP$$

Assume that the free energy of the system in the initial state is G_1, and that in the final state, it is G_2, when the pressure of the system changes from P_1 to P_2.

So, the change in free energy is given by

$$\Delta G = G_2 - G_1$$

$$= \int_{P_1}^{P_2} V\,dP \quad (3.57)$$

For 1 mole of ideal gas

$$PV = RT$$

So,

$$V = \frac{RT}{P} \quad (3.58)$$

Putting the value of V from Equation (3.58) in Equation (3.57) :

$$\Delta G = \int_{P_1}^{P_2} \frac{RT}{P}\,dP$$

or,

$$\Delta G = RT \int_{P_1}^{P_2} \frac{1}{P}\,dP$$

or,

$$\Delta G = RT\left[\ln P\right]_{P_1}^{P_2} \quad \text{or,} \quad \Delta G = RT\left[\ln \frac{P_2}{P_1}\right]$$

For n moles of ideal gas,

$$\Delta G = nRT \ln \frac{P_2}{P_1} \quad (3.59)$$

Criteria of the Spontaneity of a Chemical Reaction

For a chemical reaction to be spontaneous, ΔG must be negative. ΔG will be negative if:

- ΔH is negative and ΔS is positive.
- Both ΔH and ΔS are positive, and the magnitude of TΔS > ΔH.
- Both ΔH and ΔS are negative, and the magnitude of ΔH > TΔS.

TABLE 3.1 Spontaneous and Nonspontaneous Reactions

ΔH	*ΔS*	*– TΔS*	*ΔG*	*Comment*
– ve	+ ve	– ve	– ve at all T	Spontaneous at all T
– ve	– ve	+ ve	– ve at low T	Spontaneous at low T
			+ ve at high T	Nonspontaneous at high T
+ ve	+ ve	– ve	– ve at high T	Spontaneous at high T
			+ ve at low T	Nonspontaneous at low T
+ ve	– ve	+ ve	+ ve at low T	Nonspontaneous at all T

Exothermic reactions are favored by a decrease in temperature and endothermic reactions are favored by an increase in temperature.

Criteria of the Spontaneity of a Chemical Equilibrium

The system is said to be in a state of equilibrium if $\Delta G = 0$.

3.13 CLAPEYRON–CLAUSIUS EQUATION

Consider a system consisting of only 1 mole of a substance, which exists in two phases A and B. The temperature and pressure of the system is T and P, respectively. Suppose free energy per mole of the substance in two phases A and B is G_A and G_B. The system is in equilibrium, so there is no free energy change, that is

$$G_A = G_B \tag{3.60}$$

Let the temperature of the system be raised to $T + dT$, so that pressure becomes $P + dP$. Suppose the new free energies per mole of A and B phases are $G_A + dG_A$ and $G_B + dG_B$, respectively. As the two phases are still in equilibrium, so

$$G_A + dG_A = G_B + dG_B \tag{3.61}$$

We know, $$G = H - TS = E + PV - TS \tag{3.62}$$

$\therefore$ $$dG = dE + PdV + VdP - TdS - SdT \tag{3.63}$$

But, $$dq = dE + dW = dE + PdV \tag{3.64}$$

and $$dq/T = dS \text{ or } TdS \tag{3.65}$$

From Equations (3.63), (3.64) and (3.65), we get

$$dG = VdP - SdT \tag{3.66}$$

In the various equilibria considered here, work done is due to volume change only, so Equation (3.66) may be applied to phase A as well as B.

$\therefore$ $$dG_A = V_A dP - S_B dT \tag{3.67}$$

and $$dG_B = V_B dP - S_B dT \tag{3.68}$$

where V_A and V_B are the molar volumes of phase A and B, respectively, and S_A and S_B are their molar entropies.

From Equations (3.66), (3.67) and (3.68), we get

$$V_A dP - S_A dT = V_B dP - S_B dT$$

or, $$\frac{dP}{dT} = \frac{S_B - S_A}{V_B - V_A} = \frac{\Delta S}{V_B - V_A} \tag{3.69}$$

where ΔS = molar entropy change, and $V_B - V_A$ is the change in volume when 1 mole of substance changes from A to B.

If molar heat of transition from phase A to B is q, then

$$\Delta S = q/T \tag{3.70}$$

From Equation (3.69) and (3.70), we get

$$\frac{dP}{dT} = \frac{q}{T(V_B - V_A)} \tag{3.71}$$

This is the Clapeyron–Clausius equation.

Applications

1. The molar heat of vaporization (L_V) can be calculated by using the integrated Clapeyron–Clausius equation, that is $\ln \frac{P_2}{P_1} = \frac{L_V}{R}\left[\frac{T_2 - T_1}{T_2 T_1}\right]$.
2. The latent heat of fusion can be calculated provided the vapor pressure at two temperatures of a solid phase in equilibrium with its liquid phase is known.
3. Using the integrated Clapeyron–Clausius equation, it is also possible to calculate the effect of pressure on the boiling point. Hence, if the boiling point of a liquid at one pressure is known, then the other pressure can be calculated.

3.14 THE THIRD LAW OF THERMODYNAMICS

This law, stated by Nernst in 1906, states that "at absolute zero, the entropy of a perfectly crystalline substance is taken as zero."

At absolute zero, the kinetic energy is zero, and the substances are in rigid state, thereby providing them complete order. Due to the total lack of disorder, the entropy of pure elements and compounds at absolute zero is zero. This law is true only for the substances that exist in perfect crystalline form at 0 K. However, if there are imperfections at 0 K, then entropy will be larger than zero.

3.15 PHYSICAL EQUILIBRIUM

Some common physical equilibria are the solid-liquid equilibrium (ice $\rightleftharpoons$ water), the liquid-gas equilibrium (water $\rightleftharpoons$ water vapors), the solid-gas equilibrium, and the dissolution of solids and gases in liquids.

General Characteristics of Physical Equilibrium

Physical equilibria have these characteristics:

- For the solid-liquid equilibrium there is only one temperature at which the two phases can coexist at a given pressure. This is called the *melting point* or *freezing point*.
  At equilibrium, the rate of melting = the rate of freezing.
- For the liquid-gas equilibrium, there is only one temperature at which the two phases can coexist at a given pressure. This is called the *boiling point*.
  At equilibrium, the rate of evaporation = the rate of condensation.
- Certain solids on heating directly change from solid into vapor state.
- For dissolution of solids in liquids, the solubility is constant at a given temperature.
- For dissolution of gases in liquids, the concentration of a gas in a liquid at a given temperature is directly proportional to the pressure of the gas over the solution. Dissolution of gases in water is always exothermic.
- Physical equilibrium is stable and dynamic in nature. At equilibrium, two opposite processes occur at the same rate.
- The measurable properties of the physical system in equilibrium remain constant.
- Physical equilibrium can only be attained in a closed system.

3.16 CHEMICAL EQUILIBRIUM

The stage of a reversible reaction carried out in a closed vessel at which the forward and backward reactions proceed with the same speed is known as *chemical equilibrium.* Chemical equilibrium is designated as a state of dynamic equilibrium. This is illustrated by taking a general example of a reaction:

$$A + B \rightleftharpoons C + D$$

Characteristics of Chemical Equilibrium

The characteristics include the following:

- Chemical equilibrium takes place in a closed container.
- At equilibrium, the rate of the forward reaction is exactly equal and opposite to the rate of the backward reaction.
- The concentration of the reactants and products remains constant at the equilibrium state.

- The equilibrium can be attained either from the reactants or the products side.
- Equilibrium is not affected by the presence of a catalyst.
- Chemical equilibrium is dynamic in nature.

3.17 THE LAW OF MASS ACTION

The law of mass action states that under a given set of conditions, the rate of chemical reaction is directly proportional to the product of the concentrations or active masses of the reacting substances.

Explanation

Let us consider a general reversible reaction:

$$aA + bB \rightleftharpoons cC + dD$$

The rate of forward reaction = $K_f\,[A]^a\,[B]^b$.

The rate of backward reaction = $K_b\,[C]^c\,[D]^d$

where K_f and K_b are rate constants.

At equilibrium, the rate of forward reaction is equal to the rate of backward reaction.

So, $$K_f[A]^a[B]^b = K_b[C]^c[D]^d$$

or, $$\frac{K_f}{K_b} = \frac{[C]^c\,[D]^d}{[A]^a\,[B]^b}$$

or, $$K_c = \frac{[C]^c\,[D]^d}{[A]^a\,[B]^b}.$$

where, $$K_c = \frac{K_f}{K_b}$$

K_c is called as the equilibrium constant. It has a constant value for a given reaction at a specific temperature.

For a gaseous reaction, the term active masses are replaced by partial pressures. Thus,

$$K_p = \frac{[p_C]^c \times [p_D]^d}{[p_A]^a \times [p_B]^b}$$

where p_A, p_B, p_C, p_D, and so on are the partial pressures of A, B, C, and D, respectively. If value of K_c and K_p are large, then it indicates that the amount of product is large, whereas the reactant is small.

Relationship between K_p and K_c

According to the gas equation

$$PV = nRT$$

or,
$$P = \frac{n}{V} . RT = CRT$$

Now,
$$K_p = \frac{[p_C]^c \times [p_D]^d}{[p_A]^a \times [p_B]^b}$$

$$= \frac{[C_C . RT]^c \times [C_D . RT]^d}{[C_A . RT]^a \times [C_B . RT]^b}$$

[By substituting the value of P]

$$= \frac{C_C^c \times C_D^d}{C_A^a \times C_B^b} \times RT^{(c+d)-(a+b)}$$

$$= K_c \times (RT)^{\Delta n}$$

Δn = Number of gaseous moles of products – Number of gaseous moles of reactants.

$$= (c+d) - (a+b)$$

$\therefore$
$$K_p = K_c \times (RT)^{\Delta n}$$

1. When $\Delta n = 0$; $K_p = K_c$
   For example, $H_2 (g) + I_2 (g) \rightleftharpoons 2HI (g)$.
   Here, $\Delta n = 2 - 2 = 0$.
2. When $\Delta n > 0$, $K_p > K_c$.
   the number of moles of products is more than that of the reactants,
   For example, $PCl_5 \rightleftharpoons PCl_3 + Cl_2$; $\Delta n = 1$
3. When $\Delta n < 0$, $K_p < K_c$,
   the number of moles of products is less than that of reactants,
   For example, $2SO_2 + O_2 \rightleftharpoons 2SO_3$; $\Delta n = -1$

Units of K_p and K_c

Because partial pressures are measured in terms of atmospheres, therefore units of K_p will be $(atm)^{\Delta n}$. Similarly, because the concentrations are expressed in moles/liter, units of K_c will be $(moles/liter)^{\Delta n}$. K_p and K_c will be a pure number when $\Delta n = 0$.

Characteristics of Equilibrium Constant (K)

Following are the characteristics of equilibrium constant K:

- Its value is independent of the original concentration of reactants, presence of the catalyst, and change of pressure.
- Its value is independent of the direction from which equilibrium is attained.
- Its value is independent of the nature and number of steps in the reaction as long as stoichiomentry is unchanged.
- It has a definite value at a given temperature and changes with temperature.
- The larger its value, the greater is the reaction toward the product side.
- If the reaction is reversed, the value of equilibrium constant is also inversed,

  For example, $H_2(g) + I_2(g) \rightleftharpoons 2HI(g), K_c$

  $$2HI(g) \rightleftharpoons H_2(g) + I_2(g); K_c' = \frac{1}{K_c}$$

- If the chemical equilibrium is divided by 2, the value of the equilibrium constant for the new equation is the square root of the previous value of K.

  $$H_2(g) + I_2(g) \rightleftharpoons 2HI(g), K_c$$

  $$\frac{1}{2} H_2(g) + \frac{1}{2} I_2(g) \rightleftharpoons HI(g), K_c' = \sqrt{K_c}$$

3.18 TYPES OF CHEMICAL EQUILIBRIA

There are two types of chemical equilibria:

Homogeneous Equilibria: When all the reactants and products of an equilibrium reaction are present in the same phase (that is gaseous or liquid), it is called a homogeneous equilibrium. It has two types:

- **Homogeneous reaction in gaseous phase:** When all the reactants and products are in gaseous state, it is called a homogeneous reaction in gaseous phase.

$$H_2\,(g) + I_2\,(g) \rightleftharpoons 2HI\,(g)$$

- **Homogeneous reaction in liquid phase:** When all the reactants and products are in liquid state, it is called a homogeneous reaction in liquid phase.

$$CH_3COOH\,(l) + C_2H_5OH\,(l) \rightleftharpoons CH_3COOC_2H_5\,(l) + H_2O\,(l)$$

Heterogeneous Equilibria: When the reactants and products of an equilibrium reaction are present in two or more phases, it is called a heterogeneous equilibrium.

$$CaCO_3\,(s) \rightleftharpoons CaO\,(s) + CO_2\,(g)$$

3.19 THE LE-CHATELIER PRINCIPLE

When a system at equilibrium is disturbed by changing the pressure, temperature, or concentration, then the system will tend to adjust itself to minimize the effect of that change as much as possible.

Explanation

Effect of change of temperature: In an endothermic reaction, an increase in temperature favors the forward reaction, whereas a decrease in temperature favors the backward reaction. In an exothermic reaction, low temperature favors the forward reaction, and high temperature favors the backward reaction.

$$N_2 + O_2 + \text{Heat} \rightleftharpoons 2NO\ (\text{Endothermic})$$

$$N_2 + 3H_2 \rightleftharpoons 2NH_3 + \text{Heat}\ (\text{Exothermic})$$

Effect of change of pressure: Increase of pressure shifts the equilibrium toward the side in which the number of gaseous moles decreases.

$$N_2\,(g) + 3H_2\,(g) \rightleftharpoons 2NH_3\,(g)$$

Increase of pressure favors the forward reaction. Note that the change in pressure has no effect on equilibrium if the total volume of the reactants and the products is the same.

Effect of change of concentration: A decrease in the concentration of any of the product or an increase in the concentration of any of the reactants shift the equilibrium in the forward direction.

Effect of a catalyst: A catalyst increases the rate of forward and backward reaction to the same extent.

Applications

- Synthesis of ammonia (Haber's process):

$$N_2 + 3H_2 \rightleftharpoons 2NH_3 + 23\text{ Kcal}$$

High pressure, low temperature, and high concentration of N_2 and H_2 favor the forward reaction.

- Formation of SO_3 (contact process):

$$2SO_2 + O_2 \rightleftharpoons 2SO_3 + 45\text{ Kcal}$$

High pressure, low temperature, and high concentration of SO_2 and O_2 favor the forward reaction.

- Formation of NO (Birkland–Eyde process):

$$N_2 + O_2 \rightleftharpoons 2NO - 43\text{ Kcal}$$

High temperature and high concentration of N_2 and O_2 favor the forward reaction. Because, reaction takes place without change in volume, pressure has no effect on equilibrium.

3.20 IONIC EQUILIBRIUM

Chemical substances are classified as electrolytes and nonelectrolytes based on their behavior in an aqueous solution or on fusion. Substances that are good conductors of electricity in their molten state or in their aqueous solution are called *electrolytes* (*e.g.*, NaCl, $CuSO_4$, acids, and bases). On the other hand, a substance whose aqueous solution or melt do not conduct electricity is called a *nonelectrolyte* (*e.g.*, urea, glucose, and starch). Electrolytes in their aqueous solution or in molten state dissociate into ions. These ions are responsible for the conduction of current through them. However, all the electrolytes are not ionized to the same extent. Depending upon the extent of their ionization, these are classified as *strong* or *weak electrolytes.*

Strong Electrolytes

Strong electrolytes ionize almost completely into their ions when dissolved in water. For example, NaCl, HCl, H_2SO_4, KOH, NaOH, $CuSO_4$, and so on.

Weak Electrolytes

Weak electrolytes ionize to a small extent when dissolved in water, for example, HCOOH, NH_4OH, and so on.

3.21 OSTWALD'S DILUTION LAW

Consider a solution of a binary electrolyte that dissociates as follows:

$$BA \rightleftharpoons B^+ + A^-$$

One mole of BA is dissolved in VL of water, and, at equilibrium, a fraction α is dissociated into ions.

	BA	$\rightleftharpoons$	B^+	+	A^-
Initial concentration	1		—		—
Amount at equilibrium	$(1-\alpha)$ mol		α mol		α mol
Equilibrium concentration	$\left(\frac{1-\alpha}{V}\right)$ mol L^{-1}		$\frac{\alpha}{V}$ mol L^{-1}		$\frac{\alpha}{V}$ mol L^{-1}

By applying the law of mass action, we get

$$\frac{[B^+][A^-]}{[BA]} = \frac{\frac{\alpha}{V}\cdot\frac{\alpha}{V}}{\frac{1-\alpha}{V}} = \frac{\alpha^2}{(1-\alpha)V} = K$$

or

$$K = \frac{\alpha^2}{(1-\alpha)V} = \frac{\alpha^2 C}{(1-\alpha)} \qquad \left[\because C = \frac{1}{V}\right]$$

This equation is known as Ostwald's dilution formula.

where C is the concentration of electrolyte in mol L^{-1}, and K is the constant called the ionization or dissociation constant of the electrolyte.

For a weak electrolyte, α is small as compared with 1, so the term $(1-\alpha)$ may be taken as approximately equal to 1.

So,

$$\frac{\alpha^2}{V} = K$$

or,

$$\alpha = \sqrt{KV} = \sqrt{\frac{K}{C}} \qquad \text{(when } \alpha \text{ is small)}$$

The degree of ionization of a weak binary electrolyte is proportional to the square of the dilution (the volume in L of solution containing 1 mol of solute) or inversely proportional to the square root of concentration (number of mol L^{-1}). The ionization of weak electrolytes increases on dilution.

3.22 ACIDS AND BASES

Arrhenius Concept

According to Arrhenius, an acid is a substance that gives H^+ ions in an aqueous solution, and all the acidic properties are due to these H^+ ions.

$$HCl \rightleftharpoons H^+ + Cl^-$$

$$H_2SO_4 \rightleftharpoons 2H^+ + SO_4^{2-}$$

$$CH_3COOH \rightleftharpoons H^+ + CH_3COO^-$$

A base may be defined as a substance that gives OH^- ions in an aqueous solution, and all the basic properties are due to these OH^- ions.

$$NaOH \rightleftharpoons Na^+ + OH^-$$

$$Ca(OH)_2 \rightleftharpoons Ca^{2+} + 2OH^-$$

The strength of an acid or base in the Arrhenius concept is based on their rates of ionization or the equilibrium constants of acids and bases. An acid is strong when it produces a large concentration of H^+ ions in its aqueous solution. On this basis, HCl, HNO_3, H_3PO_4, and H_2SO_4 are strong acids, because they ionize strongly to give high H^+ ion concentrations, whereas HCN, CH_3COOH, and so on are weak acids, because they ionize feebly in an aqueous solution and yield very low H^+ ion concentration. Similarly, NaOH and KOH are strong bases, whereas $Ca(OH)_2$ and $Mg(OH)_2$ are weak bases.

Bronsted-Lowry Concept

In 1923, Bronsted and Lowry advanced the protonic concept for acids and bases.

- An acid is a substance that has a tendency to give up a proton to any other substance, whereas a base is a substance that has a tendency to accept a proton from any other substance. In other words, an acid is a proton donor, and a base is a proton acceptor.

$$\underset{\text{Acid}}{A} \rightleftharpoons \underset{\text{Proton}}{H^+} + \underset{\text{Base}}{B^-}$$

Thus all ions with positive charge are called *acids*, whereas all ions with negative charges are called *bases*.

- The proton or H^+ ion in an aqueous solution does not exist as H^+ but as hydrated or solvated ion $H^+.H_2O(H_3O^+)$ is called hydronium or oxonium ion. Thus, for an acid to show its acidic character, the presence of solvent is absolutely essential.

$$\underset{\text{Acid}}{HCl} + \underset{\text{Solvent}}{H_2O} \rightleftharpoons H_3O^+ + Cl^-$$

- When an acid donates a proton, the portion of acid that remains is the conjugate base of the acid. Similarly when a base accepts a proton, the entity produced is the conjugate acid of the base.

$$\underset{\text{Acid}}{CH_3COOH} + \underset{\text{Base}}{H_2O} \rightleftharpoons \underset{\text{Conjugate base}}{CH_3COO^-} + \underset{\text{Conjugate acid}}{H_3O^+}$$

Acid + Base ⟶ Conjugate base + Conjugate acid

CH_3COO^- is the conjugate base of the acid (acetic acid), whereas H_3O^+ is the conjugate acid of base (H_2O).

- If a chemical entity can behave both as an acid and as a base, the entity is called amphiprotic or amphoteric (*e.g.*, water).

$$\underset{\text{Acid}}{CH_3COOH} + \underset{\text{Base}}{H_2O} \rightleftharpoons \underset{\text{Base}}{CH_3COO^-} + \underset{\text{Acid}}{H_3O^+}$$

$$\underset{\text{Acid}}{H_2O} + \underset{\text{Base}}{NH_3} \rightleftharpoons \underset{\text{Base}}{OH^-} + \underset{\text{Acid}}{NH_4^+}$$

- If an acid has a strong tendency to donate protons to water molecules, it is called a strong acid.

$$\underset{\text{Strong acid}}{HNO_3 + H_2O} \rightleftharpoons \underset{\text{Weak base}}{H_3O^+ + NO_3^-}$$

- If an acid has a weak tendency to donate protons to water molecules, it is called a weak acid.

$$\underset{\text{(Weak acid)}}{CH_3COOH + H_2O} \rightleftharpoons \underset{\text{(Strong base)}}{H_3O^+ + CH_3COO^-}$$

- If a base has a strong tendency to accept protons from water molecules, it is a strong base. On the other hand, a weak base has a weak tendency to accept protons from water molecules.

$$\underset{\text{(Weak base)}}{NH_3 + H_2O} \rightleftharpoons \underset{\text{(Strong acid)}}{NH_4^+ + OH^-}$$

- The conjugate acid of a strong base is always a weaker acid than the conjugate acid of a weak base.

$$\underset{\text{(Weak acid)}}{H_2O} + \underset{\text{(Weak base)}}{H_2O} \rightleftharpoons \underset{\text{Strong acid}}{H_3O^+} + \underset{\text{Strong base}}{OH^-}$$

The strength of an acid is determined by the relative ease with which its molecule or ion can lose a proton. A strong acid loses protons easily. Therefore, hydrochloric acid has a very strong tendency to lose protons and is a strong acid. Similarly, the strength of a base is determined by the ease with which protons are accepted by it.

Lewis Concept

According to the Lewis theory (1923), an acid is defined as any substance that can accept a pair of electrons, whereas a base is any substance that can donate a pair of electrons.

A Lewis base may be of the following types:

- All negative ions, for example, OH^-, CN^-, CH_3COO^-, Cl^-, Br^-, and so on.
- Neutral molecules having one or more lone pairs of electrons, for example, $\ddot{N}H_3$, $R\ddot{N}H_2$, $H_2\ddot{O}$:, and so on.

A lewis acid may be of following types:

- All cations, for example, Cu^{2+}, Ag^+, Ca^{2+}, and so on.
- Molecules with electron-deficient atoms, for example, $FeCl_2$, $ZnCl_2$, SO_3, $AlCl_3$, and so on.
- Molecules with atoms that can accommodate more electrons in vacant *d*-orbitals in the valence shell, for example, $SiCl_4$, SiF_4, and so on.
- Molecules with multiple bonds between atoms of different electro-negativities, for example, CO_2, SO_2, and so on.

3.23 pH AND pOH

Sorensen suggested an expression that defines the pH of a solution as the negative logarithm of the hydrogen ion concentration expressed in moles/liters.

$$pH = -\log_{10}[H^+] = \log_{10}\frac{1}{[H^+]} \tag{3.72}$$

or,

$$[H^+] = 10^{-pH}$$

In other words, the pH of a solution is equal to the negative power to which 10 must be raised to express the H^+ ion concentration in moles per litre. So, the higher the hydrogen ion concentration, the lower is its pH value.

Similarly, the hydroxyl-ion concentration can be represented as, pOH

$$pOH = -\log[OH^-] \tag{3.73}$$

Thus in aqueous neutral solutions, when $[H^+][OH^-] = 10^{-14}$, taking logs, we get

$$\log[H^+] + \log[OH] = \log 10^{-14} = -14$$

or,

$$-\log[H^+] - \log[OH] = 14$$

So,

$$pH + pOH = 14 \tag{3.74}$$

In neutral solution, $[H^+] = [OH^-] = 10^{-7}$

pH of neutral solution (water) = 7

Any solution with pH lower than seven is an acid, and a solution with pH above seven is an alkaline.

pH of acidic buffer:

$$pH = -\log K_a + \log \frac{[\text{Salt}]}{[\text{Acid}]}$$

$$= pK_a + \log \frac{[\text{Salt}]}{[\text{Acid}]}$$

pH of basic buffer:

$$pOH = -\log K_b + \log \frac{[\text{Salt}]}{[\text{Base}]}$$

$$= pK_b + \log \frac{[\text{Salt}]}{[\text{Base}]}$$

3.24 SOLUBILITY

Solubility is defined as the concentration of solute in its saturated solution in a given solvent at a particular temperature. It depends on the following factors:

Nature of the solvent: The solubility of a substance in a solvent depends on the nature of solvent. Like dissolves like. For example, nonpolar solutes are soluble in nonpolar solvents, and polar solutes are soluble in polar solvents.

Nature of the solute: Solubility depends upon the solute-solvent interaction. The greater the interaction, the greater the solubility, for example the solubility of NaCl in water is more than sugar in water.

Temperature: Most of the solids become more soluble with increasing temperature in a particular solvent. However, the reverse trend is observed in gases.

Pressure: The solubility of gases in liquid varies greatly with pressure. According to Henry's law, the mass of a gas dissolved per unit volume of a solvent is directly proportional to the pressure of the gas in equilibrium with the solution at a particular temperature.

3.25 THE SOLUBILITY PRODUCT

The ionic product of a saturated electrolyte solution is called the *solubility product* and is denoted by K_{sp}.

Consider the dissociation of a sparingly soluble binary electrolyte, A_xB_y as

$$A_xB_y \rightleftharpoons xA^{y+} + yB^{x-}$$

Applying the law of mass action:

$$K = \frac{[A^{y+}]^x [B^{x-}]^y}{[A_x B_y]}$$

In saturated solution:

$$[A_xB_y] = K' \text{ (constant)}$$

Thus, $$[A^{y+}]^x [B^{x-}]^y = K [A_xB_y] = KK' = K_{sp}.$$

$[A^{y+}]$ and $[B^{x-}]$ are the ionic concentrations in a saturated solution. Thus, the solubility product of an electrolyte may be defined as "the maximum product of the concentrations of its constituent ions each raised to the power equal to its ions present in the equation representing the solution of one molecule of the electrolyte."

Case I. When the ionic product is equal to the solubility product, the solution is just saturated.

Case II. When the ionic product is less than K_{sp}, the solution is unsaturated.

Case III. When the ionic product is greater than K_{sp}, the solution is super saturated (the solid is precipitated).

Relation Between Solubility and the Solubility Product

Consider the general form of electrolyte A_xB_y, which dissociates:

$$\underset{S}{A_xB_y} = \underset{xS}{x[A^{y+}]} + \underset{yS}{y[B^{x-}]}$$

Let S mole/liter be the solubility of the electrolyte, then concentration

$$[A^{y+}] = x \,.\, S \text{ mole/L}$$

$$[B^{x-}] = y \,.\, S \text{ mole/L}$$

$$\therefore \quad K_{sp} = [A^{y+}]^x [B^{x-}]^y$$
$$= [x \,.\, S]^x [y \,.\, S]^y$$
$$= x^x \,.\, y^y \,.\, S^{x+y}$$

$$K_{sp} = x^x \,.\, y^y \,.\, S^{x+y}$$

For example, $$BaCl_2 \rightleftharpoons Ba^{2+} + 2Cl^-$$

Because $$x = 1 \text{ and } y = 2$$

$$\therefore \quad K_{sp} = 1^1 \,.\, 2^2 \,.\, S^{1+2} = 4S^3.$$

3.26 THE COMMON ION EFFECT

The common ion effect defined as "the supression of the dissociation of a weak electrolyte by the addition of a strong electrolyte having some common ion." For example, the addition of NH_4Cl (strong electrolyte) to a solution of NH_4OH (weak electrolyte) results in the decrease in dissociation of NH_4OH.

$$NH_4OH \rightleftharpoons \boxed{NH_4^+} + OH^-$$

$$NH_4Cl \rightarrow \boxed{NH_4^+} + Cl^-$$

Common ion

Application of the Common Ion Effect

Precipitation of soap: Soap is a sodium or potassium salt of higher fatty acid. To precipitate soap, we add saturated NaCl solution in the soap solution.

Formation of common salt: When HCl gas is passed through sea water, crystallization of NaCl takes place due to the common ion effect.

Second group sulphides and third group hydroxides: The common ion effect is also applicable to the precipitation of second group sulphides and third group hydroxides.

SOLVED NUMERICAL PROBLEMS

1. *When a system moves from state 1 to state 2 by a certain path, q and W are found to be + 120 cal, and + 70 cal, respectively. Calculate ΔE =?*

Sol. We know $\Delta E = q - W$

or $E_2 - E_1 = q - W = 120 - 70 = +50$ cals

2. *Calculate the difference in ΔH and ΔE for combustion of CH_4 at 25°C.*

Sol. $CH_4(g) + 2O_2(g) \longrightarrow CO_2(g) + 2H_2O(l)$

$$\Delta H = \Delta E + \Delta n RT$$

$$\Delta n = 1 - (1 + 2) = -2$$

$$\therefore \quad \Delta H = \Delta E - 2RT$$

$$\Delta H - \Delta E = -2 \times 2 \times 298 \text{ cals} = -1192 \text{ cals}$$

3. *ΔG and ΔH values for a reaction at 300 K are –16.0 K cal and –10 K cal, respectively. Calculate the free energy at 330 K.*

Sol. We know entropy change (ΔS)

$$\Delta S = \frac{\Delta H - \Delta G}{T} = \frac{-10{,}000 \text{ cal} + 16{,}000 \text{ cal}}{300} = 20 \text{ cal deg}^{-1}$$

$\therefore$ At 330 K, $\Delta G = \Delta H - T\Delta S$

$$= -10{,}000 \text{ cal} - 330 \times 20 \text{ cal}$$

$$= -16{,}000 \text{ cal} = -16.6 \text{ K cal}$$

4. *Calculate the change in free energy at 25°C for the reaction*

$$CO\,(g) + \frac{1}{2} O_2\,(g) \longrightarrow CO_2\,(g)\ \Delta H = -67.37 \text{ K cal}$$

and the change in entropy accompanying the process is – 20.7 cal deg^{-1} mol^{-1}.

Sol. $\Delta G = \Delta H - T\Delta S$

$$= -67.37 \text{ K cal mol}^{-1} - 298 \times (-20.7 \text{ cal deg}^{-1} \text{ mol}^{-1})$$

$$= -67.37 \text{ K cal mol}^{-1} + 6{,}169 \text{ cal mol}^{-1}$$

$$= -67.37 \text{ K cal mol}^{-1} + 6{,}169 \text{ K cal mol}^{-1}$$

$$= -61.061 \text{ K cal mol}^{-1}$$

5. *When 1 g of liquid naphthalene ($C_{10}H_8$) solidifies, 149 joules of heat is evolved. Calculate the enthalpy of fusion of naphthalene.*

Sol. Enthalpy of fusion of naphthalene

$$= 149 \text{ J/g} \times (10 \times 12 + 8 \times 1) \text{ g mol}^{-1}$$

$$= 149 \text{ J/g} \times 128 \text{ g mol}^{-1} = 19{,}072 \text{ J mol}^{-1}$$

$$= 19.072 \text{ k J mol}^{-1}$$

6. *The enthalpies of formation of C_2H_2 (g) and C_6H_6 (g) at 298 K are 230 and 85 kJ $mole^{-1}$, respectively. Calculate the enthalpy change for the reaction.*

Sol. $3C_2H_2\,(g) \longrightarrow C_6H_6\,(g)$

$$\Delta H = \Delta H_{C_6H_6} - 3\Delta H_{C_2H_6}$$

$$= 85 - (3 \times 230) = -605 \text{ k J mole}^{-1}$$

7. *Calculate the free energy and entropy change per mole when liquid water boils at 1 atmosphere, given, for water ΔH_{vap} = 2.0723 k J/g.*

Sol. $\Delta H_{vap} = 2.0723 \text{ k J/g} \times 18 \text{ g mol}^{-1}$ [$\because$ Mol. wt. of water = 18 amu]

$$= 2.0723 \times 18 \text{ k J mol}^{-1} = 37.30 \text{ k J mol}^{-1}$$

Entropy change, $\Delta S = \frac{\Delta H_{vap}}{T_b} = \frac{37.30}{373\ K}\ kJ\ mol^{-1}$

$= 0.1\ kJ\ mol^{-1}\ K^{-1} = 100\ J\ mol^{-1}\ K^{-1}$

Free energy change, $\Delta G = \Delta H - T\Delta S = 37.30\ kJ\ mol^{-1} - 373\ K \times 0.1\ kJ\ mol^{-1}\ K^{-1}$

$= 37.30\ kJ\ mol^{-1} - 37.30\ kJ\ mol^{-1} = 0$

8. *Calculate the change in entropy when 5 moles of an ideal gas expand from a volume of 4 liters to 40 liters at 27°C.*

Sol. We know, $dS = nR \ln \frac{V_2}{V_1} = nR \times 2.303 \log_{10} \frac{V_2}{V_1}$

$= 5 \times 1.987 \times 2.303 \times \log \frac{40}{4} = 22.89\ cal\ degree^{-1}$

9. *10 g of calcium carbonate on thermal decomposition absorbs 4.5 K cal of heat energy. Calculate the heat of reaction.*

Sol. 10 g $CaCO_3$ = (10/100) mol = 0.1 mol

Now, $CaCO_3 + H_2O \rightarrow CaCO_3\ (aq.)$

To find ΔH of the reaction,

$$\Delta H = 4.5\ K\ cal\ /\ 0.1\ mol = 45\ K\ cal\ mol^{-1}$$

10. *At NTP, 2.8 litres of oxygen were mixed with 19.6 liters of hydrogen. Calculate the increase in entropy. Assume ideal gas behavior.*

Sol. Here $n_A = \frac{2.8\ L}{22.4\ L\ mol^{-1}} = 0.125\ mol$

and $n_B = \frac{19.6\ L}{22.4\ L\ mol^{-1}} = 0.875\ mol$

∴ $X_A = \frac{0.125}{0.125 + 0.875} = 0.125$

and $X_B = \frac{0.875}{0.125 + 0.875} = 0.875$

∴ $\Delta S_{mixing} = -R\ [n_A \ln X_A + n_B \ln X_B]$

$= -R\ 2.303\ [n_A \log X_A + n_B \log X_B]$

$= -8.314\ J\ K^{-1}\ mol^{-1} \times 2.303$

$[0.125\ mol \log 0.125 + 0.875\ mol \log 0.875]$

$= -8.314\ J\ K^{-1} \times 2.303\ [-0.1129 - 0.0507]$

$= 3.132\ J\ K^{-1}$

11. *For 1 mole of an ideal gas, T = f (P, V); show that dT is a perfect differential.*
Sol. For 1 mole of ideal gas,

$$PV = RT \text{ or } T = \frac{PV}{R} \tag{3.75}$$

Differentiating with respect to V at constant P,

$$\left[\frac{\partial T}{\partial V}\right]_P = \frac{P}{R}$$

Differentiating again with respect to P at constant V,

$$\frac{\partial^2 T}{\partial V \,.\, \partial P} = \frac{1}{R} \tag{3.76}$$

Now differentiating Equation (3.75) with respect to P at constant V,

$$\left[\frac{\partial T}{\partial P}\right]_V = \frac{V}{R}$$

Differentiating again with respect to V at constant P,

$$\frac{\partial^2 T}{\partial V \,.\, \partial P} = \frac{1}{R} \tag{3.77}$$

From Equations (3.76) and (3.77), we get

$$\frac{\partial^2 T}{\partial V \,.\, \partial P} = \frac{\partial^2 T}{\partial P \,.\, \partial V}$$

Hence, $d\,T$ is a perfect differential.

12. *ΔG accompanying a given process is –85.77 k J at 25°C and –83.68 at 35°C. Calculate the change in enthalpy (ΔH) for the process at 30°C.*
Sol. $\Delta G = \Delta H - T\Delta S$
At 25°C,

$$-\,85.77 \text{ k J} = \Delta H - 298 \text{ K } (\Delta S)$$

or,

$$\Delta S = \frac{(\Delta H + 85.77 \text{ k J})}{298 \text{ K}}$$

At 35°C,

$$-\,83.68 \text{ k J} = \Delta H - 308 \text{ K } (\Delta S)$$

or,

$$\Delta S = \frac{(\Delta H + 83.68 \text{ k J})}{308 \text{ K}}$$

$$\therefore \quad \frac{\Delta H + 85.77 \text{ kJ}}{298 \text{ K}} = \frac{\Delta H + 83.68 \text{ kJ}}{308 \text{ K}}$$

$$\Delta H = -148 \text{ kJ}$$

13. *Calculate ΔS, ΔA, and ΔG for the vaporization of 2 moles of liquid benzene at its boiling point of 80.2°C. Assume ideal gas behavior for benzene vapor, given the latent heat of vaporization (L_V) = 101 cal g^{-1}, and molecular weight of benzene = 78.*

Sol. $\Delta H = \Delta E = 2 \text{ mol} \times 101 \text{ cal g}^{-1} \times 78 \text{ g mol}^{-1}$

$$= 7{,}878 \text{ cal.}$$

$$\Delta S = \frac{\Delta H}{T} = \frac{7{,}878 \text{ cal}}{350.2 \text{ K}} = 22.50 \text{ cal K}^{-1}.$$

$$\Delta G = \Delta H - T\Delta S$$

$$= 7{,}878 \text{ cal} - 350.2 \text{ K} \times 22.50 \text{ cal K}^{-1} = 0$$

and $\qquad \Delta A = \Delta G = 0$

14. *Five moles of a monoatomic ideal gas are compressed reversibly and adiabatically. The initial volume is 6 dm^3, and the final volume is 2 dm^3. The initial temperature is 27°C.*

(i) What would be the final temperature in this process?

(ii) Calculate W, q, and ΔE for the process, given C_V = 20.91 J K^{-1} mol^{-1}, and γ = 1.4

Sol. (*i*) For an adiabatic reversible process,

$$T_1 V_1^{\gamma-1} = T_2 V_2^{\gamma-1}$$

$$T_2 = \left(\frac{V_1}{V_2}\right)^{\gamma-1} . T_1$$

Again, $\qquad \gamma - 1 = \frac{C_P}{C_V} - 1 = \frac{C_P - C_V}{C_V} = \frac{R}{C_V}$

Here, $\qquad T_1 = 300 \text{ K}, V_1 = 6 \text{ dm}^3, V_2 = 2 \text{ dm}^3$

$$T_2 = 300 \times \left(\frac{6}{2}\right)^{\frac{8.314}{20.91}} = 464.3 \text{ K}$$

(*ii*) The process is adiabatic.

So, $\qquad q = 0$

$$\Delta E = -W = -n\, C_V (T_2 - T_1)$$

$$= -5 \times 20.91 \,(464.3 - 300) = -17.18 \text{ kJ / mol}$$

15. *One mole of an ideal gas expands isothermally and reversibly from 5 dm³ to 10 dm³ at 300 K. Calculate q, W, ΔE, ΔH, ΔG, and ΔA.*

Sol. For isothermal and reversible process,

$$\Delta E = \Delta H = 0 \text{ and } \Delta A = \Delta G = -2.303\, nRT \log \frac{V_2}{V_1}$$

$$\Delta A = \Delta G = 2.303 \times -1 \text{ mol} \times 8.314 \text{ JK}^{-1} \text{ mol} \times 300 \text{ K} \log \frac{10}{5}$$

$$= -1.729 \text{ J mol}^{-1}.$$

For isothermal and reversible expansion

$$q = W = -\Delta G = 1{,}729 \text{ J mol}^{-1}$$

16. *Compute the free energy change when 5 moles of an ideal gas expands reversibly and isothermally at 27°C from an initial volume of 50 L to 1,000 L.*

Sol. For isothermal and reversible change,

$$\Delta G = 2.303\, nRT \log \frac{V_1}{V_2}$$

$$= 2.303 \times 5 \text{ mole} \times 8.314 \text{ JK}^{-1} \text{ mol}^{-1} \times 300 \text{ K} \log \frac{50}{1000}$$

$$= 2.303 \times 5 \times 8.314 \times 300 \text{ J} \log \frac{1}{20}$$

$$= 2.303 \times 5 \times 8.314 \times 300 \text{ J} \times (-1.3010)$$

$$= -37{,}366 \text{ J} = -37.366 \text{ k J}$$

17. *The heat of neutralization of NH_4OH and CH_3COOH is –51.46 k J/mol. Calculate ΔH°_{ion} of NH_4OH (for strong acid and strong base ΔH_{neut} = – 57.1 k J/mol).*

Sol. $\Delta H = \Delta H_{ion} + \Delta H_{neut}$

$$\therefore \quad \Delta H^\circ_{ion} = 57.1 \text{ k J/mol} - 51.46 \text{ k J/mol}$$

$$= 5.64 \text{ k J/mol}$$

18. *If V = f (P, T), then show that dV is an exact differential for an ideal gas.*

Sol. For 1 mole of an ideal gas,

$$PV = RT \text{ or } V = \frac{RT}{P}$$

$$\left(\frac{\partial V}{\partial P}\right)_T = -\frac{RT}{P^2} \qquad \left[\because \frac{\partial}{\partial P}\left(\frac{1}{P}\right) = -\frac{1}{P^2}\right]$$

$$\frac{\partial^2 V}{\partial P . \partial T} = -\frac{R}{P^2} \qquad (3.78)$$

$$\left(\frac{\partial V}{\partial T}\right)_P = \frac{R}{P}$$

$$\frac{\partial^2 V}{\partial T.\partial P} = -\frac{R}{P^2} \qquad (3.79)$$

From Equations (3.78) and (3.79) we get

$$\frac{\partial^2 V}{\partial P.\partial T} = \frac{\partial^2 V}{\partial T.\partial P}$$

Hence, dV is an exact differential.

19. *Show that $W_{rev} > W_{irr}$ for an ideal gas.*

Sol. Let P_1, V_1, T_1, and P_2, V_2, T_2 be the initial and final states, respectively of the system:

$$W_{rev} = nRT \ln\left[1-\left(1-\frac{P_1}{P_2}\right)\right] = nRT\left[\frac{P_1}{P_2}-1\right]$$

$$W_{irr} = P_2(V_2 - V_1) = P_2\left[\frac{nRT_2}{P_2} - \frac{nRT_1}{P_1}\right] = nRT\left[1-\frac{P_2}{P_1}\right],$$

$$\because \quad T_1 = T_2 = T$$

$$W_{rev} - W_{irr} = nRT\left[\left(\frac{P_1}{P_2}-1\right)-\left(1-\frac{P_2}{P_1}\right)\right] = \frac{nRT}{P_1P_2}(P_1-P_2)^2$$

Because the R.H.S of the above equation is always positive,

$$\Rightarrow \qquad W_{rev} - W_{irr} > 0 \Rightarrow W_{rev} > W_{irr}$$

20. *Two moles of HI were heated in a scaled tube at 440°C, until the equilibrium state was reached. HI was found to be 22% dissociated. Calculate the equilibrium constant for the dissociation reaction.*

Sol.	$2HI$	$\rightleftharpoons$	H_2	+	I_2
Initial	2 moles		0		0
At equilibrium	$2 - 2 \times 0.22$		0.22		0.22 mole
	$= 1.56$				
Equilibrium concentration	$\frac{1.56}{2}$		$\frac{0.22}{2}$		$\frac{0.22}{2}$ mole/L

$$\therefore \qquad K_C = \frac{[H_2][I_2]}{[HI]} = \frac{\frac{0.22}{2} \times \frac{0.22}{2}}{\left[\frac{1.56}{2}\right]^2} = 0.0119$$

21. *K_C for the reaction $N_2 + 3H_2 \rightleftharpoons 2NH_3$ is 0.5 mole^{-2} liter^2 at 400 K. Find K_P, given that R = 0.082 liter atm. degree^{-1} mole^{-1}.*

Sol. The given equilibrium is

$$N_2(g) + 3H_2(g) \rightleftharpoons 2NH_3(g)$$

$$K_P = K_C(RT)^{\Delta n}$$

$$= 0.5\,(0.0821 \times 400)^{-2}$$

$$= 4.636 \times 10^{-4}$$

22. *The solubility of a sparingly soluble salt, silver chromate Ag_2CrO_4, is 7.5 × 10^{-5} mol L^{-1}. Assuming complete dissociation, calculate the solubility product of the salt.*

Sol. $$Ag_2CrO_4 \rightleftharpoons 2Ag + CrO_4^{2+}$$

$$K_{sp} = 2^2 \times 1^1\,(S)^{2+1} = 4S^3$$

Hence, $$K_{sp} = 4\,(7.5 \times 10^{-5}\text{ mol L}^{-1})^3 = 1.6875 \times 10^{-12}\text{ mol}^3\text{ L}^{-3}.$$

23. *The solubility of AgCl in water at 25°C is 0.00179 g/L. Calculate its solubility product at 25°C.*

Sol. Solubility, $$S = 0.00179\text{ g/L} = 0.0179/170\text{ mol/V}$$

$$= 1.053 \times 10^{-5}\text{ mol/L}$$

∴ Solubility product of AgCl

$$K_{sp} = 1^1 \times 1^1 \times (S)^{1+1} = S^2$$

$$= (1.053 \times 10^{-5})^2$$

$$= 1.1087 \times 10^{-10}\text{ mol}^2\text{ L}^{-2}$$

24. *The solubility product of lead bromide is 8 × 10^{-5}. If the salt is 80% dissociated in saturated solution, find the solubility of the salt.*

Sol. $$PbBr_2 \rightleftharpoons Pb^{2+} + 2Br^-$$

Solubility S S 2S

$$K_{sp} = 1^1 \times 2^2\,(S)^{1+2} = 4S^3$$

∴ $$S = \left(\frac{K_{sp}}{4}\right)^{\frac{1}{3}} = \left(\frac{8 \times 10^{-5}}{4}\right)^{\frac{1}{3}}$$

$$= 0.02714\text{ mole/L}$$

Solubility when salt is 100% dissociated

$$= 2.714 \times 10^{-2}\text{ M}$$

∴ Solubility for 80% dissociation

$$= \frac{80}{100} \times 2.714 \times 10^{-2} = 1.73 \times 10^{-2}\ M$$

Solubility in gm per liter = Molarity × Molecular weight

$$= 1.73 \times 10^{-2} \times 367 = 6.349\ gm/liter$$

25. *Consider the following esterification reaction.*

$$CH_3COOH + C_2H_5OH \rightleftharpoons CH_3COOC_2H_5 + H_2O$$

1 mole of the acid and 1 mole of the alcohol are mixed at a temperature of 25°C. At equilibrium, 0.667 moles of the acid have reacted. Calculate the equilibrium constant, K_C. How much ester would be obtained if 2 moles of the acid were mixed with 1 mole of alcohol under identical conditions?

Sol.	CH_3COOH +	$C_2H_5OH \rightleftharpoons$	$CH_3COOC_2H_5$ +	H_2O
Initial	1 mole	1 mole	—	—
Equilibrium	1 – 0.667	1 – 0.667	0.667	0.667 mole
Concentration	$\frac{0.333}{V}$	$\frac{0.333}{V}$	$\frac{0.667}{V}$	$\frac{0.667}{V}$

$$\therefore \qquad K_C = \frac{[CH_3COOC_2H_5][H_2O]}{[CH_3COOH][C_2H_5OH]} = \frac{\left(\frac{0.667}{V}\right)\left(\frac{0.667}{V}\right)}{\left(\frac{0.333}{V}\right)\left(\frac{0.333}{V}\right)} = 4.00$$

When 2 moles of acid are mixed with 1 mole of alcohol, let x mole of acid react.

$$[CH_3COOH] = \frac{2-x}{V}, [C_2H_5OH] = \frac{1-x}{V}, [CH_3COOC_2H_5] = [H_2O] = \frac{x}{V}$$

$$\therefore \qquad K_C = 4.0 = \frac{\frac{x}{V} \cdot \frac{x}{V}}{\frac{2-x}{V} \cdot \frac{1-x}{V}} = \frac{x}{2 - 3x + x^2}$$

$$3x^2 - 12x + 8 = 0$$

$$\therefore \qquad x = \frac{12 \pm \sqrt{144 - 96}}{8} = \frac{12 \pm \sqrt{48}}{8} = 2.366 \text{ or } 0.634$$

0.634 mole of ester would be formed, because the other value, $x = 2.366$, is not permissible.

26. *Find the pH of buffer solution of 0.2 mole of acetic acid per litre and 0.10 mole of sodium acetate per litre. The dissociation constant of acetic acid is 1.8 × 10^{-5}.*

Sol. The pH of acidic buffer,

$$= -\log K_a + \log \frac{[\text{Salt}]}{[\text{Acid}]} = -\log 1.8 \times 10^{-5} + \log \frac{[0.1]}{[0.2]}$$

$$= 4.4737$$

27. *Calculate the pH of a buffer solution containing 0.2 M NH_4OH and 0.02 M NH_4Cl (pK_b = 4.7).*

Sol.
$$\text{pOH} = \text{p}K_b + \log \frac{[\text{Salt}]}{[\text{Base}]}$$

$$= 4.7 + \log \frac{[0.02]}{[0.2]} = 3.7$$

∴
$$\text{pH} = 14 - 3.7 = 10.3$$

EXERCISES

1. Derive the Gibbs-Helmholtz equation.
2. Derive the Clausius-Clapeyron equation and discuss its applications.
3. State and explain the second law of thermodynamics. Explain the term entropy.
4. State Hess's law, and mention its important applications with suitable examples.
5. What is Kirchhoff's equation? State its significance.
6. Derive the Kirchhoff equation. Write down the Gibbs-Helmholtz equation.
7. Write a short note on entropy.
8. State the law of mass action, and apply it to the reaction:

$$N_2\,(g) + 3H_2\,(g) \rightleftharpoons 2NH_3\,(g) + Q$$

9. (*a*) What is Le-Chatelier's principle?
   (*b*) What is the effect of pressure on the system?

$$PCl_5 \rightleftharpoons PCl_3 + Cl_2$$

10. Explain the solubility product.
11. Write short notes on
    (*a*) The common ion effect (*b*) The solubility product
12. Explain the effect of pressure and temperature on the following industrial reactions:

$$N_2 (g) + 3H_2 (g) \rightleftharpoons 2NH_3 (g) + 22{,}000 \text{ cal}$$

$$N_2 (g) + O_2 (g) \rightleftharpoons 2NO (g) + 43{,}200 \text{ cal}$$

MULTIPLE CHOICE QUESTIONS

1. Which one of the following is not an intensive property?
   (*a*) pressure (*b*) volume
   (*c*) temperature (*d*) time
2. Evaporation of water is
   (*a*) an exothermic process
   (*b*) an endothermic process
   (*c*) a process where no heat change occurs
   (*d*) a process accompanied by chemical reaction
3. The first law of thermodynamics is expressed by which of the following?
   (*a*) $q = \Delta H - W$ (*b*) $\Delta E = q - W$
   (*c*) $\Delta H = q + W$ (*d*) $\Delta E = \Delta H + P\Delta V$
4. Which of the following is true in an endothermic reaction?
   (*a*) $\Delta G > 0$ (*b*) $\Delta H < 0$
   (*c*) $\Delta G < 0$ (*d*) $\Delta H > 0$ and $\Delta S > 0$
5. The variation of the heat of reaction with temperature is known as what?
   (*a*) Vant Hoff isotherm (*b*) Vant Hoff isochore
   (*c*) Kirchhoff's equation (*d*) None of these
6. One liter-atmosphere of work is equal to
   (*a*) 22.4 cal (*b*) 24.2 cal
   (*c*) 0.821 cal (*d*) 814 cal
7. Bond energy of a substance
   (*a*) is always positive
   (*b*) is always negative
   (*c*) is either positive or negative
   (*d*) depends upon the physical state of the system
8. The temperature of the system decreases in an
   (*a*) adiabatic expansion (*b*) isothermic expansion
   (*c*) isothermal compression (*d*) adiabatic compression
9. The heat of neutralization of a strong acid and a strong base is nearly equal to
   (*a*) 10 k J/mol (*b*) 10 cal/mol
   (*c*) 57 cal/mol (*d*) – 57 k J/mol
10. In an isothermal expansion of an ideal gas,
   (*a*) $\Delta E = 0$ (*b*) $W = 0$
   (*c*) $\Delta V = 0$ (*d*) $q = 0$

11. In $2HI = H_2 + I_2$, the forward reaction is affected by a change in
 (*a*) catalyst (*b*) pressure
 (*c*) volume (*d*) temperature
12. Which of the following conditions will shift the equilibrium of an exothermic reaction to the right?
 (*a*) Lowering (*b*) Increasing
 (*c*) Keeping fixed temperature (*d*) None of these
13. The solubility of Ca $(OH)_2$ is $\sqrt{3}$. What is its solubility product?
 (*a*) $12\sqrt{3}$ (*b*) 3
 (*c*) 27 (*d*) $3\sqrt{3}$
14. In $CaF_2 = Ca_2^+ + 2F^-$, solubility $= x$. What is the solubility product?
 (*a*) $2x^2$ (*b*) $2x^3$
 (*c*) $4x^3$ (*d*) $3x^2$

Chapter 4

PHASE RULE

4.1 INTRODUCTION

The phase rule put forward by Williard Gibbs (1876) is the most useful generalization concerning the equilibria in heterogeneous systems. The phase rule is a relation between the number of components (C), the number of phases (P), and the variable parameters (F) of a heterogeneous system in equilibrium. It is expressed as

$$F + P = C + 2$$

$$F = C - P + 2 \tag{4.1}$$

The phase rule may be defined as, "in a heterogeneous system in equilibrium the number of degrees of freedom plus the number of phases equal the number of components plus 2."

4.2 EXPLANATION OF TERMS

Phase (P)

A *phase* is that part of a system that is chemically and physically uniform throughout. A phase is defined as any homogeneous and physically distinct part of a system that is separated from other parts of a system by definite boundary surfaces. For example:

- One liquid layer constitutes one phase, whether it is pure or a mixture.
- Two liquid layers in contact represent two phases because there is a definite surface of separation between them (*i.e.*, benzene and water).
- A gas or a mixture of gases constitutes one phase, because the system is homogeneous (*i.e.*, N_2 and H_2).
- Every solid constitutes a separate phase except in case of a solid solution (*i.e.*, sulphur).
- All homogeneous solutions are one-phase systems.
- A heterogeneous system is composed of two or more phases.

Component (C)

The number of components is defined as the least number of independent chemical constituents necessary to express the composition of every possible phase of the system. For example:

- The water system consists of one component; the composition of each of the three phases (solid, liquid, and vapor) can be expressed in terms of component H_2O. Any system consisting of a single chemical individual is always a one-component system.
- The sulphur system consists of four phases: rhombic, monoclinic, liquid, and vapor. Because the chemical composition of all phases is S, it is a one-component system.
- The solution of NaCl in H_2O is a two-component system.
- The decomposition of $CaCO_3$:

$$\underset{\text{Solid}}{CaCO_3} \rightleftharpoons \underset{\text{Solid}}{CaO} + \underset{\text{Gas}}{CO_2}$$

  This system constitute three phases: solid $CaCO_3$, solid CaO, and gaseous CO_2. $CaCO_3$ is equivalent to CaO + CO_2. The two components may be taken as CaO and CO_2, so that the composition of the $CaCO_3$ phase can be represented as x CaO + x CO_2, that of calcium oxide as yCaO + $o$$CO_2$, and that of CO_2 as oCaO + $y$$CO_2$.
- The rusting of iron:

$$Fe\ (s) + H_2O\ (g) \rightleftharpoons FeO\ (s) + H_2\ (g)$$

  This system consists of three phases. Out of four constituents, any three may be considered as components to explain the system.
- The dissociation of ammonium chloride in vacuum:

$$NH_4Cl\ (s) \rightleftharpoons NH_3\ (g) + HCl\ (g)$$

  The proportions of NH_3 and HCl are equivalent, and hence the composition of both phases (solid and gas) can be expressed in terms of NH_4Cl alone. Hence, the number of component is one. However, if NH_3 or HCl is in excess, the system becomes a two-component system.

- In the dissociation reaction,

$$CuSO_4 \,.\, 5H_2O\ (s) \rightleftharpoons CuSO_4 \,.\, 3H_2O\ (s) + 2H_2O\ (g)$$

  the composition of each phase can be represented by the simplest components, $CuSO_4$ and H_2O. Hence, it is two-component system.
- Decomposition of PCl_5

$$PCl_5\ (s) \rightleftharpoons PCl_3\ (l) + Cl_2\ (g)$$

The equilibrium is expressed by three phases but need only two components. The third component is automatically fixed at a low temperature. At a high temperature, all the compounds remain in a gaseous state, and thus it can be explained by only one component.

If C′ is the number of chemical constituents in the system, r is the number of restrictions imposed on the independent variation of these constituents; as a general rule the number of components C is given by $C = C' - r$. In the preceding case there are three constituents and one restriction (equivalence of CaO and CO_2), hence the number of components is $C = 3 - 1 = 2$.

The components of a system are not synonyms with the consitutents of the system. All the constituents need not be components, but all the components must be included in the constituents. When no reaction takes place, the number of components is equal to the number of constituents.

Degree of Freedom (F)

The number of degree of freedom or variance of a system is defined as the number of variable factors, such as temperature, pressure, and concentration, which must be specified so that the condition of a system at equilibrium may be completely defined.

In other words, the number of degrees of freedom is the number of factors that can be varied independently without altering the number of phases. Systems possessing one, two, three, and so on, degrees of freedom are said to be univariant, bivariant, trivariant, and so on respectively. For example:

- In a water system, the three phases—ice, liquid water, and water vapor—can be in equilibrium only at particular temperature and pressure. If any of these two factors is altered, one of the phases disappears. Hence, there is no independent variable, or it is a nonvariant system, or the number of degrees of freedom is zero.
- For a system consisting of water in contact with its vapor, if we fix either the temperature or pressure, the system is defined completely. Hence, degree of freedom is one, or the system is univariant.
- For a system consisting of the water vapor phase only, we must state the values of both temperature and pressure to define the system completely. Hence, the system is bivariant or has two degrees of freedom.

Applications of the Phase Rule

- The phase rule is applicable to both physical and chemical equilibria.
- The phase is applicable to macroscopic systems.
- The phase indicates that different systems with the same degree of freedom behave similarly.
- The phase helps us predict the behavior of a system under different sets of variables.

Phase Diagram

A diagram that shows the conditions of equilibrium between phases of a substance is called a *phase diagram*. Phase diagrams are of considerable commercial and industrial significance, particularly for semiconductors, ceramics, steels, and alloys.

4.3 ONE-COMPONENT SYSTEMS

A one-component system is defined as a system in which it is possible to express the composition of all phases in terms of one substance. We will now discuss some important one-component systems.

Water System

The water system consists of three phases (ice, water, and water vapor) and is the simplest and most typical example of a one-component system.

In this case, $C = 1$

$$F = 1 - P + 2 = 3 - P$$

or,

$$F + P = 3$$

So, the sum of the number of phases and degrees of freedom should always equal 3, if the system is in equilibrium. The diagram for the water system is shown schematically in Fig. 4.1.

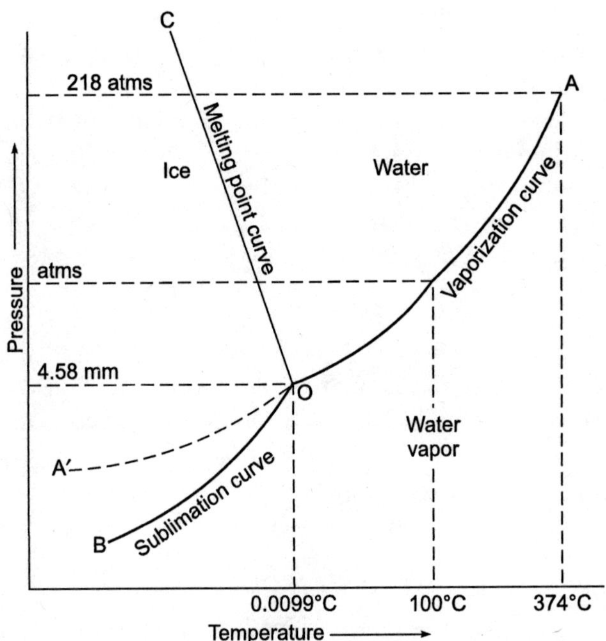

Fig. 4.1 Phase diagram for the water system.

Areas: AOB, AOC, and BOC are the fields of existence of the vapor, liquid, and ice phases, respectively. Within these single-phase areas, it is necessary to state both the temperature and pressure of the phase because to locate any point in an area, both the temperature and pressure coordinates are required. So the system is bivariant. This also follows from the phase rule equation:

$$F = 3 - P = 2.$$

Boundary lines: Separating the areas are lines OA, OB, and OC, which shows that the two phases are in equilibrium. When two phases are in equilibrium, it is sufficient to state either the temperature or the pressure. Hence, any point on the boundary lines has one degree of freedom or is univariant. This also follows from phase rule equation:

$$F = 3 - P = 3 - 2 = 1.$$

- The curve OA, which separates the liquid from vapor region, is called the vapor pressure curve for water, and the vapor pressure increases with the rise in temperature. There is an upper limit for OA terminating at A, its critical state (374°C); above this temperature, the distinction between the liquid and vapor vanishes.
- The curve OB is the sublimation curve of ice. It registers the vapor pressure under which ice is in equilibrium with its vapor at different temperatures.

This is the line of demarcation between ice and vapor. The point B has a natural limit at – 273°C, beyond which the two phases merge into each other.

- The curve OC is the freezing point curve indicating the equilibria at different termperatures between ice and water. The line has a negative slope, that is, the melting point is lowered with the increase in pressure.

Triple point: The point O where the three lines meet, all the three phases *i.e.*, ice, water and vapor should coexist. This is called a triple point. Triple point has P = 4.58 mm and t = 0.0099°C. Because there are three phases,

$$F = 3 - P = 0$$

that is the system is nonvariant at triple point. If either the temperature or the pressure is changed, the three phases would not coexist.

It is, however, possible to cool water below the freezing point without solidification. The liquid below its freezing point is said to remain in a supercooled state. This supercooled state is not stable and is called a metastable state. The dotted curve OA′ is the vapor pressure curve of supercooled water, which is a continuation of AO. The line OA′ is above OB, the vapor pressure curve of ice indicating that the metastable supercooled water has a higher vapor pressure than ice. As a principle, the vapor pressure of a metastable phase is greater than that of a stable phase.

Sulphur System

Sulphur system is somewhat complicated because of its polymorphic nature. Broadly, sulphur has four phases: rhombic sulphur (S_R), monoclinic sulphur (S_M), liquid sulphur (S_L), and sulphur vapor (S_V).

The phase diagram for the system is shown in Fig. 4.2. A system with all the four phases cannot exist. In such a case,

$$F = 1 - 4 + 2 = -1$$

The degree of freedom becomes negative, which has no meaning. There are four possibilities of three phases coexisting at equilibrium. From the phase rule,

$$F = C - P + 2 = 3 - P \text{ [}\because C = 1\text{, as it is an one-component system]}$$

Areas: There are four areas— ABEF, DCEF, BCE, and ABCD— where only one phase, that is S_R, S_L, S_M, and S_V exists, respectively. Within these single phase areas, the system is bivariant because both the temperature and pressure coordinates are required to locate any point. This also follows from the phase rule equation:

$$F = 3 - P = 3 - 1 = 2$$

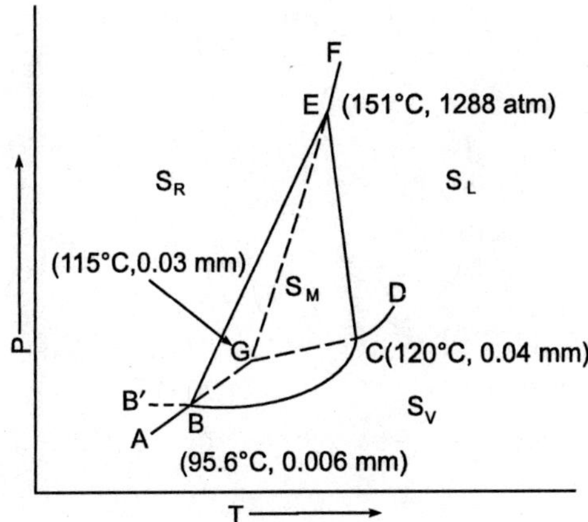

Fig. 4.2 The sulphur system.

Curves: There are six stable curves— AB, BC, CD, BE, CE, and EF— indicating stable equilibria of the two phases side by side. There are also four metastable curves— BG, CG, EG, and BB′, which are continuations indicating metastable equilibria between two phases. The curves are univariant. This also follows from the phase rule equation

$$F = 3 - P = 3 - 2 = 1$$

For a given value of pressure, temperature is automatically fixed on a curve and vice versa:

- Curve AB is the vapor pressure curve of solid rhombic sulphur ($S_R \rightleftharpoons S_V$). The rhombic sulphur is stable up to B (95.6°C, 0.006 mm), above which the rhombic crystals change to monoclinic form. Point B is known as a transition temperature ($S_R \rightleftharpoons S_M$). The melting point of rhombic sulphur is at G. The curve BG is the metastable vapor pressure curve of rhombic sulphur.
- Curve BC is the vapor pressure curve of monoclinic sulphur ($S_M \rightleftharpoons S_V$). The melting point of monoclinic sulphur is at C (120°C).
- Curve CD is the vapor pressure curve of liquid sulphur ($S_L \rightleftharpoons S_V$). CG is the metastable vapor pressure curve of the liquid sulphur (supercooled).
- The curve BE gives the conditions of equilibrium of the rhombic and monoclinic forms ($S_R \rightleftharpoons S_M$). The curve is sloping away from the pressure axis, indicating that the transition temperature of S_R and S_M increases with increasing pressure.
- The curve CE shows the equilibrium between S_M and S_L. CE slopes to the right, thus indicating density of liquid sulphur less than that of the monoclinic solid.

- Curve EF is the melting point curve of S_R ($S_R \rightleftharpoons S_L$). The dotted curve EG is the metastable vapor pressure curve of supercooled monoclinic sulphur.

Triple points: There are three stable triple points:

B (S_R, S_M, S_V)
C (S_M, S_L, S_V)
E (S_R, S_M, S_L)

and one metastable triple point: G (S_R, S_L, S_V).

For these points, $C = 1, P = 3$

So by the phase rule, $F = C - P + 2 = 1 - 3 + 2 = 0$

Therefore, these are invariant points with no degrees of freedom. Variation of any one of the variables, that is temperature or pressure, causes disappearance of one of the three phases. The diagram thus completely represents the behavior of sulphur under any given condition of temperature and pressure.

4.4 TWO-COMPONENT SYSTEMS

A two-component system may be defined as "a system in which it is possible to express the composition of all phases present in the system in terms of two constituents." For a two-component system, the phase rule equation may be put in the form

$$F = 2 - P + 2 = 4 - P$$

or

$$F + P = 4$$

In a two-component system, when P = 1, the degree of freedom (F) has the highest value:

$$F = 4 - 1 = 3$$

Because the maximum number of degrees of freedom in a two-component system is three, the phase behaviors of a binary system may be represented by a 3D diagram of pressure, temperature, and composition, which cannot be conveniently shown on a paper.

The Phase Rule for Two-Component Alloy Systems

In general, the phase rule equation is

$$P + F = C + E$$

where E is the number of external variables such as temperature, pressure, and in unusual cases, electric field and magnetic field. A solid-liquid equilibrium of an alloy has practically no gas phase, and the effect of pressure is negligible on this type of equilibrium. Thus, keeping the pressure constant of a system in which vapor phase is not considered is known as a condensed system.

As such, the phase rule becomes

$$F = C - P + 1$$

This is known as the *condensed phase rule.* The solid-liquid equilibria are represented on temperature-composition diagrams.

4.5 EUTECTIC SYSTEM

A binary system consisting of two substances, which are miscible in all proportions in liquid phase but do not react chemically, is known as a eutectic system. For example:

- Pure lead melts at 327°C.
- Pure tin melts at 232°C.

Alloys of lead and tin melt at low temperatures (183°C). The alloy so formed is a eutectic alloy. The eutectic mixture is a solid solution of two or more substances having the lowest melting point of all the possible mixture of components. Alloys of low melting point are generally eutectic mixtures.

Eutectic Point

The *eutectic point* (*i.e.*, lowest melting point) refers to two or more solid substances capable of forming solid solutions with lowering the melting point and the minimum melting point attainable corresponding to the eutectic mixture (*i.e.*, lowest melting point).

The Bismuth-Cadmium Alloy System

This two-component system has four phases: Bi (*s*) + Cd (*s*), Bi + Cd solution, Bi (*s*) + Cd (*l*), and Cd (*s*) + Bi (*l*). Because the gaseous phase is absent, one variable —pressure— is neglected, and so the condensed phase rule is applicable.

$$F = C - P + 1 = 2 - P + 1$$

or,

$$F = 3 - P$$

The system can be represented by a temperature-composition diagram (Fig. 4.3).

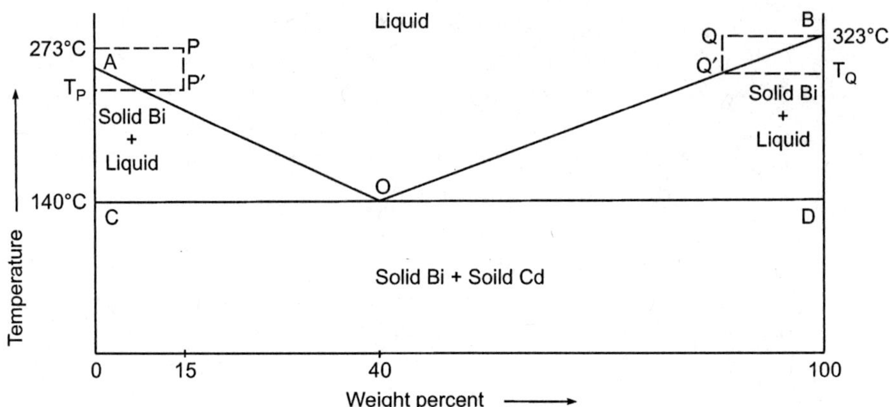

Fig. 4.3 T-C diagram of the Bi-Cd system.

Curve AO (freezing point curve of Bi): Melting point of Bi falls gradually on adding Cd, along AO until the lowest point O (140°C) is reached. At O, Bi will no further go into solution. Along AO, degree of freedom,

$$F = 3 - P = 3 - 2 = 1, \text{ that is univariant.}$$

Curve BO (freezing point curve of Cd): Point B is the melting point of pure Cd (323°C). Along BO, the melting point gradually falls on the addition of Bi until the lowest point O is reached. Melting point of Cd does not fall thereafter. Along BO, the system is also univariant.

$$F = 3 - P = 3 - 2 = 1.$$

Eutectic point O: The two curves AO and BO meet at O where three phases, that is, solid Bi, solid Cd, and their solution coexist. The system is invariant.

$$F = 3 - P = 3 - 3 = 0$$

The point O represents a fixed composition (40% Cd and 60% Bi) and is called eutectic point *i.e.*, at 140°C (lowest melting point for mixture of Bi and Cd) and the composition is called eutectic composition.

Area AOB represents the solution of Bi-Cd. It is called a liquidous curve that is above it only liquid exists.

Iron-Carbon System

The iron-carbon system is the theoretical basis for ferrous metallurgy. Pure iron on cooling from the liquid state solidifies at 1535°C, and the cooling curve then shows three breaks: 1400°C, 910°C, and 768°C. There are, therefore, four solid phases: α-iron stable below 768°C, β-iron stable from 768—910°C, γ-iron stable from 910—1400°C, and δ-iron stable from 1400—1535°C.

In this system, one of the components is a metal (*i.e.*, Fe), and the other is a nonmetal (*i.e.*, C). The phase diagram of this system is shown in Fig. 4.4.

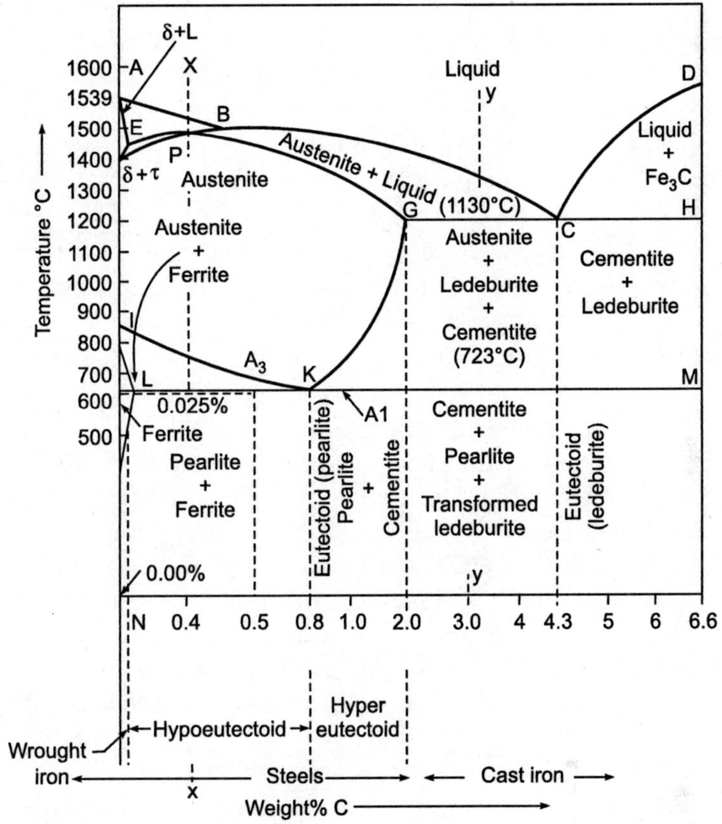

Fig. 4.4 Equilibrium diagram for the Fe-C system.

Here, the upper limit of carbon content is 6.7% only, which corresponds to the compound Fe_3C (cementite). It is evidently then, a two-component system, and the various phases as indicated above are as follows:

1. The curve ABCD is the liquidus line, above which there is only one liquid phase, consisting of Fe and dissolved C.
2. The curve AEPGCH is the solidus line, below which various Fe-C carbon compositions are completely solid. Evidently, the regions between the liquidus line and the solidus line represent mixtures of solid and liquid.
3. With increasing C content, the melting point of the carbon alloy is progressively depressed until alloy with 4.3% C melts at 1130°C (as compared with 1537°C for pure iron). As the C content increases (upto 6.67%), when only Fe_3C (cementite) is present, the melting point again rises.

4. Within the area AEB, the alloy exists as δ-iron and liquid. δ-iron has a body-centered cubic structure. The region DCH consists of cementite and liquid. The region BCGP is for austenite and liquid, and the region PGKI consists of γ-iron (*i.e.*, only austenite), which has a face-centered cubic structure.
5. The transformation of one solid phase into another occurring at certain critical temperatures indicated by lines LKM, IK, and GK are called the upper critical temperature lines.
6. The compositions are as follows:
   - Up to 0.088% C are commercially pure iron.
   - 0.088—2% C represent steel.
   - Above 2% C belong to a general class of cast iron.

     Steels are further subdivided into the following:
   - Hypoeutectoid steel (up to 0.8%)
   - Hypereutectoid steel (0.8—2%)
7. Austenite is not stable below the upper critical lines IK and GK:
   - For compositions containing less than 0.8% C, austenite on cooling begins to transform into ferrite, and the C-content of the remaining austenite increases along the line A_3 until point K (0.8% C) is reached.
   - For composition between 0.8—2.0%, cooling results in the separation of cementite, and the composition of the remaining austenite varies along the line GK, until again the point K is reached.
   - At point K (eutectoid point), austenite is transformed into pearlite (eutectoid mixture of cementite and ferrite). This is called eutectoid transformation. At K, three phases (austenite, ferrite, and cementite) coexist in equilibrium, and the point is invariant.
8. At 1130°C, the eutectic transformation (point C) takes place. The eutectic liquid (with 4.3% C) on cooling freezes to ledeburite (eutectoid of cementite and austenite). Further cooling transforms the eutectic austenite to cementite gradually, and again at the eutectoid point K, the remaining austenite is transformed into pearlite (723°C).
9. A steel of the eutectoid composition (*i.e.*, with 0.8% C) consists of pearlite only; whereas a steel with a C-content greater than 0.8% will consist of pearlite and cementite.
10. Now consider the cooling of an iron-carbon alloy containing 3% C, represented on the diagram by the line *yy*. On cooling, freezing begins at about 1270°C, the point of intersection of *yy* with ABC. On further, cooling austenite begins to separate out of the liquid. As cooling proceeds further, the amount of solid increases and its composition varies along the liquidus line PG, whereas the amount of liquid decreases along the liquidus line BC. At 1130°C, the mixture will consist of austenite containing 2% C and a liquid of eutectic composition. Further removal of heat causes no drop in temperature, until all liquid solidifies to ledeburite. At this point, cast iron will consist of primary

eutectic plus ledeburite. Still further cooling causes both primary and eutectic to decompose to cementite along the line GK. When the eutectic temperature (723°C) is reached, the residual austenite will transform to pearlite. At the temperature below 723°C, all ledeburite transforms into a pearlite and cementite mixture, which will thus be the final microstructure of cast iron.

EXERCISES

1. State Gibb's phase rule explaining the terms involved in it with suitable examples.
2. Define the phase rule. Describe the sulphur system with the help of a neat phase diagram.
3. Define phase, component, and degree of freedom. Give an account of Bismuth-Cadmium system.
4. State and explain the phase rule. Describe the phase diagram of the water system.
5. What is meant by a phase diagram? With the help of a phase diagram, explain the terms triple point and eutectic point.
6. With the help of a neat phase diagram, describe a lead-silver system.
7. State Gibb's phase rule, and explain the terms involved.
8. Draw and explain the phase diagram for a sulphur system.
9. What is the difference between triple point and eutectic point? Discuss the phase diagram of the water system.
10. State the phase rule. Explain the significance of each term with an example.
11. Draw a schematic phase diagram of the different phases (solid, liquid, and vapor) of a substance. What is a triple point?
12. (*a*) Draw the phase diagram for a sulphur system, and show the triple points with their temperature and pressure values.

    (*b*) What is the number of components in the system?

    $$CaCO_3\,(s) \rightleftharpoons CaO\,(s) + CO_2\,(g)$$

    (*c*) What is a eutectic point?

MULTIPLE CHOICE QUESTIONS

1. Which of the following is a two phase system?

   (*a*) Water and ether (*b*) Sand in water

   (*c*) Kerosene in water (*d*) All are correct

2. For a water system, the maximum number of degrees of freedom is
   (*a*) 0 (*b*) 3
   (*c*) 2 (*d*) 4
3. Gibb's phase rule is
   (*a*) F – P = C + 2 (*b*) F + P = C + 2
   (*c*) F – C = P + 2 (*d*) P + C = F + 2
4. How many phases are present in a saturated salt solution?
   (*a*) 1 (*b*) 2
   (*c*) 3 (*d*) 0
5. At the triple point of a water system, the system is
   (*a*) invariant (*b*) univariant
   (*c*) bivariant (*d*) none
6. In a Bi-Cd system, the composition at the eutectic point is
   (*a*) 60% Bi, 40% Cd (*b*) 70% Bi, 30% Cd
   (*c*) 4% Bi, 96% Cd (*d*) 2.4% Bi, 97.6% Cd
7. A system consisting of O_2 and N_2 gases is
   (*a*) a one-phase system (*b*) a two-phase system
   (*c*) a three-phase system (*d*) none of these
8. The condensed phase rule is applicable for
   (*a*) a water system (*b*) a sulphur system
   (*c*) a Bi-Cd system (*d*) none of these

Chapter 5

Chemical Kinetics and Catalysis

5.1 INTRODUCTION

Chemical kinetics is the branch of physical chemistry that deals with a study of the speed of chemical reactions. Such studies also enable us to understand the mechanism by which the reactions occur. This reveals not only the influence of different factors on the progress of the reaction, but it also throws light on the mechanism through which reactant molecules are transformed into products. The knowledge of the rate of reactions is very valuable for the success of an industrial process to give maximum yield.

5.2 RATE OF REACTION

The *rate of reaction*, that is the velocity of a reaction, is the amount of chemical change occurring per unit time. The rate is generally defined as the decrease in the concentration of the reactant or an increase in the concentration of the products per unit time. Mathematically, the rate of reaction can be expressed as

$$\text{Rate of reaction} = \frac{\text{Amount of reactant consumed}}{\text{Time interval}}$$

$$= \frac{\text{Amount of product produced}}{\text{Time interval}}$$

Consider a simple hypothetical reaction of the type

$$A \longrightarrow B$$

The rate of the reaction at any given time will be given by

$$\text{Rate of reaction } (r) = \frac{dx}{dt}$$

$$r = -\frac{d[A]}{dt} = +\frac{d[B]}{dt}$$

where [A] and [B] indicate the concentration of A and B, respectively.

The rate of reaction is invariably a positive quantity. The negative sign merely indicates that the concentration of the reactant is falling with the increase in time, whereas the positive sign indicates the increase in concentration of the product with the increase in time.

The numerical value of $\frac{dx}{dt}$ can be obtained from a plot of the concentration of a reactant (or product) versus time (Fig. 5.1).

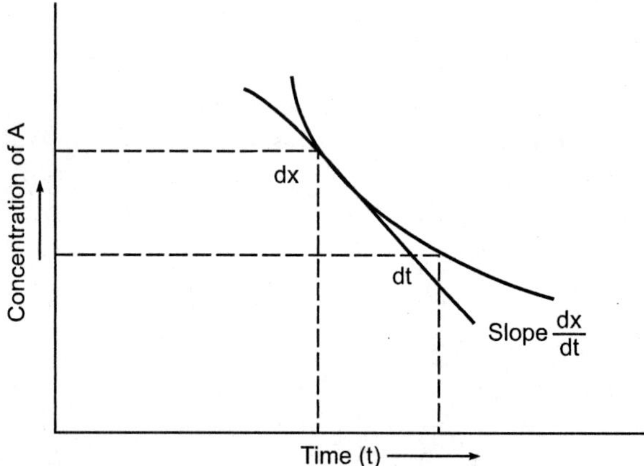

Fig. 5.1 Plot of the concentration of a reactant [A] vs time (t).

The time is usually expressed in seconds. The rate will have units of concentration divided by time. The concentrations are taken in gm-moles/liter, hence the rate is moles per liter per second.

5.3 FACTORS INFLUENCING THE REACTION RATE

The factors influencing the reaction rate:

Nature of reactants and products: Rates of reactions are considerably influenced by the nature of reactants and products. In general, reactions involving greater number of breaking and formation of bonds are slow at room temperature than those which involve lesser number of bond cleavage and formation. Consider the following two reactions:

$$2NO + O_2 \longrightarrow 2NO_2 \tag{5.1}$$

$$CH_4 + 2O_2 \longrightarrow CO_2 + 2H_2O \tag{5.2}$$

Because reaction (5.1) involves breaking one bond in NO and forming two new bonds in NO_2, and reaction (5.2) involves breaking four bonds in CH_4 and forming two bonds in CO_2 and four bonds in H_2O, reaction (5.1) is much faster than the (5.2) at the same temperature.

Effect of temperature: Most of the chemical reactions are accelerated by an increase in temperature. Usually at temperatures near room temperature, the rate of most chemical reactions increases by a factor of 2 or 3 for a 10°C rise in temperature.

Concentration of reactants: The rate of a reaction is proportional to the concentration of reactants. With an increase in time, the concentration of reactants decreases, and also the rate of reaction decreases.

Effect of catalyst: In general, a catalyst increases the rate of reaction at a given temperature. Further, a catalyst is specific in its action, that is it may affect the rate of one particular reaction only.

Surface area of reactants: This is of significance only if the reaction is heterogeneous in nature. The larger surface area of solid reactants and catalysts tends to increase the rate of a reaction.

Radiations: The rates of certain reactions increase by the absorption of photons of certain radiations. Such reactions are called *photochemical reactions*.

5.4 RATE CONSTANT

Consider the following reaction:

$$A \longrightarrow B$$

$$\text{Rate of reaction, } r = \frac{dx}{dt} = -\frac{d[A]}{dt} = K[A]$$

[A] represents the concentration of reactant A, and K is the proportionality constant called the velocity constant, rate constant, or specific reaction rate at a given temperature. It may be defined as the fraction of A that reacts per unit time.

If $[A] = 1$, then $r = K$.

Thus, at a given temperature and when the concentration of the reactant is unity, the rate constant of a reaction involving a single reactant is equal to the rate of the reaction.

In the life of a chemical reaction, the concentration and the rate are everchanging. The specific rate constant is the only constant factor directly related to the mechanism of the reaction.

5.5 THE ORDER OF REACTION

The *order of reaction* is defined as the total number of reactants whose concentrations change during the course of a reaction. It is the sum of exponents of the molar concentration of the reactants in the rate equation of the reaction:

$$mA + nB \longrightarrow \text{Products}$$

$$\text{Rate, } \frac{dx}{dt} = K\,[A]^m\,[B]^n$$

$$\text{Order of reaction} = m + n$$

Thus, the order of reaction can be defined as the sum of the power of the concentration terms:

1st order reaction: Decomposition of N_2O_5:

$$N_2O_5 \xrightarrow{\Delta} N_2O_4 + \frac{1}{2}O_2, \quad \frac{dx}{dt} = K\,[N_2O_5]$$

Order of reaction = 1.

2nd order reaction: Dissociation of HI:

$$2HI \longrightarrow H_2 + I_2, \quad \frac{dx}{dt} = K\,[HI]^2$$

Order of reaction = 2.

In some heterogeneous or surface reactions, the rate has been independent of the concentration. These are zero-order reactions.

$$\frac{dx}{dt} = K\,[A]^0 \text{ or } \frac{dx}{dt} = K$$

Characteristics of Order of Reaction

The order of reaction's characteristics:

- The order of reaction cannot be obtained from a simple balanced equation, but it can be known by experimental means.
- The order depends on the conditions of the reaction.
- The order may be a whole number, fraction, or zero.
- Theoretically, reactions of higher order are possible, but it is doubtful if reactions higher than the third order exist.

5.6 THE MOLECULARITY OF A REACTION

The *molecularity of a reaction* is defined as the minimum number of atoms, molecules, or ions of the reactants taking part in the rate-determining step. The reactions are said to be unimolecular, bimolecular, or trimolecular depending on whether one, two, or three molecules are involved during the reaction. To decide the molecularity of a reaction, only the single actual rate-determining step of a stoichiometric equation is used. For example:

Unimolecular reaction: $N_2O_5 \xrightarrow{\Delta} N_2O_4 + \frac{1}{2}O_2$

Bimolecular reaction: $2HI \longrightarrow H_2 + I_2$

Trimolecular reaction: $2NO + O_2 \longrightarrow 2NO_2$.

Characteristics of the Molecularity of Reaction

Following are the characteristics of the molecularity of reaction:

- Molecularity refers to a one-step reaction (*i.e.*, rate-determining step) of a complex reaction involving a number of steps.
- Molecularity is always a whole number and is never a fraction or zero.
- Molecularity can be determined from a simple balanced equation and is thus a theoretical concept.
- Molecularity reveals some fundamental facts about the reaction mechanism.

5.7 PSEUDO-UNIMOLECULAR REACTIONS

A reaction of the type, $A + B \longrightarrow$ Products is theoretically bimolecular and of the second order overall, with a rate law:

$$-d[C_A]/dt = k[C_A][C_B]$$

If the concentration of one of the reactants (say B) is much larger than that of the other, then it is clear that during the reaction, the concentration of B does not alter appreciably, that is the concentration of B remains essentially constant. The rate law now becomes

$$-d[C_A]/dt = k'[C_A]$$

where k' is a new constant known as the pseudo-first order rate constant and sometimes the pseudo-unimolecular rate constant. Thus, the reaction is apparently a first order in A and a zero order in B; the overall reaction is of first order. For example:

Hydrolysis of ethyl acetate

$$CH_3COOC_2H_5 + H_2O \xrightarrow{H^+} CH_3COOH + C_2H_5OH$$

In this reaction, one of the reactants (*i.e.*, water) is present in large excess, and the concentration of water does not alter appreciably during the course of the reaction. Such types of reactions, although they are bimolecular and following the kinetics of first order, are called pseudo-unimolecular reactions.

5.8 KINETIC EQUATIONS OF DIFFERENT ORDERS

Following are the kinetic equations of the different orders:

Zero order reaction: A reaction is said to be a zero order reaction when the rate is independent of the concentration of any reactant of the system. Consider a reaction:

$$A \longrightarrow \text{Products}$$

Let a = Initial concentration of A in moles/liter

x = Change in concentration at time 't'

$a - x$ = Concentration left at time 't'

Rate of reaction,

$$r = \frac{dx}{dt} = -\frac{d[A]}{dt} = k[A]^0 = k$$

or,

$$\frac{dx}{dt} = k(a-x)^0 = k$$

or,

$$x = kt + I$$

At $t = 0, x = 0$ So $I = 0$

Hence, $x = kt$

or, $k = \dfrac{x}{t}$

This is the expression for the rate constant of a zero-order reaction.

Unit of k: Concentration time^{-1} or, mol L^{-1} s^{-1}.

Examples:

- Photochemical combination of H_2 and Cl_2 over water surface:

$$H_2(g) + Cl_2(g) \xrightarrow{h\nu} 2HCl(g)$$

- Decomposition of NH_3 in the presence of Mo or W is also a zero-order reaction:

$$2NH_3 \xrightarrow{Mo} N_2 + 3H_2$$

- Thermal decomposition of HI on a gold surface.

First order reaction: The reactions whose rate is determined by the change of only one concentration term are known as first order reactions.

Consider the reaction, $A \longrightarrow$ Products

Let a = Initial concentration of A in moles/liter

x = Change in concentration at time 't'

$a - x$ = Concentration left at time 't'

Rate of reaction, $r = \dfrac{dx}{dt} = -\dfrac{d[A]}{dt} = k[A]$

or, $-\dfrac{d(a-x)}{dt} = k(a-x)$

or, $\dfrac{dx}{dt} = k(a-x)$ or, $\dfrac{dx}{a-x} = k \, . \, dt$

On integration, $\displaystyle\int_0^x \frac{dx}{a-x} = \int_0^t k \, . \, dt$ [$\because$ At $t = 0, x = 0$]

or, $\left[-\ln(a-x)\right]_0^x = k \, . \, t$

or, $[-\ln(a-x) + \ln a] = k \, . \, t$

or, $\ln \dfrac{a}{a-x} = k \, . \, t$

or, $$k = \frac{1}{t} . \ln \frac{a}{a-x} = \frac{2.303}{t} \log \frac{a}{a-x} \tag{5.3}$$

This is called a kinetic equation of a first-order reaction.

Unit of k: time^{-1}, for example, h^{-1}, m^{-1}, s^{-1}

Half-life period: Half-life period is the time taken for the completion of half of the reaction.

If $$x = \frac{a}{2} \text{ at } t = t_{0.5}$$

$$K = \frac{2.303}{t_{0.5}} \log \frac{a}{a - \frac{a}{2}} = \frac{2.303}{t_{0.5}} \log 2$$

$$K = \frac{2.303 \times 0.301}{t_{0.5}}$$

or, $$t_{0.5} = \frac{0.6932}{K} \tag{5.4}$$

Thus, for a first-order reaction, the time taken to reduce the concentration to one half of its original value is constant. Otherwise, the half-life period of a first-order reaction is independent of the initial concentration of the reacting substances.

Second order reaction: The reaction in which the reaction rate is determined by the variation of two concentration terms is known as a second-order reaction. Two types of second-order reaction are possible.

- The rate is proportional to the square of the concentration of the same reactant.

  Consider the reaction 2 A ⟶ Products

$$-\frac{d[A]}{dt} = K[A]^2$$

a = Initial concentration of A

x = Change in concentration at time 't'

$a - x$ = Concentration left at time 't'

∴ $$-\frac{d(a-x)}{dt} = K(a-x)^2$$

or, $$\frac{dx}{dt} = K(a-x)^2 \quad \text{or,} \quad \frac{dx}{(a-x)^2} = K \, . \, dt$$

On integration, $\int_0^x \frac{dx}{(a-x)^2} = \int_0^t K.dt$

or, $$\left[\frac{1}{a-x}\right]_0^x = K.t \quad \text{or,} \quad \frac{1}{a-x} - \frac{1}{a} = K.t$$

or, $$\frac{x}{a(a-x)} = K.t \quad \text{or,} \quad K = \frac{1}{t}.\frac{x}{a(a-x)} \tag{5.5}$$

This is the expression for the rate constant of second-order reaction when the concentrations of both the reactants are equal.

Unit of K: Concentration^{-1} time^{-1}, that is L mol^{-1} s^{-1}.

Half-life period:

If $$x = \frac{a}{2} \text{ at } t = t_{0.5}$$

$$K = \frac{a/2}{t_{0.5}.a(a-a/2)} = \frac{1}{t_{0.5}.a} \quad \text{or,} \quad t_{0.5} = \frac{1}{Ka} \tag{5.6}$$

Hence, the half-life period of a second-order reaction is inversely proportional to the initial concentration of the reactant.

- The rate is proportional to the product of the concentration of the two reactants.

  Consider the reaction; A + B ⟶ Products.

  Let the initial concentration of A and B be 'a' and 'b' mole per unit volume, respectively. After time 't', the x mole of each has been converted to the products. The rate is given by

$$\frac{dx}{dt} = K(a-x)(b-x)$$

or, $$\frac{dx}{(a-x)(b-x)} = K.dt$$

or, $$\frac{1}{a-b}\left[\frac{1}{b-x} - \frac{1}{a-x}\right]dx = K.dt$$

On integration,

$$\int_0^x \frac{1}{a-b}\left[\frac{1}{b-x} - \frac{1}{a-x}\right]dx = \int_0^t k.dt$$

$$\frac{1}{a-b}\left[-\ln(b-x)+\ln(a-x)\right]_0^x = \text{K} \,.\, t$$

or, $$\frac{1}{a-b}\left[-\ln(b-x)+\ln(a-x)\right]_0^x = \text{K} \,.\, t$$

or, $$\frac{1}{a-b}\left[\ln\frac{(a-x)}{a}-\ln\frac{(b-x)}{b}\right] = \text{K} \,.\, t$$

or, $$\frac{1}{a-b}\ln\frac{b\,(a-x)}{a\,(b-x)} = \text{K} \,.\, t$$

or, $$\text{K} = \frac{1}{t\,(a-b)}\ln\frac{b\,(a-x)}{a\,(b-x)}$$

$$\text{K} = \frac{2.303}{t\,(a-b)}\log\frac{b\,(a-x)}{a\,(b-x)} \tag{5.7}$$

is the rate constant expression for the second-order reaction, when the concentrations of both the reactants are different.

Third order reaction: Reactions whose rate is determined by the variation of three concentration terms are known as third-order reactions. Only a few third-order reactions are definitely known. The simplest case of a third-order reaction will be of the type

$$3\text{A} \longrightarrow \text{Products}$$

So that, $$\frac{dx}{dt} = \text{K}\,(a-x)^3, \quad \text{or,} \quad \frac{dx}{(a-x)^3} = \text{K} \,.\, dt$$

On integration, $$\text{K} = \frac{1}{2t}\left[\frac{1}{(a-x)^2}-\frac{1}{a^2}\right] \tag{5.8}$$

The half-life period of such a process would be

$$t_{0.5} = \frac{3}{2\text{K}a^2} \tag{5.9}$$

The half-life period of a third order reaction is inversely proportional to the square of the initial concentration of reactant(s).

Higher order reaction: Reactions of fourth and higher orders are very rare because in such reactions four or more reacting molecules should meet simultaneously before any reaction can occur. This seems to be improbable, and no reaction is definitely known to have a higher order.

In general, a reaction of the nth order, where all the initial concentrations are the same, has a reaction rate,

$$nA \longrightarrow \text{Products},$$

$$\frac{dx}{dt} = K\,(a - x)^n$$

On integration, $$K = \frac{1}{t\,(n-1)}\left[\frac{1}{(a-x)^{n-1}} - \frac{1}{a^{n-1}}\right] \qquad (5.10)$$

This equation is applicable for all orders except when $n = 1$.

The half-life period of an nth order reaction may be given by

$$t_{0.5} = \frac{2^{n-1}-1}{K.(n-1)} \cdot \frac{1}{a^{n-1}}$$

$$= \frac{K'}{a^{n-1}} \qquad (5.11)$$

Thus, the half-life period is inversely proportional to the $(n - 1)$th power of the initial concentration.
Unit of K: (concentration)$^{-n+1}$ time^{-1}.

5.9 DETERMINATION OF THE ORDER OF REACTION

There are several methods for determining the order of simple reactions:

Method of integration: In this method, a known amount of reactants are mixed, and the progress of the reaction is determined by analyzing the reaction mixture from time to time. The experimental data thus obtained is substituted in different kinetic equations in their integrated forms until a constant value for the specific reaction rate is obtained.

$$K = \frac{2.303}{t}\log\frac{a}{a-x} \qquad \text{(for first-order reaction)}$$

$$K = \frac{1}{t} \times \frac{x}{x\,(a-x)} \qquad \text{(for second-order reaction)}$$

$$K = \frac{1}{2t} \times \frac{x\,(2a-x)}{a^2\,(a-x)^2} \qquad \text{(for third-order reaction)}$$

For complex reactions, the method may lead to fallacious results.

Method of half-life period: The half-life period of a reaction of the nth order is

$$t_{0.5} = \frac{K'}{a^{n-1}}$$

If a and a' are the initial concentrations in two experiments of the same reaction, then

$$\frac{t_{0.5}}{t'_{0.5}} = \left(\frac{a'}{a}\right)^{n-1}$$

Taking the logarithms, we get $\log \frac{t_{0.5}}{t'_{0.5}} = (n-1) \log \frac{a'}{a}$

or, $$n - 1 = \frac{\log t_{0.5}/t'_{0.5}}{\log a'/a} \quad \text{or,} \quad n = 1 + \frac{\log t_{0.5}/t'_{0.5}}{\log a'/a} \tag{5.12}$$

Plotting the concentrations against time $t_{0.5}$, $t'_{0.5}$ can be easily obtained, so the order n for the reaction can be calculated.

Graphical method: We know that for the first-order reaction:

$$t = \frac{2.303}{K} \log a - \frac{2.303}{K} \log (a - x)$$

$$t \propto \log (a - x)$$

Similarly for second-order reaction:

$$t = \frac{1}{K(a-x)} - \frac{1}{Ka}$$

$$t \propto \frac{1}{(a-x)}$$

So in general, for the nth order reaction, $t \propto \frac{1}{(a-x)^{n-1}}$.

In this method, for a particular experiment at different intervals of time, if

- By plotting $\log (a - x)$ versus time, a straight line indicates a first-order reaction.
- By plotting $\frac{1}{a-x}$ versus time, a straight line indicates second-order reaction.

Ostwald's isolation method: This method introduced by Ostwald (1902) is applicable only in case of reactions having two or more reactants. A large excess of reactants are to be taken except one. Then the order of the reaction is determined with respect to the isolated reactant (*i.e.*, taken in the small amount). This series of experiments was carried out. The sum of the individual order, when each reactant is taken separately in a small amount, is the overall order of reaction. For example:

$$\underset{C_A}{2FeCl_3} + \underset{C_B}{SnCl_2} \longrightarrow 2FeCl_2 + SnCl_4$$

- $-\dfrac{dC_A}{dt} \propto [C_A]^2$, when $SnCl_2$ is in large excess.
- $-\dfrac{dC_B}{dt} \propto [C_B]$, when $FeCl_3$ is in large excess.

$$-\frac{dC}{dt} = K\,[FeCl_3]^2\,[SnCl_2]^1$$

Order of reaction = 2 + 1 = 3.

This is a third-order reaction. For complex reactions, the method may lead to fallacious results.

5.10 ACTIVATION ENERGY

There are two theories:

Collision theory: According to the collision theory, the reaction occurs when the reactant molecules (say A and B) come very close to each other or collide. All collisions do not lead to chemical reactions. Only those collisions give rise to chemical reaction in which the molecule acquires energy greater than the activation energy. Thus only those collisions result in product formation in which the colliding molecules are associated with a certain minimum amount of energy. This minimum energy, which the molecule should possess so that their mutual collisions result in a chemical reaction, is called *threshold energy*. The collisions that result in the formation of a product are called *effective collisions*. Collisions among molecules possessing energy less than the threshold energy are not effective collisions and do not result in the formation of products. The additional energy required by the molecule to attain the threshold energy is called *activation energy.*

Activation energy = Threshold energy – energy of colliding molecules

The threshold energy is always greater than the activation energy.

Transition state theory: According to Arrhenious, the molecules must acquire a certain energy before undergoing chemical change. The molecules involved in a chemical reaction must be raised to a higher energy level before the reaction can occur; otherwise, the molecules must be in an activated state before they can react. This may be made clear with the aid of a schematic diagram as in Fig. 5.2. The average energy of the reactant is represented by E_A, and the product is represented by E_B. There is a minimum energy level denoted by E_X to which the reactant molecule must be raised to enable it to undergo a chemical change. The excess or additional energy $(E_X - E_A)$, which the reactant must acquire to undergo transformation, is the activation energy E_1. Molecules with energy E_X or above are said to be in an activated state; such molecules are only fit for chemical reaction. The intermediate product with the partially formed bond is called an *activated complex* or *transition state*, and the energy of activation is the energy required to form the activated complex or intermediate. After the chemical change, the product will have average energy E_B. Hence, an amount of energy $(E_X - E_B)$ will be given out.

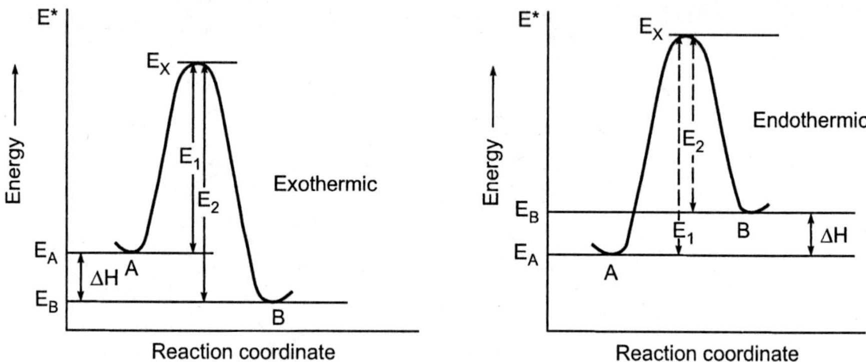

Fig. 5.2 Activation energy.

- If E_2 is greater than E_1, the reaction is exothermic.
- If E_2 is less than E_1, the reaction is endothermic.

This concept that the molecules must possess the activation energy before they can undergo the chemical change has now been universally accepted for all reactions.

Arrhenious Equation

According to Arrhenious, $K = Ae^{-E_a/RT}$ (5.13)

where, K = Rate constant E_a = Activation energy

R = Universal gas constant A = Frequency factor

T = Absolute temperature

Taking logarithm, $\ln K = \ln A - \frac{E_a}{RT}$

If K_1 and K_2 are rate constants at temperatures T_1 and T_2, respectively,

$$\ln K_1 = \ln A - \frac{E_a}{RT_1} \quad (5.14)$$

$$\ln K_2 = \ln A - \frac{E_a}{RT_2} \quad (5.15)$$

On subtracting Equation (5.14) from (5.15), we get

$$\ln \frac{K_2}{K_1} = -\frac{E_a}{R}\left[\frac{1}{T_2} - \frac{1}{T_1}\right]$$

or,

$$\ln \frac{K_2}{K_1} = -\frac{E_a}{R}\left[\frac{T_1 - T_2}{T_1 T_2}\right]$$

$$\log \frac{K_2}{K_1} = \frac{E_a}{2.303\,R}\left[\frac{T_2 - T_1}{T_1 T_2}\right]$$

Because $R = 1.987 \text{ cal } K^{-1} \text{ mole}^{-1}$

$$E_a = \frac{4.567\,(\log K_2 - \log K_1)\,T_2 T_1}{T_2 - T_1} \quad (5.16)$$

This is the equation for evaluating the activation energy by determining the rate constants of a reaction at two different temperatures.

5.11 CATALYSTS

A *catalyst* is defined as a foreign substance that alters the velocity of a chemical reaction without itself undergoing any permanent chemical change. The phenomenon is called *catalysis.*

Positive Catalyst

A catalyst that increases the rate of reaction is called a *positive catalyst.* For example:

$$2KClO_3 \xrightarrow{MnO_2} 2KCl + 3O_2$$

Negative Catalyst

A catalyst that decreases the rate of reaction is called a *negative catalyst*. For example, alcohol retards the oxidation of chloroform into poisonous phosgene.

$$CHCl_3 + \frac{1}{2}O_2 \xrightarrow{CH_3OH} COCl_2$$

Autocatalysis

The phenomenon in which one of the products formed during the reaction acts as a catalyst for the reaction is called *autocatalysis*. For example, the hydrolysis of an ester becomes faster after some time due to the formation of H^+ or carboxylic acid:

$$RCOOR' + H_2O \longrightarrow RCOOH + \underset{\text{Autocatalyst}}{R'OH}$$

Induced Catalysis

The phenomenon in which one reaction influences the rate of the reaction, which does not occur under ordinary conditions, is called *induced catalysis*. For example, the reduction of $HgCl_2$ by oxalic acid is slow but becomes faster if reduction is made in a mixture of $KMnO_4$ and $HgCl_2$, where both are reduced. The reduction of $KMnO_4$ induces the reduction of $HgCl_2$.

Homogeneous Catalysis

The catalytic reaction in which the catalyst and the reactants are in the same physical phase is called *homogeneous catalysis*. For example, the oxidation of sulphur dioxide with nitric oxide as catalyst:

$$2SO_2\,(g) + O_2\,(g) \xrightarrow{NO\,(g)} 2SO_3$$

Heterogeneous Catalysis

The catalytic reaction in which the catalyst and the reactants are in a different physical phase is called *heterogeneous catalysis*. For example, the oxidation of sulphur dioxide in presence of finely divided platinum:

$$2SO_2\,(g) + O_2\,(g) \xrightarrow{Pt\,(s)} 2SO_3$$

5.12 THE CRITERIA OF CATALYSTS

The catalytic reactions are characterized by the following criteria:

- The catalyst remains unchanged in mass and chemical composition at the end of a reaction. However, there may be a change in the physical state (color).
- A minute quantity of catalyst can produce an appreciable effect on the speed of a reaction.

$$2H_2 + O_2 \xrightarrow{Pt} 2H_2O$$

1 mg Pt is required for a 2.5 liter mixture of H_2 and O_2.

- A catalyst has a selective action, like a key can open a particular lock.
- A catalyst cannot start a reaction but only increases or decreases its speed. A catalyst simply hastens the attainment of the equilibrium.
- A catalyst is most active at a particular temperature called the optimum temperature.
- The addition of a small amount of a foreign substance, which is not catalytically active, increases the activity of the catalyst. Such substances that catalyze the catalyst are called *promoters*. For example:
  Haber's process of manufacturing ammonia.

$$N_2 + 3H_2 \xrightarrow[\text{Mo (Promoter)}]{\text{Fe (Catalyst)}} 2NH_3$$

Here, Molybdenum acts as the promoter.

- The activity of a catalyst is inhibited or completely destroyed by the presence of traces of a substance called a *catalytic poison, anticatalyst*, or *inhibitor*.
  For example: The contact process of manufacturing sulphuric acid, As_2O_3 destroys the catalytic activity of platinum. As_2O_3 acts as a catalyst poison.

5.13 THE MECHANISM OF CATALYTIC ACTION

To explain the mechanism of catalyst, consider the following two theories:

Unstable intermediate compound formation theory: The theory was forwarded by Element and Desormes in 1806. Accordingly, the catalyst forms a very reactive and unstable intermediate compound with reactants that immediately react with other reactants to yield the products of the reaction and liberate the catalyst in its original chemical composition.

$$A + K = AK \text{ (intermediate compound)}$$

$$AK + B = \underset{\text{Product}}{AB} + \underset{\text{Catalyst}}{K}$$

Some catalytic reactions are given here:

- Catalytic action of NO in the manufacture of H_2SO_4 by chamber's process:

$$\underset{\text{Catalyst}}{2NO} + O_2 = \underset{\text{Intermediate}}{2NO_2}$$

$$2SO_2 + 2NO_2 = 2SO_3 + [2NO]$$

$$[2NO] + 2SO_2 + O_2 = 2SO_3 + [2NO]$$

- In Friedel-craft's reaction, the action of anhydrous $AlCl_3$ is explained as follows:

$$C_6H_5COCl + \underset{\text{Catalyst}}{AlCl_3} = \underset{\text{Intermediate}}{C_6H_5COCl\,.\,AlCl_3}$$

$$C_6H_5COCl\,.\,AlCl_3 + C_6H_6 = \underset{\text{Product}}{C_6H_5COC_6H_5} + HCl + AlCl_3.$$

Adsorption or contact theory: This theory was postulated by Faraday (1883) and later received by many others. It explains the action of heterogeneous catalysis. According to the theory:

- The surface of the solid catalyst possesses some isolated active spots with residual affinity or free unsatisfied valency forces.
- Due to these free unsatisfied valency forces on the catalyst surface, the molecules of the gaseous reactants get adsorbed in the unimolecular thickness layer.
- The adsorbed molecules react due to their close proximity, forming products. The latter then fly off leaving the surface for fresh action.
- The chemical action is accelerated due to an increased concentration of the reacting substances on the surface of the solid catalyst, and no definite intermediate compound formation takes place.
- The forces that keep the molecules of reactants intact with a catalyst also attract reacting molecules. The distorted molecules of a catalyst, under more strain, are more reactive.

This theory successfully explains the following facts:

- The catalyst is more efficient in a finely divided state. With the increase of disintegration, the free surface area is increased, the free valencies or active spots (responsible for the adsorption of reactant molecule), increases in number, and consequently, the activity of the catalyst is also enhanced.

- The enhanced activity of a rough-surfaced catalyst is explained by postulating that a rough surface possesses cracks, peaks, corners and so on, and consequently have a large number of active spots, which in turn must be catalytically more active.
- The action of promoters is explained by assuming that a loose compound is formed between the catalyst and the promoter, which possesses more adsorption capacity than the pure catalyst only.
- The action of catalyst poisons is probably due to the preferential adsorption of poisons on the active spots of the catmalyst, and thus reduces the number of free active spots available for the adsorption of the reacting molecules. For example, the iron catalyst used in the synthesis of ammonia (Haber's process) is poisoned by H_2S:

$$N_2 + 3H_2 \xrightarrow[\text{Poisoned by } H_2S]{\text{Fe}} 2NH_3$$

$$Fe + H_2S \longrightarrow FeS + H_2$$

- The affinity of the catalyst for the reactant molecules is totally responsible for adsorption, and adsorption, occurs only when the intensity of the affinity of the catalyst for the reactants is quite high. In other words, the adsorption depends on the nature of both the adsorbent (catalyst) and the adsorbate (reactants). So, different catalysts cannot possess the same affinity for the same reactants. Hence, the action of the catalyst is also specific.

5.14 ACID-BASE CATALYSTS

Some reactions are catalyzed by acids or bases. This phenomenon is called an *acid-base catalysis*. Following are examples of acid-base catalyst:

- **Inversion of cane sugar in presence of a dilute mineral acid:**

$$\underset{\text{Sucrose}}{C_{12}H_{22}O_{11}} + H_2O \xrightarrow{[H^+]} \underset{\substack{\text{Glucose} \\ \text{(Dextrorotatory)}}}{C_6H_{12}O_6} + \underset{\substack{\text{Fructose} \\ \text{(Laevorotatory)}}}{C_6H_{12}O_6}$$

- **Hydrolysis of an ester:**

$$\underset{\text{Ethyl acetate}}{CH_3COOC_2H_5} + H_2O \xrightarrow{[H^+]} CH_3COOH + C_2H_5OH$$

- **Hydrolysis of an amide:**

$$RCONH_2 + H_2O \xrightarrow{[OH^-]} \underset{\text{Ammonium salt}}{RCOONH_4}$$

- **Decomposition of nitramide:**

$$NH_2NO_2 \xrightarrow{[OH^-]} N_2O + H_2O$$

In acid-base catalysis, H^+ or OH^- ions act as the catalyst. According to the modern concept of acid base catalysis:

- The reaction that is catalyzed by an acid is also catalyzed by substances that have the tendency to loose a proton.
- The reaction that is catalyzed by a base is also catalyzed by substances that have the tendency to gain a proton.

5.15 ENZYMES OR BIOLOGICAL CATALYSTS

Enzymes are nonliving complex nitrogeneous organic substances produced by living organisms. They have the capacity of bringing about many chemical reactions such as hydrolysis, oxidation, reduction, and so on. They are highly specific. Each enzyme can catalyze a specific reaction. The catalytic activity of enzymes, like catalysts, is due to their capacity to lower the activation energy for a particular reaction. For example:

- Starch is catalyzed by diastase to maltose:

$$\underset{\text{Starch}}{2(C_6H_{10}O_5)_n} + nH_2O \xrightarrow{\text{Diastase}} \underset{\text{Maltose}}{nC_{12}H_{22}O_{11}}$$

Again maltose is hydrolyzed by the enzyme maltase to glucose:

$$\underset{\text{Maltose}}{C_{12}H_{22}O_{11}} + H_2O \xrightarrow{\text{Maltase}} \underset{\text{Glucose}}{2C_6H_{12}O_6}$$

- The enzyme invertase converts cane sugar into glucose and fructose:

$$C_{12}H_{22}O_{11} + H_2O \xrightarrow{\text{Invertase}} \underset{\text{Glucose}}{C_6H_{12}O_6} + \underset{\text{Fructose}}{C_6H_{12}O_6}$$

Zymase converts glucose and fructose into alcohol:

$$C_6H_{12}O_6 \xrightarrow{\text{Zymase}} 2C_2H_5OH + 2CO_2$$

- The enzyme urease converts urea into ammonium carbonate:

$$NH_2CONH_2 + 2H_2O \xrightarrow{\text{Urease}} (NH_4)_2CO_3$$

- The enzyme ptyalin, present in the human mouth saliva, converts starch to glucose:

$$\underset{\text{Starch}}{(C_6H_{10}O_5)_n} + nH_2O \xrightarrow{\text{Ptyalin}} \underset{\text{Glucose}}{nC_6H_{12}O_6}.$$

5.16 MECHANISM OF ENZYME ACTION

The enzymes can increase the rate of biochemical reactions by factors ranging from 10^6–10^{12}. Moreover, they are highly selective and specific. They act only on certain molecules called *substrates*, while leaving the rest of the system unaffected. The high turnover numbers suggest that the substrate molecule cannot be very tightly bound to the enzymes; if they were, they might block the active sites. Reaction would then be slow because the active site is not quickly cleared out. Most enzyme systems behave as though they were in equilibrium between the substrate (S) and the active site (E):

$$E + S \underset{K_2}{\overset{K_1}{\rightleftharpoons}} ES \qquad (5.17)$$

The symbol ES represents a species in which the substrate is attached in some way to the enzyme. This enzyme-substrate complex then reacts to create product (P) and free the active site (E):

$$ES \xrightarrow{K_2} E + P \qquad (5.18)$$

The substrate molecules move off and onto the active site very rapidly compared with the rate at which they undergo reaction to form products (P), that is the equilibrium (Equation 5.17) is rapidly established between the enzyme and the substrate. On studying the effect of increasing the concentration of substrate, we found that the rate of product formation increases because the rate depends on the concentration of ES:

$$\text{Rate} = \frac{\Delta[P]}{\Delta t} = K_2[ES] \qquad (5.19)$$

Because a sizable fraction of the active sites are occupied, further increase in the concentration of S, does not result in the same degree of increase in [ES], and consequently, the rate also does not increase as much.

The lock and key model was developed by Emil Fischer (1894) to explain the specificity of enzymes. An enzyme is ordinarily a very large protein molecule that contains one or more active sites, where the reactions with the substrate take place; the rest of the molecule maintains the 3D integrity of the network. According to Fischer's hypothesis, the active site has a rigid structure similar to a lock. A substrate molecule has the complementary structure that causes it to fit and function like a key (Fig. 5.3). As the substrate molecules are drawn into the active site, they are somehow activated in that they are capable of extremely rapid reaction. This activation may result from the withdrawal or donation of electron-density at a particular bond by the enzyme. In addition, in the process of filling into the active site, the molecule may be distorted by the portion of the enzyme protein chain near the active site and thereby made more reactive. If co-enzymes are involved, the enzyme may promote reaction by binding both the co-enzymes and substrate to the same site or adjacent sites, thereby keeping the reactants in close proximity.

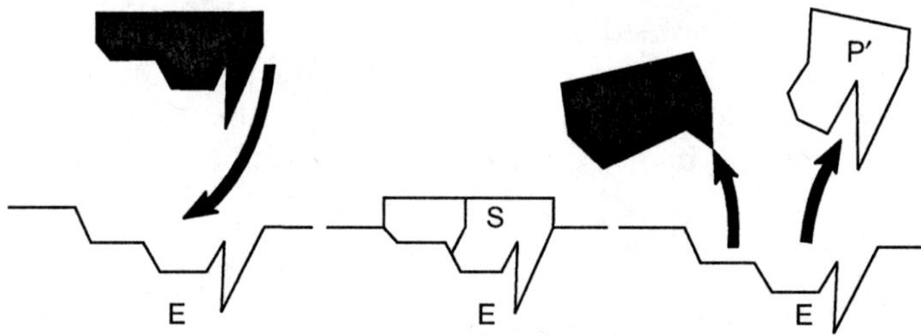

Fig. 5.3 The lock-and-key model for the formation of the enzyme-substrate complex. After the substrate (S) is bound to the enzyme (E), one or more products (P) may form and separate from the enzyme.

Enzymes are highly specific in action. Fig. 5.4 gives an illustration of how the geometry of an active site will only fit one type of substrate in most cases.

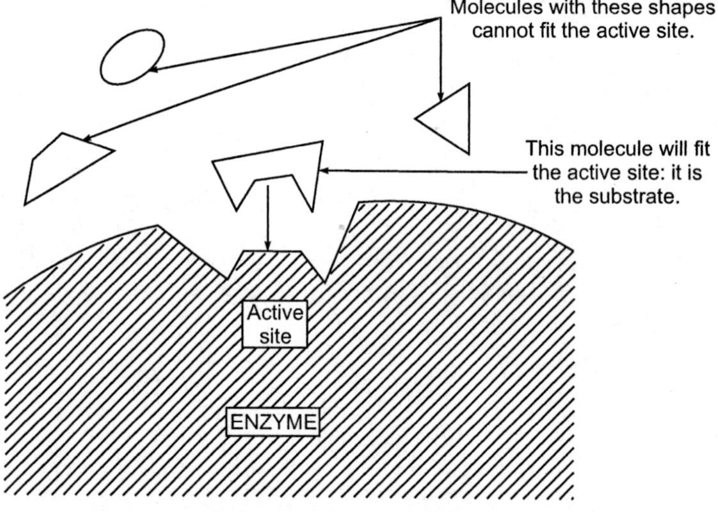

Fig. 5.4 Selectivity of an enzyme action.

5.17 INDUSTRIAL APPLICATIONS OF A CATALYST

The industrial applications of a catalyst are numerous. The presence of a catalyst is very useful in many industrially important reactions that are either very slow or take place at very high temperatures. Hence, it is essential to use a suitable catalyst to decrease the cost of production. A few examples of industrial applications of some catalysts are given in the following table.

Reactions	*Catalyst Required*
1. Haber's process of manufacturing ammonia $N_2 + 3H_2 \longrightarrow 2NH_3$	Finely divided Fe and Mo as promoter
2. Manufacture of "vanaspati" or vegetable ghee by hydrogenation of vegetable oil.	Ni
3. Contact process of manufacturing sulphuric acid	Pt
4. Manufacturing methyl alcohol from water gas $CO + H_2 + H_2 \longrightarrow CH_3OH$	ZnO and Cr_2O_3 as promoter
5. Acetic acid from acetaldehyde $2CH_3CHO + O_2 \longrightarrow 2CH_3COOH$	V_2O_5
6. Manufacturing chlorine by Deacon's process $4HCl + O_2 \longrightarrow 2H_2O + 2Cl_2$	$CuCl_2$
7. Ostwald's process of manufacturing nitric acid $4NH_3 + 5O_2 \longrightarrow 4NO + 6H_2O$ $2NO + O_2 \longrightarrow 2NO_2$ $4NO_2 + 2H_2O + O_2 \longrightarrow 4HNO_3$	Platinized asbestos
8. Manufacturing hydrogen by Bosch's process $\underbrace{CO + H_2}_{\text{Water gas}} + H_2O \longrightarrow CO_2 + 2H_2$	Fe_3O_4 and Cr_2O_3 as promoter

5.18 ORGANOMETALLIC COMPOUNDS IN CATALYSIS

An organometallic compound is defined as a compound that contains a direct carbon-metal bond.

$$-\overset{|}{\underset{|}{C}}-M$$

where, M = K, Na, Li, Ca, Mg, Pb, Hg, and so on.

The more electropositive the metal, the more ionic the carbon-metal bond. The organometallic compounds of intermediate ionic character, such as those of Li and Mg, are of greater interest to the organic chemist. These are relatively easy to prepare and handle and are useful synthetic reagents acting as catalysts.

Tetra ethyl lead (TEL):

$$(C_2H_5)_4\text{—}Pb \qquad \begin{array}{c} C_2H_5 \\ | \\ H_5C_2\text{—}Pb\text{—}C_2H_5 \\ | \\ C_2H_5 \end{array}$$

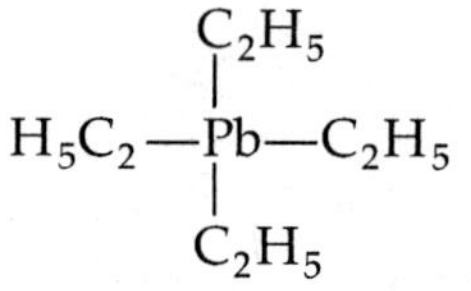

It is the most important organometallic compound that is used as an antiknock for gasoline used in automobile engines. TEL (1 to 3 ml. per gallon) serves as an antiknock. The rate of combustion reaction of gasoline hydrocarbons increases sharply under high compression, which causes engine knocking. Over-rapid combustion of motor fuel under high pressure produces free hydrocarbon radicals and knocking is due to the chain reactions set up by these radicals. The antiknocking property of TEL is probably explained by the fact that it produces ethyl free radicals that act as radical traps for the radicals (R) produced from gasoline, and chain reactions are terminated.

$$(C_2H_5)_4\,Pb \xrightarrow{\Delta} 4C_2H_5^{\bullet} + Pb^{\bullet}$$

$$\underset{\text{Gasoline}}{R^{\bullet} + {}^{\bullet}C_2H_5} \longrightarrow \underset{\text{Alkane}}{R\text{—}CH_2\,.\,CH_3}$$

In internal combustion engines, TEL acts as a *negative catalyst* for the combustion of gasoline.

Wilkinson catalyst:

$$RhCl\,(PPh_3)_3 \qquad \begin{array}{ccc} Cl & & PPh_3 \\ & Rh & \\ Ph_3P & & PPh_3 \end{array}$$

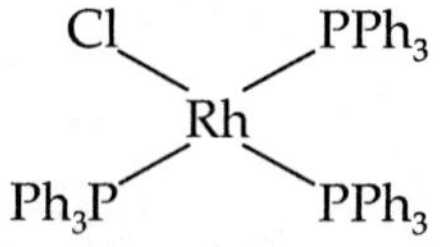

Chlorotris (triphenyl phosphine) rhodium

is used in the hydrogenation of alkenes.

Di butylphosphine substituted cobalt carbonyls:

$$(PBu)_2Co(CO)_2 \qquad \begin{array}{ccc} BuP & & CO \\ & Co & \\ BuP & & CO \end{array}$$

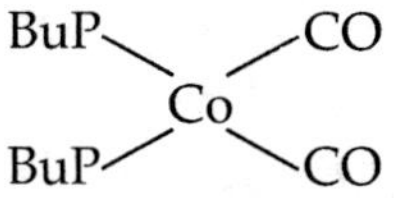

is used in the hydroformylation of alkenes.

Chloro carbonyl di (triphenyl phosphine) irridium:

$$IrCl\,(CO)\,(PPh_3)_2 \qquad \begin{array}{ccc} Cl & & PPh_3 \\ & Ir & \\ CO & & PPh_3 \end{array}$$

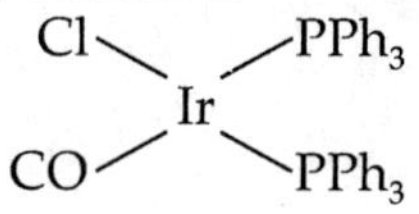

Ziegler-Natta catalyst:

Tri ethyl aluminium in titanium chloride ($AlEt_3$ in $TiCl_4$) used in alkyl polymerization and oligomerization.

Tetra ethoxy phosphine nickel:

$Ni\,[P\,(OEt)_3]_4$

$$\begin{matrix} (OEt)_3P & & P(OEt)_3 \\ & Ni & \\ (OEt)_3P & & P(OEt)_3 \end{matrix}$$

used in the 1, 4-hexadiene synthesis by Dupont Company France.

SOLVED NUMERICAL PROBLEMS

1. *The rate constant of a first-order reaction is 3.5×10^{-2} minute^{-1}. Calculate the time taken for half the initial concentration to react.*

Sol. Half-life of a first-order reaction $= \dfrac{0.693}{K}$.

Thus, $$t_{0.5} = \frac{0.693}{3.5 \times 10^{-2}} = 19.8 \text{ minutes.}$$

2. *Half-time change for a first-order reaction is 40 minutes. What is the rate constant of the reaction at the same temperature?*

Sol. Rate constant, $K = \dfrac{0.693}{t_{0.5}} = \dfrac{0.693}{40} = 0.0173\ \text{min}^{-1}$

3. *The catalyzed decomposition of hydrogen peroxide in an aqueous solution, which is of the first order, is followed by titrating equal volumes of samples of the solution with potassium permanganate at the stated time as follows:*

Time (min)	0	5	15	25	45
cm^3 of $KMnO_4$	37.0	29.8	19.6	12.3	5.0

Calculate the rate constant of the reaction.

Sol. For a first-order reaction, $K = \dfrac{2.303}{t} \log \dfrac{a}{a-x}$

Here $a = 37.0\ \text{cm}^3$

$a - x = 29.8, 19.6, 12.3$ and $5.0\ \text{cm}^3$ at $t = 5, 15, 25$ and 45 minutes, respectively.

$$\therefore \quad K = \frac{2.303}{5\text{ min}} \log \frac{37.0}{29.8} = 4.329 \times 10^{-2}\ \text{min}^{-1} \text{ (At } t = 5 \text{ min)}$$

$$= \frac{2.303}{15\text{ min}} \log \frac{37.0}{19.6} = 4.236 \times 10^{-2}\ \text{min}^{-1} \text{ (At } t = 15 \text{ min)}$$

$$= \frac{2.303}{25\text{ min}} \log \frac{37.0}{12.3} = 4.407 \times 10^{-12}\ \text{min}^{-1} \text{ (At } t = 25 \text{ min)}$$

$$= \frac{2.303}{45 \text{ min}} \log \frac{37.0}{5.0} = 4.447 \times 10^{-2} \text{ min}^{-1} \text{ (At } t = 45 \text{ min)}$$

As the different values of K are nearly same, so the reaction is of the first-order.

The average value of K = 4.355×10^{-2} min^{-1}.

4. *In a second-order reaction, where the initial concentration of the reactants is the same, half of the reactants are consumed in 60 minutes. If the specific reaction rate is 5.2×10^{-3} mol^{-1} L minute^{-1}, what is the initial concentration of the reactants?*

Sol. For a second-order reaction: $K = \frac{x}{t \cdot a \cdot (a - x)}$

At $\quad t = 60$ min, $x = 0.5a$

and $\quad K = 5.2 \times 10^{-3}$ mol^{-1} L min^{-1}

$$\therefore \quad 5.2 \times 10^{-1} \text{ L min}^{-1} = \frac{0.5\,a}{60 \text{ min} \times a \times 0.5a} = \frac{1}{60a}$$

$$\therefore \quad \text{Initial concentration} = a = \frac{1}{60 \times 5.2 \times 10^{-3}} = 3.2 \text{ mol L}^{-1}$$

5. *Show that for a first-order reaction, the time required for 99.9% completion of the reaction is 10 times that required for 50% completion.*

Sol.

$$\frac{99.9\%}{50\%} = \frac{\frac{2.303}{K} \log \frac{100}{100 - 99.9}}{\frac{2.303}{K} \log \frac{100}{100 - 50}} = 10$$

6. *The specific reaction rates of a chemical reaction at 273 K and 303 K are respectively 2.45×10^{-5} and 162×10^{-5}. Calculate the activation energy of the reaction.*

Sol. The Arrhenius equation shows that

$$\log \frac{K_2}{K_1} = \frac{E_a}{2.303\,R} \left[\frac{T_2 - T_1}{T_1 T_2} \right]$$

Substituting the values, $T_1 = 273$ K, $T_2 = 303$ K, $K_1 = 2.45 \times 10^{-5}$, $K_2 = 162 \times 10^{-5}$, R = 1.987 cal K^{-1} mol^{-1}, we get

$$\log \frac{162 \times 10^{-5}}{2.45 \times 10^{-5}} = \frac{E_a}{2.303 \times 1.987} \left[\frac{303 - 273}{273 \times 303} \right]$$

or,

$$1.8203 = \frac{E_a}{2.303 \times 1.987} \times \frac{30}{273 \times 303}$$

Hence, activation energy, E_a = 22,968 cal mol^{-1}.

7. *In a second-order reaction, the half-life was found to be 30 minutes when the initial concentration was 0.1 M. Find the rate constant.*

Sol. For a second-order reaction, $t_{0.5} = 1/Ka$

given, $t_{0.5}$ = 30 minutes, a = 0.1 M

$$\therefore \quad 30 = \frac{1}{K \times 0.1}$$

$$\text{or} \quad K = \frac{1}{3.0} = 0.333 \text{ mole lit}^{-1} \text{ min}^{-1}$$

8. *Consider a gaseous decomposition reaction: A → products at 500°C and at an initial pressure of 350 torr. The rate of the reaction was 1.07 torr s^{-1} when 5% of the decomposition was over, and the rate was 0.76 torr s^{-1} when 20% decomposition was over. Determine the order of the reaction.*

Sol. Let a be the initial concentration of reactant A and n be the order of the reaction.

$$\therefore \quad (a - x_1) = 0.95\, a \text{ and rate, } r_1 = 1.07 \text{ torr s}^{-1}$$

$$(a - x_2) = 0.80\, a \text{ and rate, } r_2 = 0.76 \text{ torr s}^{-1}$$

So,

$$1.07 \text{ torr s}^{-1} = k\,(0.95\, a)^n \tag{5.20}$$

$$0.76 \text{ torr s}^{-1} = k\,(0.80\, a)^n \tag{5.21}$$

Dividing (5.20) by (5.21),

$$\frac{1.07 \text{ torr s}^{-1}}{0.76 \text{ torr s}^{-1}} = \left(\frac{0.95\, a}{0.80\, a}\right)^n$$

$$\text{or,} \quad \left(\frac{95}{80}\right)^n = \frac{107}{76}$$

Taking log on both sides, we get

$$n\,[1.9777 - 1.9031] = [2.0294 - 1.8808]$$

$$\text{or,} \quad n\,(0.0746) = 0.1486$$

$$\text{or,} \quad n = \frac{0.1486}{0.0746} = 2$$

Hence, the order of reaction is two.

9. *For the reaction, $2NO + Cl_2 \rightleftharpoons 2NOCl$, the following mechanism has been proposed.*

Step I : $NO + Cl_2 \underset{k_{-1}}{\overset{k_1}{\rightleftharpoons}} 2NOCl_2$

Step II : $NO + NOCl_2 \xrightarrow{K_2} 2NOCl$

Show that the overall rate of the reaction is

$k\,[NO]_2\,[Cl_2]$, *where* $k = \dfrac{k_1 . k_2}{k_{-1}}$, *assuming that* $k_2\,[NO] << k_{-1}$.

Sol. Applying steady state approximation to the concentration of $NOCl_2$, we get

$$\frac{d[NOCl_2]}{dt} = 0 = k_1\,[NO]\,[Cl_2] - k_{-1}\,[NOCl_2] - k_2\,[NO]\,[NOCl_2]$$

$$\therefore \qquad [NOCl_2] = \frac{k_1\,[NO][Cl_2]}{k_{-1} + K_2\,[NO]} \qquad (5.22)$$

Now, the overall rate of reaction,

$$\frac{d[NOCl]}{dt} = k_2\,[NO]\,[NOCl_2] \qquad (5.23)$$

$\therefore$ From Equations (5.22) and (5.23), we get

$$\frac{d[NOCl]}{dt} = \frac{k_1 . k_2\,[NO][NO][Cl_2]}{k_{-1} + k_2\,[NO]}$$

$$= \frac{k_1 . k_2}{k_{-1}}\,[NO]^2\,[Cl_2],\ \text{if } k_2\,[NO] << k_{-1}$$

$$= k\,[NO]^2\,[Cl_2],\ \text{where } k = \frac{k_1 . k_2}{k_{-1}}$$

EXERCISES

1. What is meant by the order of reaction? Derive an expression for the rate constant of a second-order reaction.
2. Define the following terms:
   (*a*) Reaction rate
   (*b*) Order of reaction
   (*c*) Molecularity of reaction
   (*d*) Activation energy
   (*e*) Activated complex
   (*f*) Rate constant

3. (*a*) Differentiate between order and molecularity of a reaction.
   (*b*) Reactions of third and higher orders are usually not very common—Why?
   (*c*) Increase of temperature invariably increases the rate of reaction—Why?
   (*d*) The overall kinetics of a reaction involving several steps is controlled by the kinetics of the slowest step.
   (*e*) How do you account for the fact that an enzyme reaction has an optimum pH at which its activity is maximum?
4. Derive the kinetic expression of a second-order reaction, and show that under certain conditions it gives first-order kinetics.
5. Deduce an expression for the rate constant of a first-order reaction and its half-life period.
6. Explain and illustrate the following:
   (*a*) Auto catalysis (*b*) Promoters
   (*c*) Negative catalysis (*d*) Enzyme and catalysis
   (*e*) Acid-base catalysis
7. (*a*) What is a catalyst?
   (*b*) Cite an example of a negative catalysis.
   (*c*) What are enzyme reaction? Illustrate with an example.
8. (*a*) Differentiate between order and molecularity of a reaction. Give two examples showing how they are same and how they are different.
   (*b*) What is energy of activation? How can it be determined?

MULTIPLE CHOICE QUESTIONS

1. The rate of a reaction does not depend upon
   (*a*) Pressure (*b*) Temperature
   (*c*) Concentration (*d*) Catalyst
2. The rate constant of a reaction depends upon
   (*a*) temperature (*b*) initial concentration
   (*c*) time of reaction (*d*) extent of reaction
3. The unit of rate constant for a second-order reaction is
   (*a*) mol liter sec. (*b*) mol^{-1} $liter^{-1}$ sec^{-1}
   (*c*) mol $liter^{-1}$ sec^{-1} (*d*) mol^{-1} liter sec^{-1}
4. For a reaction of second-order kinetics, its $t_{0.5}$ is
   (*a*) $\alpha\ a$ (*b*) $\alpha\ a^3$
   (*c*) $\alpha\ a^2$ (*d*) $\alpha\ a^{-1}$
5. The half-life of a first-order reaction depends on the
   (*a*) concentration of reactants (*b*) concentration of products
   (*c*) time (*d*) temperature

6. The unit of rate constant for a first-order reaction is
   (*a*) s^{-1}
   (*b*) mol $liter^{-1}$ s^{-1}
   (*c*) mol s^{-1}
   (*d*) liter $mole^{-1}$ s^{-1}
7. 75% of the first-order reaction was completed in 32 minutes. 50% of the reaction was completed in
   (*a*) 24 min
   (*b*) 8 min
   (*c*) 16 min
   (*d*) 4 min
8. A zero-order reaction is one
   (*a*) in which reactants do not react
   (*b*) in which one of the reactants is in large excess
   (*c*) whose rate is not affected by time
   (*d*) whose rate increases with time
9. A substance that retards the rate of reaction is called
   (*a*) poison
   (*b*) negative catalyst
   (*c*) auto catalyst
   (*d*) promoter
10. A catalyst
    (*a*) increases the activation energy
    (*b*) decreases the activation energy
    (*c*) alters the reaction mechanism
    (*d*) increases the average K.E. of reactants
11. A biological catalyst is
    (*a*) an amino acid
    (*b*) $C_6H_{12}O_6$
    (*c*) N_2
    (*d*) enzymes
12. An auto catalyst is
    (*a*) a catalyst for catalyst
    (*b*) one that starts reaction
    (*c*) one of the products of a reaction that acts as catalyst
    (*d*) one that retards the rate of reaction
13. The activity of a catalyst is enhanced by the presence of a substance called
    (*a*) super-catalyst
    (*b*) promoter
    (*c*) catalytic poison
    (*d*) none of the above
14. The unit of rate constant for a zero-order reaction is
    (*a*) liter sec^{-1}
    (*b*) liter mol^{-1} sec^{-1}
    (*c*) mol $liter^{-1}$ sec^{-1}
    (*d*) mol sec^{-1}
15. For a single step reaction, $X + 2Y \longrightarrow$ Products, the molecularity is
    (*a*) 0
    (*b*) 2
    (*c*) 3
    (*d*) 1

Chapter 6

ELECTROCHEMISTRY

6.1 INTRODUCTION

Electrochemistry is a science that deals with the transformation of chemical energy into electrical energy, and vice versa. The process of transforming electrical energy into chemical energy is called *electrolysis*. The apparatus used for electrolysis is called an *electrolytic cell* in which electrical energy is used to cause a chemical reaction to take place. Faraday found out that there were quantitative relations between the amount of electricity flowing through the electrolyte and the amount of chemical change occurring at the electrodes. These relations came to be known as Faraday's law.

Laws of Electrolysis

The two laws discovered by Faraday are stated as follows:

First law of electrolysis: The law states that the amount of substance deposited or evolved at any electrode is proportional to the quantity of electricity passed through the electrolyte.

Suppose Wg of the substance is deposited at an electrode when Q coulombs of electricity are passed.

Then $$W \propto Q$$

$$\therefore \quad W = Zct \tag{6.1}$$

When Z is a constant called the electrochemical equivalent, it is the quantity of substance produced when 1 coulomb of electricity is passed that is, a unit current (1 amp) is passed for 1 second. c is the current in amperes, and t is the time in seconds. Every substance possesses a definite electrochemical equivalent (Z). An electrochemical equivalent may be defined as the amount of substance deposited by 1 ampere of current passing for 1 second (*i.e.*, 1 coulomb).

Second law of electrolysis: According to this law, if the same quantity of electricity is passed through different electrolytes, the different amounts of chemical changes produced are all chemically equivalent.

Mathematically, $W \propto E$,

where, W = Weight of substance deposited or liberated
E = Chemical equivalent weight of the substance deposited or liberated

Thus if W_1 and W_2 are the weights of two different substances liberated by the passage of the same amount of electricity, and E_1 and E_2 are their respective chemical equivalent weights, then

$$\frac{W_1}{E_1} = \frac{W_2}{E_2} \tag{6.2}$$

For example, consider the passage of electricity through the solutions of copper sulphate and silver nitrate, connected in series, so that the same amount of electricity is passed through them. Then, from the second law we get

$$\frac{\text{Wt. of copper deposited}}{\text{Equivalent wt. of copper}} = \frac{\text{Wt. of silver deposited}}{\text{Equivalent wt. of silver}}$$

6.2 ELECTROCHEMICAL CELL

An electrochemical cell is a device in which a redox reaction is used to get electrical energy. This type of cell is also commonly called a *galvanic cell* or *voltaic cell* (named after L. Galvani and A. Volta).

Galvanic Cell

A galvanic cell is a device in which the free energy of a chemical process is converted into electrical energy. When a system consists of two electrodes dipped into an electrolyte solution and electrodes connected by some metallic conductor outside the electrolyte, this produces an electric current in the external circuit called a Galvanic cell. The example of an electrochemical or galvanic cell is a Daniell cell.

Daniell Cell

A Daniell cell is a zinc-copper cell that can be represented by

$$Zn \mid ZnSO_4 \mid\mid CuSO_4 \mid Cu$$

where vertical lines indicate phase boundaries. The cell is shown in Fig. 6.1.

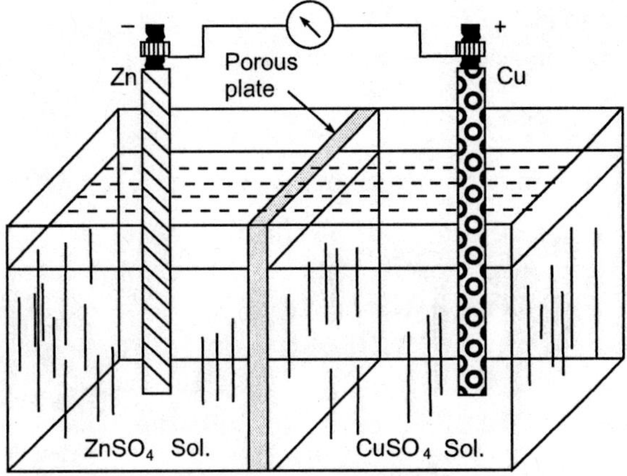

Fig. 6.1 A Daniell cell.

In a Daniell cell, a zinc rod partially immersed in a $ZnSO_4$ solution and a copper rod in a $CuSO_4$ solution are the two electrodes. The two solutions are separated by a porous barrier that allows the ions to pass through but prevents mass diffusion. A current begins to flow through the outer circuit when the two electrodes are joined with the help of a copper wire.

At the anode: The zinc electrode where zinc gives up its electrons, that is oxidation occurs, is the anode or negative electrode of the cell.

$$Zn \rightleftharpoons Zn^{++} + 2e^-$$

At the cathode: The electrons that were released at the anode move along the external circuit and are used up at the cathode or the positive electrode (copper electrode), where copper ions take up electrons to give copper metal. Reduction takes place at the cathode.

$$Cu^{++} + 2e^- \rightleftharpoons Cu$$

Because the electrons flow from the zinc rod toward the copper rod, the electrical energy is produced. Electricity is nothing but the flow of electrons, and the direction of current is opposite to that of the flow of electrons.

The cell reaction is $Zn + Cu^{++} \longrightarrow Zn^{++} + Cu$.

Representation of Galvanic cell

The cell is represented by keeping in view the following points:

- It is customary to write the anode of the cell on the left and the cathode to the right.
- In a cell, there are two electrodes. Each of the electrodes in contact with its ions is called a single electrode or half cell. In the Daniell cell, the half cells are Zn/Zn^{++} and Cu^{++}/Cu.
- A salt bridge is indicated by two vertical lines separating the two half cells. Thus the Daniell cell can be represented as

$$Zn \mid Zn^{++}_{(conc.\,C_1)} \mid\mid Cu^{++}_{(conc.\,C_2)} \mid Cu$$

Concentration may be mentioned in bracket.

The flow of the current and the chemical changes at the electrode continue as long as the circuit is closed. The chemical change ceases when the circuit is left open and there is no flow of current.

Charge Anomaly

At the anode, electrons enter into the external circuit, that is it sends electrons to the external circuit. Thus, it is considered to be negatively charged. On the other hand, the cathode receives electrons from the external circuit; thus, the cathode should have a positive charge.

Some scientists consider the anode to be positively charged. The discrepancy may be explained as, for example, the Daniell cell: the Zn^{2+} is released from the anode; hence, it appears positively charged as viewed from the solution. The zinc rod sends electrons to the external circuit through a wire; thus, it (anode) appears negatively charged as viewed from the wire.

Salt Bridge

If two electrolyte solutions at different concentrations of a galvanic cell form a heterogeneous system so that the cations and anions diffuse with different speeds then the potential difference is set up at the junction of two electrolyte solutions called *liquid junction potential*. It is necessary to avoid, eliminate, or minimize the liquid junction potential as it interferes in the accurate determination of the electromotive force of the cell (emf).

Two electrolyte solutions in galvanic cells are separated using a salt bridge, which is a device to minimize or eliminate the liquid junction potential. A saturated solution of salts such as KCl, KNO_3, NH_4Cl, NH_4NO_3, and so on in agar-agar gel is used in a salt bridge.

Role of a Salt Bridge

A salt bridge contains a high concentration of ions namely K^+ and NO_3^-, at the junction with the electrolyte solution. Thus, a salt bridge carries the whole current across the boundary; moreover the K^+ and NO_3^- ions have the same speed or transport number. Hence, a salt bridge with the uniform and same mobility of cations and anions minimizes the liquid junction potential.

Potassium chloride in a salt bridge cannot be used in the presence of electrolyte solutions of silver, mercurous, or thalus salt because it may cause precipitation of the corresponding chloride. A porous partition may also be used in place of the salt bridge. This permits the flow of ions without allowing the two solutions to mix.

6.3 REVERSIBLE CELL

A cell is considered to be working under reversible conditions if it satisfies the following conditions:

- If an opposing emf equal to the cell emf is applied to the cell, no current flows in the circuit and thus no chemical reaction occurs.
- If an opposing emf infinitesimally smaller than the emf of the cell is applied to the cell, only an infinitesimally small current should flow from the cell and thus an infinitesimally small reaction occurs.
- If an opposing emf infinitesimally greater than that of the cell is applied, an infinitesimally small current should flow in the opposite direction, and an infinitesimally small reaction in the reverse direction should occur. The cell will return to its original state.

The Daniell cell is an example of a reversible cell. The cell reaction is

$$Zn + Cu^{++} \rightleftharpoons Zn^{++} + Cu$$

The emf of the cell is 1.10 volts.

When

- the opposing emf is equal to the cell emf, there is no current.
- the opposing emf is infinitesimally smaller than the cell emf, the reaction proceeds in a forward direction.
- the opposing emf is infinitesimally greater than the cell emf, the reaction proceeds in a backward direction.

A cell that does not fulfill the preceding conditions is said to be irreversible. Thus, an irreversible cell has a tendency to change spontaneously from one state to another but not to change exactly in the reverse direction.

For example, consider a cell consisting of zinc and copper electrodes dipped into a solution of sulphuric acid.

- If the emf of the external source is slightly smaller than that of the cell, zinc dissolves with the evolution of hydrogen.

$$Zn\,(s) + H_2SO_4\,(aq) \rightleftharpoons ZnSO_4\,(aq) + H_2\,(g)$$

- If the emf of the external source is slightly greater than that of the cell, copper dissolves with the evolution of hydrogen.

$$Cu\,(s) + H_2SO_4\,(aq) \rightleftharpoons CuSO_4\,(aq) + H_2\,(g)$$

These two types of reactions illustrate that the cell is not reversible because in each case, different sets of products are formed.

6.4 STANDARD CELL

A cell may be considered a standard cell if it fulfills the following conditions:

- A constant value of emf that is reproducible.
- No change in emf with the passage of time.
- The temperature coefficient of the cell is very low.

Standard cells are not used for a supply of current but always as a source of a constant emf. The most commonly used standard cell is the Weston cell.

Weston Standard Cell

The Weston cell shown in Fig. 6.2 is prepared in a H-shaped glass vessel: one arm containing cadmium amalgam that contains 12.5% Cd by weight is the negative electrode, and the other arm containing mercury in contact with a paste of mercury and mercurous sulphate (Hg_2SO_4) is the positive electrode. Some crystals of $CdSO_4, \frac{8}{3} H_2O$ being added over the negative electrode. The vessel is then filled up with a saturated solution of $CdSO_4$ and sealed. The electrodes are connected to the external circuit through two platinum wires sealed at the bottom of two electrodes. Thus, the Weston cell is a mercury-cadmium cell that can be represented as,

$$Cd\,(Hg)\,CdSO_4, \frac{8}{3} H_2O \left| \underset{\text{(Satd. soln.)}}{CdSO_4} \right| \underset{\text{(Solid)}}{Hg_2SO_4} \Big| Hg$$

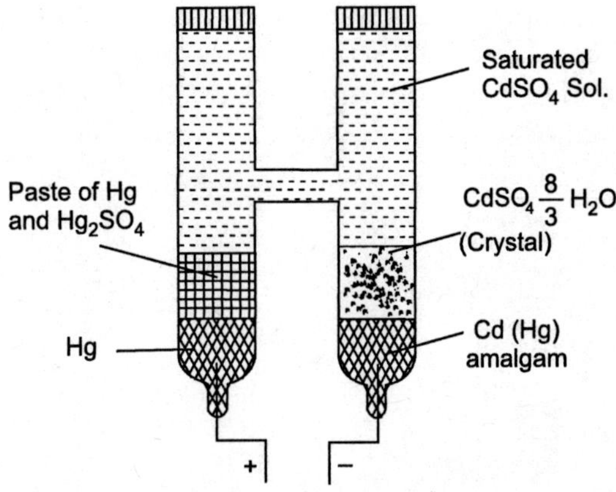

Fig. 6.2 A standard cell (Weston cell).

The following reactions occur at the two electrodes:

At the anode: $Cd \rightleftharpoons Cd^{2+} + 2e^-$

At the cathode: $Hg_2^{2+} + 2e^- \rightleftharpoons 2Hg$

The net cell reaction is

$$Cd + Hg_2SO_4 \rightleftharpoons 2Hg + CdSO_4$$

The purpose of the solid crystals of $CdSO_4, \frac{8}{3}H_2O$ is to keep the electrolyte saturated at all temperatures.

At 20°C, the emf is E = 1.01830 volts and the

$$\text{temperature coefficient is } \frac{dE}{dt} = -0.00004 \text{ volts degree}^{-1}$$

The cell works for many years at a stretch without any voltage changes. A standard cell is never used for a supply of current, but always as a source of constant emf.

6.5 ELECTRODE POTENTIAL

The tendency of an electrode to lose or gain electrons is called *electrode potential.* Electrode potential is called oxidation potential if oxidation takes place at the electrode. It is called reduction potential if reduction takes place at the electrode.

- If a metal is in contact with a solution of its own salt, the positive ions in the metal come into equilibrium with those in the solution, leaving behind an equivalent number of electrons on the metal. Thus, the metal acquires a negative charge as it is left with an excess number of electrons. A number of positive metallic ions are formed in the solution. For example, Zn in $ZnSO_4$ solution (Fig. 6.3 (*a*)).

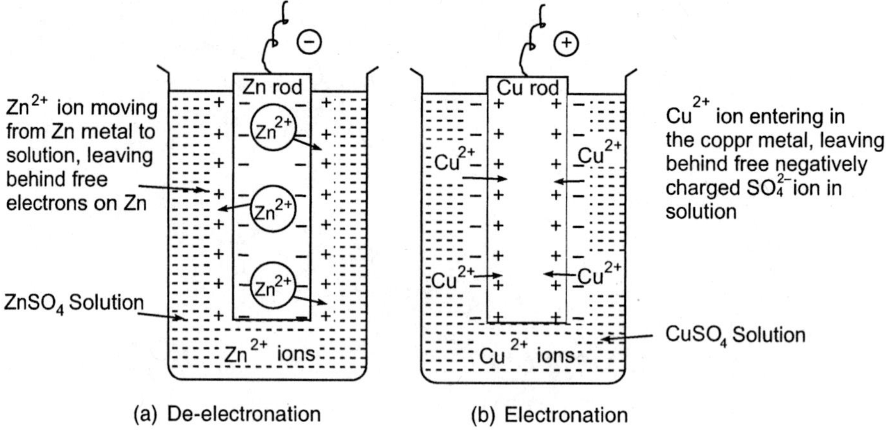

Fig. 6.3 Electrode potential.

For a metal of valency '*n*' dipping in a solution:

$$M \rightleftharpoons M^{n+} + ne^- \quad \text{(oxidation } i.e., \text{ de-electronation)}$$

- If the positive metallic ions from the solution enter the metallic lattices, then the metal acquires a positive charge.

$$M^{n+} + ne^- \rightleftharpoons M \quad \text{(Reduction } i.e., \text{ electronation)}$$

For example in the case of Cu in $CuSO_4$ solution [Fig. 6.3 (*b*)], the rate of reaction depends on

- The nature of metal
- The concentration of metal ions in the solution
- The temperature

When a metal is placed in the solution of its salt, oxidation or reduction takes place resulting in a dynamic equilibrium. Negative or positive charges developed on the metal attracts the positively or negatively charged ions in the solution. Thus, a short layer of positive or negative ions is formed around the metal. This layer is called the Helmholtz electrical double layer. A difference of potential is set up between the metal and the solution. At equilibrium, the potential difference becomes a constant value and is called the electrode potential. The contribution of each electrode to the emf of a cell is called the single electrode potential.

If the oxidation potential (tendency to lose electrons) of an electrode is + x volt, then its reduction potential (tendency to gain electrons) will be – x volt.

Mathematical expression:

Consider a general redox reaction:

$$M^{n+} + ne^- \rightleftharpoons M$$

For a reversible reaction, the free energy change (ΔG) is given by

$$\Delta G = RT \ln K + RT \ln \frac{[\text{Product}]}{[\text{Reactant}]}$$

$$= \Delta G^\circ + RT \ln \frac{[\text{Product}]}{[\text{Reactant}]} \qquad (6.3)$$

where K is the equilibrium constant, and ΔG° is the standard free energy change. Equation (6.3) is known as the Vant Hoff reaction isotherm. Electrical energy produced in a reversible cell is equal to the decrease in free energy change:

$$-\Delta G = nFE \quad \Delta G^\circ = -nFE^\circ,$$

E = Electrode potential, E° = Standard electrode potential,

F = Faraday constant

Now Equation (6.3) becomes

$$-nFE = -nFE^\circ + RT \ln \frac{[M]}{[M^{n+}]}$$

$$= -nFE^\circ + RT \ln \frac{1}{[M^{n+}]} \quad (\therefore \text{ Concentration of metal is unity})$$

or,

$$-nFE = -nFE^\circ - RT \ln [M^{n+}]$$

or,

$$E = E^\circ + \frac{RT}{nF} \ln [M^{n+}]$$

$$= E^\circ + \frac{2.303RT}{nF} \log [M^{n+}] \qquad (6.4)$$

This expression is known as the Nernst equation for the electrode potential:

At 25°C, $$\frac{2.303RT}{nF} = \frac{0.0592}{n}$$

∴ $$E = E^\circ + \frac{0.0592}{n} \log [M^{n+}].$$

Measurement of the Electrode Potential

It is impossible to know the absolute value of the single electrode potential. The difference in potential between two electrodes can be measured potentiometrically by combining them to form a complete cell. In other words, we can only determine the relative value of the electrode potential, if we can fix arbitrarily the potential of any one electrode. For this purpose, the potential of a standard hydrogen electrode (SHE) has been arbitrarily fixed at zero, and all other electrode potentials are expressed in comparison with this value. The hydrogen electrode consists of a platinum electrode in contact with $1MH^+$ ion concentration, and hydrogen gas at atmospheric pressure is constantly bubbled over it. This electrode is represented as shown in Fig. 6.4.

$$\text{Pt, } H_{2\,(1\text{ atm})}\,;\, H^+_{\;(1\text{ M})}$$

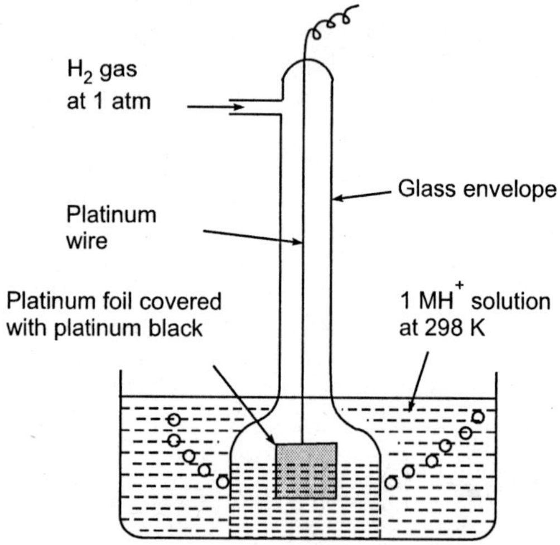

Fig. 6.4 Normal hydrogen electrode.

Thus, to measure the electrode potential of the Zn electrode dipping in 1 M $ZnSO_4$, this electrode is to be coupled with SHE through a salt bridge (Fig. 6.5). The voltmeter reading directly gives the electrode potential because the potential of SHE is zero. As the zinc electrode acts as the anode (*i.e.*, electron releasing) in this case, so the electrode potential, 0.76 volt is the oxidation potential. So, the reduction potential of the zinc electrode is – 0.76 volt. According to the modern conventions, if on coupling an electrode with SHE, a reduction occurs at the given electrode, the electrode potential (reduction) is given a positive sign; if oxidation occurs at the given electrode, the electrode potential (reduction) is given a negative sign.

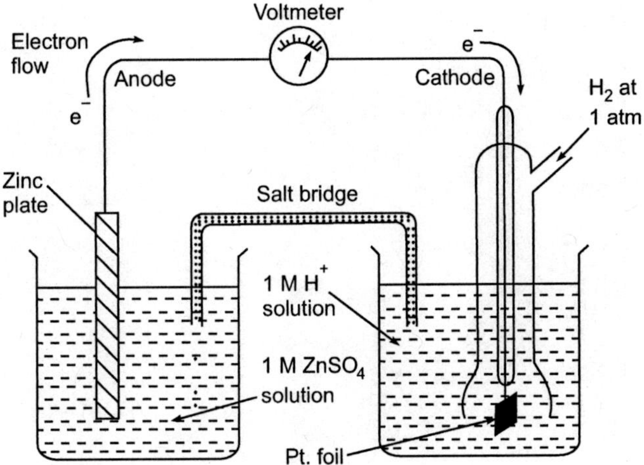

Fig. 6.5 Measurement of electrode potential of zinc electrode (Zn^{2+}/Zn). Here, the Zn electrode acts as the anode with respect to SHE, so the electrode potential of Zn is given a negative sign (namely –0.76 volt).

6.6 ELECTROCHEMICAL SERIES

The standard electrode potential of a large number of electrodes has been determined using the standard hydrogen electrode as the reference electrode, for which the electrode potential has been arbitrarily fixed as zero. It is customary to express the electrode potentials as the reduction potentials. The various electrodes have thus been arranged in the order of their increasing values of standard reduction potential (Table 6.1). This arrangement is called the *electrochemical series.*

TABLE 6.1 Standard Electrode Potentials (Reduction) at 25 °C (Electrochemical Series)

Reduction Half Reaction	*Standard Reduction Potential E° (in volts)*
$Li^+ + e^- \longrightarrow Li$	– 3.05
$K^+ + e^- \longrightarrow K$	– 2.93
$Ca^{2+} + 2e^- \longrightarrow Ca$	– 2.90
$Na^+ + e^- \longrightarrow Na$	– 2.71
$Mg^{2+} + 2e^- \longrightarrow Mg$	– 2.37
$Al^{3+} + 3e^- \longrightarrow Al$	– 1.66

$Zn^{2+} + 2e^- \longrightarrow Zn$	– 0.76
$Cr^{3+} + 3e^- \longrightarrow Cr$	– 0.74
$Fe^{2+} + 2e^- \longrightarrow Fe$	– 0.44
$Ni^{2+} + 2e^- \longrightarrow Ni$	– 0.23
$Sn^{2+} + 2e^- \longrightarrow Sn$	– 0.14
$Pb^{2+} + 2e^- \longrightarrow Pb$	– 0.13
$F^{3+} + 3e^- \longrightarrow F$	– 0.04
$H^+ + e^- \longrightarrow H$	0.00 (Reference)
$Cu^{+2} + 2e^- \longrightarrow Cu$	+ 0.34
$Ag^+ + e^- \longrightarrow Ag$	+ 0.80
$Pt^{4+} + 4e^- \longrightarrow Pt$	+ 0.86
$Au^+ + e^- \longrightarrow Au$	+ 1.69
$F_2 + 2e^- \longrightarrow 2F^-$	+ 2.87

The electrochemical series provides valuable information regarding the following:

Relative ease of oxidation or reduction: A system with a high reduction potential has a great tendency to undergo reduction, so we can easily interpret the behavior of different elements. For example, the standard reduction potential of F_2/F^- is highest (*i.e.*, + 2.87 volts), so F^- ions are very easily reduced to F_2, or, conversely, F_2 is oxidized with great difficulty to F^- ions.

Similarly, the standard reduction potential of the Li^+/Li system is the least (*i.e.*, – 3.05 volts), so Li is reduced with great difficulty to Li, or, conversely, Li is very easily oxidized to Li^+.

Replacement tendency: The higher value of the reduction potential shows a great tendency to assume the reduced form. So, a known electrode potential indicates the relative replacement tendency. For example, it can be known whether Cu will displace Zn from solution, or vice versa.

$$E°Cu^{2+}/Cu = +0.34 \text{ V}$$

$$E°Zn^{2+}/Zn = -0.76 \text{ V}$$

Cu^{2+} has a greater tendency to acquire the Cu form than Zn^{2+} has for acquiring the Zn form. So the reaction will occur in the direction

$$Zn + Cu^{2+} \longrightarrow Zn^{2+} + Cu$$

Predicting spontaneity of redox reactions: The spontaneity of a redox reaction can be predicted from the emf (E) value of the complete cell reaction.

The positive value of E of a cell reaction indicates that the reaction is spontaneous. If the value of E is negative, the reaction is not feasible. In general, an element with a lower reduction potential can displace another metal with a higher reduction potential from its salt solution spontaneously.

Calculation of equilibrium constant: The standard elecrode potential

$$E^\circ = \frac{RT}{nF} \ln K_{eq} = \frac{2.303\,RT}{nF} \log K_{eq}$$

$$\therefore \qquad \log K_{eq} = \frac{nF \times E^\circ}{2.303\,RT} = \frac{nE^\circ}{0.0592\ V} \ (\text{At } 25^\circ\text{C})$$

So from the value of the standard electrode potential (E°) for a cell reaction, its equilibrium constant can be evaluated.

6.7 EMF OF A REVERSIBLE CELL

The force that causes the flow of electrons from one electrode to the other and thus results in a flow of current is called the electromotive force, usually represented by emf, (as stated earlier). It is generally measured in volts. This force arises because of the different potentials possessed by two electrodes.

The electrode potential arises when a metal rod is kept in contact with its ions. In addition to the electrode potentials of the two electrodes, that is half cells, an additional potential arises because of the contact between the electrolytic solutions and is known as the liquid junction potential (E_{LJ}). The emf of a reversible cell or electrochemical cell is given by

$$E_{cell} = E_L + E_R + E_{LJ}$$

where, E_L = Electrode potential of the left electrode

E_R = Electrode potential of the right electrode

E_{LJ} = Electrode potential of the liquid junction

The emf is also defined as the difference of potential that causes the current to flow from an electrode at a higher potential to an electrode at a lower potential. If all the reactants and products are present in their standard state at 25°C, the emf is called the *standard emf.*

The Standard emf of Electrochemical Cell

The emf of a cell is the algebraic sum of its electrode potentials. Any electrochemical cell is based on a redox reaction that can be split up into two half-reactions:

- Oxidation half-reaction
- Reduction half-reaction

It is well known that the electrode on which oxidation occurs is written on the left side, and the electrode on which the reduction occurs is written on the right side.

Standard emf, cell = Standard oxidation potential of L.H.S. electrode + Standard reduction potential of R.H.S. electrode.

As, oxidation potential = –reduction potential, the preceding expression becomes

Standard emf, cell = Standard reduction potential of R.H.S. electrode – Standard reduction potential of L.H.S. electrode

$$E^\circ_{cell} = E^\circ_{right} - E^\circ_{left}$$

Measurement of EMF

The emf is the difference of potential that causes the current to flow from an electrode at a higher potential to an electrode at a lower potential and is expressed in volts. Thus, the simplest possible method to measuring the emf of a cell is to join the two electrodes to a voltmeter and take the reading directly. But the emf of a cell cannot be determined in this way because of the following objections:

- Because the voltmeter draws some current from the cell, the emf of the cell will change.
- Due to the appreciable flow of current, a part of the emf is used up in overcoming the internal resistance of the cell.

To overcome these objections, the potentiometric method, which is based on Poggendorff's compensation principle, is used. The principle involved is shown in Fig. 6.6. AB is the potentiometer slidewire of a uniform cross-section with high resistance. A storage cell 'C' is connected across the terminals of slidewire AB, so that the potential drops from A to B.

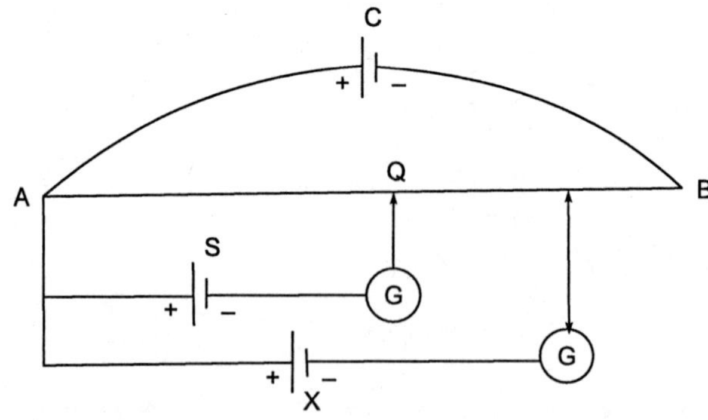

Fig. 6.6 Measurement of EMF of a cell.

Now the cell 'X,' whose emf is required is connected to 'A.' The other terminal of cell X is connected through a galvanometer (G) to a sliding contact P. This is moved along the slidewire until there is no deflection in the galvanometer. This means that the emf of the cell 'X' just balances the drop of the potential between 'A' and 'P.' The contact point is assumed to be at 'P' when there is no deflection. Next, a standard cell 'S' of known emf is connected to A replacing cell 'X.' The experiment is repeated. The sliding contact is again moved along the wire AB, until the point 'Q' is reached, when there is no deflection.

If, E_s is emf of standard cell and

E_x is emf of unknown cell

$E_s \propto$ Length AQ

$E_x \propto$ Length AP

$$\therefore \quad \frac{E_x}{E_s} = \frac{\text{Length AP}}{\text{Length AQ}} \text{ or, } E_x = \frac{\text{Length AP}}{\text{Length AQ}} \times E_s$$

The two lengths are known, and because E_s, the emf of the standard cell, is known, E_x is easily determined. The most widely used standard cell is the Weston cell.

6.8 THE EMF OF A CELL AND THE FREE ENERGY CHANGE

The electrical energy produced in a cell is the product of the amount of electricity that passes through the cell and the emf of the cell:

$$\text{Electrical energy} = \text{Quantity of electricity} \times \text{EMF}$$

If E is the emf of the cell,

$$\text{Electrical energy} = n\text{FE}$$

where, n = Moles of electrons transferred

F = Faraday constant

It was earlier accepted that the electrical energy produced from the cell is quantitatively equal to the heat of the reaction of the chemical process occurring therein.

But according to Gibb's and Helmholtz, the electrical energy produced in a reversible cell is equal to the decrease in the free energy change of the process.

$$-\Delta G = n\text{FE} \qquad (6.5) \; [\Delta G = \text{Change in free energy}]$$

From the Gibb's Helmholtz equation:

$$\Delta G = \Delta H + T\left[\frac{\partial}{\partial T}(\Delta G)\right]$$

Hence, $$-nFE = \Delta H + T\left[\frac{\partial}{\partial T}(\Delta G)\right] = \Delta H + T\left[\frac{\partial}{\partial T}(-nFE)\right]$$

or, $$nFE = -\Delta H + TnF\left(\frac{\partial E}{\partial T}\right)$$

or, $$E = -\frac{\Delta H}{nF} + T\left(\frac{\partial E}{\partial T}\right) \tag{6.6}$$

- If $\left(\frac{\partial E}{\partial T}\right)$ is very small, as in the case of a standard cell, the electrical energy is equal to the change in enthalpy.
- If $\left(\frac{\partial E}{\partial T}\right)$ is positive, the electrical energy is more than the enthalpy change. The additional energy is absorbed from the surroundings.
- If $\left(\frac{\partial E}{\partial T}\right)$ is negative, the electrical energy is less than the enthalpy change. The energy is lost as heat to the surroundings.
- At absolute zero, the equation becomes

$$nFE = -\Delta H, \text{ when } T = 0° \text{ A}$$

Thus, at absolute zero, the decrease in enthalpy is equal to the electrical energy obtainable from the cell. The value of ΔG can be known from Equation (6.5) and ΔH from Equation (6.6) knowing the values of $\frac{\partial E}{\partial T}$ and E. The value of ΔS (entropy change) can be calculated by using the following:

$$\Delta G = \Delta H - T\Delta S. \tag{6.7}$$

6.9 DETERMINATION OF THE ACTIVITY COEFFICIENT FROM THE CELL EMF

The activity coefficient can be computed from emf measurements. Consider the cell:

$$(Pt)H_2\,(1\text{ atm}) \mid HCl\,(m) \mid AgCl\,(s)\,.\,Ag \quad (m = \text{molality of solution})$$

The cell reaction is $AgCl + \frac{1}{2}H_2 \longrightarrow Ag + H^+ + Cl^-$.

$\therefore$ The cell emf $E = E° - \frac{RT}{F}\ln a_{H^+}\,.\,a_{Cl^-} = E° - \frac{RT}{F}\ln a_{\pm}^2$

where $a_{\pm}$ is the mean activity of HCl.

If we add $\left(\frac{2RT}{F}\right)$ ln m to each side:

$$E + \frac{2RT}{F} \ln m = E^\circ - \frac{2RT}{F} \ln a + \frac{2RT}{F} \ln m$$

or,
$$\left(E + \frac{2RT}{F} \ln m\right) - E^\circ = -\frac{2RT}{F} \ln \frac{a}{m} = -\frac{2RT}{F} \ln r$$

r is the mean activity coefficient of the hydrogen ion and chloride ion in solution. If a plot is drawn between $\left(E + \frac{2RT}{F} \ln m\right)$ vs the function of molality and extrapolated to $m = 0$ (*i.e.*, to infinite dilution), then the ordinate gives the value of E°. At infinite dilution, r is unity and the right side of the preceding equation becomes zero. With the knowledge of r, the mean activity of the solution can be obtained:

$$a = r \cdot m$$

The method, which can be used with other electrolytes also, is employed both when finding the normal emf or when finding the electrode potential as also the mean activity coefficient.

6.10 REFERENCE ELECTRODES AND DETERMINATION OF pH

The *reference electrode* is an electrode of standard potential, with which we can compare the potentials of other electrodes.

Reference electrodes are classified into two types:

Primary Reference Electrode: Standard hydrogen electrode is the primary reference electrode. The emf of such a cell has arbitrarily been fixed at zero.

A hydrogen electrode may be employed to find the pH value of an unknown solution. The solution, whose pH is to be determined, is taken in a vessel, and a platinum electrode is half dipped in it. A slow current of hydrogen gas at 1 atmosphere is bubbled through the solution. The platinum catalyzes the reaction:

$$H^+(aq) + e^- \rightleftharpoons \frac{1}{2} H_2(g)$$

This reaction develops a definite potential at the electrode, depending on the H^+ ion concentration of the solution under test.

$$E = E^\circ - \frac{2.303\,RT}{nF}\log\frac{[H_2]^{\frac{1}{2}}}{[H(aq)]}$$

$$= 0 - \frac{0.0592\text{ V}}{1}\log\frac{[H_2]^{\frac{1}{2}}}{[H(aq)]} \qquad \text{(At 25°C)}$$

$$= -0.0592\text{ V}\log\frac{1}{[H^+(aq)]} = -0.0592\text{ V pH.}$$

The preceding half-cell so formed is connected to a standard or normal hydrogen electrode (*i.e.*, having a solution of 1N HCl). The two solutions are separated by a salt bridge to eliminate the liquid junction potential. The emf of a full cell is then determined by using a potentiometer. With the emf of the reference electrode being zero, the observed emf directly supplies the emf of the half-cell containing the solution that is being tested (Fig. 6.7).

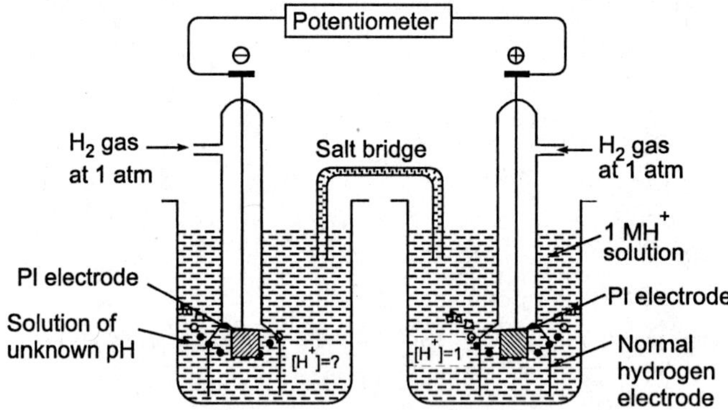

Fig. 6.7 Measurement of pH by the hydrogen electrode.

∴ $$E_{cell} = E_{right} - E_{left}$$

or, $$E_{cell} = 0 - (-0.0592\text{ V pH})$$

$$= 0.0592\text{ V pH}$$

$$pH = \frac{E_{cell}}{0.0592\text{ V}}$$

It is quite cumbersome to set up a hydrogen electrode. It cannot be used in solutions containing redox systems.

Secondary Reference Electrode: In view of the difficulties encountered with the primary reference electrode, the secondary reference electrodes are used. They include the calomel electrode, glass electrode, quinhydrone electrode, and so on.

The Calomel Electrode

The calomel electrode is the mercury-mercurous chloride electrode used as a secondary reference electrode for potential measurements.

Construction

The calomel electrode consists of a tube at the bottom of which a small amount of mercury is placed. It is covered with a paste of solid mercurous chloride Hg_2Cl_2 (Calomel). A solution of potassium chloride is then placed over the paste. A platinum wire, dipping into the mercury layer, is used for making the electrical contact. The side tube is used for making the electrical contact with a salt bridge. The saturated calomel electrode is formulated as shown in Fig. 6.8.

$$Hg, Hg_2Cl_2\,(s), KCl\,(\text{satd. solution})$$

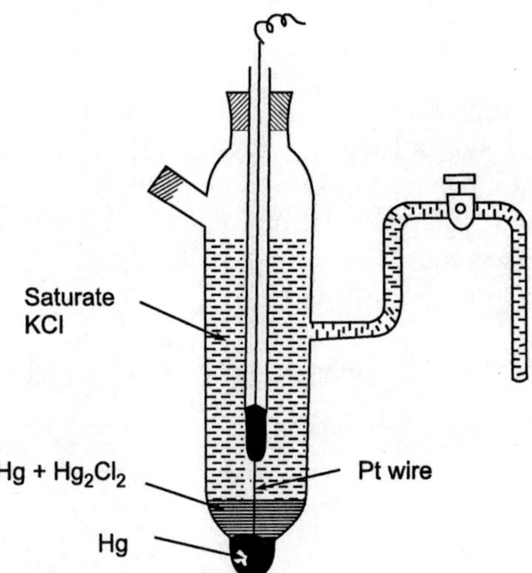

Fig. 6.8 Saturated calomel electrode.

The electrode reaction taking place in the half-cell is

$$\frac{1}{2}\,Hg_2Cl_2\,(s) + e^- \rightleftharpoons Hg + Cl^-$$

The electrode can be coupled with the hydrogen electrode containing the solution of the unknown pH.

$$Pt,H_2\,(1\,\text{atm})\,|\,H^+c = ?\;||\;Hg_2Cl_2\,(s)\,|\,Hg^+$$

The emf of the cell:

$$E_{cell} = E_{right} - E_{left}$$

$$= 0.2422 \text{ V} + 0.0592 \text{ VpH}$$

$$pH = \frac{E_{cell} - 0.2422 \text{ V}}{0.0592 \text{ V}}$$

The calomel electrode is simple to construct.

The Glass Electrode

The glass electrode is the universally employed electrode for pH measurement.

Construction

The glass electrode is made of a special glass with a relatively low melting point and a high electrical conductivity. It is a corning glass containing Na_2O (22%), CaO (6%), and SiO_2 (72%). The assembly of the glass electrode (Fig. 6.9) consists of a thin glass bulb filled with 0.1 M HCl, and a silver wire coated with silver chloride is immersed in it. Here, Ag | AgCl acts as an internal reference electrode. The glass electrode is represented as

$$Ag \mid AgCl\,(s) \mid 0.1\,M\,HCl \mid glass$$

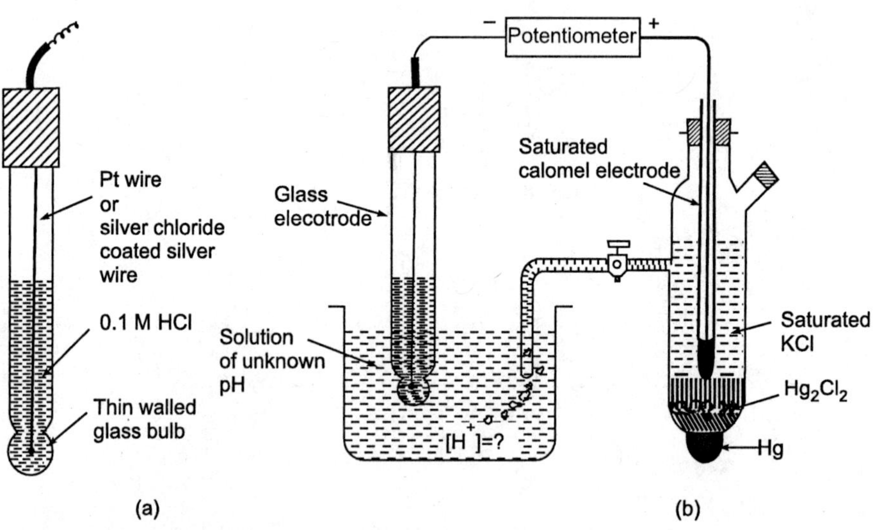

Fig.6.9 (a) Glass electrode (b) determination of pH.

HCl in the bulb gives a constant H^+ ion concentration.

Thus, it is a silver-silver chloride electrode that is reversible with respect to chloride ions.

For the measurement of pH, the glass electrode is immersed in the solution whose pH is to be determined. It is then combined with a reference-saturated calomel electrode. The cell assembly is represented as

Ag | AgCl | 0.1 M HCl | Glass | solution of unknown pH || saturated calomel electrode

The emf of the cell at 298 K is given as

$$E_{cell} = E_{right} - E_{left}$$

$$= 0.2422 \text{ V} - \left[E°_G + 0.0592 \text{ V pH}\right]$$

$$pH = \frac{0.2422 \text{ V} - E_{cell} - E°_G}{0.0592 \text{ V}}$$

The $E°_G$ value of a glass electrode is first measured by dipping the electrode in buffer solutions of known pH values. Instruments are available in which the galvanometer readings are directly scaled to read pH values. These instruments are called *pH meters.*

The glass electrode can be used up to a pH of 13 but becomes sensitive to Na^+ ions above pH = 9, resulting in an alkaline error.

The Quinhydrone Electrode

The quinhydrone electrode contains equimolar quantities of quinone and hydroquinone. The working of this electrode is based upon the fact that quinone is reduced to hydroquinone ions:

O=C₆H₄=O $+ 2H^+ + 2e^- \rightleftharpoons$ HO–C₆H₄–OH

Quinone (Q) Hydroquinone (H_2Q)

The cell reaction can be represented as

$$Q + 2H^+ + 2e^- \rightleftharpoons H_2Q$$

When a platinum electrode is immersed in a solution containing quinhydrone, a potential is established, the value of which is given by

$$E_Q = E_Q^\circ - \frac{2.303\,RT}{2F}\log\frac{[H_2Q]}{[Q][H^+]^2}$$

If quinone and hydroquinone are taken in equimolar concentrations, then

$$[Q] = [H_2Q]$$

Now, the preceding equation becomes

$$E_Q = E_Q^\circ - \frac{2.303\,RT}{2F}\log\frac{1}{[H^+]^2}$$

$$= E_Q^\circ - \frac{2.303\,RT}{F}\log\frac{1}{[H^+]}$$

$$= E_Q^\circ - 0.0592\text{ V pH} = 0.6994\text{ V} - 0.0592\text{ V pH}$$

Construction

The quinhydrone electrode can very easily be set up by adding a pinch of quinhydrone powder to the experimental solution with stirring until the solution is saturated and a slight excess remains undissolved. Then, the indicator electrode, usually of bright platinum, is inserted in the solution.

For determining pH value, this half-cell is combined with any other reference electrode, that is the saturated calomel electrode, and the emf of the cell is determined potentiometrically (Fig. 6.10).

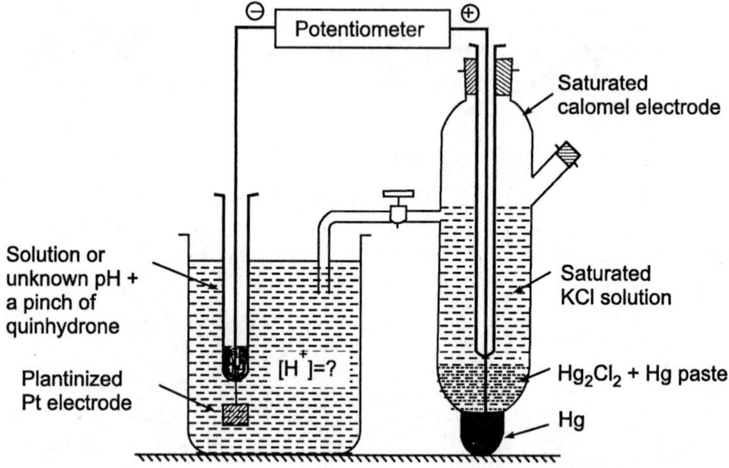

Fig. 6.10 Determining the pH of a solution using quinhydrone electrode.

The cell may be represented as

$$\text{Pt} \mid \text{H}_2\text{Q, Q, H}^+_{\text{(unknown)}} \mid\mid \text{KCl}_{\text{(satd.)}}, \text{Hg}_2\text{Cl}_2\ (s) \mid \text{Hg}^+.$$

$$E_{\text{cell}} = E_{\text{right}} - E_{\text{left}} = 0.2422\ \text{V} - (0.6994\ \text{V} - 0.0592\ \text{V pH})$$

$$\text{pH} = \frac{0.6994\ \text{V} - 0.2422\ \text{V} + E_{\text{cell}}}{0.0592\ \text{V}}$$

The quinhydrone electrode is very useful in the titration of an acid by an alkali.

6.11 BATTERIES (COMMERCIAL CELLS)

Some galvanic cells are commonly used as sources for supplying of electric energy for different commercial purposes. A *battery* is defined as an electrochemical cell or often several electrochemical cells connected in series that can be used as a source of direct electric current at a constant voltage. Batteries are of two types:

Primary cells: The primary cells are those in which the cell reaction is not reversible. In such a cell, electrical energy is obtained at the expense of chemical reactivity only as long as the active materials are present. Dry cells used in flash-lights, radios, and so on are primary cells.

Secondary cells: In the secondary cells (or storage cells or accumulators), the chemicals used up can be brought back to the original state by charging the exhausted cells with a current from an external source. The cell is thus reproduced in its original state and is again fit to supply energy as before. The process can be repeated over and over again. These cells can then be used for storing electrical energy, hence the name.

Leclanche or Dry Cell

In this primary cell, a zinc plate is the negative electrode, and the carbon rod is the positive electrode (Fig. 6.11). The zinc plate usually forms the outer covering of the cell. The electrolyte is a paste of MnO_2, NH_4Cl, and $ZnCl_2$, and the space between the central carbon electrode and the zinc plate is completely filled with the paste.

At the anode, the oxidation process is

$$\text{Zn} + 2\text{MnO}_2 + \text{H}_2\text{O} \longrightarrow \text{Zn}^{++} + \text{Mn}_2\text{O}_3 + 2\text{OH}^- \qquad (6.8)$$

At the cathode, the reduction process is

$$2\text{MnO}_2 + \text{H}_2\text{O} + 2e^- \longrightarrow \text{Mn}_2\text{O}_3 + 2\text{OH}^- \qquad (6.9)$$

The net cell reaction is

$$\text{Zn} + 2\text{MnO}_2 + \text{H}_2\text{O} \longrightarrow \text{Zn}^{++} + \text{Mn}_2\text{O}_3 + 2\text{OH}^-$$

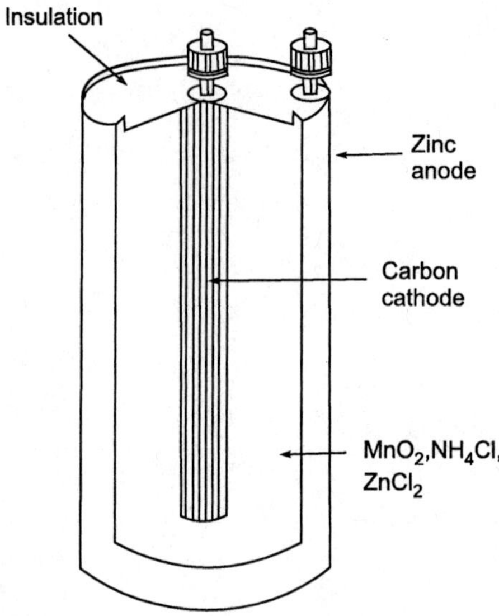

Fig. 6.11 Dry cell.

The resulting OH^- ions react with NH_4Cl to liberate NH_3, which combines with Zn^{2+} ions to precipitate sparingly soluble Zn $(NH_3)_2Cl_2$.

$$2NH_4Cl + 2OH^- \longrightarrow 2NH_3 + 2Cl^- + 2H_2O$$

$$Zn^+ + 2NH_3 + 2Cl^- \longrightarrow Zn\,(NH_3)_2\,Cl_2$$

These are the secondary reactions and are not involved in the electrode reaction. Hence, they also do not contribute to the potential of the cell. The exhausted dry cell cannot be brought back to the original state. The emf of the cell depends upon the reactions (Equation 6.8) and (Equation 6.9). Dry cell is cheap to make and gives a voltage of about 1.5 V.

Dry cells are used in flash-lights, transistor radios, calculators, and so on.

Alkaline Version Dry Cell

In such a dry cell, NH_4Cl in the electrolyte is replaced by KOH to avoid the corrosion of Zn can (negative electrode).

In this battery, zinc in powdered form is mixed with KOH to get a gel. The graphite rod is surrounded by a paste containing MnO_2. The cell reactions are

Anode: $Zn\,(s) + 2OH^-\,(aq) \longrightarrow Zn\,(OH)_2\,(s) + 2e^-$

Cathode: $2\,MnO_2\,(s) + H_2O\,(l) + 2e^- \longrightarrow Mn_2O_3\,(s) + 2OH^-\,(aq)$

The net cell reaction is

$$Zn\,(s) + 2MnO_2\,(s) + H_2O\,(l) \longrightarrow Zn\,(OH)_2\,(s) + Mn_2O_3\,(s)$$

In some cases steel can is also used. Alkaline version dry cells are longer lasting and expensive as well. These are used in camera exposure controls, calculators, watches, etc.

Nickel-Cadmium (NiCad) Battery

This battery consists of a cadmium anode and a cathode composed of a paste of NiO (OH) (*s*). The cell reactions is

Anode: $Cd\ (s) + 2OH^-\ (aq) \longrightarrow Cd\ (OH)_2\ (s) + 2e^-$

Cathode: $2NiO\ (OH)\ (s) + 2H_2O\ (l) + 2e^- \longrightarrow 2Ni\ (OH)_2\ (s) + 2OH^-\ (aq)$

The net cell reaction is

$$2NiO\ (OH)\ (s) + Cd\ (s) + 2H_2O\ (l) \longrightarrow Cd\ (OH)_2\ (s) + 2Ni\ (OH)_2\ (s)$$

The reaction can be readily reversed because the reaction products Ni $(OH)_2$ and Cd $(OH)_2$ adhere to the electrode surfaces.

The nickel-cadmium battery is a portable, rechargeable cell, and its cell voltage is constant at about 1.4 V. It can be left for long periods of time without any deterioration because no gases are produced during the discharge (or charge).

Nickel-cadmiums batteries are used in calculators, transisters, mobile phones, cordless electronic shavers, telephones, and so on.

Lead-Storage Cell (Lead Accumulator)

A storage cell is a secondary cell that can operate both as a voltaic cell and an electrical cell. When operating as a voltaic cell, it supplies electrical energy and as such becomes run down. It must then be recharged. When being recharged, the cell operates as an electrolytic cell. Thus, the storage cell has the advantage of its ability to work both ways, to receive electrical energy, and to supply it.

The common example of a storage cell is the lead-acid storage cell. It essentially consists of two lead electrodes, one of which is covered with an adherent layer of PbO_2 (positive electrode) on its surface.

The electrolyte is 20% H_2SO_4 solution of specific gravity 1.15 at room temperature. The cell is represented as

$$Pb \mid PbSO_4\,(s), H_2SO_4\,(aq) \mid PbO_2\,(s), Pb$$

- At the anode (negative electrode), lead dissolves to form Pb^{++} ions, which react with sulphate ions to be precipitated as sulphate:

$$Pb \longrightarrow Pb^{++} + 2e^-$$

$$Pb^{++} + SO_4^{--} \longrightarrow PbSO_4\,(s)$$

$$Pb + SO_4^{--} \longrightarrow PbSO_4 + 2e^-$$

- At the cathode (positive electrode), the reactions are as follows:

$$PbO_2 \longrightarrow Pb^{++++} + 2O^{--}$$
$$4H^+ + 2O^{--} \longrightarrow 2H_2O$$
$$Pb^{++++} \longrightarrow Pb^{++}$$
$$Pb^{++} + SO_4^{--} \longrightarrow PbSO_4$$

$$PbO_2 + 4H^+ + SO_4^{--} + 2e^- \longrightarrow PbSO_4 + 2H_2O$$

The plumbic ion is reduced to Pb^{++} ions, and ultimately lead sulphate is formed in the cathode as well.

The net cell reaction during discharging, that is when the accumulator gives current, is

$$PbO_2 + Pb + 2H_2SO_4 \longrightarrow 2PbSO_4 \downarrow + 2H_2O$$

Note that lead sulphate is precipitated at both the electrodes. The voltage of each cell is about 2.0 volts at a concentration of 21.4% H_2SO_4 at 25°C. The lead cell commonly used in automobiles is a combination of six such cells in series to form a battery with an emf of 12 volts.

When both the anode and the cathode become covered with $PbSO_4$, the cell ceases to function as a voltaic cell. The cell is charged by passing a current from an external emf, which reverses the chemical process:

$$PbO_2 + Pb + 2H_2SO_4 \underset{\text{charge}}{\overset{\text{discharge}}{\rightleftharpoons}} 2PbSO_4 + 2H_2O$$

During charging, the electrodes of the cell are restored to their original conditions. Also note that during the discharging operation, the concentration of acid (H_2SO_4) decreases, whereas the concentration of acid increases during the charging operation.

Accumulators are widely used in laboratories, power stations, telephones, motors, and cars, trains, electric clocks, radios, doorbells, and so on.

6.12 FUEL CELLS

Fuel cells are really electrochemical cells in which the chemical energy of easily available fuel such as hydrogen, methane, and so on, is converted into electrical energy by their oxidation at suitable anodes. At the cathodes of the cell, oxygen is dissolved to form OH^- ions. The essential process in a fuel cell is

$$\text{Fuel} + \text{oxygen} \longrightarrow \text{Oxidation products} + \text{electricity}$$

The cells have low voltages but are capable of supplying direct current for considerable periods.

One of the simplest and most successful fuel cells is the *hydrogen-oxygen* fuel cell. It consists essentially (Fig. 6.12) of an electrolytic solution such as a 25% KOH solution and two inert porous electrodes. Hydrogen and oxygen gases are bubbled through the anode and cathode compartments, respectively, where the following reactions take place:

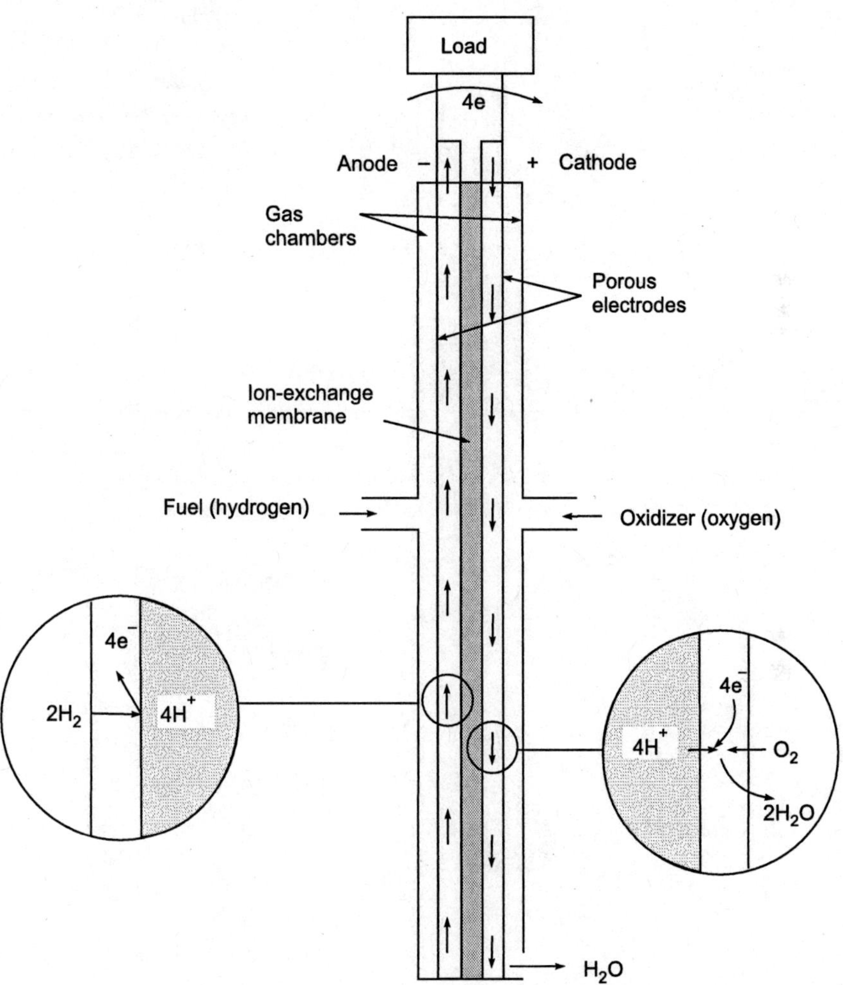

Fig. 6.12 Hydrogen-oxygen fuel cell.

At the anode: $2H_2\ (g) + 4OH^-\ (aq) \longrightarrow 4H_2O\ (l) + 4e^-$

At the cathode: $O_2\ (g) + 2H_2O\ (l) + 4e^- \longrightarrow 4OH^-\ (aq)$

$$2\ H_2\ (g) + O_2\ (g) \longrightarrow 2H_2O\ (l)$$

The standard emf of the cell E° = 1.23 V; the experimental value is 1.15 V. The only product discharged by the cell is water. Usually a large number of these cells are stacked together in series to make a battery called *fuel cell battery* or *fuel battery*.

Hydrogen-oxygen fuel cells are used as an auxiliary energy source in space vehicles (Apollo space craft), submarines, military vehicles, and so on. The product water proved to be a valuable source of fresh water for the astronauts.

A few examples of other fuel cells are shown in the following table.

Fuel Cell	*Electrode*	*Electrolyte*	*Anode Process*	*E(expt.) V*
CH_4......O_2	Pt/Pt	H_2SO_4	$CH_4 + 2H_2O \longrightarrow 8H^+ + CO_2 + 8e^-$	0.6
N_2H_4......O_2	Ni/C	KOH	$N_2H_4 \longrightarrow N_2 + 4H^+ + 4e^-$	1.28
CO......O_2	Cu/Ag etc.	KOH	$CO + H_2O \longrightarrow CO_2 + 2H^+ + 2e^-$	1.22

Such cells provide power at a comparatively lower cost, and less attention is required for their maintenance.

SOLVED NUMERICAL PROBLEMS

1. *Calculate the standard electrode potential of Cu^{2+}/Cu, if the electrode potential at 25°C is 0.296 V, when $[Cu^{2+}]$ = 0.015 M.*

Sol.

$$E = E^\circ + \frac{0.0592}{n} \text{ V} \log [M^{n+}] = E^\circ + \frac{0.0592}{2} \text{ V} \log [Cu^{2+}]$$

$\therefore$

$$E^\circ = E - 0.0296 \text{ V} \log [Cu^{2+}]$$

$$= 0.296 \text{ V} - 0.0296 \text{ V} \log 0.015$$

$$= 0.296 \text{ V} - 0.0296 \text{ V} (-1.8239)$$

$$= 0.296 \text{ V} + 0.054 \text{ V} = 0.350 \text{ V}$$

2. *Calculate the emf of a Daniell cell at 25°C, when the concentration of $ZnSO_4$ and $CuSO_4$ are 0.001 M and 0.1 M, respectively. The standard potential of the cell is 1.1 V.*

Sol.

$$E = E^\circ + \frac{0.0592}{2} \text{ V} \log \frac{[Cu^{2+}]}{[Zn^{2+}]}$$

$$= 1.1 \text{ V} + 0.0296 \text{ V} \log \frac{0.1 \text{ M}}{0.001 \text{ M}}$$

$$= 1.1 \text{ V} + 0.0296 \text{ V} \log 100$$

$$= 1.1 \text{ V} + 0.0592 \text{ V} (2) = 1.1592 \text{ V}$$

3. *Calculate the single electrode potential for copper metal in contact with a 0.15 M Cu^{2+} solution. E° for copper = + 0.34 V. [R = 8.314 JK^{-1} mol^{-1}, T = 298 K].*

Sol.

$$E = E° + \frac{2.303 \text{ RT}}{n\text{F}} \log [\text{Cu}^{2+}]$$

$$= 0.34 \text{ V} + \frac{2.303 \times 8.314 \text{ JK}^{-1} \text{ mol}^{-1} \times 298 \text{ K}}{2 \times 96{,}500 \text{ C mol}^{-1}} \log [0.15]$$

$$= 0.3156 \text{ V} \qquad [\text{ J/C} = \text{V}]$$

4. *Calculate the standard emf of the Daniell cell.*

Sol. Daniell cell is, Zn | $Zn^{2+}_{(a\,=\,1)}$ || $Cu^{+2}_{(a\,=\,1)}$ | Cu

$$E°_{cell} = E°_{right} - E°_{left}$$

$$= E°_{Cu^{2+}/Cu} - E°_{Zn/Zn^{2+}} = 0.3370 - (-\,0.7630)$$

$$= 1.1000 \text{ V}$$

5. *Calculate the standard emf of the cell:*

Cd | $Cd^{2+}_{(a\,=\,1)}$ || $Cu^{2+}_{(a\,=\,1)}$ | Cu.

Sol.

$$E°_{cell} = E°_{right} - E°_{left}$$

$$= E°_{Cu^{2+}/Cu} - E°_{Cd/Cd^{2+}}$$

$$= 0.3370 - (-\,0.4003) = 0.7373 \text{ V}$$

6. *Calculate the standard free energy change for the following cell at 25°C:*

$$Ni, Ni^{2+}, Ag^{+}, Ag$$

Given at 25°C, $E°_{Ag^{+}, Ag} = 0.80 \text{ V}$

$E°_{Ni^{2+}, Ni} = -\,0.24 \text{ V}$

Sol.

$$E°_{cell} = E°_{right} - E°_{left}$$

$$= 0.80 \text{ V} - (-\,0.24 \text{ V}) = 1.04 \text{ V}$$

Standard free energy change,

$$\Delta G° = -\,n\text{FE}°$$

$$= -\,2 \times 96500 \times 1.04 \text{ VC mol}^{-1}$$

$$= -\,200720 \text{ J mol}^{-1}$$

7. *Write down the reactions at the following two electrodes as well as the full cell reactions.*
   (i) $Fe \mid Fe^{2+} \mid\mid Fe^{2+} - Fe^{3+} \mid Pt$
   (ii) $Zn \mid ZnO_2^{2-}, OH^- \mid Hg^0\,(s) \mid Hg$.

Sol. (i) Reduction half reaction: $2Fe^{3+} + 2e^- \longrightarrow 2Fe^{2+}$

Oxidation half reaction: $Fe - 2e^- \longrightarrow Fe^{2+}$

Overall cell reaction: $2Fe^{3+} + Fe \longrightarrow 3Fe^{2+}$

(ii) Reduction half reaction: $Hg^{2+} + 2e^- \longrightarrow Hg$

Oxidation half reaction: $Zn - 2e^- \longrightarrow Zn^{2+}$

Overall cell reaction: $Hg^{2+} + Zn \longrightarrow Hg + Zn^{2+}$

8. *From the standard reduction potentials:*
   (i) $Ce^{3+} + 3e^- \longrightarrow Ce,\ E° = -2.48\ V$
   (ii) $Ce^{4+} + e^- \longrightarrow Ce^{3+},\ E° = 1.61\ V$

   Calculate the reduction potential for the half-cell Pt/Ce, Ce^{4+}.

Sol.

$$Ce^{3+} + 3e^- \longrightarrow Ce, \quad E_1° = -2.48\ V; \quad \Delta G_1 = -3F(-2.48)$$

$$Ce^{4+} + e^- \longrightarrow Ce^{3+}, \quad E_2° = 1.61\ V; \quad \Delta G_2 = -1F(1.61)$$

$$Ce^{4+} + 4e^- \longrightarrow Ce, \quad E_3° = ?; \quad \Delta G_3 = -4FE\ E_3°$$

For the overall reaction,

$$\Delta G_3 = \Delta G_1 + \Delta G_2$$

or,

$$-4FE_3° = -3F(-2.48) - 1F(1.61)$$

∴

$$E_3° = \frac{-(7.44 - 1.61)}{4} = -1.4575\ V$$

9. *Consider the following galvanic cell:*
   $Cd \mid Cd^{2+}$ (1M) $\mid\mid H^+$ (aq) (1M) $\mid H_2$ (g) Pt
   Write down the overall cell reaction indicating the anodic and cathodic half-reactions.

Sol. Anodic half reaction: $Cd \longrightarrow Cd^{2+} + 2e^-$

Cathodic half reaction: $2H^+ + 2e^- \longrightarrow H_2 \uparrow$

Overall cell reaction: $Cd + 2H^+ \rightleftharpoons Cd^{2+} + H_2$

10. *Calculate the emf of a Daniell, cell at 25°C, when the concentration of $ZnSO_4$ and $CuSO_4$ are 0.001 M and 0.1 M, respectively. The standard potential of the cell is 1.2 V.*

Sol. Zn (s) | Zn^{2+} (0.001 M) || Cu^{2+}(0.1 M) | Cu (s)

$$E_{cell} = E°_{cell} + \frac{0.0592\ V}{2} \log \frac{[Cu^{2+}]}{[Zn^{2+}]}$$

$$= 1.1 \text{ V} + 0.0296 \text{ V} \log\left(\frac{0.1}{0.001}\right)$$

$$= 1.1 \text{ V} + 0.0296 \text{ V } (2)$$

$$= (1.1 + 0.0592) \text{ V} = 1.1592 \text{ V}$$

11. *Write down Nernst's equation, and calculate the reduction potential for the reduction of O_2 at pH = 7 given a partial pressure of O_2.*

$[pO_2] = 0.20$ bar and $E° = 1.229$ V at pH = 7.

Sol. The reduction reaction involved is

$$2H^+ + \frac{1}{2}O_2\ (g) + 2e^- \rightleftharpoons H_2O$$

∴ Nernst's equation at 25°C is

$$E = E° - \frac{0.0592 \text{ V}}{2} \log \frac{1}{[H^+]^2 [pO_2]^{\frac{1}{2}}} \qquad \because\ [H_2O] = 1$$

$$= 1.229 \text{ V} - 0.0296 \text{ V} \log \frac{1}{(10^{-7})^2 (0.2)^{\frac{1}{2}}}$$

[Since, at pH = 7, $[H^+] = 10^{-7}$]

$$= 1.229 \text{ V} - 0.0296 \text{ V } [14 - 0.4472]$$

$$= 1.229 \text{ V} - 0.4011 \text{ V} = 0.828 \text{ V}$$

12. *The emf of the following cell is 0.445 V:*

Pt, H_2 (1 atm.) | $H^+_{(unknown)}$ || $KCl_{(satd.)}$ | Hg_2Cl_2 | Hg

Calculate the pH of the solution.

$E°_{KCl(satd.)} = 0.2415$ V

Sol. Emf of the cell is

$$E = E_R - E_L = E_{Ref} - \frac{RT}{F} \ln H^+$$

At 25°C,

$$E = E_{Ref} - 0.059 \log aH^+$$

$$= E_{Ref} + 0.059 \text{ pH}$$

$$\text{pH} = \frac{E - E_{Ref}}{0.059} = \frac{0.445 - 0.2415}{0.059} = 3.45$$

EXERCISES

1. Derive an expression for the emf of a reversible cell. What are the applications of electrode potentials?
2. What is a single electrode potential? Derive an expression for the emf of a reversible cell.
3. Describe a typical galvanic cell, and show how the chemical energy is converted to electrical energy. What is the relation between the energy and the emf of a cell?
4. Explain the different applications of the emf series.
5. Describe the construction and working of a Daniell cell.
6. What is meant by a single electrode potential? How is it measured experimentally?
7. Represent the standard Weston cell.
8. What is meant by a single electrode potential and the total emf of a cell? How are single electrode potentials determined?
9. Derive Nernst's equation for a single electrode potential.
10. Write a short note on electrochemical series.
11. Describe the construction of a lead-acid battery with the reactions occurring during discharge.
12. (*a*) Discuss the working principles of primary batteries.
    (*b*) Distinguish between a cell and a battery. Give the classification of cells with examples.
13. Define a fuel cell. How does a fuel cell differ from a battery?
14. (*a*) What are the reactions taking place when the storage cell is supplying electrical energy?
    (*b*) Write down the reactions involved in a dry cell.

MULTIPLE CHOICE QUESTIONS

1. The potential of a standard hydrogen electrode is taken as
   (*a*) 1 V (*b*) 0 V
   (*c*) 8 V (*d*) 10 V
2. The electrode potential is the tendency of metal
   (*a*) to gain electrons (*b*) to lose electrons
   (*c*) either to lose or gain electrons (*d*) none of these
3. A galvanic cell converts
   (*a*) electrical energy into chemical energy
   (*b*) chemical energy into electrical energy
   (*c*) electrical energy into heat energy
   (*d*) chemical energy into heat energy

4. Which metal will dissolve in the galvanic cell Cu | Cu^{2+} || Ag^{+} | Ag?
   (*a*) Cu (*b*) Ag
   (*c*) Both (*d*) None
5. The passage of electricity in the Daniell cell, when Zn and Cu electrodes are connected, is from
   (*a*) Cu to Zn in the cell (*b*) Cu to Zn outside the cell
   (*c*) Zn to Cu in the cell (*d*) Zn to Cu outside the cell
6. The galvanic cell, Zn + $Cu^{2+} \longrightarrow Zn^{2+}$ + Cu, may be represented as
   (*a*) Cu | Cu^{2+} || Zn^{2+} | Zn (*b*) Pt | Cu^{2+} || Zn^{2+} | Zn
   (*c*) Zn | Zn^{2+} || Cu^{2+} | Cu (*d*) Zn | Cu^{2+} || Zn^{2+} | Cu
7. For a cell reaction to be spontaneous (ΔG = Free energy change, E° = Reduction potential)
   (*a*) E° is – ve (*b*) E° is + ve
   (*c*) ΔG is + ve (*d*) ΔG and E° are + ve
8. The standard reduction potentials of metal electrodes A, B, C, and D are + 0.14 V, + 0.34 V, – 0.74 V, and – 0.4 V, respectively. Which of the following is the best reducing agent?
   (*a*) A (*b*) B
   (*c*) C (*d*) D
9. Which of the following relations between the standard free energy change ($\Delta G°$) and E° is correct?
   (*a*) $\Delta G° = nFE°$ (*b*) $\Delta G° = -nFE°$
   (*c*) $\Delta G° = nE°$ (*d*) None of these
10. Which one of the following electrodes require the use of a special electronic potentiometer?
   (*a*) Hydrogen electrode (*b*) Calomel electrode
   (*c*) Glass electrode (*d*) Quinhydrone electrode
11. When a lead storage battery is charged, it acts as
   (*a*) a fuel cell (*b*) an electrolytic cell
   (*c*) a galvanic cell (*d*) a concentration cell
12. The approximate voltage of a dry cell is
   (*a*) 2.0 V (*b*) 1.2 V
   (*c*) 6 V (*d*) 1.5 V
13. An example of a simple fuel cell is
   (*a*) a lead storage battery (*b*) a Leclanche cell
   (*c*) a $H_2 - O_2$ cell (*d*) All of these
14. A reaction that takes place at the anode is
   (*a*) ionization (*b*) reduction
   (*c*) oxidation (*d*) hydrolysis
15. When the cell reaction attains a state of equilibrium, the emf of the cell is
   (*a*) zero (*b*) positive
   (*c*) negative (*d*) not definite

Chapter 7

SOLID STATE

7.1 INTRODUCTION

Solids are characterized by their definite shape and also their considerable mechanical strength and rigidity. The rigidity is due to the absence of the translatory motion of the structural units of the solid. These units remain fixed to a mean position about which they may vibrate. The forces of attraction between these units are quite large. In general, solids are classified as crystalline or amorphous:

Crystalline solids exist in well-defined crystalline forms, with their atoms, molecules, or ions arranged in a certain definite geometrical pattern, throughout the 3D network of the crystal. Thus, in crystalline solids, there is a long range order in their structures. Sulphur, sugar, sodium chloride, diamond, and so on are examples of crystalline solids.

Amorphous solids have completely random particle arrangement. Thus, they lack the regular arrangement of atoms or molecules, and they also lack a long range order in their structures. In fact, amorphous solids can be regarded as supercooled liquids with small structural units and a short range order. Examples of amorphous solids are glass, rubber, plastics, pitch, and so on. Table 7.1 shows the difference between crystalline and amorphous solids.

TABLE 7.1 Difference Between Crystalline and Amorphous Solids

Crystalline Solids	*Amorphous Solids*
A regular arrangement of particles in an orderly manner.	A completely random particle arrangement.
Sharp melting points.	Melt over a range of temperature and do not have sharp melting points.
Physical properties are different in different directions. Hence, crystals are anisotropic.	The same physical properties in all the directions, and hence amorphous solids are isotropic.
Definite crystal symmetry.	Do not have symmetry.
True solids.	Not true solids but supercooled liquids.

7.2 THE BAND THEORY OF SOLIDS

The Band theory of solids may be considered as the molecular orbital theory of covalent bands extended to solids. The salient features of band theory are the following:

- Solids are made up of giant molecules, in which a very large number of spherical atoms are arranged in a regular close-packed pattern in the form of crystals.
- When the atoms are brought together in solids, the atomic orbitals of the valence shells interact forming molecular orbitals.
- There are approximately 10^{23} atoms in 1 cm^3 of a solid; during interaction between different atoms, they influence each other to such an extent that their outer shells (empty as well as partially filled) constitute one single system of electrons, common to entire array of atoms in crystal, whereas the inner shells remain intact and are not affected.
- When n energy levels of identical atoms interact, each energy level gets split up into n number of infinitesimal energy levels. As a result of this, each original energy level becomes a band of very closely spaced levels of extremely small energy differences. The individual energies within an energy band are so close together that for all purposes, the energy band may be considered to be continuous. An electron in a solid crystal can thus occupy any of these large numbers of energy levels, within the band. Thus, a *band* is a group of infinitesimal energy levels in a solid/crystal.
- Depending upon the composition of a solid/crystal, the bands are of the following two types:

Overlapping Band: The higher band overlaps the lower band to some extent [Fig. 7.1(*a*)].

For example, the valence 2*s*-band overlaps with the empty 2*p*-band in beryllium ($1s^2\ 2s^2\ 2p^0$).

Nonoverlapping or Separate Bands: [Fig. 7.1 (*b*)].

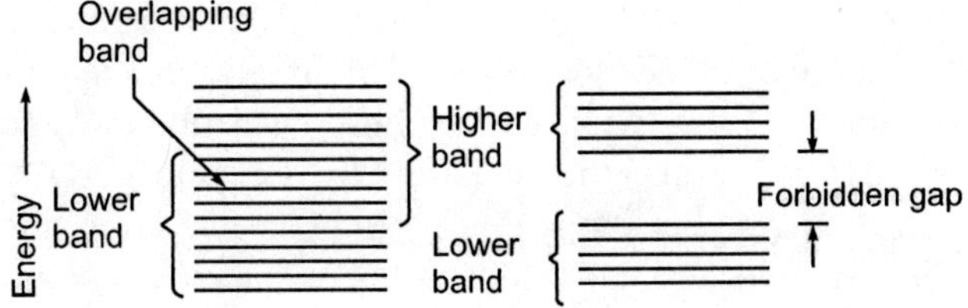

Fig. 7.1 (a) Overlapping bands, and (b) the forbidden gap.

Examples are lithium ($1s^2\ 2s^1$) and sodium ($1s^2\ 2s^2\ 2p^6\ 3s^1$).

- The band of energy levels occupied by the valence electrons is called the *valence band*. This may be occupied by partially or completely with electrons.
- A band that is either partly or completely vacant is called a *conduction band*. The electrons present in the conduction band are called *conduction electrons*, which are free electrons, and after an electron enters the conduction band, it moves freely.
- The gap between the valence band and the conduction band is called the *forbidden gap*. No electron can exist in the forbidden gap. The width of the forbidden gap is called the *energy gap* (E_g).
- In conductors (metallic elements), either the valence bands and conduction bands overlap, or the valence band is only partly full. Because the filled and unfilled molecular orbitals are not separated by significant gap, perturbation can occur readily.

The Band Model of Magnesium

The atomic number of Mg is 12. The electronic configuration is $1s^2\ 2s^2\ 2p^6\ 3s^2$, which shows that 1*s*, 2*s*, 2*p*, and 3*s* bands are all completely filled with electrons. Hence, magnesium may be expected to be an insulator. But magnesium is a good conductor of electricity. It has empty 3*p* bands. The filled 3*s* band and empty 3*p* band are so close that they virtually overlap. Therefore, some electrons in the filled 3*s* band can move into the empty 3*p* band creating vacancies both in the 3*s* and 3*p* band. These bands constitute the conduction band. The electrons in these bands are mobile and are responsible for the electrical conductivity of magnesium metal.

The electrical conductivity of a metal decreases with a rise in temperature because increased thermal vibrations of the metal atoms causes scattering of the electrons, thereby obstructing their free flow.

- In insulators (nonmetallic elements), there is a large band gap between the filled valence bands and the empty conduction band. Therefore, electrons cannot be promoted from the valence band to the conduction band, where they could move freely. However, at high temperatures, some of the valence band electrons acquire sufficient energy to overcome the forbidden gap and enter the conduction band. These electrons in the conduction band are now free to move, and the insulators exhibit conductivity with increased temperature. The higher the temperature of an insulator, the higher is its conductivity. In other words, conductivity of an insulator increases with a rise in temperature.
- A solid is a semiconductor (a material with electrical conductivity in between those of the insulator and conductor), if it has an almost filled valence band, an almost empty conduction band, and a very narrow energy gap (E_g) of the order of 1 eV.

Fig. 7.2 shows the energy bands of (*a*) an insulator, (*b*) a conductor, and (*c*) a semiconductor.

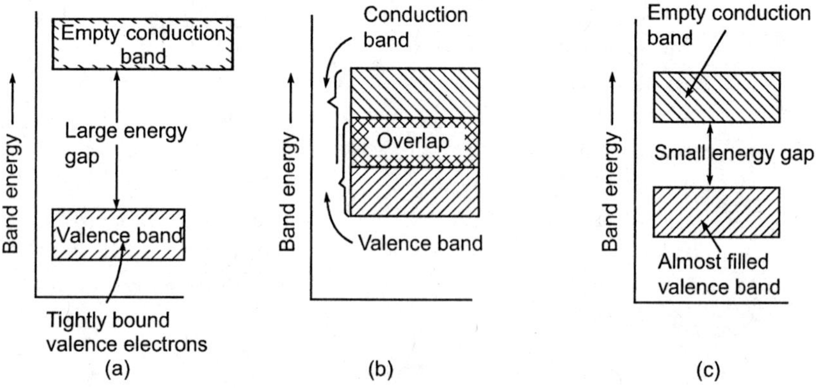

Fig. 7.2 Energy band in (a) insulators, (b) conductors, and (c) semiconductors.

The energy gaps of Si and Ge are 1.0 and 0.7 eV, respectively. Hence, silicon and germanium are semiconductors. At 0 K, all semiconductors are insulators. As the temperature is increased, some valence electrons gain enough thermal energy to jump over the energy gap (E_g) into the conduction band. Hence, the conductivity of a semiconductor increases with the rise in temperature.

7.3 TYPES OF SEMICONDUCTORS

The semiconductors are either of the following types:

- *Intrinsic semiconductors* are extremely pure (above 99.9999% purity). Pure silicon and germanium are intrinsic semiconductors. Pure silicon and germanium are tetravalent, that is they have four valence electrons. In their crystals, each atom is held in a covalent bond with four of its neighbors. Two valence electrons are shared by two atoms. Therefore, at low temperatures, energy bands are filled with electrons and do not conduct electricity.

  When the temperature is increased, due to thermal energy, a few electrons are freed from the bond and can move throughout the crystal. When an electron leaves a band, the vacancy created is called a *hole*. Electrically, the hole is equivalent to a positive charge. When an electric field is applied across an intrinsic semiconductor at room temperature, electrons in the conduction band move to the anode; whereas the positively charged holes in the valence band move to the cathode. Hence, current in an intrinsic semiconductor consists of simultaneous movement of the conduction band electrons and the valence band holes in the opposite directions. Thus, electrons and holes are the charge carriers, and they are equal in number.
- *Extrinsic semiconductors* are basically intrinsic semiconductors, whose conducting properties have been improved by adding extremely small amounts of specific substitutional impurities (called *doping agent* or *dopants*). The addition of a doping agent reduces the energy gap (E_g), thereby allowing more electrons to flow from the valence band to the empty conduction band. Usually, 1 part of a doping agent is added to 10^6 parts of the parent element (Si, Ge, etc.). By appropriate doping, the conductivity of an intrinsic semiconductor may be increased by 10,000 times. Based on the nature of the doping agent added, the extrinsic semiconductors are classified into two types:

  ***n*-type Extrinsic Semiconductor:** V group elements, such as P, As, and Sb, have five valence electrons. When an atom of this group occupies a site in the crystal lattice of Si or Ge, four of its electrons enter bonding, and the fifth electron that is in excess to octet arrangement remains. This electron can enter the conduction band relatively easier. An impurity atom acts as a donor, and doping increases the number of electrons. An impurity atom donates the conduction electron without leaving a hole in the valence band. Because the negatively charged electrons are the charge carriers, the semiconductor is called an *n*-type semiconductor.

  ***p*-type Extrinsic Semiconductor:** III group elements, such as boron and aluminium, have three valence electrons. When the silicon or germanium crystal is doped with boron or aluminium, at the site, one electron is less for octet. A deficiency of electrons results in a hole. At an ordinary temperature, the positive holes introduced by the impurity atom are localized around the impurity atom. At a higher temperature, many of these holes are excited and are free to move throughout the crystal lattice. Thus, the positively charged holes are in excess and are the charge carriers. Positive holes are less mobile than free electrons, so the conductivity of a *p*-type semiconductor is inherently less than that of an *n*-type semiconductor.

7.4 CRYSTALLOGRAPHY

The branch of science that deals with the geometry, properties, and structure of crystals and crystalline substances is known as *crystallography*. A *crystal* is a solid substance with a definite geometrical shape that has flat faces and sharp edges. Crystallography depends upon the following three fundamental laws:

Steno's law of constancy of interfacial angle: According to this law, "the shape and size of crystals of a certain compound or element may vary with the conditions under which crystallization occurs, but the angle between the corresponding faces is always constant."

The crystals of a substance are bounded by plane surfaces called *faces*. These faces always intersect at an angle, called the *interfacial angle*. The interfacial angle has a characteristic value for a given crystalline solid. A crystalline substance may crystallize under different conditions to produce its crystals with faces of a variable shape and size. However, the angle of intersection of any two corresponding faces is always the same. For example, the interfacial angles of all crystals of sodium chloride are found to be 90°.

Haüy's law of rationality of indices: Haüy proposed that the regular external form of crystals was a reflection of the inner regular order of the arrangement of the building units. Each building unit, the smallest crystal, is called the *unit cell*. According to this law, the ratio between intercepts on crystallographic axes for the different faces of a crystal can always be expressed by rational numbers. The law states that, "it is possible to choose along the three coordinate axes, unit distances (a, b and c) that may be or may not be of the same length such that the ratio of the three intercepts of any plane in the crystal is given by ($xa : yb : zc$) where x, y, and z are integral numbers." In other words, "all faces cut a given axis at distances from the origin, which bear a simple ratio to one another."

In Fig. 7.3., OX, OY, and OZ are the three axes at right angles to each other. For the plane ABC, let a, b, and c be the intercepts along the axes. Suppose the other plane A'B'C' makes intercepts at $2a$, $2b$, and $3c$, respectively.

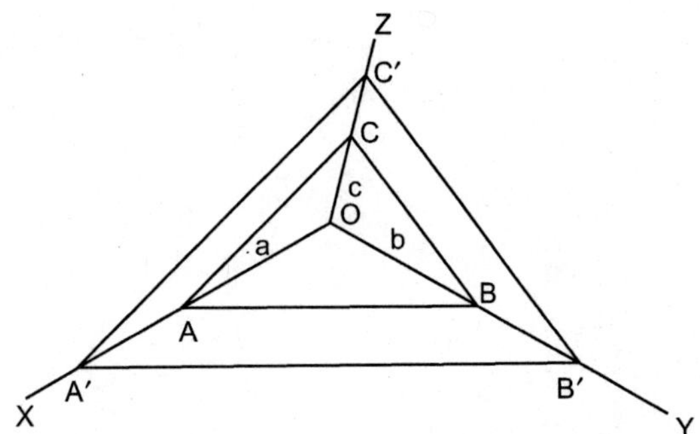

Fig. 7.3 Crystallographic axes.

The ratio of the intercepts in terms of standards is 2 : 2 : 3. The coefficients of *a*, *b*, and *c* are known as *Weiss indices* of the given plane.

If any plane is parallel to one axis, then it will cut that axis at infinity. In such a case, the use of Weiss indices is awkward. So, the Weiss indices have been replaced by the *Miller indices*. The Miller indices of a plane are obtained by taking the reciprocals of the Weiss indices and multiplying throughout by the least common multiple to obtain integers. Thus, the Miller indices of plane A′B′C′ are

$$\frac{1}{2}:\frac{1}{2}:\frac{1}{3} \text{ or } 3:3:2$$

Thus, the plane A′B′C′ is designated by Miller indices as (332). In general, a face is represented as (*hkl*).

The distance between the parallel planes in a crystal is given by

$$d_{hkl} = \frac{a}{\sqrt{h^2 + k^2 + l^2}}$$

where, a = Length of the cubic side.

h, k, and l are the Miller indices of the plane.

Law of symmetry: The law states that all crystals of the same substance possess the same element of symmetry. Following are the three types of symmetry associated with a crystal:

- A *plane of symmetry* is defined as an imaginary plane that divides the crystal into two identical halves so that one is the mirror image of the other.
- An *axis of symmetry* is defined as an imaginary line drawn through the crystal about which if the crystal is rotated through 360°, the crystal presents exactly the same appearance more than once. If a similar view

appears twice, it is called an axis of a two-fold symmetry or diad axis. If it appears three times, it is an axis of three-fold symmetry, or a triad axis, and so on (Fig. 7.4).

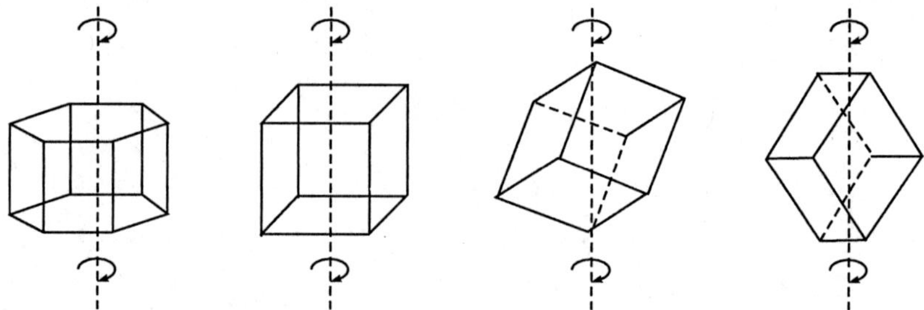

Fig. 7.4 Various axes of symmetry.

- A *center of symmetry* of a crystal is a point within the crystal so that any line drawn through it will intersect the surface of a crystal at equal distances in both directions (Fig. 7.5).

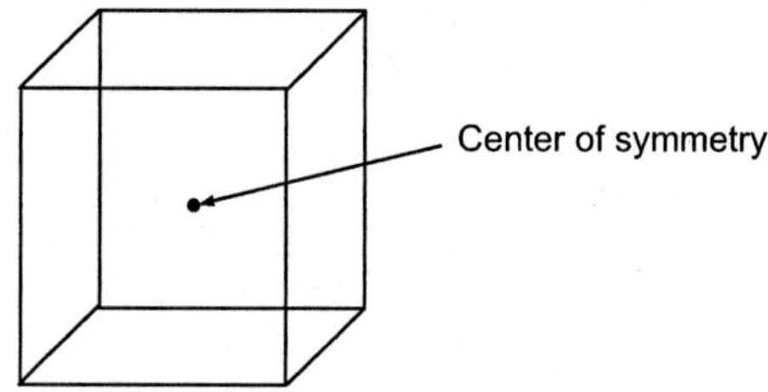

Fig. 7.5 Center of symmetry of cubic crystals.

A crystal can have one or more axes of symmetry but not more than one center of symmetry. The total number of elements of symmetry depends entirely on the nature of the crystal.

For a cubic crystal, there are 9 planes of symmetry, 13 axes of symmetry, and 1 center of symmetry.

Thus, there are 23 total symmetries. All crystals of the same substance possess the same elements of symmetry.

7.5 CRYSTAL LATTICE

A *crystal lattice* is a highly ordered 3D structure formed by its constituent atoms, molecules, or ions. The unit cell is the smallest building unit in the space of the crystal, which when repeated over and over again in 3D, results in a space lattice of the crystalline substance. The unit cell is the essential feature of the crystal structure. On the basis of the arrangement of structural units in their crystal lattices, crystalline substances can be classified into the groups shown in Table 7.2.

TABLE 7.2 Crystal Systems and Their Characteristics

System	*Axial Characteristics*	*Angles*	*No. of Space Lattices*	*Examples*
1. Cubic	$a = b = c$	$\alpha = \beta = \gamma = 90°$	3	NaCl, KCl, CaF_2, Cu, ZnS, Cu_2O
2. Tetragonal	$a = b \neq c$	$\alpha = \beta = \gamma = 90°$	2	SnO_2, TiO_2, $ZrSiO_4$
3. Orthorhombic	$a \neq b \neq c$	$\alpha = \beta = \gamma = 90°$	4	$BaSO_4$, KNO_3, K_2SO_4, $CdSO_4$, AgBr
4. Monoclinic	$a \neq b \neq c$	$\alpha = \gamma = 90°$ $\beta \neq 90°$	2	$CaSO_4 . 2H_2O$, $NaHCO_3$, $Na_2SO_4 . 10\,H_2O$, Monoclinic sulphur
5. Triclinic	$a \neq b \neq c$	$\alpha \neq \beta \neq \gamma \neq 90°$	1	$CuSO_4. 5H_2O$, $NaHSO_4$, H_3PO_3, $K_2Cr_2O_7$
6. Hexagonal	$a = b \neq c$	$\alpha = \beta = 90°$ $\gamma = 120°$	1	PbI_2, Mg, Cd, Zn, ZnO, BN, CdS, SiO_2, HgS
7. Rhombohedral (or trigonal)	$a = b = c$	$\alpha = \gamma = 90°$ $\beta \neq 90°$	1	Graphite, ICl, Al_2O_3, As, Sb, Bi

The classification is based on the magnitudes of the unit cell length and the angle of inclination between them. These crystal systems in all are made up of 14 types of crystal lattice. These crystal lattices, known as *Bravais lattices* are shown in Fig. 7.6.

7.6 CUBIC CRYSTALS

In a cubic crystal, the intercepts on the three axes are equal, and all the angles are equal to 90°. A cubic crystal can be any of the following three types:

Simple crystal lattice (SC): In a simple crystal lattice, there are lattice points at the eight corners of the unit cell. In a simple cubic structure, an atom situated at any corner of each unit is shared by a total of eight unit cells, thus, each unit cell has $\frac{1}{8}$ shares of every corner atom. So the total contribution of all the eight corner atoms to each cell is $\frac{1}{8} \times 8 = 1$ atom per unit cell of SC.

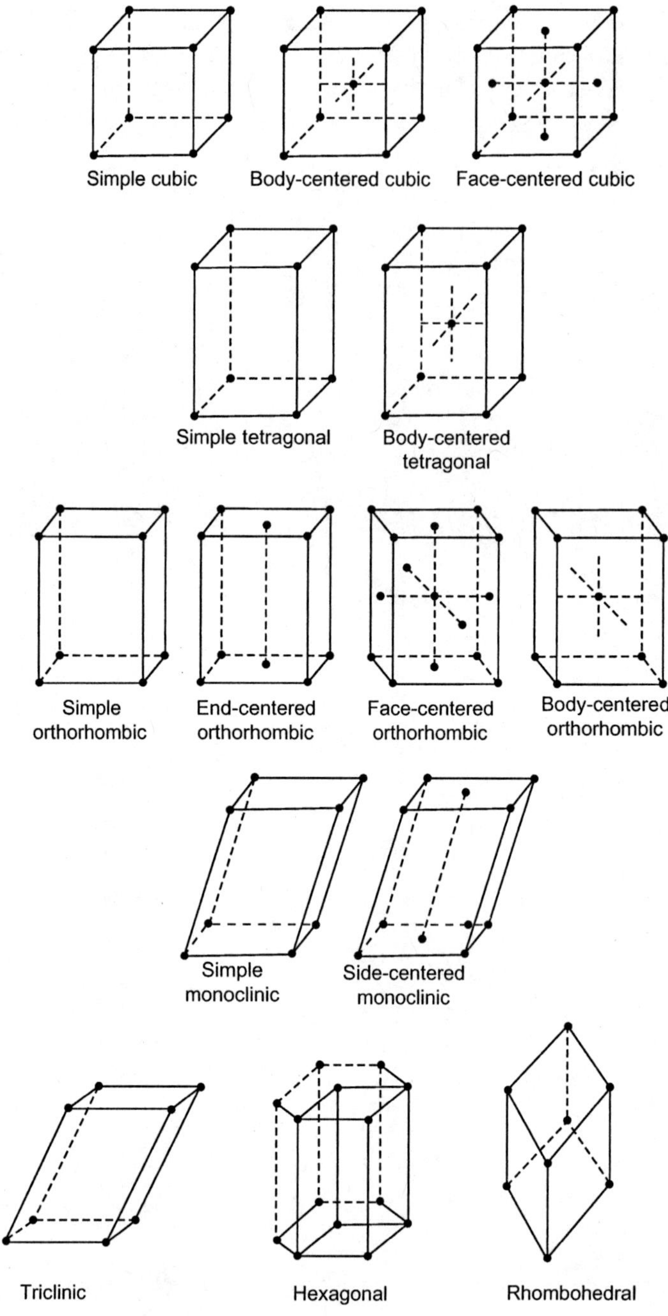

Fig. 7.6 The 14 Bravais lattices.

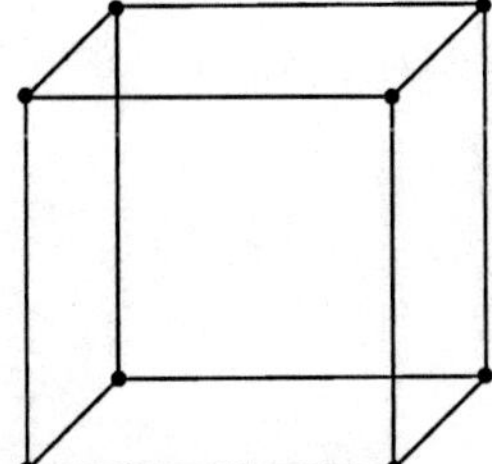

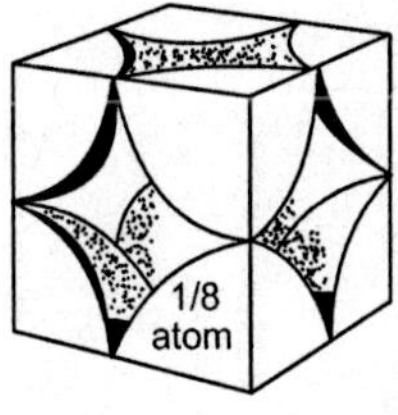

Fig. 7.7 Unit cell of a simple cubic.

Body centered crystal lattice (BCC): In body centred crystal lattice, there are lattice points at the eight corners and at the center of the unit cell. A BCC cell has one additional atom at the center, besides having one atom each at its eight corners. The atom at the center is independent of other cells, while each of the eight atoms situated at the corners are shared by a total of eight unit cells. Thus, the total number of atoms per unit cell are

$$= 1 \text{ (at the center)} + 8 \times \frac{1}{8} \text{ (at the eight corners)}$$

$$= 1 + 1 = 2 \text{ atoms per unit cell of BCC.}$$

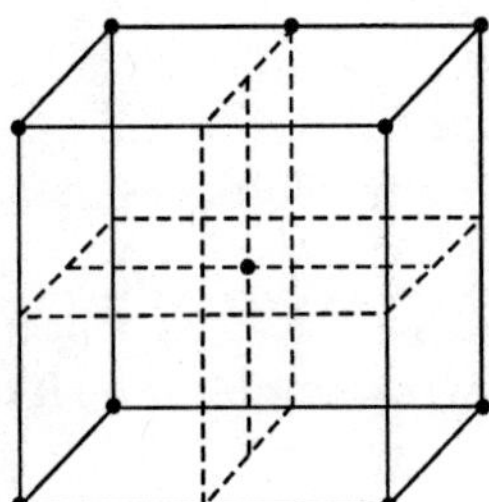

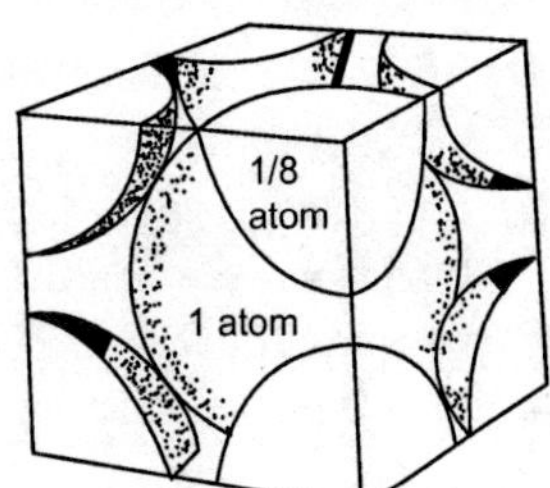

Fig. 7.8 Unit cell of body centered cubic.

Face centered crystal lattice (FCC): In a face-centered crystal lattice, there are lattice points at the center of all faces, in addition to those at the eight corners of the unit cell. A FCC has one atom, at the center of each face, besides having one atom at each corner. Thus, every atom situated at the center of a face of a unit cell is shared by two adjoining unit cells. Because there are six faces of a cube, the total number of atoms per unit cell are

$$= \frac{1}{2} \times 6 \text{ (at the center of six faces)} + 8 \times \frac{1}{8} \text{ (at 8 corners)}$$
$$= 3 + 1$$
$$= 4 \text{ atoms per unit cell of FCC}$$

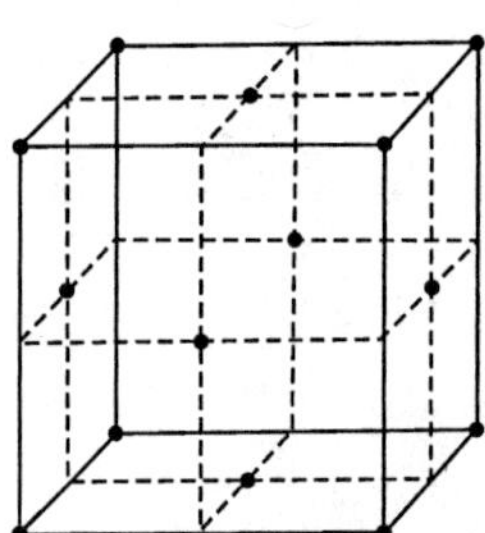
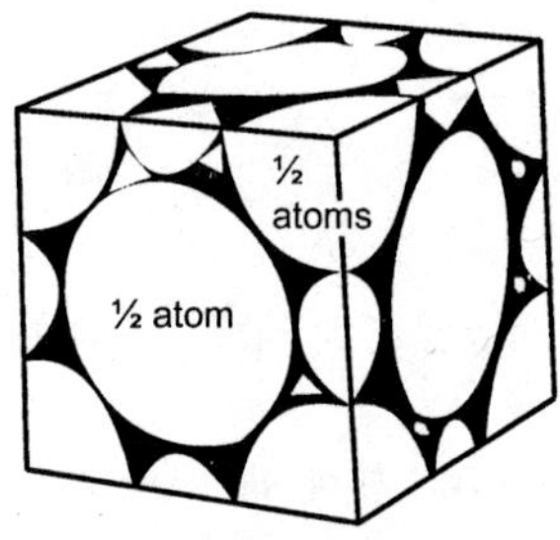

Fig. 7.9 Unit cell of a face-centered cubic.

The Coordination Number of a Cubic Lattice

The total number of nearest neighbor atoms of a particular atom in a crystal lattice is called the coordination number.

Simple-cubic cell (sc): In this type of cell, there are six atoms that are the nearest neighbors for every corner atom. Thus, there are two nearest atoms, one along the + X axis and the other along the – X axis. Similarly, there are two nearest neighbors along the ± Y axis and two along the ± Z axis. Thus, in total there are 2 + 2 + 2 = 6 nearest neighbors. Hence, the coordination number is 6 (Fig. 7.10).

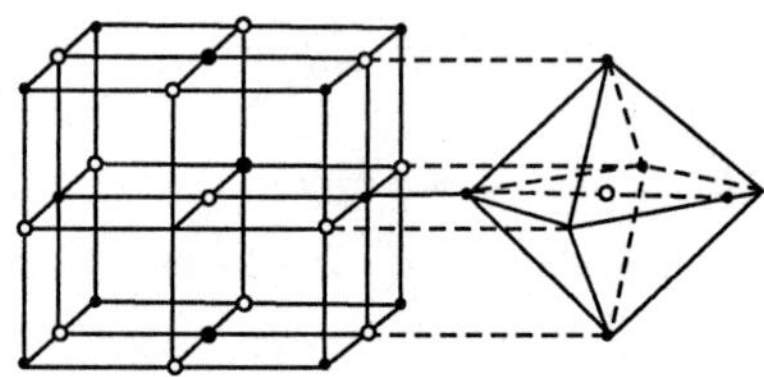

Fig. 7.10 Simple cubic.

CsCl crystallizes in *sc*-lattice.

Body-centered cell (bcc): In this type, an atom in the center or body of the cell has all the eight corner atoms as its close neighbors. Hence, the coordination number is 8 (Fig. 7.11). NaCl, Cr, Fe, Mo, and W crystallize in the *bcc* lattice.

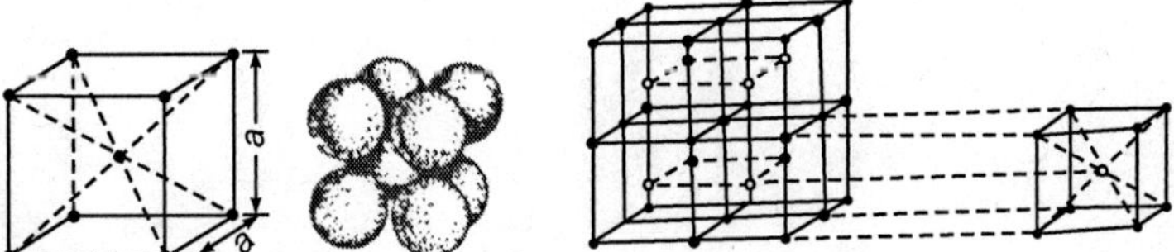

Fig. 7.11 Body-centered cubic.

Face-centered cell (fcc): In this type, each atom is in direct contact with its 12 nearest neighbors – 6 of which lie in one layer in the same plane, 3 from the layer above, and 3 from the layer below. Hence, the coordination number is 12 (Fig. 7.12). This structure is akin to the tetrahedral structure. Cu, Au, and Al crystallize in the *fcc* lattice.

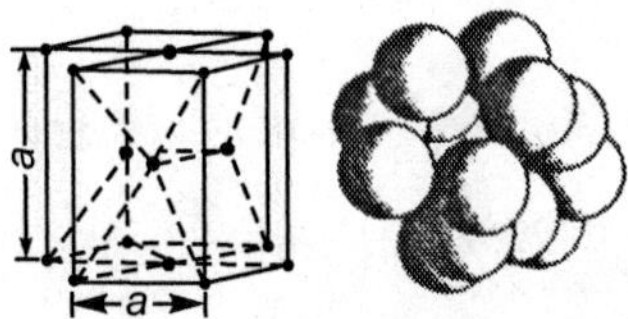

Fig. 7.12 Face-centered cubic or tetrahedral cubic.

The Atomic Radius of a Cubic Lattice

The atomic radius of a unit cell may be defined as "half the distance between the centers of two immediate neighbors in a unit cell" and is denoted by r. The distance between the centers of the two corner atoms of the cube is called the length of the cube edge and denoted by a.

Thus, for a simple cubic (*sc*) cell (Fig. 7.13):

$$\text{atomic radius } r = \frac{a}{2}$$

as we know that $a = 2r$

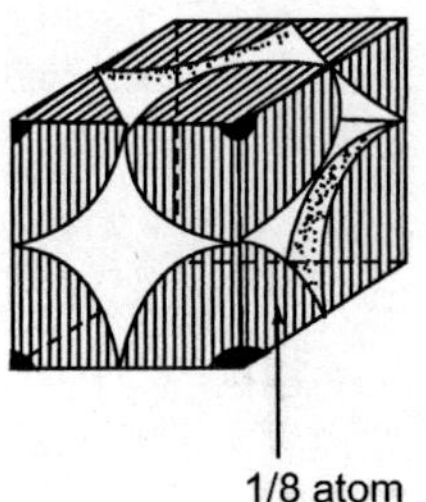

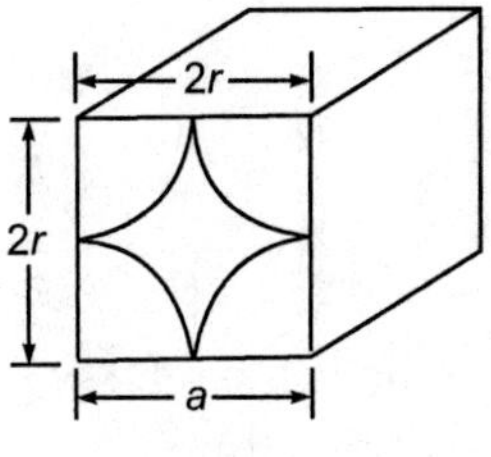

Fig. 7.13 Simple cubic cell.

For a body-centered cell (*bcc*) (Fig. 7.14):

$$(4r)^2 = (a^2 + a^2) + a^2 = 3a^2$$

$$r = \sqrt{\frac{3a}{4}}$$

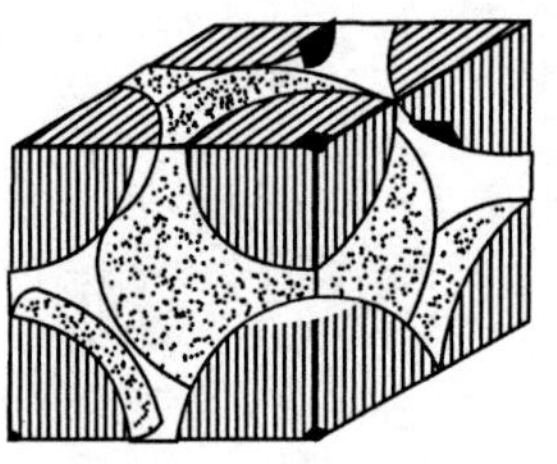

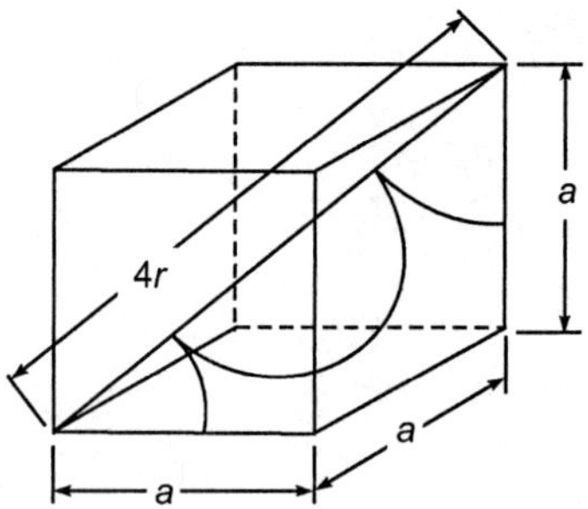

Fig. 7.14 Body-centered cubic cell.

For a face-centered cubic cell (*fcc*) (Fig. 7.15):

$$(4r)^2 = a^2 + a^2$$

or

$$16r^2 = 2a^2$$

Thus,

$$r = \frac{a}{2}\sqrt{2}$$

$$r = \frac{a}{\sqrt{8}}$$

Fig. 7.15 Face-centered cubic cell (fcc).

Radius Ratio

The radius ratio is the ratio of the cation radius to that of the anion in an ionic solid. Thus, the radius ratio

$$= \frac{\text{Radius of cation } (r^+)}{\text{Radius of anion } (r^-)}$$

Significance

The higher the radius ratio, the larger the size of the cation, and the greater its coordination number. Thus,

- When the radius ratio is greater than 0.732, a coordination number of 8 is usually observed.
- When the radius ratio is greater than 0.414, but less than 0.732, a coordination number of 6 is usually observed.
- When the radius ratio is greater than 0.225 but less than 0.414, a coordination number of 4 is often observed.

7.7 X-RAY STUDIES OF CRYSTALS

Von Laue (1912) made the brilliant suggestion that if X-rays are actually radiations of very small wavelengths and if crystals are made of the orderly arrangement of atoms, the crystals must act as 3D natural gratings for X-rays. When X-rays are incident on a crystal face, they penetrate into the crystal and are scattered by the atoms or ions.

A simple equation was developed by H. Bragg for the study of the internal structure of crystals. Consider a beam of monochromatic X-rays incident on a set of parallel and equidistant planes, called Bragg's planes, in the crystal structure. Let d be the distance between successive planes and θ be the glancing angle, that is the angle between the direction of the incident beam and the planes. In Fig. 7.16, the parallel lines (*aa′*, *bb′*, *cc′* ...) represent the successive planes. Now consider an incident ray, say APQ, which meets two planes at P and Q. A portion of the ray AP will be reflected from P along PB, so that the angle BP*a′* will also be θ. Part of this ray will penetrate and be reflected from the inner planes *bb′*, *cc′*, ... The reflection of this ray from the second plane *bb′* occurs at Q in the direction QC, PB, and QC being parallel. Draw the line QO perpendicular to the planes at Q meeting the first plane *aa′* at O. Also draw XOY perpendicular to PB and QC meeting them at X and Y; XOY represents the wave front of the reflected beam. The beams reflected from the parallel planes will generally combine in different phases and will destroy each other. Under certain conditions of wavelength and angle of incidence, the reflected waves from different planes will combine in the same phase and reinforce each other. It is evident that constructive interference will occur if the path PQY exceeds the path PX for waves scattered at P by an integral multiple of wavelength (λ). Draw OZ perpendicular to PQ meeting the latter at Z. It is easy to see

$$PX = PZ \text{ and } \angle QOZ = \angle QOY = \theta$$

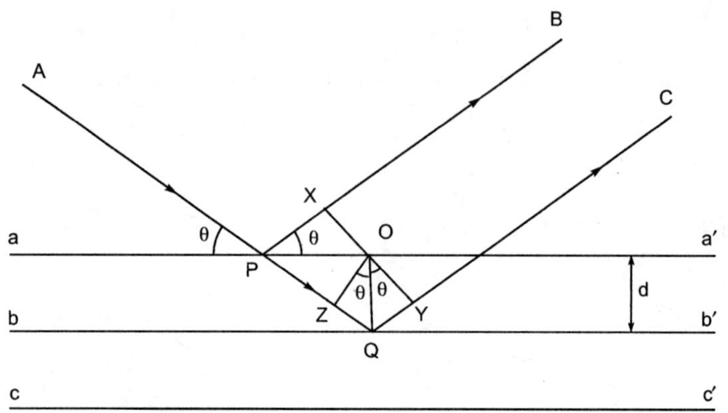

Fig. 7.16 Reflection of X-rays from lattice-planes (Bragg's law).

Hence, the excess path, (PQY – PX) = ZQY = 2QY = $2d \sin \theta$.

This, for constructive interference, must equal $n\lambda$, that is

$$n\lambda = 2d \sin \theta \tag{7.1}$$

where n is an integer 1, 2, 3,.... and so on, known as the order of reflection.

This is the famous Bragg equation, correlating the wavelength of the radiation with the spacing of the lattice planes. The reflection corresponding to $n = 1$ (for a given family of planes) is called first-order reflection; the reflection corresponding to $n = 2$ is the second-order reflection and so on.

Bragg's equation may alternatively be written as

$$\lambda = 2\left(\frac{d}{n}\right)\sin \theta$$

$$= 2d_{hkl} \sin \theta \tag{7.2}$$

where d_{hkl} denotes the perpendicular distance between adjacent planes with the indices hkl.

Determining the Crystal Structure by Bragg's Method

The apparatus used by Bragg for the investigation of a crystal is shown in Fig. 7.17. The crystal is mounted over a turntable, which can be moved over a scale.

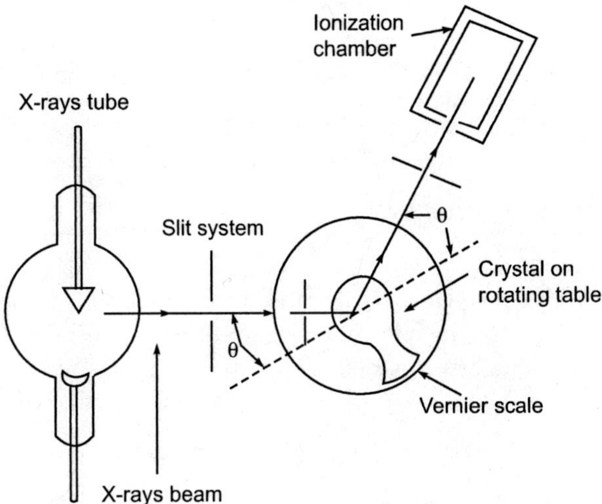

Fig. 7.17 Diagrammatic representation of Bragg's X-ray spectrometer.

A beam of X-rays of a definite wavelength generated in an X-ray tube with a molybdenum or copper anticathode passes through a slit and falls on a known face of the crystal. The X-rays reflected from the crystal pass through another slit and enter into the ionization chamber containing methyl bromide (CH_3Br) vapor.

The ionization chamber can also be moved over a circular scale independently from the crystal as shown. The walls of the chamber are insulated and connected to a source of high potential. The entry of reflected rays causes ionization so that there is a flow of current, which is measured with an electrometer. The strength of current is proportional to the intensity of the reflected X-rays. The intensities of the reflected X-rays are determined at different angles of incidence and plotted. From the graph thus obtained, angles of maximum reflection are ascertained.

As the crystal is rotated, successive maxima appear. These maxima should be governed by the equation:

$$n\lambda = 2d \sin \theta$$

For NaCl, the maximum reflections for the (100) face occur at 5.9°. The distance between the planes in the space lattice parallel to this face can now be calculated using the preceding equation.

When $\quad n = 1°$

$$\theta = 5.9°$$

and $\quad \sin \theta = 0.103$

$$\lambda = 2d\,(0.103)$$

or, $d = 4.85\ \lambda$

Thus, the distance between the reflecting planes of particles parallel to the (100) face can be calculated if we know the wavelength of the X-ray employed.

7.8 THE BORN HABER CYCLE

The Born Haber cycle is applicable in the calculation of lattice energy of ionic crystalline solids. The energy terms involved in building a crystal lattice (such as NaCl) may be taken in steps. Thus, the overall reaction

$$Na\,(s) + \frac{1}{2}\,Cl_2\,(g) \longrightarrow NaCl\,(crystal)$$

involves the following three steps:

1. The elements are converted into gaseous atoms:

$$Na\,(s) \xrightarrow[(+\Delta H_s)]{\text{Sublimation}} Na\,(g)$$

$$\frac{1}{2}\,Cl_2 \xrightarrow[(+\frac{1}{2}\Delta H_d)]{\text{Dissociation}} Cl\,(g)$$

2. The gaseous atoms are converted into the ions:

$$Na\,(g) \xrightarrow[(+I)]{\text{Ionization}} Na^+\,(g) + e^-$$

$$Cl\,(g) \xrightarrow[(\text{Addition of electron} - E)]{\text{Electron affinity}} Cl^-\,(g)$$

3. The ions are packed together forming the NaCl crystal:

$$Na^+\,(g) + Cl^-\,(g) \xrightarrow[(\text{Lattice energy})]{\text{Crystal formation}} NaCl\,(crystal)$$

where, ΔH_s = Heat of sublimation of the sodium atom
ΔH_d = Heat of dissociation of the molecular chlorine.
I = Ionization energy of the sodium
E = Electron affinity of the chlorine (g) atoms
U = Lattice energy of sodium chloride

Thus, the entropy of the formation (ΔH_f) of NaCl (crystal) is the sum of the terms going around the cycle:

$$-\Delta H_f = \Delta H_s + I + \frac{1}{2}\,\Delta H_d - E - U$$

Therefore, the Born Haber cycle for the formation of the NaCl crystal is illustrated in Fig. 7.18.

Application of the Born Haber Cycle

The most important applications of the Born Haber cycle are listed here:

- The cycle was used to calculate the electron affinities by using a known crystal structure. For example, for NaCl,

$$-\Delta H_f = \Delta H_s + I + \frac{1}{2}\Delta H_d - E - U$$

$$-381.2 = 108.4 + 495.4 + \frac{1}{2}(241.8) - E - 757.3$$

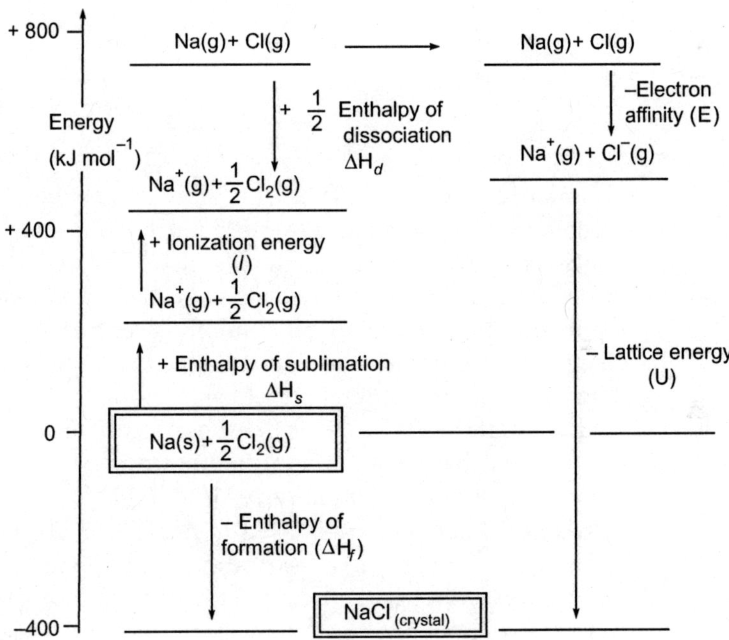

Fig. 7.18 The Born Haber cycle for the formation of a NaCl crystal.

Therefore, $E = -348.6\ \text{K Jmole}^{-1}$

- By knowing the values of electron affinity, the cycle is used to calculate the lattice energy of unknown crystals.
- Lattice energy may provide some information about ionic covalent bonding.
- The cycle is useful in analyzing and correlating the stabilities of various ionic solids.

7.9 CRYSTAL IMPERFECTIONS

An imperfection or defect refers to any deviation from the regular arrangement of atoms, molecules, or ions, thereby resulting in a change in the properties. The deviations arise because some atoms, molecules, or ions do not occupy the expected theoretical positions. Common types of imperfections in ionic crystals are listed here:

Electronic Imperfections

Electronic imperfections correspond to defects in ionic crystals, due to electrons. For example, when a trivalent impurity (*e.g.*, B, Al, Ga) or pentavalent impurity (*e.g.*, P, As) is present in Si or Ge crystal, it becomes semiconductor, due to the presence of excess holes or excess electrons respectively.

Atomic Imperfections

Atomic imperfections or point defects arise due to an irregularity in the arrangement of atoms or ions in the crystal lattice. Point defects are of two types:

Stoichiometric defects: Those in which the ratio between cations and anions is the same as in its molecular formula. Such defects include the following:

- *The Schottky defects* refers to the fact that in certain crystals, some of the lattice points are unoccupied, that is the crystals have vacancies. An equal number of cation and anion vacancies exist to maintain the electroneutrality. This type of defect results in a decrease in the overall density of the crystalline ionic substance. Ionic solids, such as NaCl and CsCl, exhibit this defect. This defect is usually observed in strongly ionic crystals that have a high coordination number, and both cations and anions are of almost similar sizes [Fig. 7.19 (*b*)].
- *In a Frenkel defect,* an ion leaves its normal site and occupies an interstitial site as shown in Fig. 7.19 (*c*). As cations are, generally, much smaller than anions, this defect is, generally, caused by the displacement of cations from the original sites to the adjacent voids.

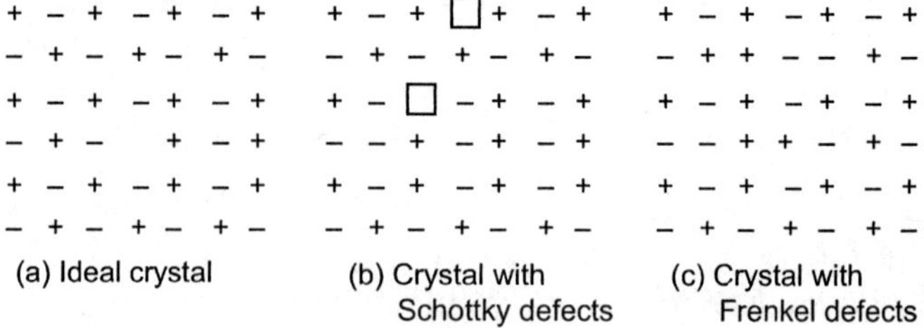

Fig. 7.19 Defects in crystals.

For example, in silver bromide (AgBr), some of the silver ions (Ag^+) are missing from their regular positions and are found squeezed between other ions. This defect is shown by the following:

- Ionic crystals, whose anions are much larger in size than cations, for example silver halides
- Ionic solids, whose crystal structure is of open type with large interstices, for example, CaF_2, ZnS, and so on.

Nonstoichiometric defects: These arise when an ionic crystal lattice contains either an excess of cations or anions so that the ionic compound as a whole does not correspond to the exact stoichiometric formula. The balance of the + ve and – ve charge is maintained by extra electrons or extra positive charges to maintain electrical neutrality. As a result, the crystal becomes irregular, that is, it contains this defect in addition to Schottky and Frenkel defects. This defect arises due to the presence of either the metal or the nonmetal in excess. When excess of metal ions are present, the defect is known as a *metal excess defect*. This defect may occur either due to anion vacancies or due to extra cations occupying the interstitial sites. The nonstoichiometric defects are therefore classified into the following types:

- *Metal ion excess defect due to vacancies* arise when some of the negative ions in the crystal lattice are absent. The absence of an anion from a crystal lattice leaves a hole, which is occupied by a trapped electron. The vacant site occupied by trapped electrons are called F-centers (Fig. 7.20).

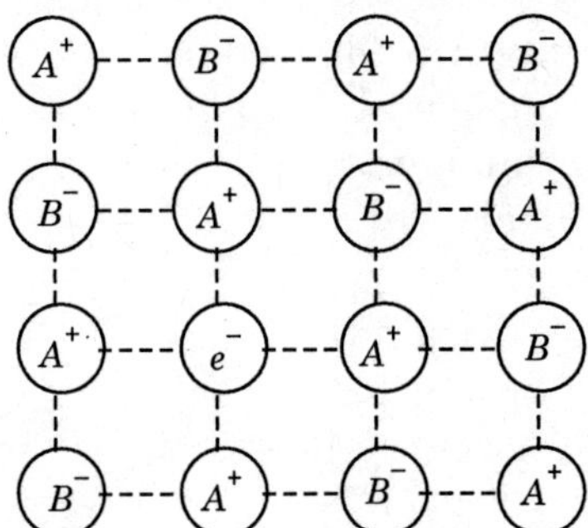

Fig. 7.20 Metal ion excess defect due to absence of anion.

- *Metal ion excess defect due to extra interstitial cation* arises when additional metal ions occupy interstitial sites, and electrical neutrality is maintained by the presence of a corresponding number of electrons in the same interstices. For example (Fig. 7.21), white zinc oxide (ZnO) turns yellow on heating, due to the loss of oxygen so that additional Zn^{2+} ions and electrons simultaneously occupy the interstices.

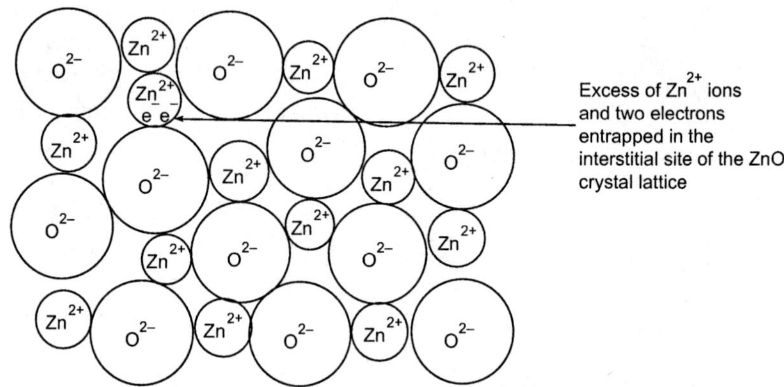

Fig. 7.21 Metal ion excess defect in ionic crystal of ZnO due to the presence of an extra Zn^{2+} ion seated in an interstice.

- *Metal ions deficiency defect due to missing cation* arises when a cation in an ionic solid is missing from its crystal lattice. This defect occurs in ionic solids in which the metal ions can exhibit variable valency. For example, FeO crystal forms a cubic close-packed structure. In an ideal structure, Fe^{2+} ions occupy all the octahedral points. However in usual practice, some of the cations (*i.e.*, Fe^{2+}) are missing from their lattice sites. Consequently, the composition of Wusite is $Fe_{0.95}O$. In this type of structure, some of the metal ions are present as Fe^{3+} so that electrical neutrality is maintained (Fig. 7.22). In a compound such as Fe_3O_4 or FeO. Fe_2O_3 (magnetite), two-thirds of Fe^{2+} ions are replaced by Fe^{3+} ions. In such a compound, all the Fe^{2+} ions occupy octahedral sites, whereas Fe^{3+} ions are equally distributed between octahedral and tetrahedral sites.

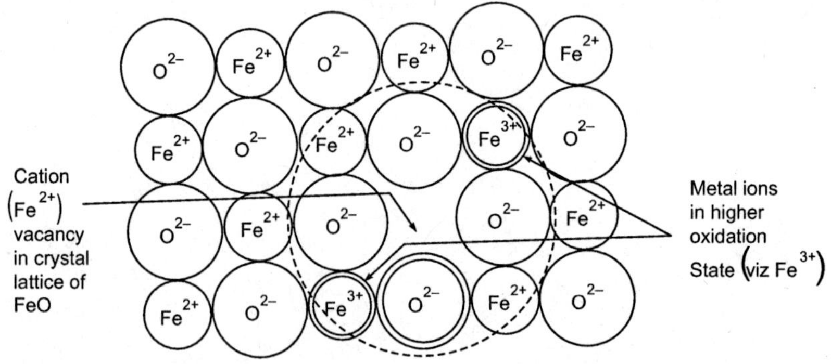

Fig. 7.22 Metal ion deficiency defect caused by missing cation.

- *Metal ions deficiency defect* due to the presence of an impurity ion of a higher valency but of identical size. For example, when a small amount of cadmium chloride ($CdCl_2$) is added to silver chloride (AgCl) to form a solid solution, some of the monovalent Ag^+ sites are occupied by the bivalent Cd^{2+}. The replacement of one Ag^+ ion site with one Cd^{2+} ion in the AgCl structure causes the removal of an adjacent Ag^+ ion also to maintain neutrality (Fig. 7.23). The presence of cation vacancies on the surfaces of those nonstoichiometric compounds imparts good catalyst characteristics.

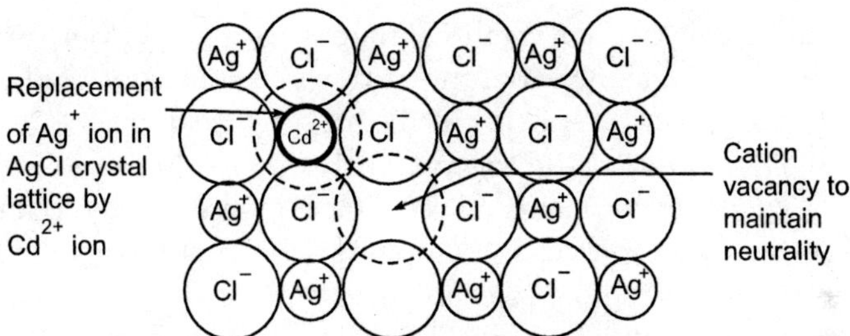

Fig. 7.23 Metal ion deficiency defect caused by replacement of the Ag+ ion in AgCl crystal lattice by a same-sized Cd^{2+} ion.

It is concluded on the whole that:

- The ionic crystal with excess metal ions conducts electricity due to the presence of free electrons or F-centers. However, the conductivity is very low, because the number of free electrons present in such crystals is very small. Due to low conductivity, these crystals are called *n*-type semiconductors because the current is carried by the electrons.
- The ionic crystals with less metal ions are *p*-type semiconductors because when an electron moves from an A^+ ion to form, A^{2+} ions, a positive hole is created, which moves from one A^+ ion to another, thereby creating a positive hole. The current is thus primarily due to the movement of positive holes.

7.10 VOIDS IN A CRYSTAL STRUCTURE

The following two types of voids are encountered in close-packed spheres in crystals:

- *Tetrahedral void* is formed between three spheres on a close-packed plane and a fourth sphere on an adjacent plane (either on top or at the bottom) fitting in the cavity space between the three spheres (Fig. 7.24).

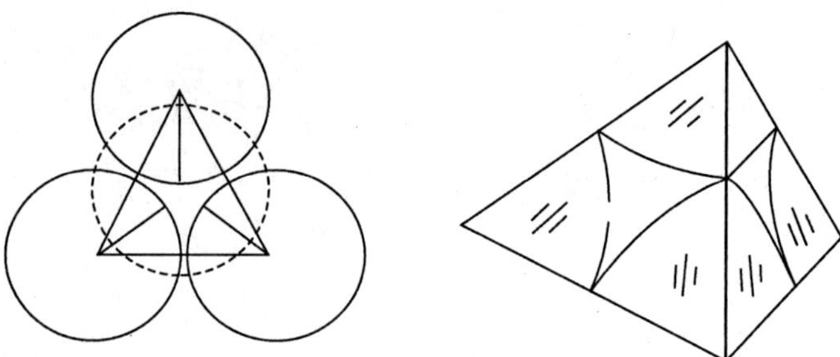

Fig. 7.24 Geometry of the tetrahedral void in a close-packing structure.

For every sphere in three dimensional array, there are two tetrahedral voids. It can be shown that the radius of tetrahedral void is 0.225 r, where r is the radius of the spheres of the close-packed array.

- *Octahedral void* is formed with three spheres on a close-packed plane and three more spheres on an adjacent close-packed plane, so that the centers of the three spheres in one plane are directly over the three triangular valleys surrounding the central valley of the first plane, and there is no sphere over the central valley (Fig. 7.25). There is one octahedral void per sphere in the 3D array. It can be shown that the radius of an octahedral void is 0.414 r, where r is the radius of the spheres of the close-packed array.

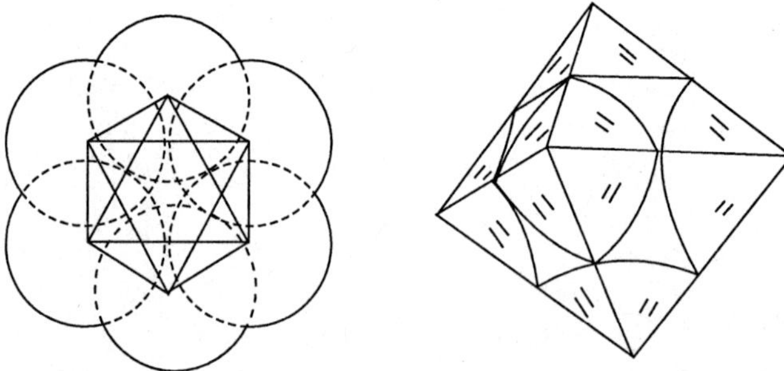

Fig. 7.25 Geometry of the octahedral void in a close-packed structure.

The voids in a crystal lattice are used to accommodate small-sized nonlattice atoms such as H, B, C, N, and so on. For example, transition metals form crystals of hydrides, borides, carbides, and nitrides with the help of interstices or voids.

7.11 LIQUID CRYSTALS

A *mesophase* is a phase intermediate between a solid and liquid. When the solid melts, some aspects of the long-range order characteristic of the solid may be retained, and the new phase may be a liquid crystal, a substance having a liquid-like imperfect long-range order in at least one direction in space but a positional or orientational order in at least one other direction. In other words, liquid crystals are liquids that can have a sufficient long-range order in them to make them behave like solids. Usually, a liquid crystal is made from molecules that are long, thin, and not very symmetrical.

For example,

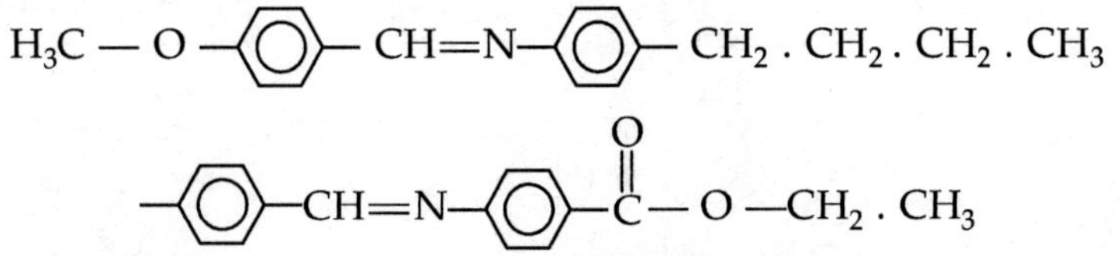

$$H_3C - O - C_6H_4 - CH{=}N - C_6H_4 - CH_2 \cdot CH_2 \cdot CH_2 \cdot CH_3$$

$$-C_6H_4-CH{=}N-C_6H_4-\overset{O}{\overset{\|}{C}}-O-CH_2 \cdot CH_3$$

The intermolecular forces in molecules of liquid crystals are strong enough to hold the molecules together but not so strong as to restrict their movement too much. The very useful property of liquid crystals is that the arrangement of the molecules can be upset by very slight changes in their surroundings. In the presence of a small electric field, the molecules rearrange themselves.

Liquid crystals are classified into following three types:

- *Smectic or soap-like liquid crystals* on heating retain long-range order and yield a smectic (soap-like) phase. They lose periodicity within the planes but retain the orientation and arrangement in equispaced planes [Fig. 7.26 (*b*)]. Thus, molecules in phase align themselves in layers. On the application of stress, one plane glides over another. Soap, such as cholesteryl oleyl carbonate, $C_{27}H_{40}O \cdot COO - (CH_2)_8\ CH = CH - (CH_2)_7 \cdot CH_3$, exhibit a smectic liquid state between 0°C and 17°C.
- *Nematic or thread-like liquid crystals* on heating lose their planar structure but retain a parallel alignment. Thus, they retain orientation but lose periodicity [Fig. 7.26 (*c*)]. In other words, the molecules lie parallel to each other but can move up, down, or sideways, or can relate along their axes. N-paramethoxy benzylidene-*p*-butylaniline changes to nematic liquid at 24°C, and this persists up to 43°C, after which it melts into isotropic liquid.

  The nematic liquid crystals do not conduct electricity when in pure form. They flow like liquids, but their mechanical, electrical, optical, and magnetic properties depend upon the direction along which these are measured.

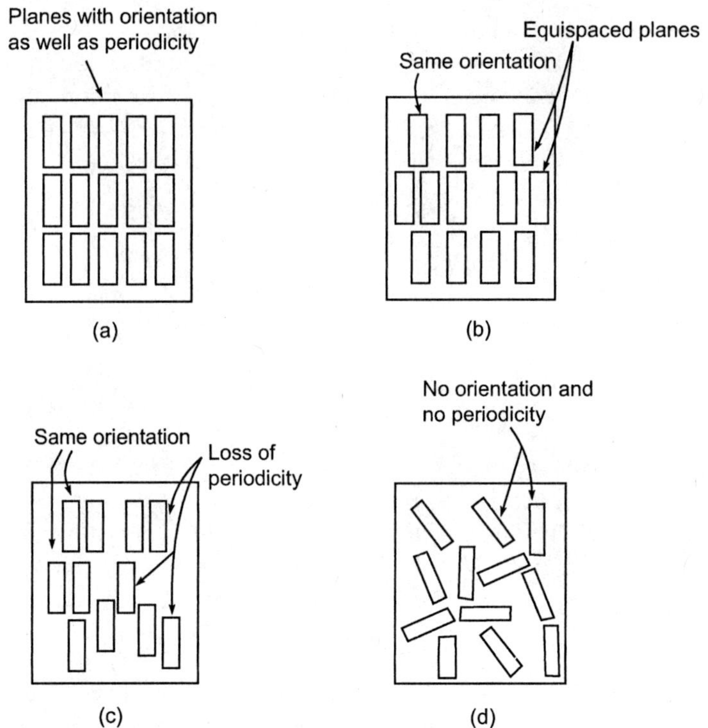

Fig. 7.26 (a) Long-chain rod-like crystalline solid possessing both orientation and periodicity (b) Smectic liquid crystals (c) Nematic liquid crystals (d) Isotropic liquid.

- *Cholesteric liquid crystals* are optically active and possess the arrangement of molecules similar to those in the nematic type. This type of liquid crystals are characterized by very high optical rotation, that is, a thousand times greater than that of its crystalline variety. Moreover, on raising the temperature, the pitch decreases. This results in a corresponding change in the wavelength of reflection. They are named so because the skeleton of these substances passes through a state similar to that of cholesterol, a steroid present in blood.

Applications of Liquid Crystals

- A liquid crystal reflects only one color when white light falls on it. If the temperature is changed, it reflects a different color light. Such liquid crystals can be used to detect even small temperature changes (as small as 0.01°C).
- The orientation of molecules in a thin film of nematic liquid crystal is easily altered by pressure and electric field. The altered orientation affects the optical properties of the film, causing the film to become opaque. If we arrange electrodes in certain patterns and an electric field is imposed on these

electrodes onto a thin film of liquid crystal, the patterns of the electrodes become visible. This is the principle used in the liquid-crystal display (LCD) in calculators and digital watches.

- Liquid crystals are used in gas-liquid chromatography because their mechanical and electrical properties lie between crystalline solids and isotropic liquids.
- Liquid crystals are employed in digital displays such as pocket calculators, digital watches, and so on, because they consume very little electrical power.
- Liquid crystals are used as a solvent during the spectroscopic study of the structure of anisotropic molecules.
- Cholesteric liquid crystals are used in thermography for detecting tumors in the human body.

7.12 SUPERCONDUCTORS

Superconductivity is a phenomenon exhibited by some materials that have a zero value of resistivity below a threshold temperature and have a certain value of resistivity above that transition temperature. This threshold temperature is called *critical temperature* (T_C). Materials in a superconducting state become diamagnetic and are repelled by magnets. The phenomenon of superconductivity was first observed by Kammerlingh Onnes in 1913, when they found that mercury became superconducting at around 4 K (at boiling point of liquid helium).

Properties of Superconductors

Superconductor properties include the following:

- They possess greater resistivity than other elements at room temperature.
- The transition temperature for different isotopes of a superconductive element decreases with the atomic mass of the isotope.
- On adding an impurity to a superconducting element, the critical temperature is lowered.
- Superconductivity is more concerned with the conduction of electrons than the atom of the element.
- During transition, neither thermal expansion nor elastic properties change.
- All electromagnetic effects disappear in the superconducting state.
- A strong magnetic field below the critical temperature of a superconductor destroys its superconducting property.
- When a not too strong magnetic field is applied to a superconductor and is cooled to a low temperature below its transition temperature, then the superconductor expels all magnetic flux from its interior. This is known as

the Meissner effect. In other words, if the magnetic field is applied after the superconductor has been cooled below its critical temperature, then the magnetic flux is excluded from the superconductor.

Preparation of Superconductors

Certain metals and metallic alloys when cooled to a temperature near the absolute zero (the lowest temperature possible is 0°K or 273°C) undergo a transition in their atomic structure, suddenly lose all their resistance to the flow of electricity, and become superconductors.

In 1987, Paul Chu (Houston University) and Wu Jr (Alabama University) reported that a ceramic compound, $YBa_2Cu_3O_{7-y}$ ($y \leq 0.5$) is superconducting up to 100 K. This compound is also known as the 1 : 2 : 3 compound due to its Y : Ba : Cu stoichiometry. This is prepared as follows:

> A mixture of 0.75 g of yttrium oxide, 2.622 g of barium carbonate, and 1.581 g of copper oxide is ground to a fine powder. This powder is taken into a porcelain crucible and heated in a furnace at 920°–930°C for 12 hours. The furnace is switched off, and the mixture is allowed to cool gradually. A sufficient amount of oxygen is absorbed by the mixture during the cooling process, and the crucible is removed when the temperature is about 100°C. The final sample obtained after cooling is a 1 : 2 : 3 compound that exhibits superconductivity well above 77 K (the boiling point of nitrogen).

Applications of Superconductivity

The phenomenon of superconductivity has several practical applications:

- Superconducting magnets capable of generating high fields with low power consumptions are being employed in scientific tests and research equipment.
- They are used for Magnetic Resonance Imaging (MRI) in the medical field as a diagnostic tool. Magnetic Resonance Spectroscopy (MRS) shows the chemical analysis of body tissues.
- Other fields in which superconductivity is used are memory or storage elements on computers, magnets for high-energy particle accelerators, electrical power transmission, high-speed switching, signal transmission for computers, and so on.

SOLVED NUMERICAL PROBLEMS

1. *Calculate the Miller indices of crystal planes whose Weiss indices are*

   (*i*) $\frac{a}{2}, \frac{2b}{3}, \infty c$ (*ii*) $\frac{2a}{3}, 2b, \frac{c}{3}$

   Sol. Weiss indices $\frac{1}{2}, \frac{2}{3}, \infty$ $\frac{2}{3}, 2, \frac{1}{3}$

   Reciprocal of Weiss indices $2, \frac{3}{2}, \frac{1}{\infty}$ $\frac{3}{2}, \frac{1}{2}, 3$

   Clear fractions 4, 3, 0 3, 1, 6

   Miller indices (430) (316)

2. *Determine the interplanar spacing between the (220) planes of a cubic lattice of length 450 pm.*

   Sol. $d_{220} = \frac{a}{\sqrt{2^2 + 2^2 + 0^2}} = \frac{450 \text{ pm}}{\sqrt{8}} = 159 \text{ pm}$

3. *The faces in a crystalline solid intersect the three crystal axes at (a, b, c), (a, 2b, c), (a, 2b, 2c), and (∞ 1, b, – c). Calculate the Weiss and Miller indices for these faces.*

Sol. Intercept	(*a*, *b*, *c*)	(*a*, 2*b*, *c*)	(*a*, 2*b*, 2*c*)	(∞, *b*, – *c*)
Weiss indices	1, 1, 1	1, 2, 1	1, 2, 2	∞, 1, – 1
Reciprocals	1, 1, 1	$1, \frac{1}{2}, 1$	$1, \frac{1}{2}, \frac{1}{2}$	0, 1, – 1
Miller indices	(111)	(212)	(211)	$(01\bar{1})$

4. *The ionic radii of Na^+ and Cl^- ions are 0.98×10^{-10} m and 1.81×10^{-10} m. Find the coordination number to each ion.*

   Sol. Radius ratio $= \frac{\text{Radius of Na}^+}{\text{Radius of Cl}^-} = \frac{0.98 \times 10^{-10} \text{ m}}{1.81 \times 10^{-10} \text{ m}} = 0.54$

   When the radius ratio is 0.54, the coordination number is 6.

5. *If the radius of Li^+ is 68 pm and that of F^- is 136 pm, predict the structure of LiF. What is the coordination number of the Li^+ ion?*

   Sol. Radius ratio $= \frac{\text{Radius of Li}^+}{\text{Radius of Cl}^-} = \frac{68 \text{ pm}}{136 \text{ pm}} = 0.5$

   The structure of LiF is scc and C.N of $Li^+ = 6$.

6. *Calculate the distance between (100) planes of a crystal, which exhibits first-order reflection at an angle of incidence equal to 30° with X-rays of wavelength 2×10^{-10} m.*

Sol. For first-order reflection

$$\lambda = 2d \sin \theta$$

$$\therefore \quad 2 \times 10^{-10} \text{ m} = 2d \sin 30° = 2d \times 0.5 = d$$

Hence, distance between planes is 2×10^{-10} m.

7. *A face-centered cubic crystal possesses an atomic radius of 174.6 pm. Calculate the interplanar spacing of (200) and (220) planes.*

Sol. Length of *fcc* edge (a) is related to atomic radius (r) as

$$a = r\sqrt{8} = 174.6 \text{ pm} \times \sqrt{8} = 493 \text{ pm}$$

$$d_{hkl} = \frac{a}{\sqrt{h^2 + k^2 + l^2}}$$

(*i*) For 200 plane: $h = 2, k = 0, l = 0$

$$d_{200} = \frac{493 \text{ pm}}{\sqrt{2^2 + 0^2 + 0^2}} = 246.5 \text{ pm}$$

(*ii*) For 220 plane: $h = 2, k = 2, l = 0$

$$\therefore \quad d_{220} = \frac{493 \text{ pm}}{\sqrt{2^2 + 2^2 + 0^2}} = 174.6 \text{ pm}$$

EXERCISES

1. What are Miller indices? Describe a method of determining the Miller indices for a plane of cubic crystal.
2. What are Bravais lattices? Assuming that the unit cell is made of the same atoms, calculate the number of atoms, after sharing in a simple cubic unit cell, a face-centered cubic unit cell, and a body-centered cubic unit cell.
3. What do space lattice and unit cell mean?
4. Explain the band theory of a solid.
5. Explain with an example the different types of semiconductors.
6. Define the terms plane of symmetry, center of symmetry, and axis of symmetry.
7. What is meant by a coordination number, and how do you correlate it with the radius ratio?
8. Explain the difference between the Schottky and Frenkel defects.
9. Discuss, in detail, imperfections in crystals.

10. What do stoichiometric and nonstoichiometric defects mean?
11. Deduce the Bragg's equation for the diffraction of X-rays.
12. Differentiate between *n*-type and *p*-type semiconductors.
13. Write short notes on the following:
    (*a*) Laws of crystallography
    (*b*) Symmetry elements
    (*c*) Liquid crystal
14. What are liquid crystals and how do they differ from solid crystals?
15. (*a*) How many sodium ions surround each chloride ion in the sodium chloride crystal?
    (*b*) Explain the Schottky defect in crystals. How is it different from the Frenkel defect?
    (*c*) What are the coordination numbers of each of the ions present in the cubic close-packed structure of CaF_2 at an ordinary temperature and pressure?

MULTIPLE CHOICE QUESTIONS

1. In a face-centered cubic arrangement (*fcc*), the number of atoms per unit cell is
   (*a*) 8 (*b*) 2 (*c*) 1 (*d*) 4
2. Bravais lattices come in
   (*a*) 7 types (*b*) 10 types (*c*) 14 types (*d*) 17 types
3. In a NaCl crystal, each Cl^- ion is surrounded by
   (*a*) four Na^+ ions (*b*) six Cl^+ ions (*c*) six Na^+ ions (*d*) four Cl^- ions
4. Which one of the following defects in the crystal lowers its density?
   (*a*) Frenkel defect (*b*) Schottky defect
   (*c*) Interstitial (*d*) F-centers
5. The coordination number of Na^+ in an NaCl crystal is
   (*a*) 4 (*b*) 6 (*c*) 12 (*d*) None of these
6. Which of the following are cubic close-packed arrangement?
   (*a*) ABCABC ... (*b*) ABAC ... (*c*) ABBA ... (*d*) ABCCB ...
7. What structural units occupy the lattice sites in a metallic crystal?
   (*a*) Atoms (*b*) Electrons
   (*c*) Positive ions (*d*) Molecular crystals
8. An example of a cubic crystal system is
   (*a*) sodium chloride (*b*) potassium chloride
   (*c*) diamond (*d*) all of these

9. The crystal system for which $a \neq b \neq c$ and $\alpha = \beta = \gamma = 90°$ is called
   (*a*) cubic (*b*) triclinic (*c*) orthorhombic (*d*) tetragonal
10. Cesium chloride has the structure
    (*a*) FCC (*b*) BCC (*c*) SC (*d*) None of these
11. Silicon is used in transistors as a
    (*a*) conductor (*b*) insulator (*c*) semiconductor (*d*) capacitor
12. A plane with intercepts 0.5 *a*, 0.25 *a*, and 1.5 *a* has the Miller indices
    (*a*) 136 (*b*) 361 (*c*) 631 (*d*) 163
13. Pure silicon doped with phosphorous is a
    (*a*) metallic conductor (*b*) *p*-type semiconductor
    (*c*) *n*-type semiconductor (*d*) insulator
14. An example of an ionic crystal is
    (*a*) NaCl crystal (*b*) graphite (*c*) ice (*d*) Cu crystal

PART 2

Chapter 8

STEREOCHEMISTRY

8.1 INTRODUCTION

Compounds that have the same molecular formula but different properties (either physical and chemical or only physical) are known as *isomers*, and the phenomenon is known as *isomerism*. When isomerism is caused by the different arrangements of atoms or groups in space, the phenomenon is called *stereoisomerism* (Greek *stereos* = occupying space). The isomers that differ in their 3D structures are known as *stereoisomers* or *stereomers*. The stereomers have the same structural formula but differ in the spatial arrangement of atoms or groups in the molecule. In other words, stereoisomerism is exhibited by compounds that have an identical molecular structure but different configurations. The stereoisomerism can be classified into two categories:

- Configurational isomerism
- Conformational isomerism

Configurational Isomerism

The stereoisomers that are nonsuperimposable and noninterconvertible by rotation around single bonds are known as *configurational isomers*, and the phenomenon is known as *configurational isomerism*. The configurational isomers can be interconverted only by breaking and making bonds. These isomers may further be of two types:

Enantiomers (Optical Isomers): When the two isomers are the mirror image of each other, they are known as *enantiomers*. This is also known as *inversional isomerism.*

Diastereomers (Geometrical Isomers): When the configurational isomers are not the mirror image of each other, they are known as *diastereomers*. These also include *geometrical isomers.*

Conformational Isomerism

The stereoisomers that are nonsuperimposable but easily interconvertible by rotation about single bonds are known as *conformational isomers, conformers, conformational enantiomers,* or *conformational diastereomers*. This type of isomerism is found in alkanes and cycloalkanes.

Stereochemistry of Carbon

The two important stereoisomerism are geometrical and optical.

8.2 GEOMETRICAL ISOMERISM

The type of isomerism due to the different geometrical arrangement of two different groups about the carbon-carbon double bond is called *geometrical isomerism* or *cis-trans isomerism.* When two similar groups lie on the same side of a double bond, the arrangement is called a *cisisomer.* When two similar groups lie on the opposite side of the double bond, the arrangement is called a *transisomer.*

Some examples of compounds showing geometrical isomerism are shown here:

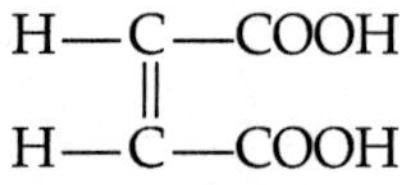

- $$\begin{array}{c} H-C-COOH \\ \| \\ H-C-COOH \end{array} \qquad \begin{array}{c} H-C-COOH \\ \| \\ HOOC-C-H \end{array}$$

  Maleic acid (cis) 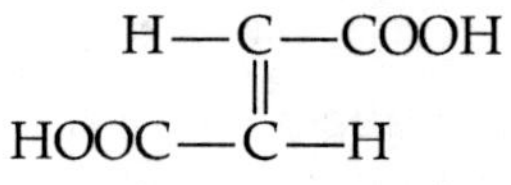 — Fumaric acid (trans)

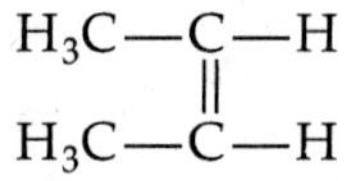

- $$\begin{array}{c} H_3C-C-H \\ \| \\ H_3C-C-H \end{array} \qquad \begin{array}{c} H_3C-C-H \\ \| \\ H-C-HC_3 \end{array}$$

  Cis-Butene-2 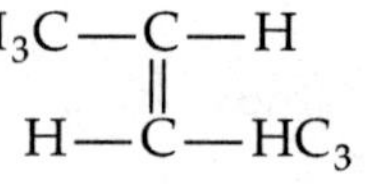 — Trans-Butene-2

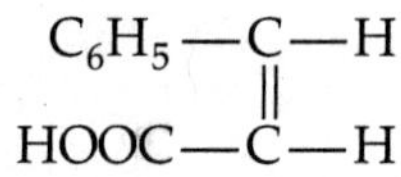

- $$\begin{array}{c} C_6H_5-C-H \\ \| \\ HOOC-C-H \end{array} \qquad \begin{array}{c} C_6H_5-C-H \\ \| \\ H-C-COOH \end{array}$$

  Cis-cinnamic acid 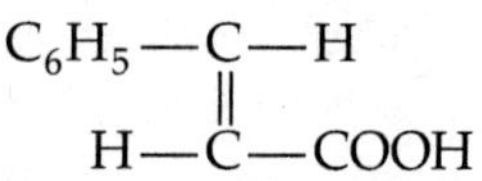 — Trans-cinnamic acid

The cause of geometrical isomerism is the restricted or hindered rotation about the carbon-carbon double bond. The cis and trans forms differ from each other in many physical properties. Trans isomers are more stable, have higher melting points, higher densities, lower solubilities, lower dipole moments, lower boiling points, and lower refractive indices than those of the corresponding cis isomers.

8.3 E/Z SYSTEM OF NOMENCLATURE

The compounds, where all the four substituents attached to the carbon atoms of double bond are different, do not obey the cis-trans nomenclature. Such geometrical isomers are designated by E/Z nomenclature. The newer method, which can be applied to all cases, is based on the Cahn-Ingold Prelog system. This system is known as the E (German word *entgegen* meaning opposite) and Z (German word *Zusammen* meaning together) system of nomenclature.

In this system, each of the two groups on each carbon atom are assigned the priority numbers (1) and (2) on the basis of the following *sequence rules*:

Rule 1: The atom of the higher atomic number is given the higher priority, that is, it is marked as 1, and the other is marked as 2. When the two atoms attached to the double bond are isotopes, the isotope with the higher mass number is given the higher priority.

Rule 2: When the two atoms attached to the double bond are of the same atomic number, then the priority can be determined by comparing the next elements in the group, and so on.

Rule 3: A doubly or triply bonded atom is considered equivalent to two or three such atoms. Thus, a carbonyl group is considered as if carbon had two single bonds to oxygen, that is

$$>C=O \text{ treated as } >C(-O)-O$$

By the application of these rules, some common substituents are given the following priority sequence:

(I, Br, Cl, F, O, N, C, H) (—C $(CH_3)_3$, —CH $(CH_3)_2$, —CH_2CH_3, —CH_3) (—$COOCH_3$, —COOH, —$CONH_2$, —$COCH_3$, —CHO)

Now on the basis of relative priorities and the position of groups/atoms, the E/Z designation is assigned as follows:

Select the atom/group with the higher priority on each doubly bonded carbon. If the atoms/groups of higher priority (denoted by 1) on each carbon are on the same side of the double bond, the isomer is assigned the configuration Z. On the other hand, if the atoms/groups of higher priority on each carbon are on the opposite sides of the double bond, the isomer is assigned the configuration E.

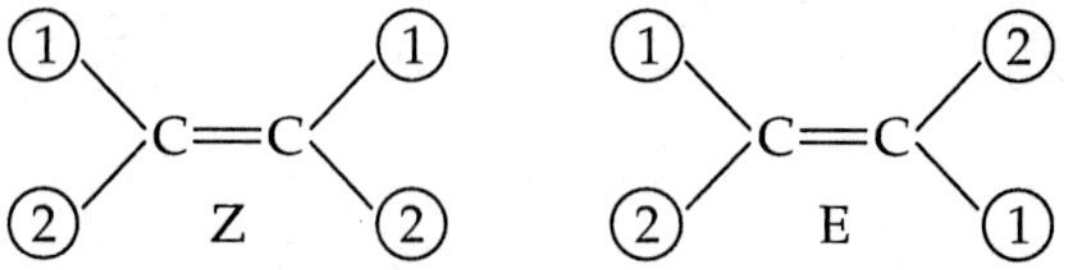

(1) C=C (1), (2) Z (2) — (1) C=C (2), (2) E (1)

For example, consider the following compound:

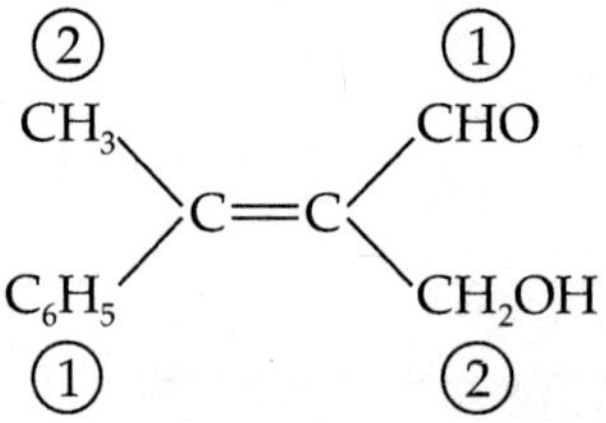

(2) CH_3, C_6H_5 (1) on one carbon; (1) CHO, CH_2OH (2) on the other: $CH_3(C_6H_5)C{=}C(CHO)(CH_2OH)$

One of the carbon atoms of the double bond carries CH_3 and C_6H_5 groups. Because in the C_6H_5 group, the first carbon is attached to two other carbons, one by a double bond and the other by a single bond, therefore, C, C, CH of phenyl gets higher priority over H, H, H of CH_3. Thus, C_6H_5 is assigned priority (1), and CH_3 is assigned priority (2). The other carbon atom of the double bond carries groups CH_2OH and CHO. Because in CH = O, C is attached to O by a double bond while in CH_2OH, C is attached to O by a single bond; therefore, O, O, H of CHO gets higher priority than O, H, H of the CH_2OH group. Thus, CHO is assigned priority (1), and CH_2OH is assigned priority (2).

Other examples of E/Z isomers are as given here:

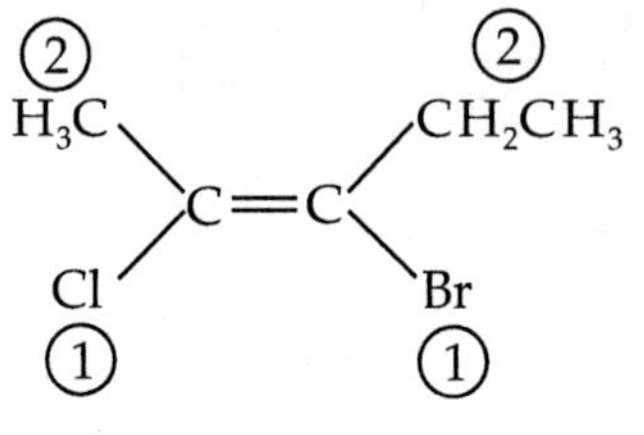

(2) H_3C, Cl (1) — C=C — (2) CH_2CH_3, Br (1)

Z-isomer

(1) H_3C, H (2) — C=C — (2) CH_3, CHO (1)

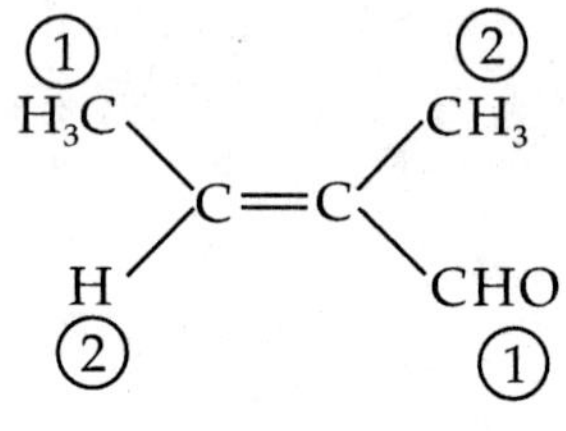

E-isomer

8.4 A GEOMETRICAL ISOMERISM IN CYCLIC COMPOUNDS

A geometrical isomerism is also possible in cyclic compounds. There can be no rotation about carbon-carbon single bonds forming a ring because rotation would break the bonds and break the ring. A requirement for a geometrical isomerism in cyclic compounds is that there must be at least two other groups besides hydrogens on the ring, and these must be on different ring carbon atoms. For example, no geometrical isomers are possible for 1, 1-dimethyl cyclopropane.

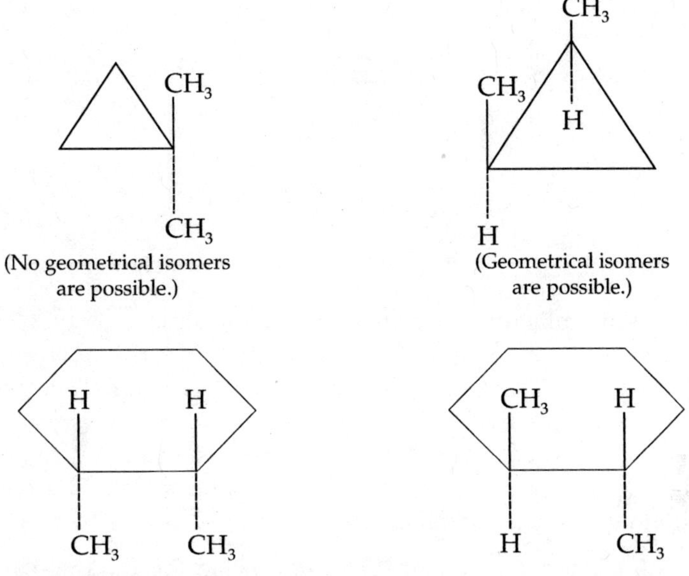

8.5 OPTICAL ISOMERISM

Certain substances have a remarkable power of rotating the plane polarized light. This phenomenon of rotating the plane polarized light is known as *optical activity*, and the compounds exhibiting this property are known as *optically active compounds*. Compounds possessing identical molecular and structural formulae and identical physical and chemical properties but differing only in optical activity are called *optical isomers* or *enantiomorphs*. The phenomenon is called *optical isomerism* or *enantiomerism*. The optical isomerism is exhibited by compounds containing asymmetric C, N, or Si atoms.

Optical Activity

According to the physical theory, an ordinary ray of light vibrates in all planes as shown in Fig. 8.1. When this light is passed through a *Nicol prism or a polarized lense, light is found to vibrate in only one plane. Such lights, that

* *Nicol prism is made of calcite or iceland spar (crystalline form of $CaCO_3$).*

vibrate only in a single direction are known as *plane polarized lights,* and the phenomenon is known as *polarization.*

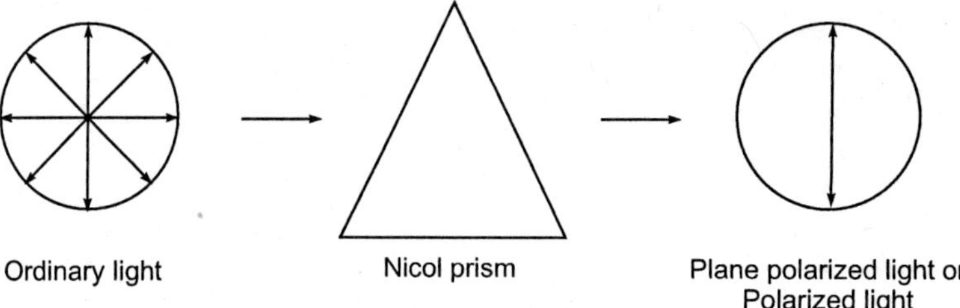

Fig. 8.1 Polarization.

When plane polarized light is passed through certain substances or their solutions, its plane of polarization is rotated either toward the right (clockwise) or toward the left (anticlockwise) by a certain angle. Such substances that rotate the plane polarized light are known as *optically active,* and the phenomenon is referred to as an *optical activity.* Those substances that rotate the plane polarized light to the right are called *dextro-rotatory,* indicated by *d* or (+), and those that rotate to the left are called *laevorotatory,* indicated by *l* or (–).

An inactive form that does not rotate the plane polarized light at all is a mixture of equal amounts of (+) and (–) forms, and is optically inactive. It is named the (±) mixture or the *Racemic mixture* (Latin, *racemic* = mixture of equal components).

The simple organic compounds that show optical activity are

$CH_3—CH(OH)—COOH$

Lactic acid

$\begin{matrix} CH_3 \\ C_2H_5 \end{matrix} \rangle CH—CH_2OH$

Iso amyl alcohol

The measurement of optical activity is reported in terms of a specific rotation that is constant. The instrument used for the purpose is a polarimeter (Fig. 8.2). The specific rotation is as much a characteristic for a particular compound as its m.p., b.p., refractive index, and so on. It is expressed by

$$\text{Specific rotation} \quad [\alpha]_D^{t°C} = \frac{\alpha_{obs}}{lxc}$$

where α_{obs} is the rotation observed by the polarimeter, *l* is the length of solution in decimeter, and *c* is the amount of substance in gm in 1 ml of solution. Thus, it may be defined as the observed rotation when the polarized light is passed through 1 decimeter (10 centimeters) of the solution with a concentration of 1 gm/ml. The sign (+) or (–) attached with the angle of rotation signifies the direction of rotation; the (–) sign indicates the rotation toward the left, and the (+) sign indicates the right.

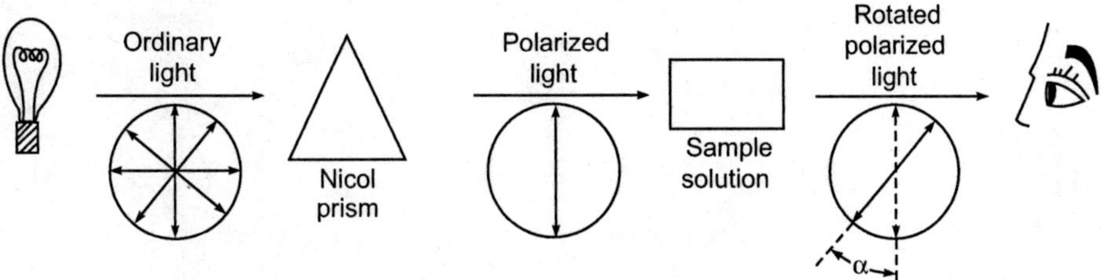

Fig. 8.2 Diagrammatic representation of a simple polarimeter in operation.

Cause of Optical Activity

For an organic compound to show optical activity, it should have an asymmetric or chiral carbon atom and not have a plane of symmetry.

8.6 CHIRALITY

The property of nonsuperimposability of an object on its mirror image is called *chirality*. Chirality in organic molecules is due to the tetrahedral nature of the carbon atom. The carbon atom attached to four different atoms or groups is called an *asymmetric chiral carbon atom* indicated by C*. Thus it can be represented as

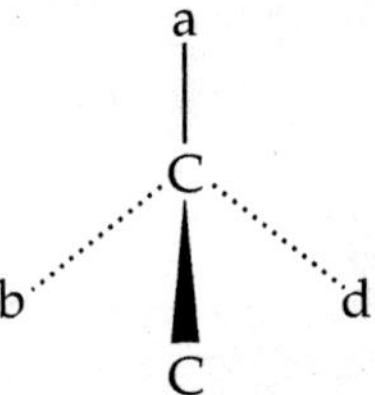

All organic compounds containing one asymmetric or chiral carbon atom are optically active. A chiral object does not have any elements of symmetry such as a plane of symmetry, a center of symmetry, or an axis of symmetry. On the other hand, achiral objects or molecules have at least one or more elements of symmetry. The most commonly observed element of symmetry is a plane of symmetry. A plane that bisects the object into two superimposable mirror image halves is called the *plane of symmetry* (Fig. 8.3). Thus, chiral objects do not superimpose on their mirror image. A chiral object and its mirror image are called *enantiomers*. For example, our two hands are enantiomers because they are mirror images of each other but are not superimposable. The two nonsuperimposable mirror images represent the two optical isomers of the compound.

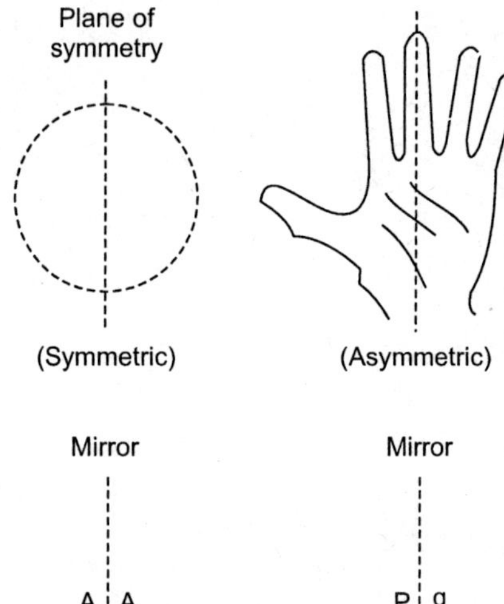

Fig. 8.3 Symmetric and asymmetric object.

Following are explanations of the two optical isomers:

Optical isomerism of lactic acid: Lactic acid [$CH_3CHOH\,.\,COOH$] contains an asymmetric central carbon atom, and it exists in two optical active forms, that are related to each other as mirror images. The two are nonsuperimposable.

Mirror

CH_3 — $H-\overset{*}{C}-OH$ — $COOH$	CH_3 — $HO-\overset{*}{C}-H$ — $COOH$
(+) or *d*-lactic acid	(+) or *l*-lactic acid

A third form called a *racemic mixture* or *modification* is obtained when equimolar quantities of *d*– and *l*– enantiomorphs are mixed. Such a mixture is optically inactive as a result of the cancellation of the equal and opposite rotations of the two isomers. The symbol ± is used to denote a racemic mixture.

Optical isomerism of tartaric acid: Tartaric acid has two asymmetric carbon atoms, each linked to the same four different groups, that is —OH, —H, —COOH, and —CH (OH) COOH. The spatial arrangement of various groups in tartaric acid can be represented in three ways. So, tartaric acid can exist in three different forms: *d*-tartaric acid, *l*-tartaric acid, and racemic tartaric acid. Meso-tartaric acid is the inactive variety as the rotation of the upper half is compensated by the rotation of the lower half (being in opposite direction). It cannot be resolved into active constituents, so it is inactive by internal compensation. Unlike dextro and laevo-varieties, it has a plane of symmetry. The molecule of a meso compound is a chiral.

Fig. 8.4 shows the spatial arrangement of various groups in isomerides of tartaric acid.

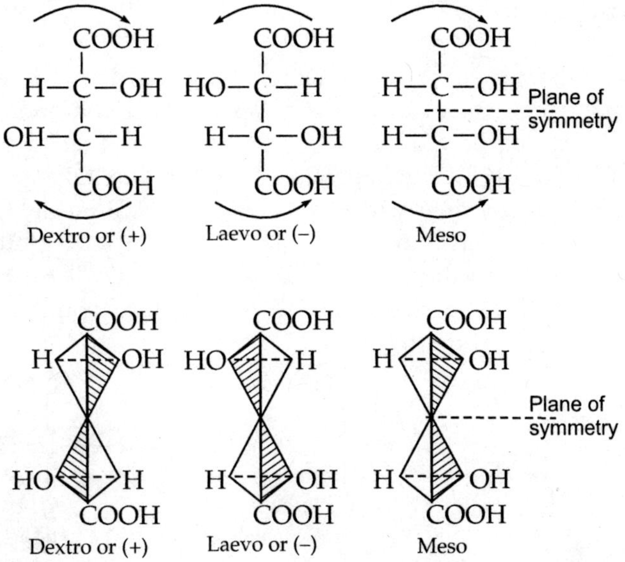

Fig. 8.4 Spatial arrangement of various groups in isomerides of tartaric acid.

8.7 FISCHER PROJECTIONS

While discussing the stereochemistry of organic compounds, it is inconvenient to always use the 3D formulae of the various organic compounds. To overcome this difficulty, Emil Fischer (1891) suggested the 2D projection formulae. In the Fischer convention, the carbon chain is always written vertically; the horizontal lines represent the groups above the plane of the paper, whereas the vertical lines represent the groups below the plane of the paper (shown by dotted lines). Furthermore, because the vertical bonds are actually below the plane of the paper, the formula may be rotated by an angle of 180° (not by 90° or 270°). The various types of representations can be exemplified in the case of lactic acid, CH_3CH (OH) . COOH.

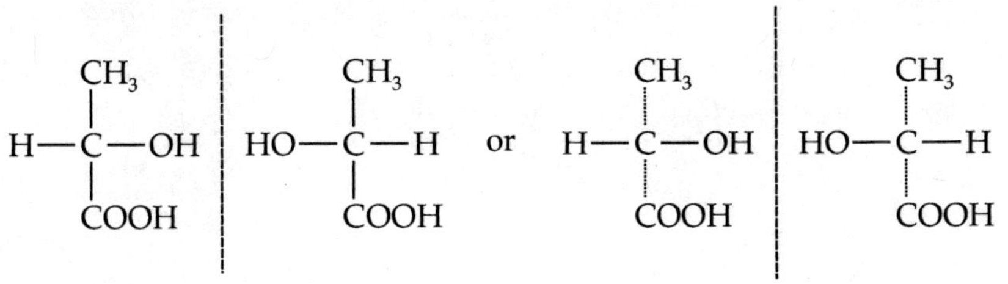

8.8 NOMENCLATURE

Following are the two systems of nomenclature for optical isomers:

Relative Configuration (D and L System)

Before 1951, there was no method for determining the absolute configuration (actual arrangement of atoms in space) of a compound, so the configuration of all the compounds were studied with respect to glyceraldehyde (relative configuration), the configuration of which was taken as an arbitrary standard. (+)-glyceraldehyde —with the —OH group on the right, the hydrogen atom on the left, the —CHO at the top and the —CH_2OH group at the bottom —was arbitrarily given the configurational symbol D. The mirror image compounds (–)-glyceraldehyde, in which the —OH group is on the left and hydrogen is on right, was given the configuration L.

```
    CHO                 CHO
     |                   |
  H—C—OH             HO—C—H
     |                   |
    CH2OH               CH2OH
D(+)-Glyceraldehyde   L(–)-Glyceraldehyde
```

(+)- and (–)-signs indicate that the two forms are dextro- and laevo-rotatory, respectively, but it must be noted that it is not necessary that all the compounds belonging to D- and L-series will be dextro- and laevo-rotatory, respectively.

Any compound that can be prepared from, or converted into, D (+)-glyceraldehyde will belong to the D-series, and similarly any compound that can be prepared from, or converted into, L (–)-glyceraldehyde will belong to the L-series (relative configuration). For example, D (+)-glyceraldehyde can be converted to glyceric acid by simple oxidation, so the configuration of glyceric acid obtained must be D.

```
     CHO                        COOH
      |          [O]             |
  H—C—OH    ——————————→     H—C—OH
      |                          |
     CH2OH                      CH2OH

D (+)-Glyceraldehyde        D (–)-Glyceric acid
```

Similarly, lactic acid obtained from D (+)-glyceraldehyde in the following way is also assigned the D configuration.

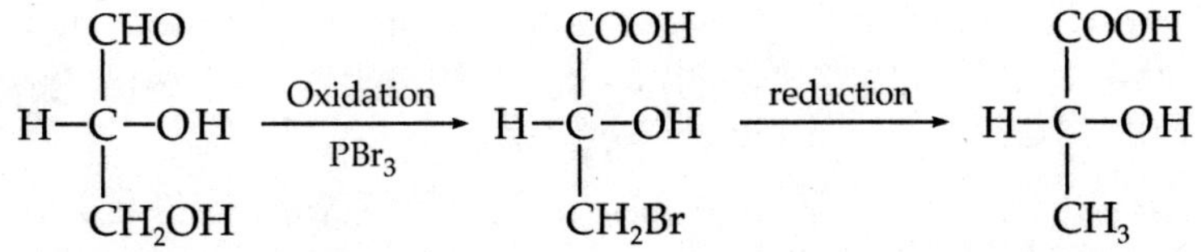

$$\underset{\text{D (+)-Glyceraldehyde}}{\begin{array}{c} CHO \\ | \\ H-C-OH \\ | \\ CH_2OH \end{array}} \xrightarrow[PBr_3]{\text{Oxidation}} \begin{array}{c} COOH \\ | \\ H-C-OH \\ | \\ CH_2Br \end{array} \xrightarrow{\text{reduction}} \underset{\text{D(−)-Lactic acid}}{\begin{array}{c} COOH \\ | \\ H-C-OH \\ | \\ CH_3 \end{array}}$$

Absolute Configuration (R and S System)

Cahn, Ingold, and Prelog devised this system of nomenclature that is based on the actual 3D formula. To specify the configuration about an asymmetric carbon **Cabde*, the groups *a*, *b*, *d*, and *e* are first assigned the order of priority determined by the sequence rules. The sequence rules to determine the order of priorities of groups are the following:

- The atoms or groups directly bonded to the asymmetric carbon are arranged in the order of decreasing atomic number and assigned priority 1, 2, 3, 4 accordingly. For example, in chlorobromofluoro methane (CHClBrF), the substituents Br (at. no. = 35), Cl (at. no. = 17), F (at. no. = 9), and H (at. no. = 1) give the order of priorities.

$$\underset{(1)}{Br} > \underset{(2)}{Cl} > \underset{(3)}{F} > \underset{(4)}{H}$$

- When two or more groups have identical first atoms attached to an asymmetric carbon, the priority order is determined by considering the atomic numbers of the second atoms. If the second atoms are also identical, the third atoms along the chain are examined. Consider three groups:

$$-\overset{1}{C}H_2-\overset{2}{H} \quad \text{methyl}$$

$$-\overset{1}{C}H_2-\overset{2}{C}H_2-\overset{3}{H} \quad \text{ethyl}$$

$$-\overset{1}{C}H_2-\overset{2}{C}H_2-\overset{3}{C}H_3 \quad n\text{-propyl}$$

In methyl and ethyl, the first atom (carbon) is identical, and therefore, atomic numbers of the second atoms H (at. no. = 1) and C (at. no. = 6) decide the priority order.

ethyl > methyl

In ethyl and *n*-propyl, the second atom (carbon) is also identical, so the third atoms (H, C) decide the priority order.

n-propyl > ethyl

- If the first atoms of the two groups have the same substituents of the higher atomic number, the one with more substituents takes priority.
  Thus — $CHCl_2$ has a higher priority than —CH_2Cl.
- A doubly or triply bonded atom is considered equivalent to two or three such atoms. Thus,

$$=A \text{ equals } \langle^{A}_{A} \quad (A, A)$$

$$\equiv A \text{ equals } -C\begin{smallmatrix}A\\A\\A\end{smallmatrix} \quad (A, A, A)$$

A phenyl group is handled as if it had one of the Kekule structures.

$$-C_6H_5 \equiv -C\begin{smallmatrix}C\\C\\C\end{smallmatrix}$$

For example, between groups —C(H)=O (O, O, H) and —CH_2OH (O, H, H), the former will have higher priority.

- When the molecule contains more than one asymmetric center, the same rules are applied to each.
  Now, for the compound $^*C_{abde}$, the order of priorities may be stated as

$$\begin{matrix} a & > & b & > & d & > & e \\ (1) & & (2) & & (3) & & (4) \end{matrix}$$

Now the tetrahedral model of the molecule is viewed from the direction opposite to the group *'e'* if lowest priority (4). The conversion rule says that:

- If the eye while moving from $a \rightarrow b \rightarrow d$ travels in a clockwise or right-hand direction, the configuration is designated R (Latin, *Rectus* = right).
- If the eye while moving from $a \rightarrow b \rightarrow d$ travels in a counter-clockwise or left-hand direction, the configuration is designated S (Latin, *Sinister* = left).

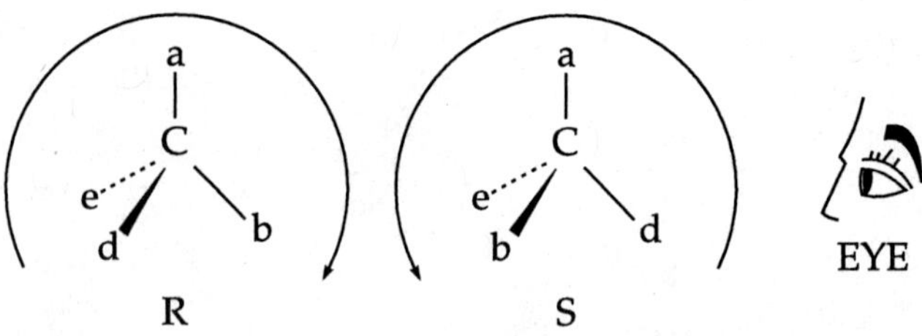

Let us now illustrate this R/S system nomenclature of enantiomers of some compounds.

Example 1: 1, 1-chlorobromoethane.
The four atoms directly bonded to the asymmetric carbon are Br (at. no. = 35), Cl (at. no. = 17), C (at. no. = 6), and H (at. no. = 1). Hence, the priority order is

$$\underset{(1)}{Br} > \underset{(2)}{Cl} > \underset{(3)}{CH_3} > \underset{(4)}{H}$$

Applying the conversion rule, the configuration of two enantiomers are assigned as

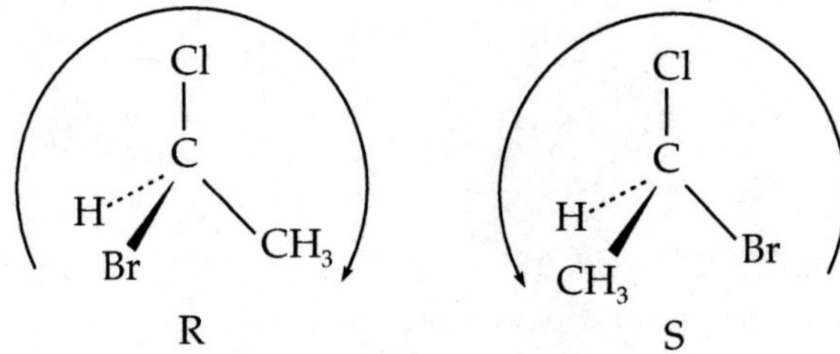

Example 2: Lactic acid [CH_3*CH (OH) COOH].
The priorities of the four groups are

$$\underset{(1)}{OH} > \underset{(2)}{COOH} > \underset{(3)}{CH_3} > \underset{(4)}{H}$$

The configurations R and S are shown as

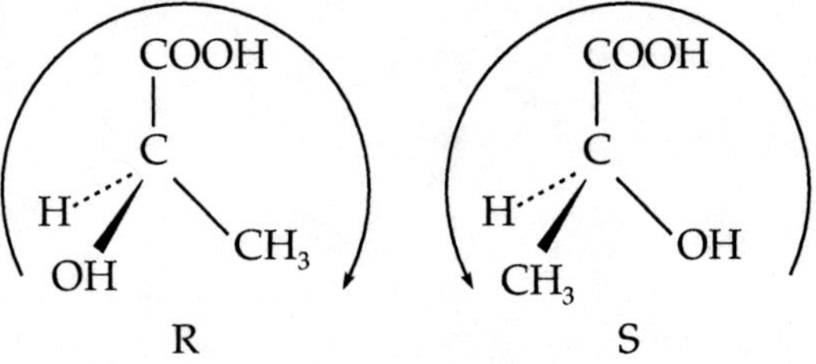

Example 3: Glyceraldehyde [CHO . *CHOH . CH_2OH].
The groups bonded to asymmetric carbon in glyceraldehyde have the priority order:

$$OH > CHO > CH_2OH > H$$

The configurations of the two isomers are

CHO
C
H
OH
CH_2OH
R

CHO
C
H
CH_2OH
OH
S

Example 4: Alanine [NH_3 . *CH . CH_3 . COOH].
The priorities of the four groups are

$$NH_3 > COOH > CH_3 > H$$

The configurations of the two isomers are

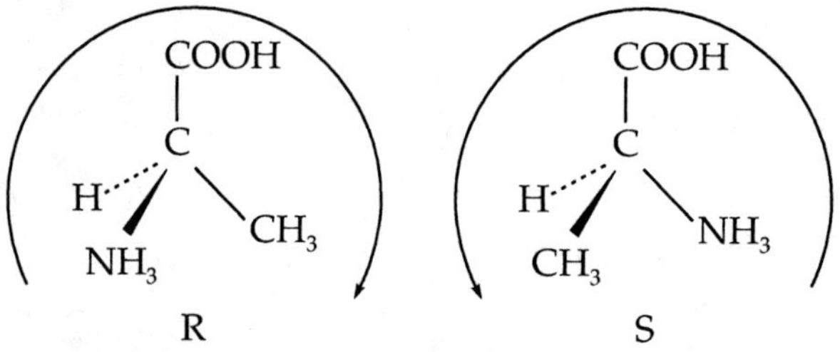

COOH, C, H, NH_3, CH_3 — R   COOH, C, H, CH_3, NH_3 — S

Example 5: Mandelic acid [C_6H_5 . *CHOH . COOH].
The priorities of the four groups attached to the asymmetric carbon atom are

$$OH > COOH > C_6H_5 > H$$

The configuration of the two isomers are

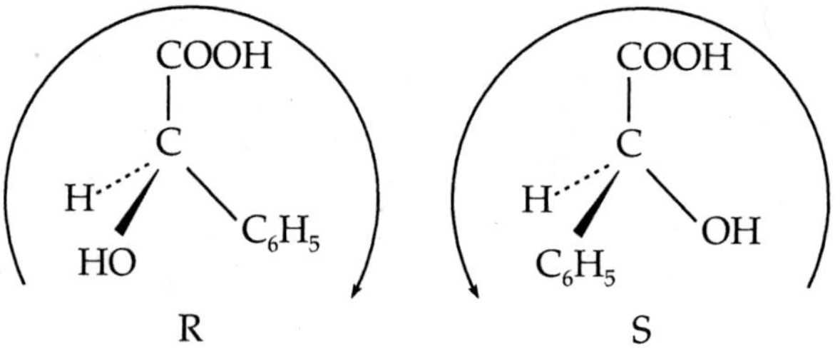

COOH, C, H, HO, C_6H_5 — R   COOH, C, H, C_6H_5, OH — S

8.9 ENANTIOMERS

Optical isomers that are nonsuperimposable mirror images of each other are called *enantiomers,* and the phenomenon is called *enantiomerism.* Chirality is the necessary and sufficient condition for the existence of enantiomers.

Some of the common examples of enantiomers are given here:

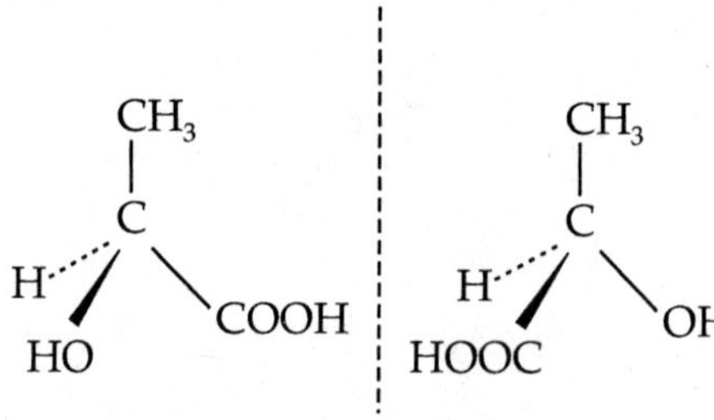

CH_3, C, H, HO, COOH | CH_3, C, H, HOOC, OH

Enantiomers of lactic acid

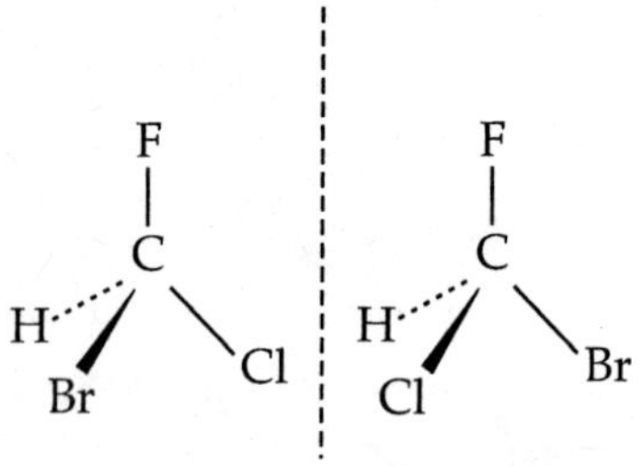

F, C, H, Br, Cl | F, C, H, Cl, Br

Enantiomers of bromo-chlorofluoro methane

Characteristics of Enantiomers

Following are the characteristics of enantiomers:

- Enantiomers have identical physical properties such as melting point, boiling point, density, refractive index, and so on. They differ only in their action on plane polarized light; one of the enantiomers rotates the plane polarized light to the right, and the other to the same magnitude but to the left. The two enantiomers are designated as dextrorotatory *d* or (+) and laevorotatory *l* or (–), respectively.
- Enantiomers have identical chemical properties except their action toward optically active compounds. For example, in the rate of esterification of (+)-lactic acid or (–)-lactic acid with the same alcohol, (+)-sec-butyl alcohol would be different.
- When the two enantiomers are mixed in equimolar quantities, it results in the formation of an optically inactive compound called racemic, *dl* or (±)-form. For example, equal amounts of (+)-lactic acid and (–)-lactic acid mixed together produce optically inactive (±) or racemic (*r*) lactic acid.

8.10 DIASTEREOMERS

The configurational isomers that are optically active isomers but not mirror images are called *diastereomers* or *diastereoisomers.*

Consider an example of 2-bromo-3-chloro butane, which can be represented in four configurational forms:

$$
\begin{array}{ccc}
 & CH_3 & \\
 & | & \\
H- & C & -Br \\
 & | & \\
H- & C & -Cl \\
 & | & \\
 & CH_3 & \\
 & \text{I} &
\end{array}
\qquad
\begin{array}{ccc}
 & CH_3 & \\
 & | & \\
Br- & C & -H \\
 & | & \\
Cl- & C & -H \\
 & | & \\
 & CH_3 & \\
 & \text{II} &
\end{array}
\qquad
\begin{array}{ccc}
 & CH_3 & \\
 & | & \\
Br- & C & -H \\
 & | & \\
H- & C & -Cl \\
 & | & \\
 & CH_3 & \\
 & \text{III} &
\end{array}
\qquad
\begin{array}{ccc}
 & CH_3 & \\
 & | & \\
H- & C & -Br \\
 & | & \\
Cl- & C & -H \\
 & | & \\
 & CH_3 & \\
 & \text{IV} &
\end{array}
$$

The forms I and III are not mirror images or enantiomers, yet they are optically active isomers and called diastereomers. Similarly, the forms II and IV are also not enantiomers but optically active isomers and are diastereomers.

Characteristics of Diastereomers

Diastereomers characteristics include the following:

- Diastereomers have different physical properties such as melting points, boiling points, densities, solubilities, refractive indices, dielectric constants, and specific rotations.
- Diastereomers other than geometrical isomers may or may not be optically active.
- Diastereomers show similar chemical properties.
- Diastereomers can be easily separated through fractional crystallization, fractional distillation, chromatography, and so on.

8.11 RESOLUTION OF RACEMIC MIXTURES

When an optically active compound is synthesized, a mixture of both *d*- and *l*-isomers in equal amounts is formed. Such a mixture is known as a *racemic mixture.* Conversion of an optically active compound into a racemic mixture is called *racemization.* Racemization can be accomplished by means of heat, light, or by conversion of the isomers into an optically inactive intermediate that reverts to the racemic mixture. The conversion of either of the optically active lactic acids into a racemic mixture by heating its aqueous solution may proceed through an enol intermediate:

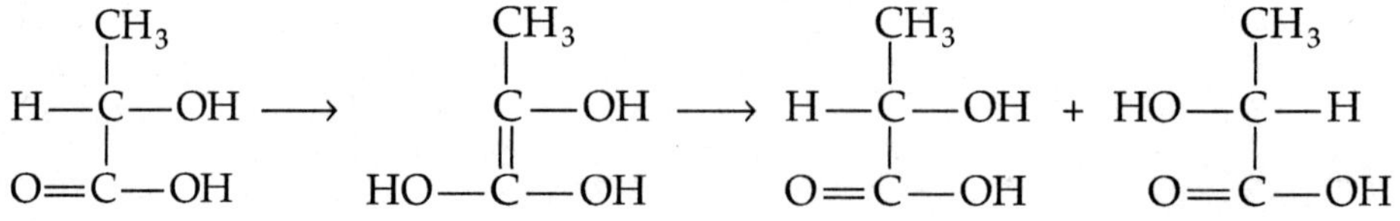

$$\underset{\substack{(+)\text{-acid}\\(2\text{ molecules})}}{\begin{array}{c}CH_3\\|\\H-C-OH\\|\\O=C-OH\end{array}} \longrightarrow \underset{\substack{\text{Enol}\\(\text{unstable})}}{\begin{array}{c}CH_3\\|\\C-OH\\\|\\HO-C-OH\end{array}} \longrightarrow \underset{(+)\text{-acid}}{\begin{array}{c}CH_3\\|\\H-C-OH\\|\\O=C-OH\end{array}} + \underset{(-)\text{-acid}}{\begin{array}{c}CH_3\\|\\HO-C-H\\|\\O=C-OH\end{array}}$$

The process of separating a racemic or *dl*-mixture into its two optically active *d*- and *l*-forms is known as *resolution.* Following are some possible methods used for the separation of racemic mixtures:

- Conversion of diastereomers
- Differential absorption
- Biochemical process
- Mechanical separation
- Differential reactivity

8.12 WALDEN INVERSION

When an atom or group directly linked to the asymmetric carbon atom is replaced, the reaction may proceed with an inversion of configuration. This phenomenon was first observed by Walden (1895). The Walden inversion may also be defined as the conversion of the (+)-form to the (–)-form, or vice versa, without resolution. Thus, (+)-malic acid may be converted to (–)-malic acid as follows:

$$\begin{array}{ccccc} COOH & & COOH & & COOH \\ | & & | & & | \\ H-C-OH & \underset{KOH}{\overset{PCl_5}{\rightleftharpoons}} & Cl-C-H & \xrightarrow{AgOH} & HO-C-H \\ | & & | & & | \\ CH_2COOH & & CH_2COOH & & CH_2COOH \\ \text{(+)-malic acid} & & \text{(–)-chlorosuccinic acid} & & \text{(–)-malic acid} \end{array}$$

The Walden inversion is a type of $S_N{}^2$ reaction.

8.13 CONFORMATIONAL ISOMERISM

The isomers that differ in their conformation are known as *conformational isomers*, and the phenomenon is known as *conformational isomerism.* This type of isomerism is observed in alkanes and cycloalkanes and their substituted derivatives.

Newmann Projections

Newmann projection formulae are adopted for projecting a molecule on a planar surface. In the Newmann projection formula, the carbon atoms nearer to the eye (*i.e.*, the front carbon) and the groups attached to it are represented by equally spaced radii, and the distant carbon atom (*i.e.*, the carbon atom farther from the eye or the rear carbon) and the groups attached to it are represented by a circle with three equally spaced radial extensions (Fig. 8.5).

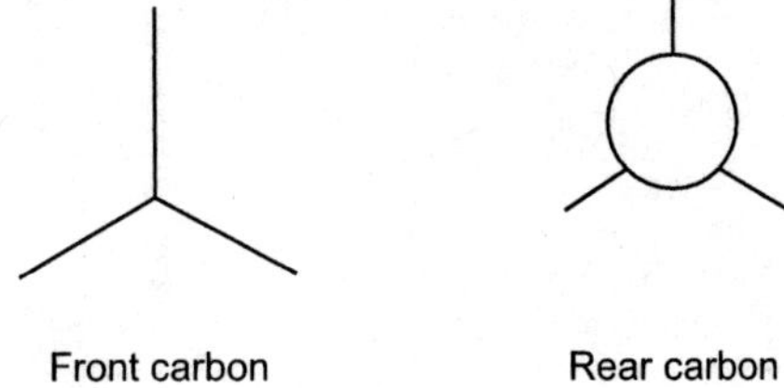

Fig. 8.5 Representation of carbon atoms in the Newmann projection formula.

Conformations of n-butane: Butane has four extreme conformations out of an infinite number of total conformations:

Anti Partially eclipsed Gauche Fully eclipsed

The energies of various conformations of butane are anti (0.0 k J mol^{-1}), skew or gauche (3.35 k J mol^{-1}), partially eclipsed (12.13 k J mol^{-1}), and fully eclipsed (15.06 k J mol^{-1}). The increasing order of stability of these conformations is

Fully eclipsed < Partially eclipsed < Gauche < Anti

Conformations of cyclohexane: Cyclohexane has many conformations, such as chair form, boat form, twist boat form, and half chair form.

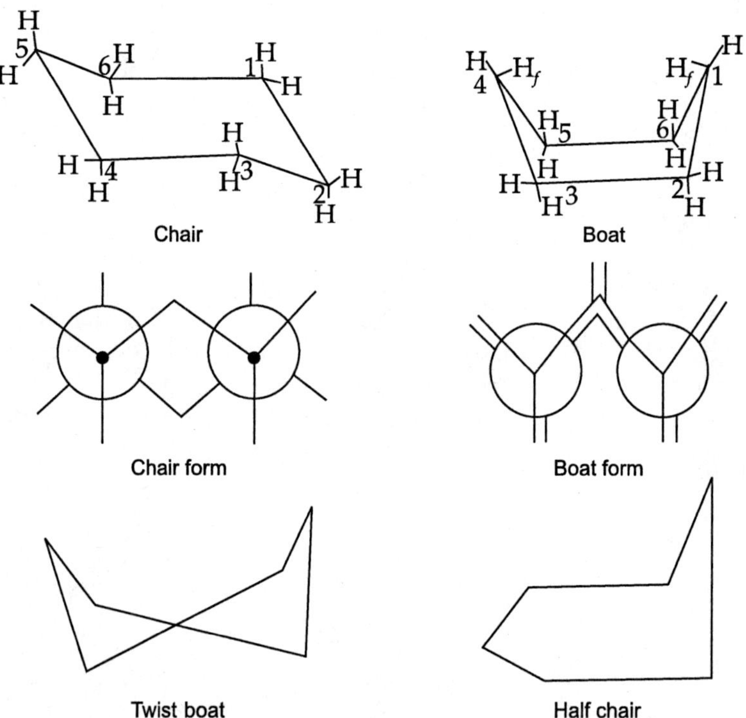

Chair Boat

Chair form Boat form

Twist boat Half chair

The chair conformation is more stable than the boat conformation because of the following two reasons:

- In chair conformation, all the H-atoms on $C_1 - C_2$, $C_2 - C_3$, $C_3 - C_4$, $C_4 - C_5$, $C_5 - C_6$, and $C_6 - C_1$ are in more stable staggered orientations with no torsional strain. On the other hand, in boat conformation, the adjacent hydrogens on $C_2 - C_3$ and $C_5 - C_6$ are in the less stable eclipsed orientation. These eclipsing interactions raise the energy of the boat form relative to the chair form.
- The two hydrogen atoms (marked H_f) are called *flagpole hydrogens.* In boat form, these are quite close to each other (1.83 Å) as compared to chair form (2.29 Å). These hydrogens strongly repel each other and introduces, the steric strain.

The two other conformations, twist boat and half chair, are unstable. The relative energies of various conformations are chair form (0.0 k J mol^{-1}), twist boat (23.0 k J mol^{-1}), boat form (29.7 k J mol^{-1}), and half chair (44.0 k J mol^{-1}).

EXERCISES

1. (*a*) What is an optical activity?
   (*b*) Describe the stereochemistry of tartaric acid.
2. Explain the necessary conditions for a molecule to show a geometrical isomerism, and provide an example.
3. Write notes on
   (*a*) Racemisation
   (*b*) Enantiomers
   (*c*) Diastereomers
4. Assign R or S configuration to each of the following compounds:

(*i*) CHO on top, H—C—OH, CH_2OH at bottom

(*ii*) COOH on top, H_2N—C—H, CH_3 at bottom

(*iii*) OH on top, HOOC—C—H, CH_3 at bottom

5. Assign E or Z configuration to each of the following compounds:

(*i*) $(CH_3)(CH_3CH_2)C{=}C(H)(CH_2{-}CHO)$

(*ii*) $(Br)(I)C{=}C(Cl)(H)$

6. Explain the optical isomerism of chiral organic compounds.
7. Discuss the conformational isomers of *n*-butane.
8. Explain the term *chirality*. What are the conditions for chirality?
9. Explain the Walden inversion and provide a suitable example.

10. Explain the difference between enantiomer and diastereomer.
11. (*a*) How is an optical activity measured?
    (*b*) What is a specific rotation?
12. (*a*) Draw the Newmann projection for the most stable conformation of *n*-butane.
    (*b*) Draw the Newmann projection of the boat conformation of cyclohexane.
    (*c*) Write the structures of all enantiomers and diastereomers of 2-Bromo-3-chlorobutane.

MULTIPLE CHOICE QUESTIONS

1. An alkane forms isomers if the least number of carbon atoms is
   (*a*) 1 (*b*) 2
   (*c*) 3 (*d*) 4
2. The lowest alkene that can show geometrical isomerism is
   (*a*) ethene (*b*) propene
   (*c*) 1-butene (*d*) 2-butene
3. A compound has 3 chiral carbon atoms. The number of possible optical isomers it can have is
   (*a*) 3 (*b*) 2
   (*c*) 8 (*d*) 4
4. *d*- and *l*-tartaric acids are
   (*a*) diastereomers (*b*) enantiomers
   (*c*) achiral molecules (*d*) tautomers
5. How many chiral carbons are present in glucose molecule CHO $(CHOH)_4$ CH_2OH?
   (*a*) 4 (*b*) 3
   (*c*) 2 (*d*) 1
6. The most stable cyclohexane is
   (*a*) boat (*b*) chair
   (*c*) skew (*d*) eclipsed
7. Optical isomerism is shown by
   (*a*) butanol-1 (*b*) butanol-2
   (*c*) 3-pentanol (*d*) 4-heptanol
8. Which of the following compounds exhibit geometric isomerism?
   (*a*) 2-methyl-1-pentene (*b*) 2-methyl-2-pentene
   (*c*) 2-hexene (*d*) 2, 3-dimethyl-2-butene

9. Optically active isomers but not mirror images are called
   (*a*) enantiomers (*b*) mesomers
   (*c*) tautomers (*d*) diastereomers
10. The most stable conformation of *n*-butane is
   (*a*) skew boat (*b*) eclipsed
   (*c*) gauche (*d*) staggered anti
11. The number of optical isomers of the compound $CH_3CHBrCHBrCOOH$ is
   (*a*) 0 (*b*) 1
   (*c*) 3 (*d*) 4
12. How many carbon atoms in the molecule $HOOC - (CHOH)_2 - COOH$ are asymmetric?
   (*a*) 1 (*b*) 2
   (*c*) 3 (*d*) None of these
13. Isomers that can be interconverted through rotation around a single bond are
   (*a*) conformers (*b*) diastereomers
   (*c*) enantiomers (*d*) positional isomers
14. Rotation of polarized light can be measured by
   (*a*) monometer (*b*) galvanometer
   (*c*) polarimeter (*d*) viscometer

Chapter 9

Electron Displacement and the Organic Reaction Mechanism

9.1 INTRODUCTION

Electron displacement plays a major role in organic reactions. Organic reactions are essentially molecular in nature. They may involve a single molecule or may take place between two or more molecules. Organic reactions proceed by the cleavage or breaking of covalent bonds. This breaking of a bond is known as *bond cleavage* or *bond fission* and has two types: homolysis and heterolysis.

Homolytic Bond Fission or Homolysis

The bond is broken in such a manner that the shared pair of electrons is equally divided between two separating atoms.

$$A:B \longrightarrow A\cdot + \cdot B$$

The species (A· or B) obtained after fission are neutral and known as free radicals.

Free Radical

A *free radical* is an atom or group of atoms with a single odd unpaired electron; for example, in the presence of sunlight, a chlorine molecule is dissociated to give chlorine free radicals.

$$Cl_2 \xrightarrow{\text{sunlight}} Cl. + .Cl$$

Characteristics of Free Radicals

Free radical characteristics:

- A free radical is an atom or a group of atoms in the form of a neutral species.
- A free radical always contains an unpaired electron.
- Free radicals are generally obtained as a result of homolytic bond fission by the help of irradiation with sunlight or by supplying heat, but they can also be prepared by reacting with another free radical:

$$CH_4 + Cl\cdot \longrightarrow CH_3\cdot + HCl$$

- Most of the free radicals are highly unstable, and most are reactive. The order of stability of alkyl free radicals is

$$\text{Tertiary} > \text{Secondary} > \text{Primary}$$

$$(CH_3)_3\,C\cdot > (CH_3)_2\,CH\cdot > CH_3CH_2\cdot$$

  The less the bond dissociation energy, the greater will be the stability.

- Free radicals are paramagnetic.
- A free radical is either sp^2 hybridized or sp^3 hybridized. In the former case, it is trigonal planar and flat in shape; in the latter case, it is pyramidal.

Heterolytic Bond Fission or Heterolysis

The bond is broken in a manner that the shared pair of electrons is completely retained by one of the separating atoms.

$$A:B \longrightarrow A^+ + :B^-$$

or

$$A^-: + B^+$$

The species obtained after heterolysis are charged and known as *ions*. The ionic intermediate carrying a positive charge at the carbon atom is known as a *carbonium ion*. The ionic species carrying a negative charge on the carbon atom is known as a *carbanion*.

Carbonium Ions

The cationic species containing a positive carbon is called as *carbonium ion*. The process is called as *carbocation*.

$$R-X \longrightarrow R^{\oplus} + X^{\ominus}$$

Following are the characteristics of carbonium ions:

- The positively charged carbon atom in a carbonium ion uses sp^2 hybrid orbitals to form three sigma bonds. The carbon atom in a carbonium ion has only a sextet of six electrons in the outermost shell.

$$H\cdot\times \overset{\displaystyle H\cdot\times}{\underset{\displaystyle \times\cdot H}{C}} \times\cdot\ddot{\underset{\cdot\cdot}{X}}\!: \longrightarrow H\cdot\times \overset{\displaystyle H\cdot\times}{\underset{\displaystyle \times\cdot H}{C^{\oplus}}} + {}^{\times}\ddot{\underset{\cdot\cdot}{X}}\!:^{\ominus}$$

Carbonium ion

- It does not contain any unpaired electrons.
- The carbonium ions are not very stable and are reactive. The order of stability of simple alkyl carbonium ions is

$$\text{Tertiary} > \text{Secondary} > \text{Primary}$$

The stability of carbonium ions is influenced by resonance and the inductive effect.

- They are diamagnetic because of the absence of unpaired electrons.
- The reactivity of these species follows the reverse order of their stability order.
- The carbonium ion is sp^2 hybridized and has trigonal planar geometry.

Carbanions

Certain organic compounds when treated with a strong base can function as an acid in the classical sense by donating a proton. The resulting ion containing a carbon atom with an unshared electron pair is called a *carbanion*. A carbon atom with negative charge during heterolytic cleavage is a carbanion.

$$[\text{Base}:]^{\ominus} + H:C\Big\langle \longrightarrow [\text{Base}:H] + :\overset{\ominus}{C}\Big\langle$$

The characteristics of carbanions include the following:

- A carbanion is a negatively charged species in which a carbon atom has an octet of electrons.

$$\mathrm{H\cdot\times\overset{\displaystyle H\atop\displaystyle\cdot\times}{\underset{\displaystyle\times\cdot\atop\displaystyle H}{C}}\times\cdot Mg\cdot\times\overset{\times\times}{\underset{\times\times}{Br}}\!\!\overset{\times}{\underset{\times}{}}} \longrightarrow \mathrm{H\cdot\times\overset{\displaystyle H\atop\displaystyle\cdot\times}{\underset{\displaystyle\times\cdot\atop\displaystyle H}{C}}\overset{\ominus}{\dot{\times}}\;\overset{\oplus}{+}Mg\cdot\times\overset{\times\times}{\underset{\times\times}{Br}}\!\!\overset{\times}{\underset{\times}{}}}$$

Carbanion

- The carbon atom in a carbanion is trivalent and has two extra electrons that are not bonded. A carbanion is isoelectronic with an amine, is sp^3 hybridized, and has a tetrahedral geometry. But the carbon atom of a conjugated carbanion is in a sp^2 hybridized state with planar structure.

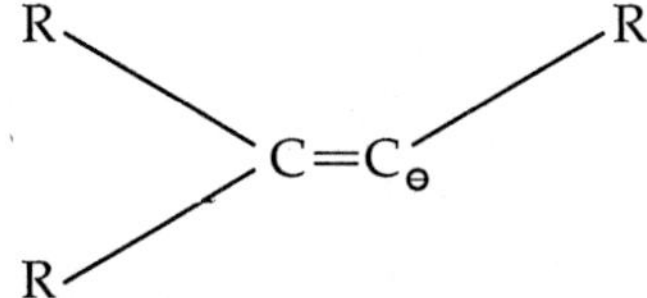

- By definition, every carbanion possesses an unshared pair of electrons and is therefore a base. When a carbanion accepts a proton, it is converted into its conjugate acid. The stability of a carbanion is directly related to the strength of the conjugate acid. The weaker the acid, the greater the base strength and the lower is stability of the carbanion. On the basis of the work of Dessy and Coworkers and the rate measurements, the order of carbanion stability is found to be

  Methyl > Primary > Secondary > Tertiary

- Carbanions are diamagnetic because they do not contain any unpaired electrons.
- Carbanions behave as nucleophiles due to the negative charge on the carbon, initiates addition and substitution reactions.

9.2 REACTIVE INTERMEDIATES

Because an organic reaction involves the homolysis, the heterolysis, or any other such process, the species such as free radicals, carbonium ions, carbanions, and so on, are frequently formed. These species are short lived, highly reactive, and often the intermediate in organic reactions. Hence, these species are generally referred to as *reactive intermediates*. There is yet another class of intermediates, the divalent carbon compounds known as *carbenes*. Carbenes are neutral carbon intermediates with two nonbonded electrons in sextet.

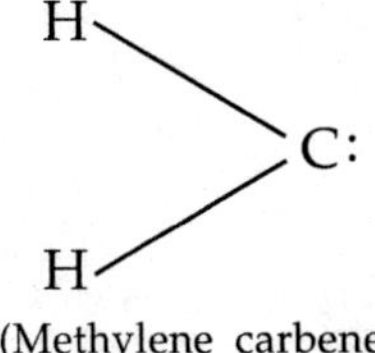

(Methylene carbene)

Carbenes do not possess any charge. They can exist in two forms: singlet (two nonbonded electrons are in the same orbital and paired) and triplet (unshared electrons are not paired). The triplet state is more stable than the singlet state. A singlet carbene may have trigonal geometry with carbon undergoing sp^2 hybridization. The paired electrons are present in one sp^2 hybrid orbital. Triplet carbenes are linear, whereas carbon undergoes sp hybridization. The two electrons in triplet state are placed in each p_y and p_z orbital, and it is linear. Carbenes are highly reactive.

9.3 ELECTRON DISPLACEMENT EFFECTS

Three effects influence the reactivity and the properties of organic compounds: an inductive effect, a resonance effect, and a steric effect.

9.4 THE INDUCTIVE EFFECT

This is the permanent displacement of electrons along a carbon chain in a compound due to the attachment of an electronegative or electropositive group at the terminal of the chain.

Consider a carbon atom chain with an electronegative atom on group X attached to its terminal:

$$C—C—C—X$$

Because the attached group is more electronegative than carbon, polarity is developed in the C—X bond. As a result, X develops a partial negative (δ^-) charge, and carbon develops a partial positive (δ^+) charge.

This carbon (C_1) attracts electrons from the neighboring carbon atom (C_2) to become neutral. This will cause C_2 to acquire a partial positive charge but of smaller extent. Similarly, C_3 will acquire a positive charge that will be still smaller.

$$\overset{\delta\delta\delta^+}{C_3} \rightarrow \overset{\delta\delta^+}{C_2} \twoheadrightarrow \overset{\delta^+}{C_1} \ggg \overset{\delta^-}{X}$$

This process of electron shift along a chain of atoms due to the presence of a polar covalent bond is called an *inductive effect*. This permanent effect becomes rapidly weaker and weaker with the length of the chain and is almost negligible beyond two carbon atoms.

The inductive effect (I effect) is of two types:

Negative inductive effect: When an electronegative atom or group (more electronegative than hydrogen) is attached to the terminal of the carbon chain in a compound, the electrons are displaced in the direction of the attached atom or group as shown previously. The groups showing – I effect (in order of decreasing – I effect) are

$$N^+(CH_3)_3 > \text{—}NO_2 > \text{—}CN > \text{—}SO_3H > \text{—}COOH > \text{—}CHO > \text{—}COR$$
$$> \text{—}F > \text{—}Cl > \text{—}Br > \text{—}I > \text{—}NH_2 > \text{—}OH > OCH_3 > C_6H_5$$

Positive inductive effect: When an electropositive atom or group (more electropositive than hydrogen) is attached to the terminal of a carbon chain in a compound, the electrons are displaced away from the attached atom or group as shown here:

$$\overset{\delta\delta\delta^-}{C_3} \leftarrow \overset{\delta\delta^-}{C_2} \twoheadleftarrow \overset{\delta^-}{C_1} \lll \overset{\delta^+}{Y}$$

The groups showing + I effect (in order of their decreasing + I effect) are

$$C_4H_9 > C_3H_7 > C_2H_5 > CH_3$$

Applications

Inductive effect is used to explain the strength of some organic acids and bases and also to explain the mechanism of organic reactions. Some of its applications are mentioned here:

Chloroacetic acid is stronger than acetic acid: The – I effect of the chlorine atom displaces the electrons toward itself in the chloroacetic acid chain. Due to this, the electrons of the O—H bond come closer to the O-atom, thereby facilitating the removal of the H-atom as H^+ ion. Similarly, dichloroacetic acid is stronger because of the – I effect of the two chlorine atoms.

Methyl acetic acid is weaker than acetic acid: In methyl acetic acid, the + I effect of the methyl group displaces the electrons in the chain in a direction away from it. Due to this, the electrons in the O—H bond are displaced slightly toward the H-atom, making its removal in the form of H^+ ion more difficult. As a result, the tendency to furnish H^+ ions is lessened in methyl acetic acid, making it weaker in comparison to acetic acid. Dimethyl acetic acid is further weakened because of the + I effect of two methyl groups, and trimethyl acetic acid is the weakest.

Methylamine is a stronger base than ammonia: Ammonia is basic in nature because of its tendency to donate the pair of electrons. The more the tendency to donate electrons, the stronger is a base. In methylamine, the + I effect of the methyl group pushes electrons toward nitrogen, which makes more electrons available at nitrogen for donation and makes it a stronger base in comparison to ammonia.

Chloramine is a weaker base than ammonia: Cl^- atom has the – I effect.

9.5 RESONANCE

The concept of resonance was introduced by Heisenberg through quantum mechanics in the 1920s and generalized by Pauling in the early 1930s. The usual representation of an organic molecule is a very rough attempt at picturing their electronic distribution in the molecule. We know that electrons in a molecule are moving continuously at a speed of about 2×10^6 mile/hr in an orbit that has a diameter of 10^{-8} in cm. So, it is clear that the representation of a molecule by its formula is not quite accurate. Thus, the electronic distribution given for a molecule is just a rough factor. The possibility for various structures in which polycentric and dicentric molecular orbitals and atomic orbitals are occupied gives rise to the phenomenon called *resonance*. According to resonance theory, the real structure of a molecule is the composite of several extreme structures (probable). Further note that each covalent bond has a certain amount of partial ionic character. Different electronic structures may be written down for every organic molecule. Now the question arises as to which form will contribute significantly toward the formation of the hybrid. For example, an ethane molecule might be considered to have the structures (I), (II), and (III), in which the positions of the atomic Kernels in the different structures are assumed to be very nearly the same:

```
    H H             H H              H H
    | |          δ-  | | δ+        δ+ δ-| |
  H:C:C:H         H:C:C:H          H:C:C:H
    | |             | |              | |
    H H             H H              H H
    (I)             (II)             (III)
```

The different structures for the ethane molecule have been written down only by the distribution of electrons. The properties of ethane are explained by considering only formula (I). But the structures (II) and (III) are considerably important in the discussion of absorption spectra. So resonance theory states that the properties of ethane should not correspond to those expected of any one of these structures but rather should be those expected of a hybrid of these structures. Consider the example of carbonyl groups as shown here:

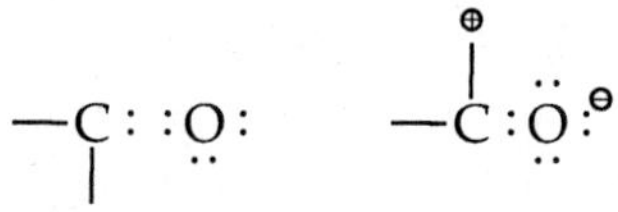

—C : : O: $\overset{\oplus}{—C}$: $\ddot{O}:^{\ominus}$

(I) (II)

The dipole moment study reveals that a 47 percent contribution of form II and the deviation of the actual properties of the carbonyl group from those to be expected of form I (or II) is large.

Evidence for persistence of a single structure of the aromatic nucleus gives the idea conveyed in the concept of resonance, rather than the dynamic structure proposed by Kekule comes from data gathered in the twentieth century. The most important structure is Kekule's structure, which opens a new gateway in realising the theory of resonance.

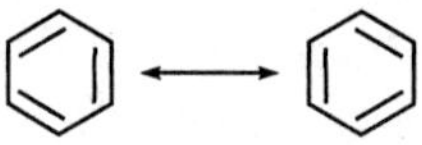

(I) ⟷ (II)

Electron and X-ray diffraction data reveal that the C—C bond distances of the aromatic ring are all the same—intermediate between a single bond and double bond distance. Also from the Raman spectral measurements, it is found that the nuclear C—C bond force constants (*i.e.*, the force required to stretch a bond a unit distance) are all the same. Again it is the value between the values characteristic of single bond and double bonds, that is the aromatic nucleus has a structure that can be called a composite of the two Kekule structures, but neither one nor both. The historic oscillation formula was suggested to explain the formation of a single orthoisomer, which is experimentally feasible. But afterwards, it was realized that these two oxidation formulas are written down simply by delocalizing the electrons over the skeleton of a regular hexagon.

Kekule's structure explains about 90 percent of the properties of benzene. However, the other forms are considered because those structures also have contributions and explain some properties of benzene. Essentially the need for the concept of resonance arises from the fact that conventional valence bond structures are inadequate to describe certain molecules.

(I) ⟷ (II) ⟷ (III) ⟷ (IV) ⟷ (V)

The linear combination of the five functions of structures is called the *resonance hybrid* of all these five structures. This hybrid cannot be represented by a single structure or the assume to be a composite of the different contributing structures or canonical structures or resonating structures. Thus, resonance may be defined as the circumstance existing whenever it is possible to write two or more reasonable electron configurations for a molecule by merely altering the position of the electrons over the same atomic skeleton. The properties of that molecule will not be those expected of any of these structures, but rather they will be expected to be a hybrid of all. A molecule or ion in which resonance can occur will always be more stable than one in which it cannot occur. The stability of the resonance hybrid is greater than that of any individual canonical structures. Resonance is sometimes referred to as mesomerism.

Resonance Energy

The resonance energy can be defined as the difference in the experimental heat of the formation of a compound, which is calculated from tabulated bond energies for its most stable valence bond structures. This is the difference in energy between the energy of the most stable of the contributing structures and the energy of the actual molecule (resonance hybrid). The larger the resonance energy, the more stable the compound.

The resonance energy is calculated by the physical measurement, such as heat of combustion, heat of hydrogenation, and so on.

The method of determining resonance energies by heat of hydrogenation was introduced by Kistiakowsky. For example, the determination of the resonance energy of benzene:

The heat of hydrogenation of benzene is 49.8 K cal/mole

$$C_6H_6 + 3H_2 \longrightarrow C_6H_{12} + 49.8 \text{ K cal/mole}$$

In case of cyclohexene, the heat of hydrogenation is 28.8 K cal/mole.

$$C_6H_{10} + H_2 \longrightarrow C_6H_{12} + 28.8 \text{ K cal/mole}$$

Benzene is regarded as cyclohexatriene. The total value of heat of hydrogenation will be

$$28.8 \times 3 = 86.4 \text{ K cal/mole}$$

The difference of heat of hydrogenation:

$$86.4 - 49.8 = 36.6 \text{ K cal/mole}$$

It is called the resonance energy or the delocalized energy of benzene.

Applications

Acidity of phenol: Phenol is a weak acid. It is weaker than the acetic acid and formic acid.

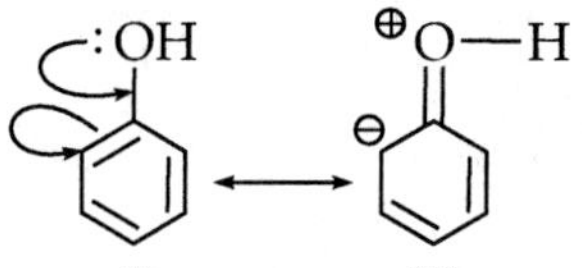

(I) (II)

The fractional positive charge is developed on an oxygen atom due to the resonance or the delocalization of a lone pair of electrons of the oxygen atom with the benzene ring. This fractional positive charge facilitates the dissociation of a proton, thereby imparting acidity to phenol. Thus, phenol is acidic in nature. However, phenol is not a very strong acid.

Basicity of amine: Aliphatic amines are stronger bases than aromatic amines. Thus, methylamine is a stronger base than aniline.

$$R\text{—}NH_2 + H_2O \rightleftharpoons RNH_3^+ + OH^- \quad K_b \simeq 10^{-5}$$

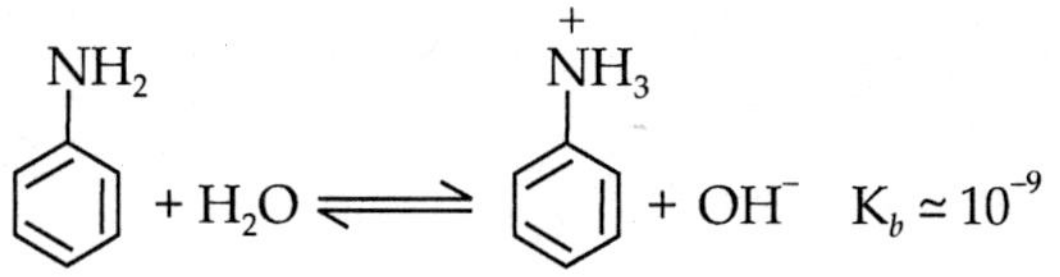

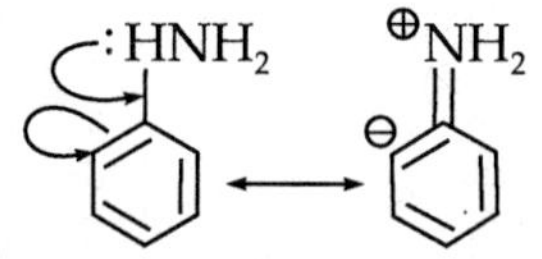

$$C_6H_5NH_2 + H_2O \rightleftharpoons C_6H_5\overset{+}{N}H_3 + OH^- \quad K_b \simeq 10^{-9}$$

(I) (II)

Resonance occurs in an aromatic amine due to the conjugation of a lone pair electrons on the seat of the hydrogen atom in the benzene ring. Due to this delocalization of electrons, the electron is not available for combination with a proton. Therefore, aromatic amines are weaker bases than aliphatic amines.

9.6 THE STERIC EFFECT

The presence of bulky groups makes the molecule behave in a different manner, which is called the *steric effect*. The steric effect is very important as a side effect in inhibiting resonance stabilization. The effect mostly occurs in aromatic bases because the resonance stabilization of an aromatic base requires the involved groups to become coplanar with the rest of the conjugated system.

A few examples of steric effect are cited here:

- N, N dimethyl aniline is six times weaker than a base.

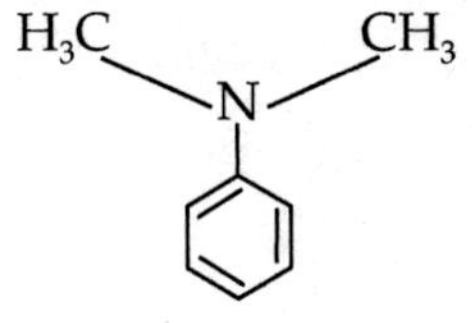

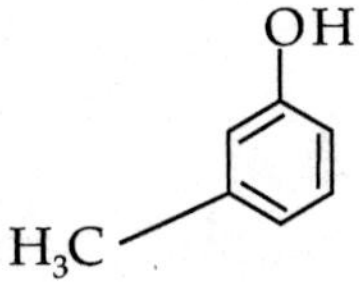

(Phenol) 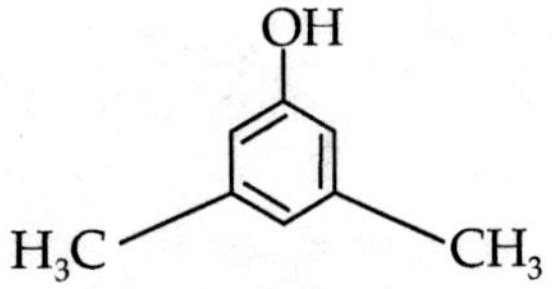 (3, 5 dimethyl phenol)

The acidity of phenol is due to the delocalization of a lone pair of electrons on the oxygen atom. The more delocalization of electrons, the more stabilization occurs due to resonance, which causes more pronounced acidity of phenol or a phenolic compound. By introducing two bulky groups at 3, 5 positions 3, 5 dimethyl phenol is obtained. Those two bulky groups offer steric hinderance, destroy coplanarity, and thereby inhibit the delocalization of electrons and resonance. Thus, the presence of two —CH_3 groups at 3, 5 positions decreases the acidity of phenol. In some cases, the rate of reaction decreases. Such a hinderance due to spatial crowding is called *steric hinderance*. It should be noted that all steric effects do not decrease the reaction rates.

9.7 HYPER CONJUGATION

Hyper conjugation is a particular kind of resonance called no bond resonance, which involves a hydrogen atom alpha to a double bond or an electronically deficient atom. This arises because the α-hydrogen atom gets activated when attached to the carbon atom adjacent to a carbonyl group, a double bond between carbon atoms, or an electron deficient atom.

Hyper conjugation is also called the Baker-Nathan effect (referring to the scientists who first proposed this phenomenon to explain the properties); the name *hyper conjugation* was given by Mulliken.

To understand this effect, consider the following canonical structures of the ethyl carbonium ion:

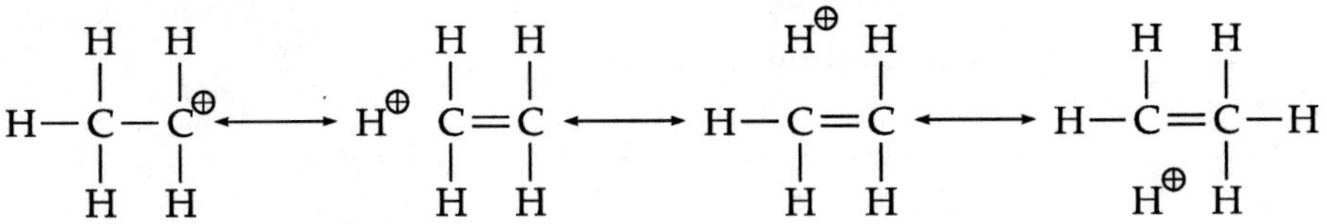

In each of these structures, the positive charge on the carbon atom has been displaced after splitting a C—H bond and making a C = C bond. The hydrogen atom carries the positive charge, but there is no bond between the hydrogen and carbon atom. Where Markonikoff's rule fails, the hyper conjugation is used.

$$CH_3-CH=CH_2+HBr \longrightarrow CH_3CHBrCH_3.$$

9.8 TYPES OF ORGANIC REACTIONS

Most organic reactions involve the conversion of one functional group into another by the attack of a reagent. Such organic compounds that are attacked by a reagent are known as a *substrate* or a *reactant.* An organic reaction may be represented as below:

$$\text{Substrate} + \text{Attacking agent} \longrightarrow [\text{Intermediate}] \rightarrow \text{Product}$$

It is convenient to classify the numerous reactions belonging to organic chemistry into the following four types:

- Substitution or displacement reactions
- Addition reactions
- Elimination reactions
- Rearrangements

Substitution or Displacement Reaction

This is the most important type of organic reaction. In a substitution or displacement reaction, an atom or group present in a compound is replaced by another atom or group without undergoing any change in the structure of the molecule or compound. The products thus obtained are known as substitution products. These reactions may involve free radicals, nucleophiles, or electrophiles as reactive intermediates. Hence, substitution reactions may be of three types depending upon the reactive species involved: free radical substitution, nucleophilic substitution, and electrophilic substitution.

Free Radical Substitution

Substitution reactions that proceed through free radical formation are known as *free radical substitution,* for example, halogenation of alkanes in the presence of light or at a temperature of 300–400 °C.

$$CH_4+Cl_2 \xrightarrow[\text{Light}]{\text{U.V.}} HCl+\underset{\text{Methyl chloride}}{CH_3Cl}$$

$$CH_3Cl + Cl_2 \xrightarrow[\text{Light}]{\text{U.V.}} HCl + \underset{\text{Methylene chloride}}{CH_2Cl_2}$$

$$CH_2Cl_2 + Cl_2 \xrightarrow[\text{Light}]{\text{U.V.}} HCl + \underset{\text{Chloroform}}{CHCl_3}$$

$$CHCl_3 + Cl_2 \xrightarrow[\text{Light}]{\text{U.V.}} HCl + \underset{\text{Carbon-tetrachloride}}{CCl_4}$$

The mechanism of a free radical substitution involves the following steps:

Step I. Chain initiation: Light energy or heat splits up the chlorine molecule into two free chlorine radicals as a result of homolytic fission.

$$Cl:Cl \xrightarrow{\text{UV light}} \underset{\text{Chlorine free radicals}}{\bullet Cl + \bullet Cl}$$

This homolytic fission is an endothermic process and requires 58 k cal/mole of energy. Hence, the free radicals produced are energy-rich chemical species.

Step II. Chain propagation: The chlorine free radical attacks the methane and abstracts the hydrogen atom from the CH_4 molecule to produce a methyl free radicals ($\bullet CH_3$) and HCl.

$$CH_4 + \bullet Cl \longrightarrow \underset{\text{Methyl free radical}}{\bullet CH_3 + HCl}$$

The methyl free radical then attacks the chlorine molecule to form an alkyl halide and a free chlorine atom.

$$\bullet CH_3 + Cl_2 \longrightarrow CH_3Cl + \bullet Cl$$

The chlorine free radical thus formed reacts again and again to create a chain reaction. Propagation steps are exothermic, so they release energy and are repeated several times.

Step III. Chain termination: The propagation of the chain continues until all the chlorine or methane is used up or until the process is terminated by any of the following reactions:

$$\bullet Cl + \bullet Cl \longrightarrow Cl_2$$

$$\bullet CH_3 + \bullet Cl \longrightarrow CH_3{-}Cl$$

$$\bullet CH_3 + \bullet CH_3 \longrightarrow CH_3{-}CH_3$$

Nucleophilic Substitution (S_N)

When a substitution reaction is brought about by a nucleophile, the reaction is called *nucleophilic substitution* (S_N), for example, the hydrolysis of alkyl halides by aqueous alkalis to furnish alcohols:

$$R-X+\bar{O}H \longrightarrow R—OH+\bar{X}$$

The nucleophilic substitution, reactions may take place mainly by two mechanisms—unimolecular, and bimolecular—which can be determined by the kinetic studies (In gold).

Unimolecular mechanism (S_{N^1}): When the rate of a nucleophilic substitution reaction is dependent only upon the concentration of the alkyl halide and is independent of the concentration of the nucleophile (*i.e.*, $\bar{O}H$, etc.), the reaction is obviously of first order and is designated as S_{N^1}. It is a two-step process, out of which first is the rate-determining step and involves the heterolysis of the compound into two ions. The cation so called carbonium ion combines with the nucleophilic reagent to form the final substituted product.

Step I: $$A—X \; S_{N^1} \; A^+ + X^-$$

Step II: $$\bar{N}u: + A^+ \xrightarrow{\text{Fast}} Nu—A$$

The rate-determining step is the first step. Because the molar concentration of one reactant (A—X) is changed, so the overall reaction is of the first order, so the mechanism is known as unimolecular and designated as S_{N^1} (substitution nucleophilic unimolecular), for example, the nucleophilic substitution of tert-butyl bromide.

$$CH_3—\underset{CH_3}{\overset{CH_3}{\underset{|}{\overset{|}{C}}}}—Br \xrightarrow{\text{Slow}} CH_3—\underset{CH_3}{\overset{CH_3}{\underset{|}{\overset{|}{C^+}}}} + Br^- \xrightarrow[\substack{H_2O \\ [-H^+]}]{\text{Fast}} CH_3—\underset{CH_3}{\overset{CH_3}{\underset{|}{\overset{|}{C}}}}—OH + Br^-$$

Carbonium ion

Bimolecular mechanism (S_{N^2}): When the rate of the nucleophilic substitution reaction is dependent both upon the concentration of the substrate and the nucleophile, the reaction is of second order, and the mechanism is known as bimolecular and designated as S_{N^2} (substitution nucleophilic bimolecular). The bimolecular mechanism is a one-step process that involves simultaneous bond breaking and bond making.

$$\bar{N}u: + A—X \longrightarrow \left[\overset{\delta+}{Nu} \ldots A \ldots \overset{\delta-}{X}\right] \longrightarrow Nu—A + \bar{X}:$$

(Nucleophile) (Transition state)

In an S_{N^2} reaction, the nucleophile attacks a carbon atom from the side opposite to that of the leaving group. The attacking nucleophile $\bar{N}u$: attacks at carbon in substrate A—X, forming a high-energy transition state in which the Nu-carbon bond is in the process of forming, and the X—carbon bond is in the process of breaking with a simultaneous secession of the leaving group $\bar{X}$:, for example, hydrolysis of alkyl halide by aqueous alkali.

$$HO^- + H-\underset{H\ \ H}{C}-Br \xrightarrow{S_N2} \left[\overset{\delta^+}{HO}\cdots\overset{H}{\underset{H\ \ H}{C}}\cdots\overset{\delta^-}{Br} \right]^+ \xrightarrow[-Br^-]{} HO-\overset{H}{\underset{H\ \ H}{C}}$$

Transition state

Electrophilic Substitution (S_E)

When a substitution reaction involves an attack by an electrophile, the reaction is referred to as an S_E (substitution electrophilic), for example, nitration of benzene with conc. HNO_3 and conc. H_2SO_4.

$$C_6H_6 + \overset{\oplus}{N}O_2 \xrightarrow[H_2SO_4]{HNO_3} C_6H_5NO_2 + H^+$$

The general mechanism of an electrophilic substitution reaction includes the following steps:

Step I: Formation of an electrophile (nitronium ion) from HNO_3:

$$HO-NO_2 + 2H_2SO_4 \longrightarrow H_3O^+ + 2HSO_4^{\ominus} + \overset{\oplus}{N}O_2$$

Nitric acid Sulphuric acid Nitronium ion (Electrophile)

Step II: Benzene (a nucleophile, *i.e.*, an electron rich molecule) is attacked by a nitronium ion forming a carbocation, which is stabilized by resonance:

$$C_6H_6 + \overset{\oplus}{N}O_2 \xrightarrow{Slow} \overset{\oplus}{C_6H_5}\begin{smallmatrix}H\\NO_2\end{smallmatrix} \quad \text{or} \quad C_6H_5^{+}(H)(NO_2)$$

Carbocation (resonance stabilized)

Step III: The carbocation then loses a proton to yield nitrobenzene:

$$C_6H_5^{+}(H)(NO_2) \xrightarrow{HSO_4^-} C_6H_5NO_2 + H_2SO_4$$

Nitrobenzene

Another example of electrophilic substitution is the sulphonation of benzene by conc. H_2SO_4 to yield benzene sulphonic acid:

$$C_6H_6 + H_2SO_4 \longrightarrow C_6H_5SO_3H + H_2O$$

The mechanism of the sulphonation involves the following steps:

Step I: Formation of electrophile:

$$2H_2SO_4 \longrightarrow H_3O^{\oplus} + HSO_4^{\ominus} + \underset{\text{(electrophile)}}{SO_3}$$

Step II: Benzene (a nucleophile, *i.e.*, an electron rich molecule) is attacked by SO_3 forming a carbocation, which is resonance stabilized.

$$C_6H_6 + SO_3 \xrightarrow{\text{Slow}} \underset{\substack{\text{Carbocation}\\\text{(resonance stabilized)}}}{C_6H_6^{+}(H)SO_2}$$

Step III: The carbocation then loses a proton to form the anion of benzene sulphonic acid:

$$C_6H_6^{+}(H)SO_3 \xrightarrow[\text{Fast}]{HSO_3^{\ominus}} C_6H_5SO_3^{\ominus} + H_2SO_4$$

Step IV: The anion of benzene sulphonic acid takes a proton to yield benzene sulphonic acid:

$$C_6H_5SO_3^{\ominus} + H_3O^{\oplus} \longrightarrow C_6H_5SO_3H + H_2O$$

Nucleophilic Aromatic Substitution

The replacement of hydrogen or a substituent by a nucleophilic reagent is known as *nucleophilic aromatic substitution*. It is important to note that nucleophilic substitution does not take place with the benzene itself but with its substituted derivatives and with naphthalene. The unreactivity of the benzene toward the nucleophilic reagents is partly due to the concentration of a negative charge above and below the plane of the ring carbon atoms and partly due to the formation of the less stable hydride ion.

$$C_6H_6 + X^- \longrightarrow C_6H_5X + H^-$$

There are three mechanisms under which all the nucleophilic aromatic substitution reactions fall: unimolecular, bimolecular, and the benzyne type.

Unimolecular substitution: The only example of unimolecular nucleophilic aromatic substitution is the uncatalyzed decomposition of the aryl diazonium cation.

$$C_6H_5 . \overset{+}{N_2} \xrightarrow{\text{Slow}} N_2 + \overset{+}{C}_6H_5 \xrightarrow[\text{Fast}]{X^-} C_6H_5 . X$$

where $\bar{X} = \bar{O}H, \bar{O}R, \bar{C}l, C\bar{N}$ and $N\bar{O}_2$

Bimolecular substitution: The general bimolecular mechanism may be represent as

$$\bar{Y} + C_6H_5Z \xrightarrow{\text{Slow}} [C_6H_5(Y)(Z)]^- \xrightarrow{\text{Fast}} C_6H_5Y + Z^-$$

The conversion of chlorobenzene into phenol:

$$C_6H_5Cl + \bar{O}H \longrightarrow [C_6H_5(Cl)(OH)]^- \xrightarrow{\text{Fast}} C_6H_5OH + Cl^-$$

Benzyne mechanism: This mechanism of the reaction has been a subject of great study, which led to the conclusion that it proceeds by an elimination of hydrogen halide (HX) from the aryl halide to give a highly reactive intermediate with a triple bond. The intermediate, known as benzyne or dehydrobenzene, now adds the nucleophilic species to form the product, for example, hydrolysis of chlorobenzene:

$$C_6H_5Cl + :\bar{N}H_2 \longrightarrow C_6H_4 + NH_3 + Cl^-$$

Benzyne

$$C_6H_4 + NH_3(l) \longrightarrow C_6H_5NH_2$$

Aniline

Addition Reaction

A reaction in which two molecules combine to yield a single molecule of the product is called an addition reaction. Addition reactions may be initiated by electrophiles, nucleophiles, or free radicals. Hence, depending upon the mechanism, addition reactions are of three types: free radical, nucleophilic, and electrophilic.

Free Radical Addition Reaction: These reactions are brought about by free radicals, for example, the addition of HBr to propene in the presence of peroxide.

$$CH_2 = CH - CH_3 + HBr \xrightarrow{\text{Peroxide}} Br—CH_2—CH_2—CH_3$$
$$(n\text{-propyl bromide})$$

The mechanism involves the following steps:

- **Step I. Initiation:** Peroxide dissociates to give alkoxy free radicals.

$$\underset{\text{Peroxide}}{RO : OR} \longrightarrow \underset{\text{Alkoxy free radical}}{2R - O^{\bullet}} \quad R—O^{\bullet} + HBr \longrightarrow R—OH + {}^{\bullet}Br$$

- **Step II. Propagation:**

$$CH_2 = CH—CH_3 + {}^{\bullet}Br \longrightarrow CH_3—\dot{C}H—CH_2Br$$
$$\text{Secondary free radical (more stable)}$$

- **Step III. Termination:**

$$CH_3—\dot{C}H—CH_2Br + HBr \longrightarrow CH_3—CH_2—CH_2Br + {}^{\bullet}Br$$
$$Br^{\bullet} + {}^{\bullet}Br \longrightarrow Br_2$$

Nucleophilic addition reaction: These reactions are brought about by nucleophiles. Aldehydes and ketones undergo nucleophilic addition reactions, for example, the addition of base catalyzed HCN to an aldehyde or ketone:

$$\underset{\text{Base}}{{}^{-}OH} + HCN \longrightarrow H_2O + \underset{\text{(Nucleophile)}}{CN^{-}}$$

$$>\overset{\delta^+}{C} = \overset{\delta^-}{O} + CN^{-} \xrightarrow{\text{Slow}} \underset{\text{(Nucleophile)}}{>\underset{CN}{\underset{|}{C}}-O^{-}} \xrightarrow{H^+} \underset{\text{(Cyanohydrin)}}{\underset{CN}{\underset{|}{C}}-OH}$$

Electrophilic addition reaction: These reactions are brought about by electrophiles. Such reactions are common in compounds containing carbon-carbon double and triple bonds. For example:

1. $\underset{\text{Ethylene}}{CH_2 = CH_2} + H_2 \xrightarrow{Ni} \underset{\text{Ethane}}{CH_3 - CH_3}$

2. $\underset{\text{Ethylene}}{CH_2 = CH_2} + HBr \longrightarrow \underset{\text{Ethyl bromide}}{CH_3—CH_2Br}$

A carbon-carbon double bond is made up of a strong σ bond, in which the electron distribution is concentrated along the line joining the nuclei of the two atoms, and a weaker π bond, in which the electron pair lies in an orbital above and below the plane of the two nuclei. The pair of electrons in the π orbital are less firmly held between the carbon nuclei and are more readily polarizable than those of the σ bond. The π orbital is readily accessible and attracts electron-deficient (electrophilic) species; it also tends to repel electron-rich (nucleophilic) species. Thus, due to the presence of π orbital, olefines (carbon-carbon double bonds) react with electrophilic reagents (X^+).

Elimination Reaction

Elimination reactions are formally the reverse of addition reactions and remove atoms or groups from the organic compound under suitable conditions to give an unsaturated molecule:

$$CH_3CH_2Br + KOH\,(alc.) \longrightarrow CH_2{=}CH_2 + H_2O + KBr$$

Elimination reactions can be classified into two categories: β-elimination and α-elimination.

- The *β-elimination reaction* involves the loss of two atoms or groups from the adjacent carbon atoms of the molecule, for example, base catalyzed dehydrohalogenation of propyl bromide.

$$\underset{\boxed{\quad\ H\qquad Br\ }}{CH_3-\overset{\beta}{C}H-\overset{\alpha}{C}H_2} \xrightarrow[\Delta]{KOH\,(alc.)} \underset{\text{Propene}}{CH_3\,.\,CH=CH_2} + KBr + H_2O$$

The reaction may proceed either by a bimolecular (E_2) or by an unimolecular (E_1) mechanism.

Bimolecular mechanism (E_2): This single-step mechanism consists of an attack of a nucleophilic base on the β-carbon atom followed by a simultaneous loss of a halide ion from the α-carbon atom.

$$CH_3-\underset{\displaystyle H}{\underset{|}{C}H}-CH_2-Br + \bar{O}H \longrightarrow \underset{\text{Transition state}}{CH_3-\underset{H\cdots\overset{\delta^-}{O}H}{\underset{|}{C}H}-CH_2\cdots\overset{\delta^-}{Br}}$$

$$\longrightarrow H_2O + CH_3-CH=CH_2 + Br^-$$

Unimolecular mechanism (E_1): The rate of certain elimination reactions is independent of the concentration of the base. Such reactions occur in two stages and involve the formation of a carbonium ion:

$$CH_3-\underset{\displaystyle H}{\underset{|}{C}H}-CH_2-Br \xrightarrow{\text{Slow}} CH_3-\underset{\displaystyle H}{\underset{|}{C}H}-\overset{\oplus}{C}H_2 + Br^-$$

$$CH_3-\underset{\substack{|\\H}}{CH}-\overset{\oplus}{C}H_2 \xrightarrow[\text{Fast}]{\bar{O}H} CH_3-CH=CH_2$$

- *α-elimination reaction* involves the loss of two atoms or groups from the same carbon atom of the molecule, for example, base catalyzed dehydrohalogenation of chloroform to form dichlorocarbene:

$$H\bar{O} + H-CCl_2-Cl \longrightarrow :CCl_2 + H_2O + Cl^-$$

Chloroform — Dichlorocarbene

Dichlorocarbene is the reactive intermediate involved in the carbylamine reaction and the Reimer-Tiemann reaction.

The reaction occurs in two stages:

$$CHCl_3 + \bar{O}H \underset{}{\overset{\text{Fast}}{\rightleftharpoons}} H_2O + \bar{C}Cl_3 \xrightarrow{\text{Slow}} :CCl_2 + Cl^-$$

Rearrangement Reaction

In these type of reactions, some atoms/groups shift from one position to another within the substrate molecule itself, giving a product with a new structure. The reactions that proceed by a rearrangement or reshuffling of the atoms/groups in the molecule to produce a structural isomer of the original substance are called *rearrangement reactions.*

$$NH_4CNO \xrightarrow{\Delta} NH_2-\overset{\overset{O}{||}}{C}-NH_2$$

Ammonium cyanate — Urea

$$CH_3-\underset{\substack{|\\CH_3}}{\overset{\substack{H\\|}}{C}}-\overset{\oplus}{C}H_2 \xrightarrow{H^- \text{Shift}} CH_3-\underset{\substack{|\\CH_3}}{\overset{\oplus}{C}}-CH_3$$

1° Carbocation (Less stable) — 3° Carbocation (More stable)

Rearrangement reactions may proceed either by an intramolecular or intermolecular change:

Intramolecular rearrangement: The rearrangements in which the migrating group is never fully detached from the system during the process of migration, are called intramolecular rearrangements. For example, butane changes to isobutane on heating in the presence of aluminium chloride:

$$\underset{\text{Butane}}{CH_3-CH_2-CH_2-CH_3} \xrightarrow[\Delta]{AlCl_3} \underset{\text{Isobutane}}{CH_3-\overset{\overset{CH_3}{|}}{C}H-CH_3}$$

Intermolecular rearrangement: The rearrangements in which the migrating group gets completely detached and is later reattached are called intermolecular rearrangements.

$$\underset{\text{N-chloroacetanilide}}{CH_3-CO-\overset{\overset{Cl}{|}}{N}-C_6H_5} \xrightarrow[\text{dil.HCl}]{\text{Warm}} \underset{p\text{-chloroacetanilide}}{Cl-C_6H_4-NH-COCH_3} + \underset{o\text{-chloroacetanilide}}{C_6H_4(Cl)-NH-COCH_3}$$

9.9 THE BECKMANN REARRANGEMENT

The Beckmann rearrangement consists of the conversion of ketoximes to N substituted amides by heating with some acidic reagents such as H_2SO_4, BF_3, P_2O_5, PCl_5, SO_3, $SOCl_2$, and so on.

$$\begin{matrix} R \\ R' \end{matrix}\!\!>C=NOH \xrightarrow{H_2SO_4} R'CONHR \text{ or } RCONHR'$$

The two different amides thus formed, can be identified by their hydrolysis to different acids and amines. Following is the mechanism:

$$R-\underset{\underset{N-OH}{\|}}{C}-R' \xrightarrow{H^+} R-\underset{\underset{N-\overset{\oplus}{O}H_2}{\|}}{C}-R' \longrightarrow R-\underset{\underset{\overset{\oplus}{N}:}{\|}}{C}-R'$$

$$\longrightarrow R-\overset{\oplus}{C}=\ddot{N}-R' \longleftrightarrow R-C\equiv\overset{\oplus}{N}-R' \xrightarrow{H_2\ddot{O}^{18}}$$

$$\begin{matrix} R \\ H_2\overset{\oplus}{O}^{18} \end{matrix}\!\!>C=N-R' \xrightarrow{-H^{\oplus}} \begin{matrix} R \\ H\ddot{O}^{18} \end{matrix}\!\!>C=N-R' \longrightarrow \begin{matrix} R \\ O \end{matrix}\!\!>C-N<\begin{matrix} H \\ R' \end{matrix}$$

9.10 THE CANNIZZARO REACTION

The disproportionation of aldehydes lacking α-hydrogen atoms (*e.g.*, HCHO, C_6H_5CHO, $R_3C.CHO$, etc.) in the presence of a strong base to form salt of an acid and a corresponding primary alcohol is known as the *Cannizzaro reaction*. The reaction best proceeds with the aromatic aldehydes:

$$\underset{\text{Formaldehyde}}{2HCHO} \xrightarrow{NaOH} \underset{\text{Methyl alcohol}}{CH_3OH} + \underset{\text{Sodium formate}}{HCOONa}$$

$$\underset{\text{Benzaldehyde}}{2C_6H_5CHO} \xrightarrow{NaOH} \underset{\text{Benzyl alcohol}}{C_6H_5CH_2OH} + \underset{\text{Sodium benzoate}}{C_6H_5COONa}$$

Following is the mechanism:

$$\underset{\text{Benzaldehyde}}{C_6H_5-\overset{}{\underset{H}{C}}=O} \underset{}{\overset{OH^-}{\rightleftharpoons}} C_6H_5-\overset{OH}{\underset{H}{C}}-O^- + \overset{O}{\underset{H}{\overset{\|}{C}}}-C_6H_5 \xrightarrow{\text{Slow}}$$

$$\underset{\text{Benzoic acid}}{C_6H_5-\overset{OH}{C}=O} + \underset{\text{Alkoxide ion}}{H-\overset{O^-}{\underset{H}{C}}-C_6H_5} \overset{\text{Fast}}{\rightleftharpoons} \underset{\text{Benzoate ion}}{C_6H_5-\overset{O^-}{C}=O} + \underset{\text{Benzyl alcohol}}{C_6H_5-\overset{H}{\underset{H}{C}}-OH}$$

In some cases, for example with HCHO, a very high concentration of alkali first forms a doubly charged anion that then transfers its hydride ion to a second molecule of aldehyde to yield acid and alkoxide anions:

$$H-\underset{\|\,O}{C}-H \overset{OH^-}{\rightleftharpoons} H-\overset{OH}{\underset{O^-}{C}}-H \overset{OH^-}{\rightleftharpoons} H-\overset{O^-}{\underset{O^-}{C}}-H + H_2O$$

$$H-\overset{O^-}{\underset{O^-}{C}}-H + \overset{H}{\underset{H}{C}}=O \longrightarrow \underset{\text{Formate ion}}{H-C\langle^{O}_{O^-}} + \underset{\text{Methoxide ion}}{H-\overset{H}{\underset{H}{C}}-O^-}$$

Crossed Cannizzaro Reaction

The Cannizzaro reaction taking place between two different aldehydes is known as a *crossed Cannizzaro reaction.*

$$C_6H_5CHO + HCHO \xrightarrow{NaOH} C_6H_5CH_2.COOH + HCOONa$$

Cannizzaro reaction sometimes fails due to sterichinderance, for example, the di-o-substituted benzaldehydes fail to react normally.

9.11 THE HOFMANN REARRANGEMENT

The conversion of an unsubstituted amide to a primary amine with one carbon atom less than the starting amide by the action of sodium hypobromite (bromine in alkali) is known as, the *Hofmann reaction* or *Hofmann rearrangement*. The overall reaction for this conversion may be represented as:

$$RCONH_2 + Br + 4NaOH \longrightarrow RNH_2 + 2NaBr + Na_2CO_3 + 2H_2O$$

where, R may be the alkyl or aryl group.

The mechanism is

$$\underset{\text{Amide}}{R-\overset{\displaystyle O}{\overset{\|}{C}}-NH_2} \xrightarrow{Br_2} \underset{\text{N-bromamide}}{R-\overset{\displaystyle O}{\overset{\|}{C}}-NHBr} \xrightarrow[(-H_2O)]{\overset{-}{O}H} \underset{\text{anion of N-bromamide}}{R-\overset{\displaystyle O}{\overset{\|}{C}}-\ddot{\overset{-}{N}}-Br} \xrightarrow{-\overset{-}{B}r}$$

$$R-\overset{\displaystyle O}{\overset{\|}{C}}-\ddot{N} \longrightarrow \underset{\text{Isocyanate}}{\overset{\displaystyle O}{\overset{\|}{C}}=\ddot{N}-R} \xrightarrow{H_2O} \underset{\text{Carbanic acid}}{HO-\overset{\displaystyle O}{\overset{\|}{C}}-\underset{H}{N}-R} \longrightarrow \underset{1^\circ\text{ Amine}}{R-NH_2} + CO_2$$

The Hofmann rearrangement involves the migration of an alkyl or aryl group with its bonding pair of electrons from the carbon to the adjacent nitrogen atoms.

9.12 THE PINACOL-PINACOLONE REARRANGEMENT

The conversion of pinacols (1, 2-glycols) to ketones or aldehydes by means of acids is known as the *pinacol-pinacolone rearrangement*.

$$CH_3-\underset{OH}{\overset{CH_3}{C}}-\underset{OH}{\overset{CH_3}{C}}-CH_3 \xrightarrow[30\%\ H_2SO_4]{\Delta} CH_3-\overset{O}{\overset{||}{C}}-\underset{CH_3}{\overset{CH_3}{C}}-CH_3$$

2, 3-dimethyl-2, 3-butanediol (Pinacol) — Methyl t-butyl ketone (Pinacolone)

The pinacol-pinacolone rearrangement is found to be a general acid-catalyzed reaction for α-glycols, and its mechanism is believed to follow the following carbonium ion path:

$$CH_3-\underset{OH}{\overset{CH_3}{C}}-\underset{OH}{\overset{CH_3}{C}}-CH_3 \overset{H^+}{\rightleftharpoons} CH_3-\underset{OH}{\overset{CH_3}{C}}-\underset{\overset{\oplus}{O}H_2}{\overset{CH_3}{C}}-CH_3 \xrightarrow{-H_2O}$$

Pinacol — Oxonium ion

$$CH_3-\underset{OH}{\overset{CH_3}{C}}-\underset{\oplus}{\overset{CH_3}{C}}-CH_3 \xrightarrow{1,2-\ddot{C}H_3\ \text{shift}} CH_3-\underset{:OH}{\overset{\oplus}{C}}-\underset{CH_3}{\overset{CH_3}{C}}-CH_3$$

Carbonium ion — Rearranged carbonium ion

$$\longleftrightarrow CH_3-\underset{\overset{\oplus}{O}-H}{\overset{||}{C}}-\underset{CH_3}{\overset{CH_3}{C}}-CH_3 \xrightarrow{-H^+} CH_3-\underset{O}{\overset{||}{C}}-\underset{CH_3}{\overset{CH_3}{C}}-CH_3$$

Pinacolone

9.13 THE ALDOL CONDENSATION

Aldehydes or ketones with one or more α-hydrogen atoms, when warmed with a trace of base, undergo self-addition reactions known as the *aldol condensation.*

$$CH_3CH{=}O + H.CH_2.CHO \overset{\text{Base}}{\rightleftharpoons} CH_3.\overset{OH}{\overset{|}{C}H}.CH_2.CHO$$

Acetaldehyde (2 moles) — Aldol

$$(CH_3)_2C{=}O + HCH_2.CO.CH_3 \overset{\text{Base}}{\rightleftharpoons} (CH_3)_2.\overset{OH}{\overset{|}{C}}.CH_2.CO.CH_3$$

Acetone (2 moles) — Diacetone alcohol

EXERCISES

1. Write notes on:
   (*a*) Inductive effect
   (*b*) Mesomeric effect
   (*c*) Steric effect
2. Explain the term *resonance*. What are the characteristic features of a resonance hybrid?
3. What are the reactive intermediates? Describe their characteristics.
4. Write a note on conditions and criteria for resonance.
5. (*a*) What are the different intermediate organic species? Discuss in detail their stabilities.
   (*b*) Write a short note on the steric effect.
6. Distinguish between the electromeric and the inductive effect, giving some suitable examples.
7. Give a detailed account of kinetics and stereochemistry of S_N1 reactions.
8. What is benzyne? Explain the mechanism of the reaction involving it.
9. What is the Beckmann rearrangement? Discuss the mechanism of the reaction.
10. Give the mechanism of the Hofmann rearrangement.
11. Write a short note on the structure and stability of the carbanion and the free radical.
12. Define and illustrate free radical, carbonium ion, and carbanion.
13. (*a*) Arrange the following in order of decreasing S_{N^2} reactivity: C_2H_5Cl, CH_3Cl, CH_3Br, $(CH_3)_2CHCl$.
   (*b*) Write the mechanism of the following reaction:

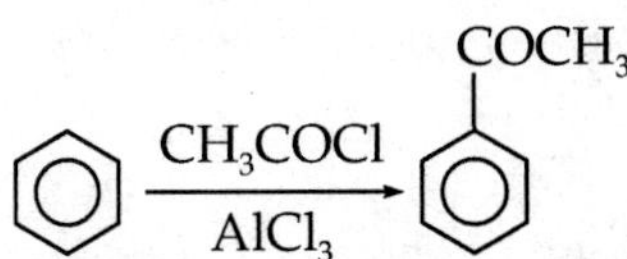

   (*c*) Convert the following in not more than two steps:

$$(CH_3)_3CBr \longrightarrow (CH_3)_2CHCH_2Br$$

MULTIPLE CHOICE QUESTIONS

1. Electromeric effect is due to
   (*a*) electronegative elements
   (*b*) double bonds
   (*c*) triple bonds
   (*d*) all of these
2. In which of the following species is the central carbon atom negatively charged?

(*a*) carbonium ion (*b*) carbanion
(*c*) free radical (*d*) carbene

3. The most stable carbanion is
(*a*) $(CH_3)_3 C^-$ (*b*) $(CH_3)_2 CH^-$
(*c*) $CH_3CH_2^-$ (*d*) $(C_2H_5)_3 C^-$

4. Phenol is less acidic than
(*a*) acetic acid (*b*) *p*-methoxy phenol
(*c*) *p*-nitrophenol (*d*) ethanol

5. The central *c*-atom of a carbonium ion possesses
(*a*) an octet of electrons (*b*) a sextet of electrons
(*c*) a duplet of electrons (*d*) no electrons

6. Which of the following contains three pairs of electrons?
(*a*) carbocation (*b*) carbanion
(*c*) free radical (*d*) none of these

7. The stability of the carbonium ion depends upon
(*a*) the substrate with which it reacts
(*b*) the polarity of the group on which it is attached
(*c*) the inductive effect of the attached group
(*d*) all are correct

8. Free radicals are highly reactive due to
(*a*) the ionic pair (*b*) an associated electron pair
(*c*) an associated odd electron (*d*) none of these

9. Due to its sextet (possessing six electrons), the carbonium ions are
(*a*) unstable (*b*) very stable
(*c*) negatively charged (*d*) none of these

10. Free radicals are paramagnetic due to
(*a*) an associated odd electron (*b*) an ionic pair
(*c*) an associated electron pair (*d*) none of these

11. Which of the following species is a nucleophile?
(*a*) $\overset{+}{N}O_2$ (*b*) $:CX_2$
(*c*) $:\ddot{N}H_2^-$ (*d*) $\dot{C}H_3$

12. In the pinacol-pinacolone rearrangement, the reactive species undergoing rearrangement is
(*a*) carbene (*b*) free radical
(*c*) carbanion (*d*) carbonium ion

13. Which of the following is an electrophile?
(*a*) RNH_2 (*b*) $: CH_2$
(*c*) H_2O (*d*) $: CN^-$

14. Which of the following is an electrophile?
    (*a*) H_2O (*b*) NH_3
    (*c*) $AlCl_3$ (*d*) $C_2H_5NH_2$
15. The shape of a carbanion is
    (*a*) linear (*b*) planar
    (*c*) pyramidal (*d*) none of these
16. The homolytic fission of a hydrocarbon results in the formation of
    (*a*) a carbonium ion (*b*) free radicals
    (*c*) carbanions (*d*) carbenes

Chapter **10**

COORDINATION CHEMISTRY

10.1 INTRODUCTION

Coordination chemistry is a branch of inorganic chemistry that deals with the study of complex addition compounds that have coordinate linkages. Such compounds are called *coordinate compounds.* A coordination compound is a complex substance that contains a central metal atom or ion surrounded by oppositely charged ions or neutral molecules. It is also defined as a compound that contains a complex ion.

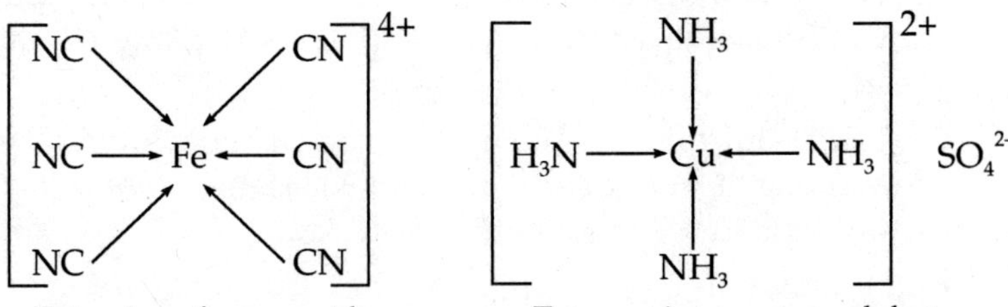

Potassium ferrocyanide   Tetrammine copper sulphate

10.2 COORDINATE COMPOUND TERMINOLOGY

The following are important terms used in a coordinate compound:

Central ion or center of coordination: The cation to which one or more neutral molecules or anions are attached.

Ligand: An atom or group of atoms that binds to the central atom through the lone pair of electrons present on its donor atom. Ligands containing one, two, three, and so on, donor atoms are known as unidentate, bidentate, tridentate, and so on respectively. Examples of various ligands are shown in Table 10.1.

TABLE 10.1 Some Typical Ligands

Type of Ligand	*Name of Ligand*	*Symbol*	*Structure*
Unidentate	Water Ammonia Cyano Chloro Bromo	aqua ammine CN^- Cl^- Br^-	H–Ö–H (with lone pairs) H—N̈—H, with H below N $:C \equiv N^-$
Bidentate	Ethylenediamine Oxalato Sulphato	en ox	$H_2N—CH_2—CH_2—NH_2$ $^-O—C(=O)—C(=O)—O^-$ SO_4^{2-}
Tridentate	Diethylenetriamine	dien	$H_2C—\ddot{N}H—CH_2$ $H_2C—\ddot{N}H_2$ $\ddot{N}H_2—CH_2$

Chelating ligand: When polydentate ligands bind to a central atom, these form a ring-like structure called *chelate* (Greek : *Chelate*-Crab's claw), and the ligand is known as chelating ligand.

Coordination sphere: The central metal atom along with the ligands surrounding it are collectively termed as the *coordination sphere.* This is usually written within a square bracket.

Coordination number or ligancy: The total number of ligands surrounding the central atom in the coordination sphere.

Oxidation number: A number that represents the electric charge on the central metal atom of a complex ion.

Calculation of the oxidation number of the metal in a complex:

1. $[Ni(CO)_4]$
   Net charge on ion

$[Ni(CO)_4]^0$
$x \quad 4 \times 0$
Oxidation number of metal:

$$x + 4(0) = 0$$
$$\therefore \quad x = 0$$

Thus nickel is present as a zero state.

2. $K_4[Fe(CN)_6]$
Net charge on ion
$[Fe(CN)_6]^{4-}$
$x \quad 6 \times (-1)$
Oxidation number of metal:

$$x + 6(-1) = -4$$
$$\therefore \quad x = +2$$

Thus, iron is present as Fe^{2+} or Fe (II).

3. $[Cu(NH_3)_4]SO_4$
Net charge on ion:
$[Cu(NH_3)_4]^{2+}$
$x \quad 4 \times 0$
Oxidation number of metal:

$$x + 4(0) = +2$$
$$\therefore \quad x = +2$$

Thus, copper is present as Cu^{2+} or Cu (II).

4. $K[PtCl_3(NH_3)]$
Net charge on ion
$[Pt\ Cl_3\ (NH_3)]^-$
$x \quad -3 \quad 3 \times 0$
Oxidation number of metal:

$$x + 3(-1) + 3(0) = -1$$
$$\therefore \quad x = +2$$

Thus, platinum is present as Pt^{2+} or Pt (II).

Charge on complex: The algebraic sum of the oxidation number of all the atoms/molecules/ions constituting the coordination sphere is the charge on the complex.

The various terms involved in complex formation are illustrated here:

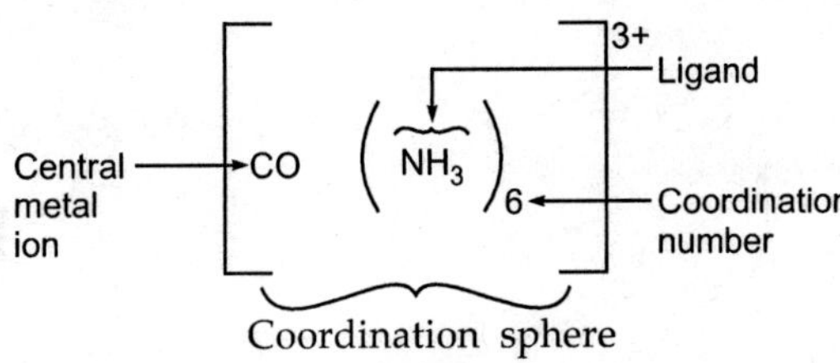

10.3 THEORY

There are four theories of formation of coordination compounds.

Werner's Coordination Theory

The fundamental postulates of Werner's theory can be summarized as follows:

- Metals possess two types of valencies, namely primary or ionizable valency and secondary or nonionizable valency.

  Primary valencies are those which a metal exhibits in the formation of its simple salts, and it corresponds to its oxidation state. The primary valency of metal ion is always satisfied by a negative ion. For example, the primary valencies of Ag, Cu, Co and Pt in the formation of their simple salts AgCl, $CuSO_4$, $CoCl_3$, and $PtCl_4$ are 1, 2, 3, and 4, respectively. The primary valencies are indicated by dotted lines because they are nondirectional in nature. Primary valencies are ionizable.

  Secondary valencies are those which a metal exhibit toward a neutral molecule or an anion during the formation of its complex ion. The secondary valency corresponds to the coordination number of the metal ion or atom. For example, in complex compound $[Cu(NH_3)_4 SO_4]$, the secondary valency of Cu^{2+} is 4.

  Secondary valencies are directional in nature and nonionizable and are shown by thick lines. Secondary valencies give a definite geometry to the complex.

- Every metal has a fixed number of secondary valencies or coordination numbers.

  Keeping in view the previously mentioned points, the amine complex of Pt^{4+} and Cu^{2+} ions may be represented as

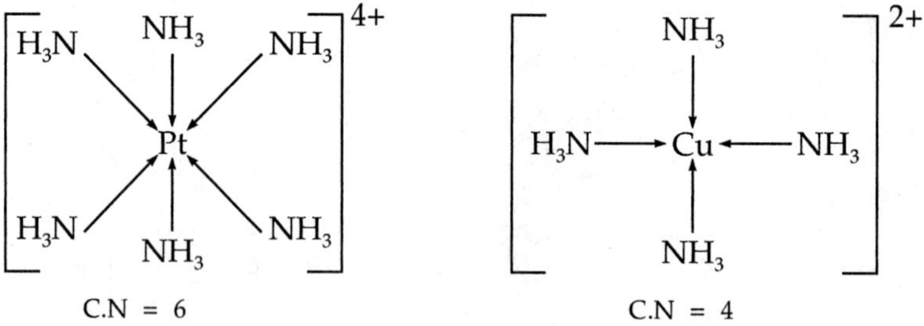

On the basis of the Werner theory, the complex $CoCl_3 . 6NH_3$ or $Co(NH_3)_6 Cl_3$ can be represented as

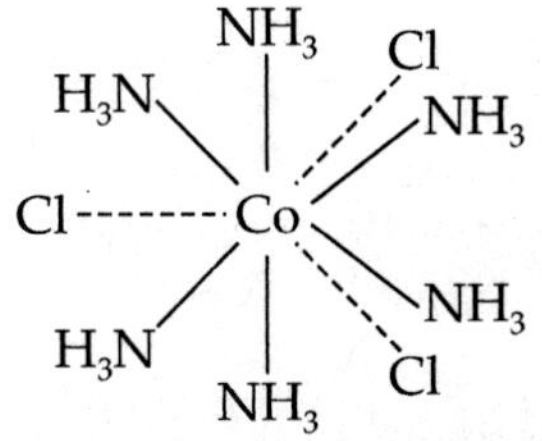

NH$_3$ Cl
H$_3$N NH$_3$
Cl-------Co
H$_3$N NH$_3$
NH$_3$ Cl

The three dotted lines indicate the three primary valencies (oxidation state) of cobalt, and the six thick lines indicate its secondary valencies (coordination number).

Sidgwick Coordination Theory

Sidgwick modified Werner's coordination theory on the basis of an electron pair bond (Lewis concept). Sidgwick observed that all the molecules or ions that coordinate to metal ions have atoms with at least one unshared electron pair in their structure, which is donated to the central metal in the formation of the bond. The atom donating the electron pair is called the *donor*, and the metal ion accepting it is called the *acceptor*. The resulting coordinate, dative, or semipolar bond is usually represented as $M \leftarrow L$, which indicates that the donor group, L (ligand), has donated both electrons to the acceptor, M (metal ion).

Sidgwick also suggested that a metal ion tends to accept sufficient electrons from ligands until the total number of electrons in the metal ion and those donated by ligands is equal to that of the next higher noble gas. This total number of electrons is called the *effective atomic number* (EAN) of the metal.

The effective atomic number of a metal in a complex is obtained by deducting the number of electrons lost in ion formation from the atomic number of that metal, and then adding the number of electrons gained by coordination. For example:

$$K_4Fe(CN)_6$$

Atomic number of iron = 26

Iron loses two electrons in forming the ferrous ion Fe^{2+}.

$\therefore$ The total number of electrons in $Fe^{2+} = 26 - 2 = 24$

Because each CN ligand contributes two electrons to the iron ion, electrons contributed by 6 CN ligands $= 6 \times 2 = 12$.

$\therefore$ The EAN of iron $= 24 + 12 = 36$

Because the number (36) corresponds to the atomic number of Krypton, according to Sidgwick, the complex will be stable.

There are certain metal ions that do not obey the EAN concept, for example, Cr^{3+}, Fe^{3+}, Ni^{3+}, Pd^{2+}, Ir^{4+}, Pt^{2+}, and so on.

Valence Bond Theory

Pauling developed the valence bond treatment of bonding in complexes, which explains satisfactorily the structure and magnetic properties of a large number of coordinate compounds. The salient features of the theory are summarised here:

- The central metal ion has a number of empty orbitals for accommodating electrons donated by the ligands. The number of empty orbitals is equal to the coordination number of the metal ion for the particular complex.
- The metal orbitals and ligand orbitals overlap to form strong bonds. The greater the extent of overlapping, the stronger will be the bond and the more stable will be the complex. For greater stability, the atomic orbitals of the metal ion hybridize to form a new set of equivalent hybridized orbitals, which overlap with the ligand orbitals to form strong chemical bonds.
- The *d*-orbitals involved in the hybridization may be either inner $(n-1)d$ orbitals or outer *nd*-orbitals. The complexes formed in these two ways are referred to as *low spin* and *high spin* complexes respectively.
- The nonbonding metal electrons occupy the inner *d*-orbitals that do not participate in hybridization and thus in bond formation with the ligand.
- Each ligand contains a lone pair of electrons.
- A covalent bond is formed by the overlap of a vacant hybridized metal orbital and a filled orbital of the ligand. The bond is also sometimes called a coordinate bond.
- If the complex contains unpaired electrons, it is *paramagnetic* in nature, whereas if it does not contain unpaired electrons, it is *diamagnetic* in nature.
- The number of unpaired electrons in the complex shows the geometry of the complex, and vice versa.
- Under the influence of a strong ligand, the electrons can be forced to pair up against the Hund's rule of maximum multiplicity.

Examples of some complexes with geometrical shapes and the type of hybridization involved are illustrated in Table 10.2.

TABLE 10.2 Geometry of Some Complexes

Type of Hybridization	*Geometrical Shape*	*Examples*
sp^3	Tetrahedral	$[Ni\ (CO)_4]$ $[Ni\ (Cl)_4]^{2-}$ $[Zn\ (NH_3)_4]^{2+}$
dsp^2	Square planar	$[Ni\ (CN)_4]^{2-}$; $[Pt\ (Cl)_4]^{2-}$ $[Pt\ (NH_3)_2\ Cl_2]$
d^2sp^3 or sp^3d^2	Octahedral	$[Co\ (NH_3)_6]^{3+}$; $[Fe\ (CN)_6]^{3-}$; $[Ni\ (NH_3)_6]^{2+}$

Crystal Field Theory

H. Bethe and Van Vleck proposed the crystal field theory according to which the metal ion is situated in an electrical field caused by surrounding molecules or ions called ligands. The ligands are either negatively charged or neutral molecules that can donate an electron pair to the metal ion and produce a field approximately equivalent to that of a set of negative point charges. Thus, it is the interaction of the *d*-orbitals of a transition metal with ligands surrounding the metal that produces crystal field effects. All the *d*-orbitals on the metal have the same energy (*i.e.*, degenerate) in the free atom. However, when a complex is formed, the ligands destroy the degeneracy of these orbitals, that is the orbitals now have different energies.

On the basis of the orientation of the lobes with respect to coordinates, five *d*-orbitals have been grouped as follows:

- d_{z^2} and $d_{x^2-y^2}$ have their lobes along the axes. They belong to the e_g set of orbitals called axial orbitals. The letter '*e*' in e_g refers to a doubly-degenerate set.
- d_{xy}, d_{yz}, and d_{zx} have their lobes lying between the axes. They are also called nonaxial orbitals and belong to the t_{2g} set of orbitals. The letter '*t*' in t_{2g} refers to a triply-degenerate set.

Under the influence of the ligands, the five degenerate *d*-orbitals of the metal ion will split into two groups of orbitals of different energy. This effect is known as *crystal field splitting* or *energy level splitting.*

Crystal Field Splitting in Octahedral Complexes

In an octahedral complex, the coordination number is 6. The central metal ion is at the center and the six ligands occupy the six corners of the octahedron. The phenomenon of splitting of *d*-orbitals in an octahedral complex is illustrated in Fig. 10.1.

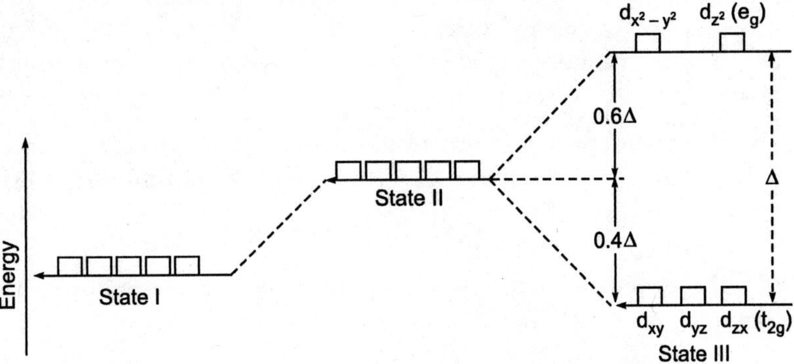

Fig. 10.1 Spliting d-orbitals in an octahedral complex.

In the free metal ion, all the five *d*-orbitals are degenerate (State I, Fig. 10.1). When the six ligands approach the central metal cation along the axes, they exert an electrostatic force of repulsion on the outer *d* electrons, that is the *d* electrons are repelled by the lone pairs of the ligands. This repulsion raises the energy of the degenerate *d* orbitals to give five excited degenerate orbitals (State II, Fig. 10.1). Because the lobes of d_{z^2} and $d_{x^2-y^2}$ orbitals (e_g set of orbitals) lie directly in the path of the approaching ligands, the electrons in these orbitals experience greater repulsion exerted by the electron clouds of the ligands than do the electrons in the d_{xy}, d_{yz}, and d_{zx} orbitals (t_{2g} set of orbitals), which are directed in space between the *x*-, *y*-, and *z*-axes. Hence, under the influence of the approaching ligands, the orbitals, that is $d_{x^2-y^2}$ and d_{z^2} are raised in energy, whereas the orbitals, that is d_{xy}, d_{yz}, and d_{zx}, are lowered in energy relative to the excited *d* levels (State III, Fig. 10.1).

The separation of five *d*-orbitals into t_{2g} and e_g sets of different energies is known as crystal field splitting. The e_g set, which is of higher energy, is doubly degenerate, whereas the t_{2g} set that is lower in energy is triply degenerate. The energy difference between e_g and t_{2g} sets is denoted by Δ_0 or 10 Dq. The magnitude of Δ_0 depends on the field strength of the ligand as well as on the metal ion. Δ_0 is called the *Crystal Field Stabilization Energy (CFSE).*

The ligands that cause only a small degree of crystal field splitting are called *weak ligands*, whereas the ligands that cause a large degree of crystal field splitting are called *strong ligands*. Strong ligands give a higher value of Δ_0, and weak ones give a lower value.

Some common ligands have been arranged in the increasing order of their splitting power as follows:

$I^- < Br^- < S^{2-} < Cl^- < NO_3^- < F^- < OH^- < EtOH < (COO^-)_2 < H_2 < EDTA < NH_3$ $<$ ethylene diamine $<$ dipyridyl $<$ o-phenanthroline $< NO_2^- < CN^- < CO$

This arrangement of ligands in the order of their abilities to split the *d*-orbitals energies is called *spectrochemical series.*

When the first, second, and third electrons enter the *d*-orbital, they occupy the three t_{2g} (lower energy) orbitals one by one. However, when the fourth and the next electron enters the *d*-orbitals, the following two alternative options are available for the electron:

- The fourth and the next electron may pair up with the electrons in the t_{2g} orbitals (contrary to the Hund's rule of maximum multiplicity).
- The fourth and the next electron may enter the higher e_g orbitals in accordance with Hund's rule of maximum multiplicity.

The exact path chosen by the electron depends upon the strength of the ligand:

Case I: In the presence of a strong ligand field such as CO, CN^-, and so on, the electrons pair up because Δ_0 is large enough to force pairing by filling the lower energy t_{2g} *d*-orbitals. Thus, in the presence of a strong ligand field, low spin (diamagnetic) complexes are formed.

$$[Fe(CN)_6]^{4-}$$

Case II: In the presence of a weak ligand field such as F^-, H_2O, and so on, because Δ_0 is not large enough to force the pairing up of electrons, the next (fourth and fifth) electron will enter the e_g orbitals. In the presence of a weak ligand field, high spin complexes (paramagnetic) are formed.

$$[Fe(H_2O)_6]^{2+}$$

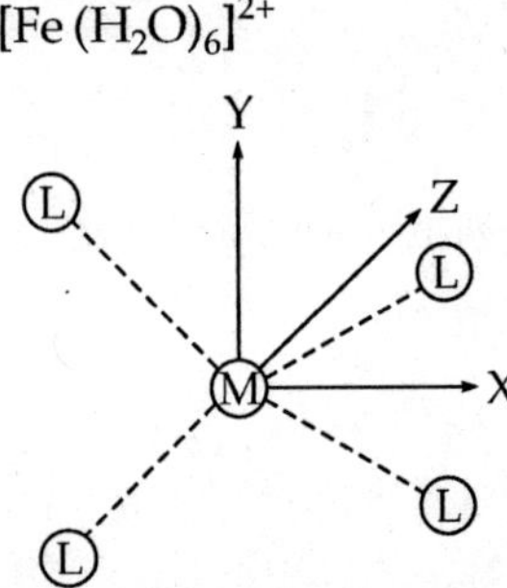

Fig. 10.2 Approach of the four ligands in a tetrahedral field.

Crystal Field Splitting in Tetrahedral Complexes

The coordination number in tetrahedral complexes is 4. In a tetrahedral field, the four ligands approach in a different fashion. Fig. 10.2 suggests that the ligands interact more with the t_{2g} (d_{xy}, d_{yz}, and d_{zx}) orbitals pointing close to the direction of the approaching ligands than the e_g ($d_{x^2-y^2}$ and d_{z^2}) orbitals lying between the ligands. The three t_{2g} orbitals are of higher energy and two e_g orbitals are of lower energy. Here, the *d*-orbitals are split into two groups but in a reverse order. This is represented in Fig. 10.3.

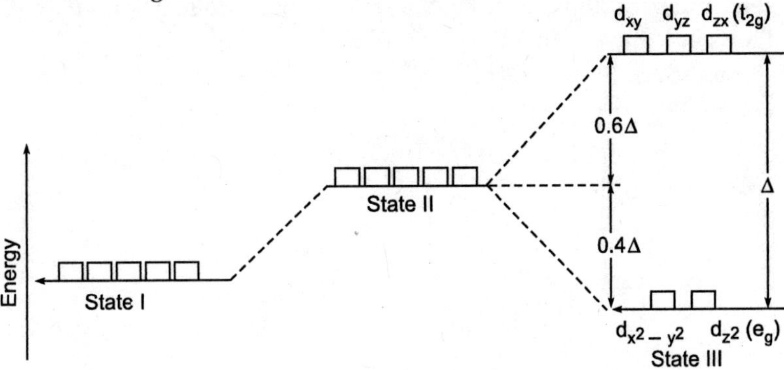

Fig. 10.3 Splitting d-orbitals in a tetrahedral complex.

The crystal field splitting in this case is denoted by Δ_t (the subscript t indicating tetrahedral complex). In the tetrahedral field, because the d-orbitals are not interacting directly with the ligand field, splitting of d-orbitals is less than that in the octahedral complexes.

$$\Delta_t < \Delta_0$$

$$\Delta_t = \frac{4}{9}\Delta_0$$

The lower value of crystal field splitting (Δ) in tetrahedral complexes may also be attributed to the lesser number of ligands (4).

Crystal Field Splitting in Square Planar Complexes

Four ligands in a square planar arrangement around the central ion would have the greatest influence on a $d_{x^2-y^2}$ orbital. So the energy of this orbital will be raised the most (Fig. 10.4).

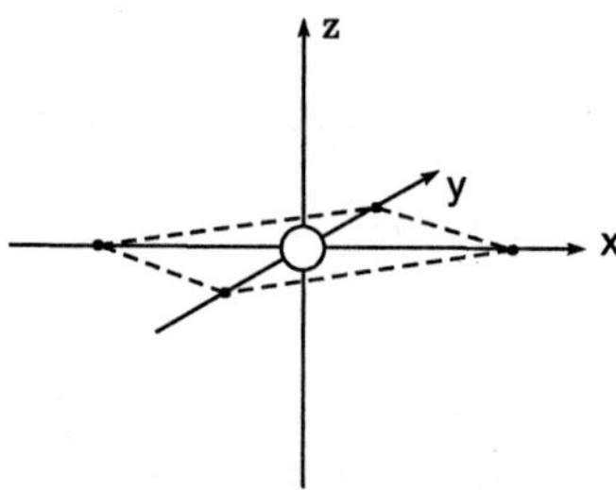

Fig.10.4 Square planar arrangement of four ligands around the central ion. The effect of the ligand field is greatest on the $d_{x^2-y^2}$ orbital.

The d_{xy} orbital lying in the same plane but between the ligands will also have a greater energy, although the effect will be less than that on the $d_{x^2-y^2}$ orbital. However, it is not possible to predict exactly the effect of the ligand field on the remaining orbitals, although it will be less than on the $d_{x^2-y^2}$ and d_{xy} orbital. However, the d_{xz} and d_{yz} pair will always be affected equally and will therefore remain degenerate. Fig. 10.5 illustrates the comparative crystal field splitting in different fields.

On the basis of crystal field theory, color and magnetic properties of complexes can be easily explained.

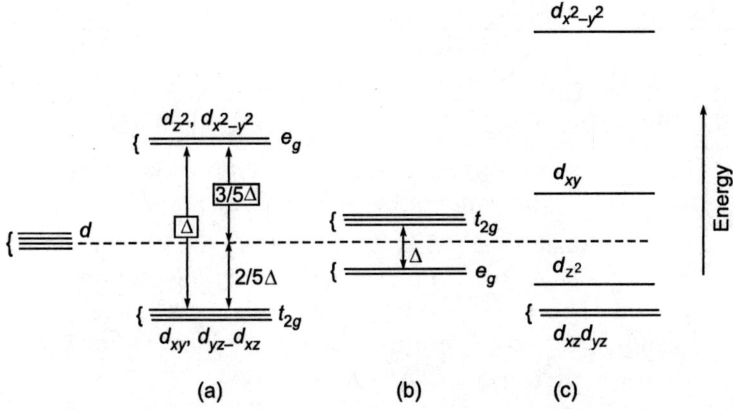

Fig.10.5 Crystal field splitting in (a) the octahedral field, (b) the tetrahedral field, (c) the square planar field.

Color

Most complexes are colored because they absorb light at one or more of the wavelengths in the visible part of the spectrum (400 to 700 mm) and reflect the others. The energy of the light in the visible region is just right to promote *d*-electrons from lower to higher energy levels (*e.g.*, from t_{2g} to e_g in an octahedral complex). The color of the reflected light is complementary to that which is absorbed.

Magnetic Properties

The magnitude of the crystal field splitting determines the magnetic properties of a complex ion. The complexes with no unpaired electrons do not exhibit magnetic properties and are therefore said to be diamagnetic. The complexes with one or more unpaired electrons show a definite value of magnetic moment and are attracted in a magnetic field. Such complexes are called paramagnetic complexes. The value of the magnetic moment (μ) and the number of unpaired electrons (n) are related as

$$\mu = \sqrt{n(n+2)}\ \text{BM}$$

Molecular Orbital Theory

According to this theory, molecular orbitals are formed by the overlap of orbitals from the ligands with the atomic orbitals of the central atom. These molecular orbitals may be either the bonding or the antibonding type.

10.4 NOMENCLATURE

International Union of Pure and Applied Chemistry (IUPAC) has framed the following rules to name the coordination compounds:

- The positive part (cation) of a coordination compound is named first and is followed by the name of the negative part (anion).
- The ligands are named first followed by the central metal. The prefixes di-, tri-, tetra-, and so on, are used to indicate the number of each kind of ligand present. The prefixes bis-(two ligands), tris-(three ligands), tetrakis-(four ligands), and so on, are used when the ligands include a number, such as dipyridil, diethylene diamine, and so on.
- In polynuclear complexes, the bridging group is indicated in the formula of the complex separating it from the rest of the complex by hyphens and using the prefix 'μ'.
- The ligands are named alphabetically. Names of the anionic ligands end in O, and those of cationic end in *ium*. Neutral ligands have their regular names except that H_2O is named as *aquo*, NH_3 as *ammine*, NO as *nitrosyl*, and CO as *carbonyl*.
- The oxidation state of the central metal is indicated in roman numbers in parentheses.
- When a complex species has a negative charge, the name of the central metal ends in *ate*. For some elements, the ion name is based on the latin name of the metal (*e.g., argentate* for silver).
- When naming geometrical isomers, the prefix *cis-* or *trans-* is used to designate similar groups at adjacent positions or opposite positions of the ligand.

  The prefix (+) or (–) [or *d* or *l*] is used to name the dextro and laevo rotatory when naming optical isomers.

  Some illustrative examples based on the preceding rules are given here:

Complex cations

$[Co(H_2O)_6]^{2+}$	Hexaaquacobalt (II)
$[Cu(NH_3)_4]^{2+}$	Tetraammine copper (II)
$[Ni(en)_3]^{3+}$	Tris (ethylenediammine) nickel (III)

Complex anions

$K_4[Fe(CN)_6]$	Potassium hexacyanoferrate (II)
$[CrF_6]^{3-}$	Hexafluoro chromate (III)
$Na[Al(OH)_4]$	Sodium tetrahydroxo aluminate (III)
$[Au(CN)_4]^{2-}$	Tetracyanoaurate (III)

Neutral complexes

$[Ni(CO)_4]$	Tetracarbonyl nickel (0)
$[Pt(NH_3)_2Cl_2]$	Diammine chloroplatinum (II)

10.5 ISOMERISM

The coordination compounds exhibit the following two types of isomerism.

Structural Isomerism

This isomerism arises due to the difference in structures. Following are the various types of structural isomerisms:

Ionization isomerism: Complex compounds with the same molecular formula but give different ions in solution on ionization. This type of isomerism arises due to the difference in the positions of the ligands because ligands in the coordination sphere do not ionize.

$[Co(NH_3)_5Br]SO_4$	$[Co(NH_3)_5SO_4]Br$
Pentaammine bromo cobalt (III) sulphate (Violet)	Pentaammine sulphato-cobalt (III) bromide (Red)

Linkage isomerism: This type of isomerism arises due to the presence of two different donor atoms in the same ligand, which may thus attach through either of the two atoms. For example,

NO_2^- may be bonded to the metal either through the nitrogen atom to act as nitro ligand (– NO_2) or through one of the oxygen atoms to act as a nitrito ligand (—ONO):

$[(NH_3)_5CoNO_2]Cl_2$	$[(NH_3)_5CoONO]Cl_2$
Pentamminenitrocobalt (III) chloride (Yellow brown)	Pentamminenitrito cobalt (III) chloride (Red)

Coordination isomerism: This isomerism occurs when both cation and anion are complex. This involves the exchange of ligands between the complex cation and the complex anion, that is between coordination spheres in a compound. For example:

$[Cr(NH_3)_6][Co(CN)_6]$	and	$[Co(NH_3)_6][Cr(CN)_6]$
Hexammine chromium (III) hexacyanocobaltate (III)		Hexamminecobalt hexacyano chromate (III)

Hydrate isomerism: This type of isomerism arises in the complexes due to different numbers of water molecules present in the coordination spheres. For example:

$[Co(H_2O)_6]Cl_3$ and $[Co(H_2O)_5Cl]Cl_2 \cdot H_2O$

Coordination position isomerism: This type of isomerism is shown by bridged complexes and involves a different placement of the ligands. This is a special type of coordination isomerism.

$$\underset{H_2N}{H_2C} - \underset{NH_2}{CH} - CH_3 \qquad \underset{NH_2}{H_2C} - CH_2 - \underset{NH_2}{CH_2}$$

1, 2-diaminopropane 1, 3-diaminopropane

Stereoisomerism

Stereoisomerism in coordination compounds is due to the difference in arrangement of ligands in space around a given metal atom or ion. It is of following two types:

Geometrical isomerism: This type of isomerism occurs in disubstituted complexes with coordination number 4 and 6 having square planar and octahedral arrangement, respectively. When two identical ligands are adjacent to each other, the isomer is known as *cis* (Latin, *cis* means same), whereas when the two identical ligands are opposite to each other, the isomer is known as *trans* (Latin, *trans* means across).

Geometrical isomerism is not possible for complexes with coordination number 2 and 3 and tetrahedral complexes with coordination number 4.

For example, the *Square planar complex*— $Pt(NH_3)_2Cl_2$— with square planar arrangement exists in two geometrical isomeric forms:

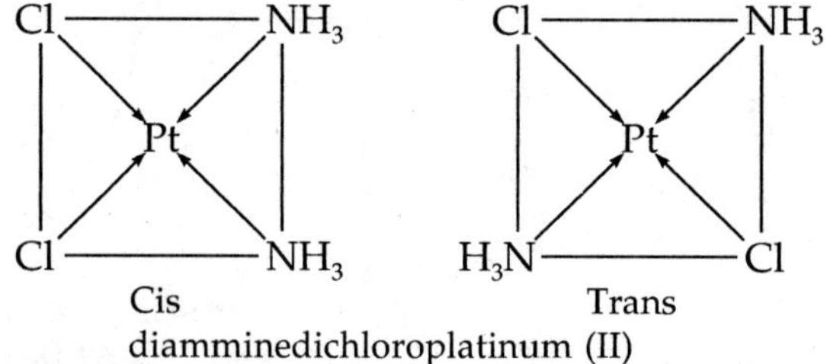

Cis Trans
diamminedichloroplatinum (II)

The *octahedral complex*—$[Co(NH_3)_4Cl_2]^+$ —ion with octahedral arrangement exists in two geometrical isomeric forms:

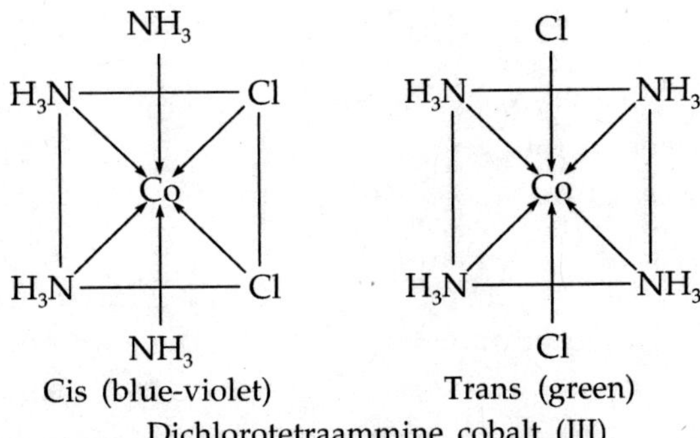

Cis (blue-violet) Trans (green)

Dichlorotetraammine cobalt (III)

Optical isomerism: The compounds with the same physical and chemical properties but differing in their action on plane polarized light are known as *optical isomers;* the phenomenon is known as *optical isomerism.* Optical isomers are just mirror images of one another, and they cannot be superimposed. Optical isomers rotate the plane of polarized light either to its left (called leavo or *l*-) or to its right (called dextro or *d*-).

Complexes with coordination number 6 and bidentate and hexadentate ligands provide examples of optical isomerism:

diamminedichloro ethylenediamine cobalt (III),
$[Co\,(en)\,(NH_3)_2\,Cl_2]^+$

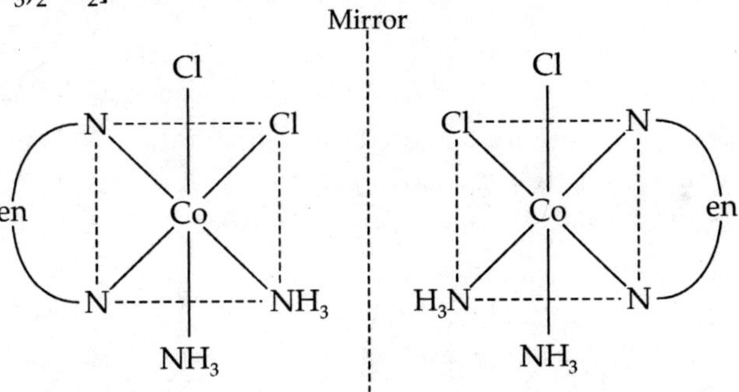

bis-dichloro-bis (ethylenediamine) cobalt (III)
$[Co\,(en)_2\,Cl_2]^+$

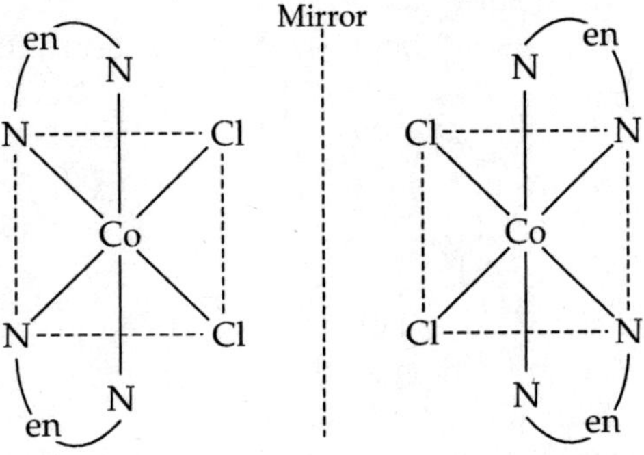

10.6 APPLICATIONS OF COORDINATION CHEMISTRY

The applications of coordination compounds in a number of general fields are summarized as follows:

Analytical chemistry: The analytical applications of coordination chemistry are numerous:

- Qualitative analysis
- Gravimetric analysis
- Volumetric analysis
- Colorimetric methods of analysis
- Solvent extraction
- Sequestration

Metal extraction: Extraction of metals such as silver and gold is carried out by forming their soluble cyanide complex.

Catalysis: Coordination compounds are used as catalysts in important commercial processes. For example:

- The Zeigler Natta catalyst ($TiCl_4$ and trialkyl aluminium) is used as a catalyst in the formation of polyethylene.
- The Wilkinson catalyst

$RhCl(PPh_3)_3$

Cl PPh$_3$
Co
Ph$_3$P PPh$_3$

Chlorotris (triphenyl phosphine) rhodium

is used in the hydrogenation of alkenes.

- The Monsanto acetic acid process : $[Rh(Cl)(Co)(PPh_3)_2]$ or $[Rh(Cl)(CO)_2]_2$ is used as a catalyst in the presence of CH_3I, I_2 or HI as activator.

EXERCISES

1. Explain crystal field splitting in octahedral complexes.
2. Explain Werner's coordination theory.
3. What are chelating complexes?
4. Explain the crystal field theory of the bonding of the coordination complex.
5. Write the IUPAC names of the following:
   - (*i*) $K_4[Fe(CN)_6]$
   - (*ii*) $K_3[Al(C_2O_4)_3]$
   - (*iii*) $[Au(CN)_4]^-$
   - (*iv*) $Li[AlH_4]$
   - (*v*) $Ca_2[Fe(CN)_6]$
   - (*vi*) $Na_2[CrOF_4]$
   - (*vii*) $Na_2[ZnCl_4]$
   - (*viii*) $K_2[NiF_6]$
6. Explain the valence bond theory of coordination complexes.
7. What do crystal field stabilization energy and pairing energy mean?
8. (*a*) Write the formula of the coordination compound, potassium tetracyanonickelate (II) and the IUPAC name of $[CoCl(NH_3)_5]Cl_2$.

(*b*) What is the type of isomerism shown by the tris (ethylene diamine) cobalt (III) ion? Draw their structures.

(*c*) On the basis of CFT, explain why $[CoF_6]^{3-}$ is paramagnetic, while $[Co(CN)_6]^{3-}$ is diamagnetic.

(*d*) What are the catalysts used in the Monsanto acetic acid process and in Wilkinson's hydrogenation process ?

MULTIPLE CHOICE QUESTIONS

1. Which of the following represents hexadentate ligand?
(*a*) 2, 2 bipyridyl (*b*) DMG
(*c*) ethylenediamine (*d*) none of these

2. The coordination number of copper in cuprammonium sulphate is
(*a*) 2 (*b*) 3 (*c*) 4 (*d*) 6

3. The IUPAC name of $[Ni\ (CO)_4]$ is
(*a*) tetracarbonyl nickel (I) (*b*) tetracarbonyl nickel (0)
(*c*) tetracarbonyl nickel (II) (*d*) tetracarbonyl nickel (III)

4. $K_3\ [Al\ (C_2O_4)_3]$ is called
(*a*) potassiumaluminooxalate
(*b*) potassiumalumino (III) oxalate
(*c*) potassiumtrioxalatoaluminate
(*d*) potassiumtrioxalatoaluminate (III)

5. EDTA has a
(*a*) monodentate ligand (*b*) bidentate ligand
(*c*) tetradentate ligand (*d*) hexadentate ligand

6. Which of the following ligands is a bidentate?
(*a*) EDTA (*b*) Ethylene diamine
(*c*) Acetate (*d*) Pyridine

7. The green pigment present in plants, chlorophyll, contains the metal
(*a*) Al (*b*) Fe (*c*) Mg (*d*) Ca

8. A solution of potassium ferrocyanide would contain ____ ions
(*a*) 2 (*b*) 3 (*c*) 4 (*d*) 5

9. The EAN of iron in $[Fe\ (CN)_6]^{3-}$ is
(*a*) 34 (*b*) 36 (*c*) 37 (*d*) 35

10. Which one of the following compounds will show optical isomerism?
(*a*) 34 (*b*) 36 (*c*) 37 (*d*) 35

11. The oxidation number of Pt in $[Pt\ (C_2H_4)Cl_3]$ is
(*a*) +1 (*b*) +2 (*c*) +3 (*d*) +4

12. Which of the following is an organometallic compound?
(*a*) Lithium methoxide (*b*) Lithium acetate
(*c*) Lithium dimethylamide (*d*) Methyl lithium

Chapter 11

CORROSION AND CONTROL

11.1 INTRODUCTION

Metals have a natural tendency to revert back to combined states. During this process, mostly oxides are formed, although in some cases sulphides, carbonates, sulphates, and so on, may result due to the presence of impurities. Any process of deterioration and loss of solid metallic material by chemical or electrochemical attack by its environment is called *corrosion*. Corrosion is the reverse process of metallurgy.

$$\text{Metal} \underset{\substack{\text{Metallurgy} \\ \text{(Reduction)}}}{\overset{\substack{\text{Corrosion} \\ \text{(Oxidation)}}}{\rightleftarrows}} \text{Matallic compound + energy}$$

Rusting of iron when exposed to atmospheric conditions is an example of corrosion. Rust is hydrated oxide ($Fe_2O_3 \, . \, xH_2O$).

11.2 TYPES OF CORROSION

Many types of corrosion occur and are described in the following sections:

Dry or Chemical Corrosion

This type of corrosion occurs mainly through the direct chemical action of atmospheric gases (O_2, halogen, H_2S, SO_2, N_2, or anhydrous inorganic liquid) with metal surfaces in immediate proximity. Three types of chemical corrosion are oxidation corrosion, corrosion by other gases, and liquid metal corrosion.

Oxidation Corrosion

This is carried out by the direct action of oxygen at low or high temperatures on metals in absence of moisture. At an ordinary temperature, metals are very slightly attacked. The exceptions are alkali metals and alkaline earth metals. At a high temperature, all metals are oxidized. The exceptions are Ag, Au, and Pt.

The reactions involved are

$$2M \longrightarrow \underset{\text{Metal ion}}{2M^{n+}} + 2ne^- \qquad \text{(Deelectronation)}$$

$$nO_2 + 2ne^- \longrightarrow \underset{\text{Oxide ion}}{2nO^{2-}} \qquad \text{(Electronation)}$$

$$2M + nO_2 \longrightarrow \underset{\text{Metal oxide}}{2M^{n+} + 2nO^{2-}}$$

Mechanism

Oxidation occurs at the surface of metal and the resulting metal oxide scale forms a barrier that restricts further oxidation. For oxidation to continue, either the metal must diffuse outward through the scale to the surface, or the oxygen must diffuse inward through the scale to the underlying metal. Both the cases are possible; however the outward diffusion of metal is generally more rapid than inward diffusion of oxygen because the metal ion is appreciably smaller than the oxygen ion and hence more mobile.

$$\text{Metal} + \text{Oxygen} \longrightarrow \text{Metal oxide}$$

When oxidation starts, a thin layer of oxide is formed on the metal surface, and the nature of this film determines further action.

- If the film is stable, it behaves as a protective coating in nature. For example, the oxide films on Al, Pb, Cu, Pt, and so on, are stable and therefore further oxidation corrosion is prohibited.
- If the film is unstable, that is the oxide layer formed decomposes back into metal and oxygen, oxidation corrosion is not possible.

$$\text{Metal oxide} \rightleftharpoons \text{Metal} + \text{oxygen}$$

For example, Ag, Au, and Pt do not undergo oxidation corrosion.

- If the film is volatile, that is the oxide layer volatilizes after formation and leaves the underlying metal surface exposed for further attack. This causes continuous corrosion that is excessive, For example, Molybdenum oxide (MoO_3).
- If the film is porous, that is, the oxide layer formed has pores or cracks, the atmospheric oxygen passes through the pores or cracks of the layer to the underlying metal surface. This causes continuous corrosion until the complete conversion of metal into its oxide.

Pilling-Bedworth Rule

The rule states that "an oxide is protective or nonporous, if the volume of the oxide is at least as great as the volume of the metal from which it is formed."

$$\text{Specific ratio} = \frac{\text{Volume of metal oxide}}{\text{Volume of metal}}$$

The smaller the specific ratio, the greater is the oxide corrosion because the oxide film formed will be porous, through which oxygen can diffuse and cause further corrosion. For example, alkali and alkaline earth metals (Li, K, Na) form oxides of a volume less than the volume of metals, which causes continuous corrosion. Metals such as aluminium form oxide, whose volume is greater than that of the metal that causes a nonporous layer, and consequently prohibits corrosion.

Corrosion by Other Gases

$$(SO_2, CO_2, Cl_2, H_2S, F_2, \text{etc.})$$

The extent of corrosion depends on the chemical affinity between the metal and the gas involved. If the film formed is protective or nonporous (*e.g.*, AgCl film resulting from the attack of Cl_2 on Ag), the extent of the corrosion decreases because the film protects the metal from further attack. If the film formed is nonprotective or porous, the metal surface is attacked, for example, the formation of volatile $SnCl_4$ by the attack of Cl_2 on Sn.

Liquid Metal Corrosion

This is due to the chemical action of flowing liquid metal at a high temperature on solid metal or alloy. Such corrosion occurs in devices used for nuclear power. The corrosion reaction is carried out either by dissolution of a solid metal by liquid metal or by internal penetration of liquid metal into solid metal. The result is a weakening of solid metal.

Wet or Electrochemical Corrosion

This type of corrosion occurs

- Where a conducting liquid is in contact with the metal
- When two dissimilar metals or alloys are dipped partially in a solution

This corrosion occurs due to the existence of separate anodic and cathodic parts, between which current flows through the conducting solution. At the anodic area, the oxidation reaction occurs thereby destroying the anodic metal either by dissolution or by the formation of compounds. Hence, corrosion always occurs at anodic parts.

At the anode: $$M \longrightarrow \underset{\text{(Metal ion)}}{M^{n+}} + ne^-$$

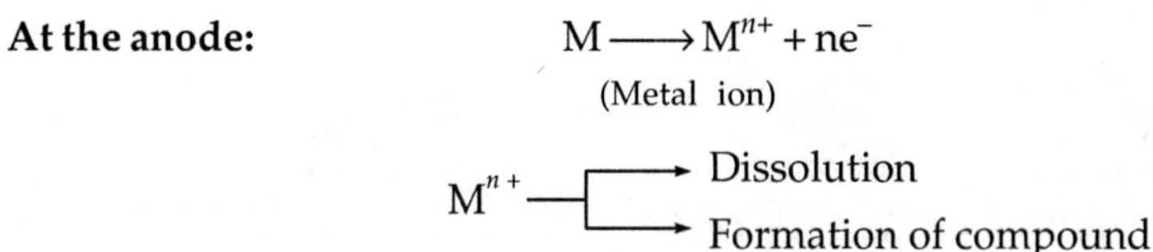

At the cathodic part, the reduction reaction (electronation) occurs. It does not affect the cathode because most metals cannot be further reduced. At the cathodic part, the dissolved constituents in the conducting medium accept the electrons forming ions (OH^-, O^{2-}). The metallic ions formed at the anodic part and the ions formed at the cathodic part diffuse toward each other through a conducting medium and form a corrosion product somewhere between the anode and the cathode.

Mechanism

Electrochemical corrosion involves the flow of electrons between the anode and the cathode. The anodic reaction involves the dissolution of metal liberating free electrons:

$$M \longrightarrow M^{n+} + ne^-$$

The cathodic reaction consumes electrons with either the evolution of hydrogen or the absorption of oxygen, which depends on the nature of the corrosive environment.

Evolution of hydrogen: This type of corrosion occurs in an acidic medium. For example, in the metal Fe, the anodic reaction is the dissolution of iron as ferrous ions with the liberation of electrons.

$$Fe \longrightarrow Fe^{2+} + 2e^- \qquad \text{(Oxidation)}$$

The released electrons flow through the metal from the anode to the cathode, whereas the H^+ ions of the acidic solution are eliminated as hydrogen gas:

$$2H^+ + 2e^- \longrightarrow H_2 \uparrow$$

The overall reaction is

$$Fe + 2H^+ \longrightarrow Fe^{2+} + H_2 \uparrow$$

This type of corrosion causes a displacement of hydrogen ions from the acidic solution by metal ions. All metals above hydrogen in the electrochemical series have a tendency to get dissolved in the acidic solution with a simultaneous evolution of H_2 gas. The anodes are large areas, whereas the cathodes are small areas.

Absorption of oxygen: For example, the rusting of iron in a neutral aqueous solution of electrolytes in the presence of atmospheric oxygen. Usually the surface of iron is coated with a thin film of iron oxide. If the film develops cracks, the anodic areas are created on the surface, whereas the metal parts act as cathodes. It shows that the anodes are small areas, whereas the rest of the metallic part forms large cathodes.

At the anode: $Fe \longrightarrow Fe^{2+} + 2e^-$ (Oxidation)

At the cathode:

The released electrons flow from the anode to the cathode through iron metal.

$$\tfrac{1}{2} O_2 + H_2O + 2e^- \longrightarrow 2OH^- \quad \text{(Reduction)}$$

$$Fe^{2+} + 2OH^- \longrightarrow Fe(OH)_2 \downarrow$$

- If oxygen is in excess, ferrous hydroxide is easily oxidized to ferric hydroxide.

$$4Fe(OH)_2 + O_2 + 2H_2O \longrightarrow 4Fe(OH)_3$$

The product called yellow rust corresponds to $Fe_2O_3 \,.\, xH_2O$.

- If the supply of oxygen is limited, the corrosion product may be black anhydrous magnetite (Fe_3O_4).

Concentration Cell Corrosion

This type of corrosion is due to an electrochemical attack on the metal surface exposed to an electrolyte of varying concentrations or of varying aerations. The most common type of concentration cell corrosion is the differential aeration corrosion that occurs when one part of metal is exposed to different air concentrations from the other part. This causes a difference in potential between the differently aerated areas. Experimentally, it has been observed that poor oxygenated parts are anodic.The differential aeration of metal causes a flow of current called the differential current.

If a metal, for example Zn, is partially immersed in a dilute solution of a neutral salt, such as NaCl, and the solution is not agitated properly, then the parts above and adjacent to the waterline are strongly aerated and become cathodic, whereas the immersed parts show a smaller oxygen concentration and become anodic. So there is a difference of potential that causes a flow of current between two differentially aerated areas of the same metal. Zinc will dissolve at the anodic areas, and oxygen will take up electrons at the cathodic areas forming hydroxyl ions.

$$Zn \longrightarrow Zn^{2+} + 2e^- \qquad \text{(Oxidation)}$$

$$\tfrac{1}{2} O_2 + H_2O + 2e^- \longrightarrow 2OH^- \qquad \text{(Reduction)}$$

Following are the facts about differential aeration corrosion:

- The less oxygenated part is the anode. Therefore, cracks serve as foci for corrosion.
- Corrosion is accelerated under the accumulation of dirt, scale, or other contaminations. This restricts the access of the oxygen resulting as an anode to promote greater accumulation. The result is localized corrosion.
- Metals exposed to aqueous media corrode under blocks of wood or glass that restricts the access of oxygen.

Stress Corrosion

This is the combined effect of static tensile stresses and the corrosive environment on a metal. The presence of tensile stress and a particular corrosive environment are necessary for stress corrosion to occur.

The corrosive agents are specific and selective, such as the following:

- Caustic alkalis and strong nitrate solution for mild steel
- Traces of ammonia for brass
- Acid chloride solution for stainless steel

This type of corrosion is marked in fabricated articles of certain alloys due to the stress caused by heavy working. Pure metals are resistant to stress corrosion. Stress corrosion is a localized electrochemical corrosion occurring along narrow paths forming anodic areas. Stress causes strain that results in localized zones of higher electrode potential. These become so chemically active that they are attacked even by a mild corrosive environment, which results in a crack that grows.

For example:

Season Cracking: This is the term applied to stress corrosion of copper alloys (brass). Pure copper is resistant to stress corrosion. Both Cu and Zn are electrochemically very active in ammonia solutions due to the formation of the stable complex ions Cu $(NH_3)_4^{2+}$ and Zn $(NH_3)_4^{2+}$, respectively. This is the cause of the dissolution of brass that ultimately results in cracks in the presence of high tensile stress.

Caustic Embrittlement: This is the most dangerous form of stress corrosion, which occurs in mild steel exposed to an alkaline medium at high stress and temperature. This kind of corrosion occurs in steam boilers and heat-transfer equipments in which the water of high alkalinity attacks mild steel plates particularly at cracks. This problem arises in high pressure boilers by using water softened with Na_2CO_3. Na_2CO_3 decomposes as

$$Na_2CO_3 + H_2O \longrightarrow 2NaOH + CO_2$$

This causes boiler water caustic. This alkaline boiler water flows into a hair-crack where water evaporates, which increases the concentration of caustic soda. The concentrated alkali dissolves iron as sodium ferroate, which decomposes as follows:

$$3Na_2FeO_2 + 4H_2O \longrightarrow 6NaOH + Fe_3O_4 + H_2 \uparrow$$

The regeneration of NaOH enhances the further dissolution of iron. The iron surrounded by dilute NaOH is the cathode, whereas the iron in contact with the concentrated caustic soda is the anodic part undergoing corrosion.

Prevention

- Using sodium phosphate as a softening agent instead of using sodium carbonate.
- Tannin or lignin is added to the boiler water, which blocks the hair cracks and prevents the caustic soda solution from coming into contact with the water.
- Adding Na_2SO_4 into boiler water to block hair cracks.

Galvanic Corrosion

When two dissimilar metals are electrically connected and exposed to an electrolyte, the metal that is higher in electrochemical series undergoes corrosion. This type of corrosion is called *Galvanic corrosion*; for example, Zinc (higher in electrochemical series) forms the anode and is attacked and gets dissolved, whereas copper (lower in electrochemical series) acts as the cathode.

Mechanism

If the solution is acidic, then corrosion occurs by the hydrogen evolution process; if the solution is neutral or slightly alkaline in nature, then corrosion occurs by the oxygen absorption process. The electrons flow from the anodic metal to the cathodic metal.

$$Zn \longrightarrow Zn^{2+} + 2e^- \quad \text{(Oxidation)}$$

Thus, corrosion occurs at the anode, that is zinc, whereas the cathode (Cu) is not corroded. For example:

- A steel pipe connected to copper plumbing
- A lead antimony solder around a copper wire

Pitting Corrosion

Pitting corrosion is a localized accelerated attack resulting in the formation of pits, holes, or cavities. Pitting corrosion therefore results in the formation of pinholes, pits, and cavities in the metal. The pitting corrosion may be due to the following:

- Metal surfaces are not homogeneous.
- External environment is not homogeneous.
- Films are not perfectly uniform.
- Crystallography directions are not equal in the reactivity.
- Environment is not uniform with respect to concentration.

Pitting is usually the result of the breakdown or cracking of the protective film on a metal at specific points. This gives rise to the formation of small anodic and large cathodic areas. In the correct environment, this produces a corrosion current. For example, stainless steel and aluminium show characteristic pitting in a chloride solution. Pitting is caused by the presence of sand, dust, scale, and other extraneous impurities on the metal surfaces. Because of the differential amount of oxygen in contact with the metal (Fig. 11.1), the small parts (underneath the impurity) become the anodic areas, and the surrounding large parts becomes the cathodic areas. Intense corrosion takes place in the anodic areas underneath the impurity. After a small pit is generated, the rate of corrosion will be increased.

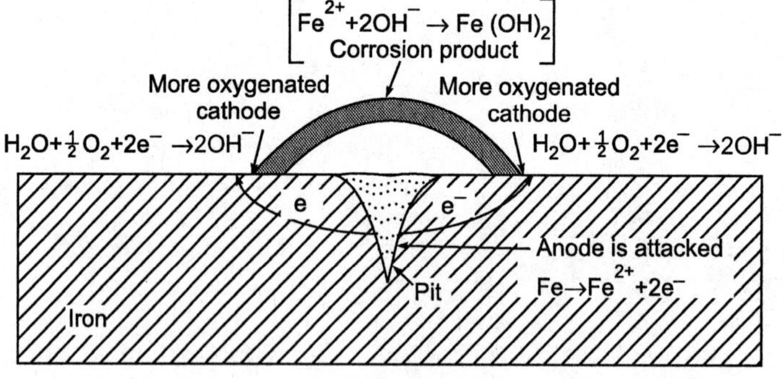

Fig. 11.1 Pitting corrosion at the surface of iron.

Waterline Corrosion

This is also known as differential oxygen concentration corrosion. In general, when water is stored in a steel tank, the maximum amount of corrosion takes place along a line just beneath the level of the water meniscus (Fig. 11.2). The area above the waterline (highly oxygenated) acts as a cathodic and is not affected by corrosion. However, if the water is relatively free from acidity, a little corrosion occurs. The problem of waterline corrosion is a matter of concern for marine engineers. This type of corrosion is prevented to a great extent by painting the sides of the ships with antifouling paints.

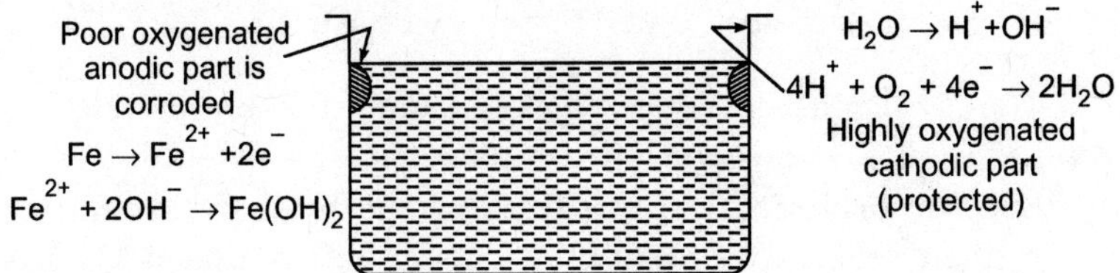

Fig. 11.2 Waterline corrosion occurs just underneath the meniscus and the water level.

Erosion Corrosion

Erosion corrosion results by the combined effect of the abrading action of vapors, gases, and liquids and the mechanical rubbing action of solids over the surface of metals. This type of corrosion is caused by the breakdown of a protective film at the spot of impingement, and its subsequent inability to repair itself under existing abrading conditions. The abrading action removes protective films from localized spots on the metal surface, resulting in the formation of differential cells at such areas and localized corrosion at anodic points of the cells. Erosion corrosion is most common in agitators, piping, condensers, tubes, and vessels in which steams of liquids or gases emerge from an opening and strike the side walls with high velocities.

Microbiological Corrosion

The result of the metabolic activity of various micro-organisms is called microbiological corrosion. The micro-organisms are either aerobic or anaerobic and develop in an environment with or without oxygen. Following are various micro-organisms that are responsible for corrosion because of their activities:

Sulphate Reducing Bacterias (Sporovobrio desulphuricous): These are responsible for the anaerobic corrosion of iron and steel. In addition to oxygen, they need sufficient amounts of sulphates for their nourishment. Their growth is maximum between pH 5–9 and temperatures of 20–30°C. The mechanism of anaerobic micro-biological corrosion of iron follows.

Anodic solution of iron:

$$8H_2O \rightleftharpoons 8H^+ + 8OH^-$$

$$4Fe + 8H^+ \rightleftharpoons 4Fe^{2+} + 8H$$

Depolarization due to activity of bacterias:

$$H_2SO_4 + 8H \rightleftharpoons H_2S + 4H_2O$$

Corrosion products:

$$Fe^{2+} + H_2S \rightleftharpoons FeS + 2H^+$$

$$3Fe^{2+} + 6OH^- \rightleftharpoons 3Fe\,(OH)_2$$

The principal corrosion products are black iron sulphide and ferrous hydroxide. The corrosion is intense and localized.

Sulphur bacterias (Thioracillus): Mostly aerobic and oxidize sulphur (present in their cells) to yield sulphuric acid, which attacks the iron. Their growth is maximum in an acidic medium (pH 0–1).

Iron and manganese micro-organisms: These are aerobic micro-organisms that live by taking into their cells iron and manganese ions, which they digest in the presence of oxygen forming insoluble hydrates of iron and manganese dioxide that are then thrown out of their bodies. Iron bacterias grow in stagnant or running water at 5–40°C and pH 4–10 and with a small amount of free dissolved oxygen.

Film-forming micro-organisms (bacterias, fungi, algae, and diatoms): These can form a micro-biological film on an iron surface. Such films are capable of maintaining concentration gradients of dissolved salts, acids, and gases on the iron surface, which leads to the formation of local biological concentration cells and consequent corrosion.

11.3 FACTORS INFLUENCING CORROSION

The rate and extent of corrosion depends on the nature of the metal and the nature of the corroding environment.

The Nature of Metal

The nature of metal is described in the following list:

Position in galvanic series: When two metals or alloys are in electrical contact in the presence of an electrolyte, the more active metal with the higher position in the galvanic series (Table 11.1) undergoes corrosion. The greater the difference in position, the faster the corrosion.

TABLE 11.1 Galvanic Series

Active (Anodic) ↓↑ Noble (cathodic)	
Active (Anodic)	1. Mg
	2. Mg alloys
	3. Zn
	4. Al
	5. Cd
	6. Al alloys
	7. Mild steel
	8. Cast iron
	9. High Ni cast iron
	10. Pb-Sn solder
	11. Pb
	12. Sn
	13. Iconel
	14. Ni-Mo-Fe alloys
	15. Brass
	16. Monel
	17. Silver solder
	18. Cu
	19. Ni
	20. Cr stainless steel
	21. 18-8 stainless steel
	22. 18-8 Mo stainless steel
	23. Ag
	24. Ti
	25. Graphite
	26. Au
Noble (cathodic)	27. Pt

A comparison of electrochemical and galvanic series is shown in Table 11.2.

TABLE 11.2 Comparison of the Electrochemical and the Galvanic Series

Electrochemical Series	*Galvanic Series*
• This series consists of metals and nonmetals.	• This series consists of metals and alloys.
• The position of a metal in this series is permanently fixed.	• Position of pure metal and when present in the form of alloy is different.
• It predicts the relative displacement tendencies.	• It predicts the relative corrosion tendencies.
• Electrode potentials are measured by dipping pure metals (without oxide film) in their salt solution of 1 M concentration.	• Corrosion of metals and alloys (without oxide film) is studied in unpolluted sea water.

The galvanic series gives real and useful information regarding the corrosion of metals and alloys.

Overvoltage: A reduction in overvoltage of the corroding metal accelerates the corrosion rate. For example, Zn in 1 N H_2SO_4 undergoes corrosion slowly because of the high overvoltage of the zinc metal (0.70 V) that reduces the effective potential to a small value. In the presence of $CuSO_4$, the corrosion rate of Zn is accelerated.

Relative areas of anodic and cathodic parts: When two dissimilar metals are in contact, the corrosion of the anodic part is directly proportional to the ratio of areas of the cathodic part and the anodic part. Corrosion is rapid and localized if the anodic area is small because the current density at a smaller anodic area is greater.

Purity of metal: Impurities in a metal generally cause a heterogeneous state forming minute electrochemical cells that result in corrosion of the anodic part, for example Zn metal with impurities (Pb or Fe). The corrosion resistance of a metal may be improved by increasing its purity.

Physical state of metal: The rate of corrosion is influenced by the physical state of the metal, such as grain size, stress, orientation of crystals, and so on. The smaller the grain size of a metal or alloy, the greater the solubility and greater the corrosion. Areas under stress even in pure metal become the anode undergoing corrosion.

Nature of surface film: In an aerated atmosphere, all metals get covered with a thin surface film of metal oxide. The ratio of the volumes of metal oxide to the metal is known as the specific volume ratio. The greater this value, the lesser the oxidation corrosion rate.

Passivity of metal: Passive metals are resistant to corrosion due to the formation of a highly protective but very thin film on the metal or alloy surface; for example, the corrosion resistance of stainless steel is due to the passivating character of the Cr present in it.

Solubility of corrosion products: In electrochemical corrosion, if the corrosion product is soluble in the corroding medium, then corrosion is rapid. If the corrosion product is insoluble, then it acts as a barrier and, suppresses further corrosion.

Volatility of corrosion products: If the corrosion product is volatile, then the underlying surface is exposed for further attack. This causes rapid and continuous corrosion; for example, MoO_3 is volatile.

The Nature of the Corroding Environment

The nature of the corroding environment is discussed in this list:

Temperature: As the temperature of the environment is increased, the reaction rate is increased, which accelerates corrosion.

Humidity of air: Critical humidity is defined as the relative humidity above which the atmospheric corrosion rate of metal increases sharply. The value of critical humidity depends on the nature of metal and the corrosion products.

The corrosion of a metal is faster in a humid atmosphere because gases (CO_2, O_2, etc.) and vapors present in the atmosphere furnish water to the electrolyte that is essential to establish an electrochemical corrosion cell. The oxide film on the metal surface has the property to absorb moisture. In the presence of this absorbed moisture, the corrosion rate is enhanced. Rain water may also wash away the oxide film from the metal surface. This leads to an enhanced atmospheric attack. The exceptions are Cr and Al.

11.4 CORROSION CONTROL

Following are the methods for controlling corrosion:

Suitable designing: The design of the material should be such that corrosion, if it occurs, is uniform and not localized. Following are the principles of suitable design:

- The contact of dissimilar metals in the presence of a corroding solution is to be avoided.
- The anodic material should have as large an area as possible when two dissimilar metals are in contact.
- When two dissimilar metals in contact have to be used, they should be as close as possible in the electrochemical series.
- An insulating filling may be used to avoid direct metal-metal electrical contact.
- The anodic metal should not be painted or coated because any break in coating would cause rapid localized corrosion.
- A suitable design should avoid the presence of cracks between adjacent parts of the structure.
- Sharp corners are poor design and should be avoided because they favor the accumulation of solids.
- The equipment should be supported on legs for free circulation of air.
- A uniform flow of corrosive liquid is desirable.
- A suitable design should prevent conditions that subject some areas of the structure to stress.

Using pure metal: Impurities in a metal cause a heterogeneous state and accelerate the corrosion rate. The corrosion resistance of a metal may be improved by increasing its purity. The corrosion resistance of a purified metal also depends on the nature of the corroding environment.

Using metal alloys: Corrosion resistance of most metals is increased by alloying them with suitable elements, for example Cr is the best suitable alloying metal for iron.

Steel containing 13 percent Cr is used in surgical equipments.

Cathodic protection: The principle is to force the metal to behave like a cathode so that corrosion does not occur. Two types of cathodic protections are possible.

- *In sacrificial protection,* the metal to be protected is connected by a wire to a more anodic metal. The more active metal loses electrons and gets corroded slowly, thereby protecting the parent cathodic metal; for example, in the galvanization process, and so on iron is protected by a covering of zinc.
  Some sacrificial anodes commonly employed are Mg, Zn, Al, and so on. Applications of this method include underground cables, water tanks, and so on.
- In *electrical cathodic protection,* an impressed current is applied in the opposite direction to nullify the corrosion current and convert the corroding metal from an anode to a cathode. The current is derived from direct sources such as a battery or rectifier on an a.c. line with an insoluble anode (graphite, platinum).

  This technique is used for long-term operations.

Modifying the environment: The rate of corrosion also depends on the corroding environment. The corrosive nature of environment can be reduced by the following:

- *Deaeration* refers to driving out dissolved oxygen by an adjustment of temperature with mechanical agitation.
- *Deactivation* involves the addition of chemicals capable of combining with oxygen in an aqueous solution.

$$2Na_2SO_3 + O_2 \longrightarrow 2Na_2SO_4$$

$$N_2H_4 + O_2 \longrightarrow N_2\uparrow + 2H_2O$$

  Using hydrated hydrazine is a suitable method because the products are N_2 (*g*) and H_2O.
- *Dehumidification* refers to the reduction of moisture content of air, for example, alumina or silica gels, which are used in an air-conditioning shop, can absorb moisture.
- *Alkaline neutralization* refers to the prevention of corrosion by neutralizing the acidic character of the corrosive environment; for example, NH_3, NaOH, lime, and so on are alkaline neutralizers.

Use of inhibitors: A corrosion inhibitor is a substance that when added in small quantities to the aqueous corrosive environment decreases the corrosion of a metal. Following are its classifications:

- *Anodic inhibitors* are adsorbed on the metal surface forming a protective film and reducing the corrosion rate, for example, chromate, phosphates, and tungstates.
- *Cathodic inhibitors* in an acidic medium creates a reaction that is the evolution of hydrogen:

$$2H^+ (aq) + 2e^- \longrightarrow H_2 (g)$$

Corrosion may be reduced either by slowing down the diffusion of the hydrated H^+ ions to the cathode or by increasing the overvoltage of the hydrogen evolution. The diffusion is decreased by organic inhibitors, whereas antimony and arsenic oxides increase the hydrogen overvoltage. In a neutral and aqueous medium, the reaction is

$$H_2O + \tfrac{1}{2} O_2 + 2e^- \rightleftharpoons 2OH^-$$

The corrosion can be controlled either by eliminating oxygen from the corroding medium or by retarding its diffusion to the cathodic area. The former is attained by adding Na_2SO_3 or by deaeration. The latter is carried out by Mg, Zn, or Ni salts, which react with OH^- ions to form the corresponding insoluble hydroxides that deposit on the cathode to form barriers.

Organic coatings: These are inert organic barriers applied on metallic surfaces for protection and also decoration. Organic coatings are of two types:

- Those that protect the metal from the environment by an impervious film.
- Organic films that contain an inhibitor.

These two types of organic coatings are combined in paints, for example, in general, there is a primary coating in paints that contains zinc chloromate and red lead, which act as powerful inhibitors.

There is another overcoat of organic coating containing ferric oxide that provides mechanical protection.

EXERCISES

1. What are concentration cell corrosion and stress corrosion? Discuss the various factors influencing corrosion.
2. (*a*) What is corrosion?
   (*b*) What are the factors that influence the corrosion?
   (*c*) Suggest some methods of corrosion control.

3. Write short notes on:
   (*a*) Waterline corrosion (*b*) Stress corrosion
4. Write a short note on cathodic protection.
5. (*a*) What is pitting corrosion?
   (*b*) Discuss any of the five methods of corrosion control.
6. Define corrosion. Discuss the mechanism of electrochemical corrosion.
7. (*a*) Name the different theories of corrosion. Describe galvanic corrosion.
   (*b*) How can corrosion be prevented or controlled?
8. (*a*) Discuss chemical corrosion in detail.
   (*b*) Describe briefly cathodic protection.
9. Explain the term cathodic protection. Indicate how metal coatings can effectively prevent corrosion.
10. (*a*) Differentiate chemical and electrochemical corrosion with suitable examples.
    (*b*) How does corrosion differ from erosion?
11. (*a*) What is chemical corrosion?
    (*b*) Discuss the role of oxygen in corrosion cells.
12. Explain the following methods of corrosion control:
    (*a*) Cathodic protection
    (*b*) Galvanization
13. (*a*) Name two metals that are noble with respect to corrosion.
    (*b*) What are the conditions for dry and wet corrosion?
14. Describe the electrochemical principles of corrosion. Illustrate with examples.
15. (*a*) Write a short note on pitting corrosion.
    (*b*) What is differential aeration corrosion? Give suitable examples.
16. Explain the mechanism of hydrogen evolution and oxygen absorption in electrochemical corrosion.
17. (*a*) Explain the rusting of iron in a neutral aqueous solution of an electrolyte in the presence of atmospheric oxygen.
    (*b*) How can the rate of corrosion be determined experimentally?
    (*c*) How is corrosion prevented by impressed current cathodic protection?
18. (*a*) State the Pilling-Bedworth rule.
    (*b*) What is galvanic corrosion?
    (*c*) How are the underground pipelines protected from soil corrosion?
    (*d*) Why does iron corrode faster than aluminium even though the oxidation potential of iron is lower than aluminium?
19. Distinguish between anodic and cathodic protection.
20. (*a*) What is a sacrificial anode? Mention its role in corrosion control.
    (*b*) Describe briefly cathodic protection.

MULTIPLE CHOICE QUESTIONS

1. Corrosion is an example of
   (*a*) oxidation (*b*) reduction
   (*c*) electrolysis (*d*) erosion
2. Which metal is less corroded?
   (*a*) pure metal (*b*) impure metal
   (*c*) impure alloys (*d*) none of these
3. When will the corrosion be rapid?
   (*a*) pH = 7 (*b*) pH > 7
   (*c*) pH < 7 (*d*) none of these
4. Rusting of iron is catalyzed by
   (*a*) Zn (*b*) Fe
   (*c*) H^+ (*d*) O_2
5. During galvanic corrosion, the more noble metal acts as
   (*a*) an anode (*b*) a cathode
   (*c*) an anode as well as a cathode (*d*) a corroding metal
6. In electrochemical corrosion,
   (*a*) the anode undergoes oxidation (*b*) the cathode undergoes oxidation
   (*c*) both undergo oxidation (*d*) none of these
7. Chemical corrosion always takes place at
   (*a*) cathodic areas (*b*) anodic areas
   (*c*) anodic and cathodic areas (*d*) in the interior of metal
8. The rate of corrosion of iron in the atmosphere depends on
   (*a*) the humidity of the air
   (*b*) the degree of atmospheric pollution
   (*c*) the frequency of rainfall
   (*d*) all are correct
9. Rusting of iron is
   (*a*) prevented by zinc coating
   (*b*) prevented if the article is connected with a wire of Mg
   (*c*) enhanced by wet air
   (*d*) retarded in the presence of dissolved salts

Chapter 12

Polymers and Polymerization

12.1 INTRODUCTION

All matter in this world is composed of extremely small units called molecules. They are too small to be seen even under the most powerful microscope and are a complex association of atoms. Molecules come in different sizes and shapes. Molecules of plastics are much larger than ordinary molecules. They are giant molecules in the form of long chains which are called polymers. The word polymer is derived from the Greek *Poly* (many) and *meros* (parts). Thus polymer means "composed of many parts or many units". *Polymer* is defined as a macromolecule of high molecular weight consisting of repeating units held together by a covalent bond. The starting material from which the molecule is formed is known as the *monomer*. The repeating unit in the polymer is called a *mer*. A polymer is a chain of repeating mers like beads in a necklace. In other words, a polymer molecule can be defined as a very long molecule in which a simple chemical unit repeats itself a large number of times.

A high polymer is one in which the number of repeating units is in excess of about 100. This number is called the *degree of polymerization,* which may be defined as the number of repeating units in the chain formed in a polymer. The length of the polymer chain is specified by the number of repeating units in the chain.

Mol. wt. of polymer = Mol. wt. of repeating unit × Degree of polymerization

Most high polymers have molecular weights between 10,000 and 1,000,000. Polymers with a low degree of polymerization are called *oligopolymers*.

12.2 NOMENCLATURE

Most of the polymers are known by their trivial names or trade names. Polymers derived from single monomers are denoted by prefixing Poly- to the name of the monomer, such as Polyethylene, Polypropylene, Polystyrene, and so on. Copolymers derived from butadiene and styrene are named Poly butadiene costyrene. Condensation polymers, such as that derived from ethylene glycol and terephthalic acid, are called poly ethylene terephthalates.

12.3 FUNCTIONALITY

The functionality of a compound depends on the number of bonding sites it possesses. A compound assumes functionality due to the presence of reactive functional groups such as —COOH, —OH, —NH_2, —SH, and so on. The number of such functional groups per molecule of the compound gives its functionality. In some cases, the functionality may arise due to the presence of double or triple bonds; for example ethylene is considered to be bifunctional because the double bond in ethylene after breaking gives two single bonds available for combination. For a substance to act as a monomer, it must have at least two reactive sites.

12.4 MOLECULAR STRUCTURE

Several molecular structures are possible depending on the functionality of monomers:

Linear polymer: The mer units are joined together end to end in single chains. Linear molecules consist of monomer units linked by primary covalent bonds, but the chains are held together by the secondary van der Waal's forces of molecular attraction. The possibility of chain movement is there (Fig. 12.1). Examples of linear polymer include:

Polyethylene, Polyvinyl chloride, and Polystyrene.

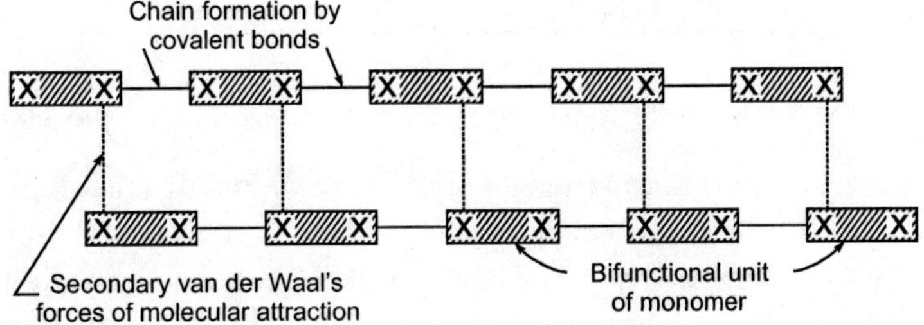

Fig. 12.1 Linear chain polymer from a bifunctional monomer.

Branched polymer: Polymers may be synthesized in which side branch chains are connected to the main chain. The movement in the branched chain molecule is restricted in comparison to the straight chain molecule. When a trifunctional monomer is mixed with a bifunctional monomer and polymerized, a branched chain polymer results (Fig. 12.2).

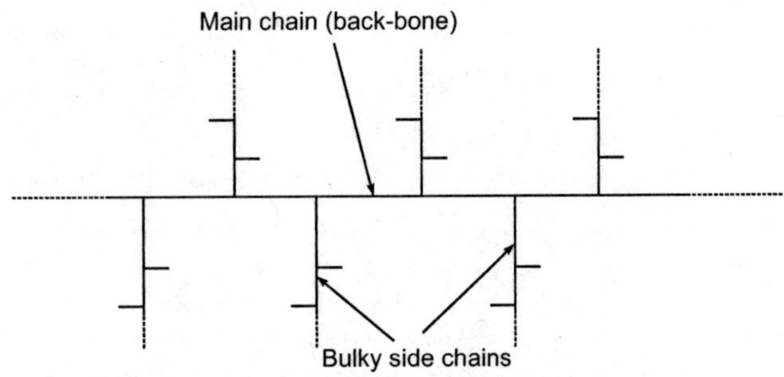

Fig. 12.2 Branched-chain polymer.

Crosslinked polymer: In crosslinked polymers, adjacent linear chains are joined one to another at various positions by covalent bonds. An example of a crosslinked polymer is an elastomer, like rubber.

Network polymers: Trifunctional mer units form 3D networks. In fact, a highly crosslinked polymer is also called a network polymer (Fig. 12.3). An example of a network polymer is Bakelite.

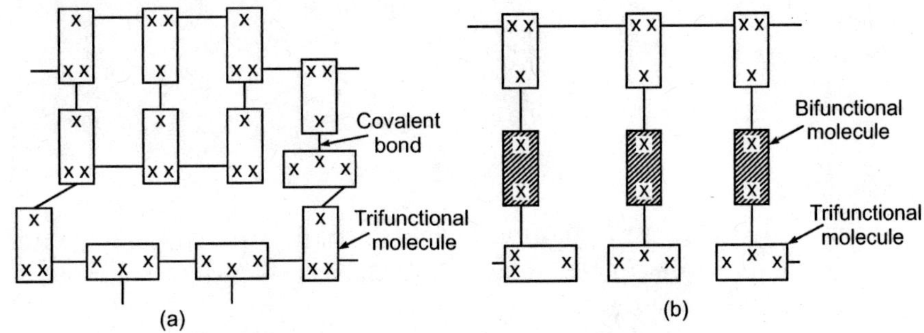

Fig. 12.3 Formation of a 3D network polymer:
(a) reaction of trifunctional molecules,
(b) reaction between bifunctional and trifunctional molecules.

12.5 CLASSIFICATION

On the basis of composition, polymers can be classified as the following:

Homopolymer: When a polymer is obtained from a single monomer unit, for example, Polystyrene, polyethylene, polypropylene, and so on.

Copolymer: When a polymer contains two monomer units in the chain. If the copolymer has three monomer units in the chain, it is referred to as a *terpolymer*. Copolymers are classified into four categories depending upon the nature of the distribution of the different monomers in the polymer chain:

- *Random copolymers* get formed by a random arrangement of monomer units in the chain.

  —A—B—A—A—B—A—B—B—B—A—A—

- *Alternating copolymer* occur when monomer units in a copolymer molecule are arranged in an alternate manner

  —A—B—A—B—A—B—A—B—A—

- A *block copolymer* consists of blocks of repeating units of one type alternating with blocks of another type.

  —A—A—A—A—B—B—B—B—A—A—A—A—

- A *graft copolymer* consists of a linear polymer chain of one type to which the side chain of a different type has been grafted.

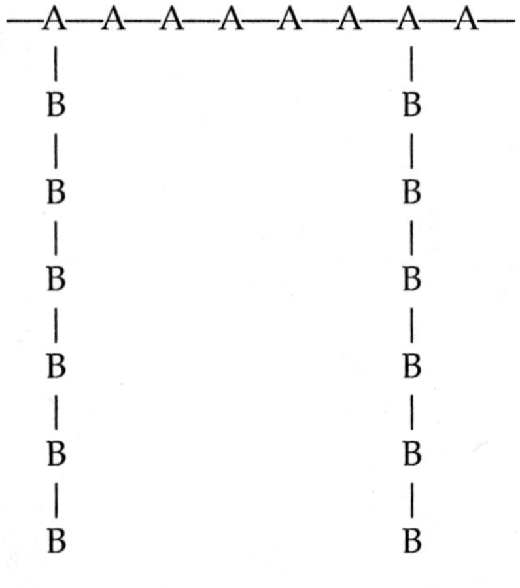

```
—A—A—A—A—A—A—A—A—
 |           |
 B           B
 |           |
 B           B
 |           |
 B           B
 |           |
 B           B
 |           |
 B           B
 |           |
 B           B
```

Tacticity

The difference in configuration because of the orientation of monomeric units in a polymer molecule in an orderly or disorderly fashion with respect to the main chain is called *tacticity*.

Isotactic polymer: Polymer in which the functional groups are all on the same side of the chain.

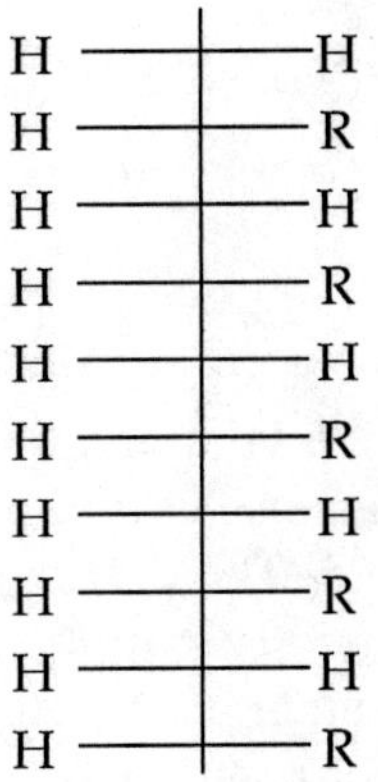

```
H ——|—— H
H ——|—— R
H ——|—— H
H ——|—— R
H ——|—— H
H ——|—— R
H ——|—— H
H ——|—— R
H ——|—— H
H ——|—— R
```

Atactic polymer: Polymer in which the arrangement of functional groups are at a random around the main chain, for example, Polypropylene.

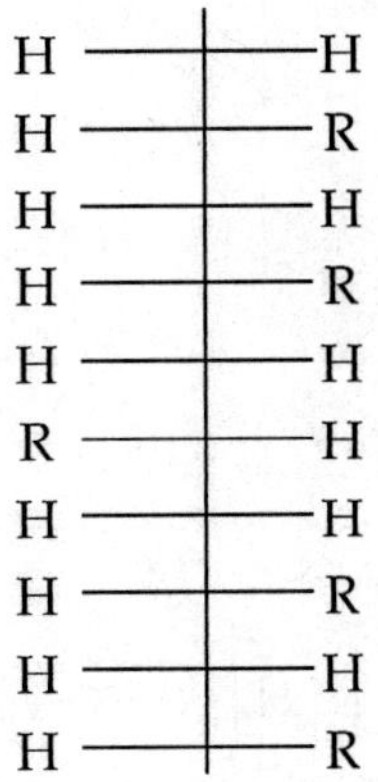

```
H ——|—— H
H ——|—— R
H ——|—— H
H ——|—— R
H ——|—— H
R ——|—— H
H ——|—— H
H ——|—— R
H ——|—— H
H ——|—— R
```

Syndiotactic polymer: Polymer in which the arrangement of side groups is in an alternating fashion called *syndiotactic polymer,* for example gutta percha.

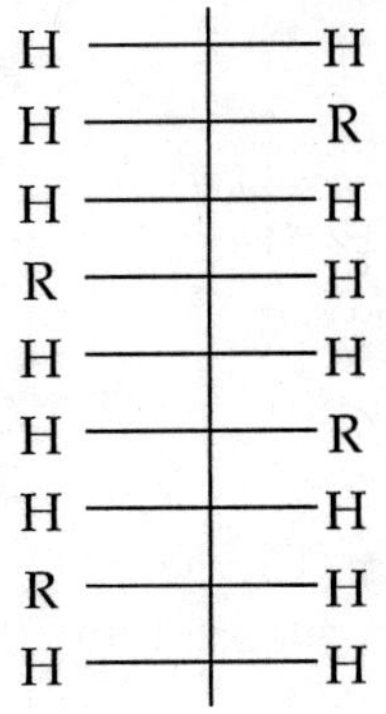

```
H ——|—— H
H ——|—— R
H ——|—— H
R ——|—— H
H ——|—— H
H ——|—— R
H ——|—— H
R ——|—— H
H ——|—— H
```

12.6 POLYMERIZATION

The process by which polymers are synthesized is called *polymerization.*

Types of Polymerization

Following are the types of polymerization:

Addition or chain growth polymerization: In this type of polymerization, the product formed is an exact multiple of the original monomeric molecule. No byproduct is formed. The product has the same elemental composition as that of the monomer. The bifunctionality is provided by the double bond present in the monomer. Compounds containing reactive double bonds can undergo this type of reaction. Such a polymerization process involving the addition of monomer units to the growing chain is known as *chain growth polymerization.*

$$n\left[\begin{array}{cc} H & H \\ | & | \\ C= & C \\ | & | \\ H & H \end{array}\right] \longrightarrow \left[\begin{array}{cc} H & H \\ | & | \\ -C- & C- \\ | & | \\ H & H \end{array}\right]_n$$

Ethylene Polyethylene

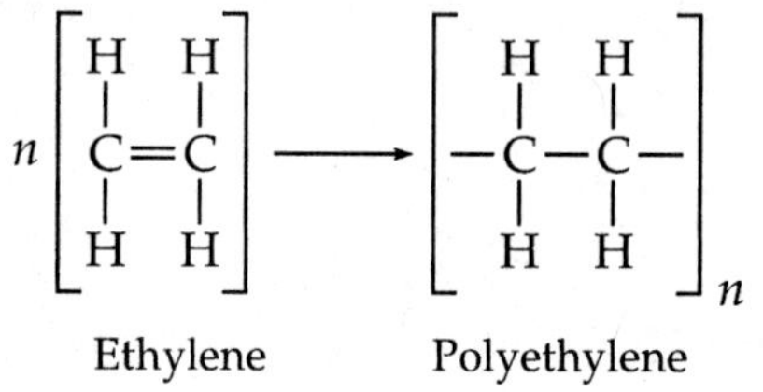

Condensation or step growth polymerization: In this type of polymerization, the polymer is formed through the reaction of functional groups. Byproducts such as H_2O and HCl are eliminated.

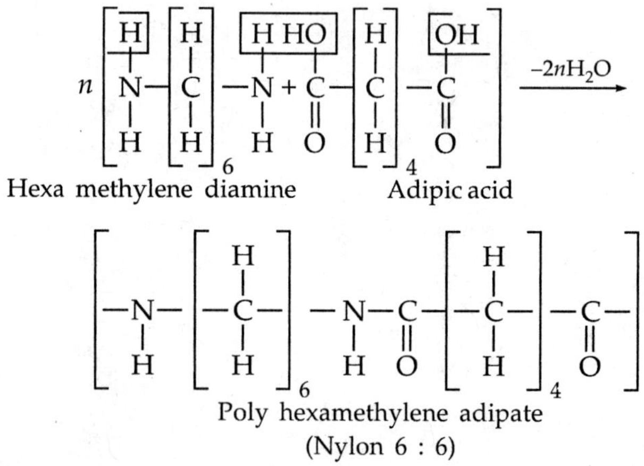

$$n\left[\begin{array}{c} H \\ | \\ N- \\ | \\ H \end{array}\left[\begin{array}{c} H \\ | \\ C \\ | \\ H \end{array}\right]_6\begin{array}{cc} H & HO \\ | & | \\ -N+ & C- \\ | & \| \\ H & O \end{array}\left[\begin{array}{c} H \\ | \\ C \\ | \\ H \end{array}\right]_4\begin{array}{c} OH \\ | \\ -C \\ \| \\ O \end{array}\right] \xrightarrow{-2nH_2O}$$

Hexa methylene diamine Adipic acid

$$\left[\begin{array}{c} \\ -N- \\ | \\ H \end{array}\left[\begin{array}{c} H \\ | \\ -C- \\ | \\ H \end{array}\right]_6\begin{array}{cc} & \\ -N- & C- \\ | & \| \\ H & O \end{array}\left[\begin{array}{c} H \\ | \\ -C- \\ | \\ H \end{array}\right]_4\begin{array}{c} \\ -C- \\ \| \\ O \end{array}\right]_n$$

Poly hexamethylene adipate
(Nylon 6 : 6)

Thus, condensation polymerization is an intermolecular combination that takes place through different functional groups present in the monomer that have affinity for each other. The process takes place in a stepwise manner, so it is called *step growth polymerization.*

Copolymerization : Copolymerization is the process in which a mixture of two or more different monomers gets polymerized to yield a product. The product is called *copolymer*.

$$nx\left[\mathrm{CH_2{=}CH{-}CH{=}CH_2}\right] + n\left[\mathrm{CH_2{=}CH(C_6H_5)}\right] \xrightarrow{\text{Copolymerization}}$$

Butadiene　　Styrene

$$\left[\left[\mathrm{{-}CH_2{-}CH{=}CH{-}CH_2{-}}\right]_x \mathrm{{-}CH_2{-}CH(C_6H_5){-}}\right]_n$$

Poly butadiene co-styrene
(Styrene-butadiene rubber)

The copolymerized product is an entirely new product containing both monomeric groups in the polymer chain in a ratio and sequence depending upon the reactivities and also on the concentration of the individual monomers.

12.7 THE MECHANISM OF ADDITION POLYMERIZATION

During addition or chain polymerization, three different types of active centers are formed: free radicals, carbonium ions, and carbanions. So the mechanisms involving these reactive species in polymerization constitute the mechanism of addition polymerization.

Free Radical Addition Polymerization

The free radical addition polymerization involves the following steps:

1. **Initiation:** This involves the dissociation of an initiator into two free radicals. Benzoyl peroxide is a commonly used initiator.

$$\underset{\text{(Initiator)}}{\mathrm{I}} \longrightarrow \underset{\text{(Free radical)}}{2\mathrm{R}^{\bullet}}$$

Then the free radical so formed adds to the first monomer molecule M

$$\mathrm{R}^{\bullet} + \mathrm{M} \longrightarrow \mathrm{R{-}M}^{\bullet}$$

2. **Propagation:** This involves the growth of a polymer chain resulting from the successive addition of other monomers to the active group.

$$R\text{—}M^{\bullet} + M \longrightarrow R\text{—}M\text{—}M^{\bullet} \longrightarrow R\ldots\ldots M^{\bullet}$$

The propagation continues until no more monomers are left for attack. The propagation steps are very fast because of the reactivity of the free radicals.

3. **Termination:** Termination can take place by either of the two means:

Coupling termination: This involves a collision between the active ends of two chains resulting in their combination.

$$R\ldots\ldots M^{\bullet} + {}^{\bullet}M\ldots\ldots R \longrightarrow R\ldots\ldots M\text{—}M\ldots\ldots R$$

Disproportionation termination: This involves the transfer of a hydrogen atom of one radical center to another radical center, resulting in the formation of two polymer units—one saturated and one unsaturated.

$$-CH_2-\underset{X}{\overset{H}{\underset{|}{\overset{|}{C}}}}{}^{\bullet} + {}^{\bullet}\underset{X}{\overset{H}{\underset{|}{\overset{|}{C}}}}-CH_2- \longrightarrow -CH_2-\underset{X}{\overset{H}{\underset{|}{\overset{|}{CH}}}} + \underset{X}{\overset{H}{\underset{|}{\overset{|}{C}}}}=\overset{H}{\overset{|}{CH}}$$

The initiator residues get incorporated into the polymer at one or both of the ends of the polymer chain. These residues do not have any appreciable effect on the properties of polymers. The term *dead polymer* signifies the cessation of the growth of the propagating radical.

Cationic Addition Polymerization

Polymerization can also be promoted by typical Friedel crafts catalysts such as aluminium chloride, boron fluoride, and other strong Lewis acids. Cationic addition polymerization involves the following steps:

Initiation: This step involves the formation of a carbonium ion, usually by the transfer of a proton.

$$Y^{+} + CH_2=\underset{X}{\underset{|}{CH}} \longrightarrow Y-CH_2-\underset{X}{\underset{|}{\overset{\oplus}{CH}}}$$

Propagation:

$$Y-CH_2-\underset{X}{\underset{|}{\overset{\oplus}{CH}}} + CH_2=\underset{X}{\underset{|}{CH}} \longrightarrow Y-CH_2-\underset{X}{\underset{|}{CH}}-CH_2=\underset{X}{\underset{|}{\overset{\oplus}{CH}}}$$

$$\xrightarrow[+nCH_2=CHX]{} Y\left[CH_2-\underset{X}{\underset{|}{CH}}\right]_n CH_2-\underset{X}{\underset{|}{\overset{\oplus}{CH}}}$$

Termination:

$$Y\!\left[\!-CH_2-\underset{\displaystyle X}{\underset{|}{CH}}-\right]_n\!CH_2-\underset{\displaystyle X}{\underset{|}{\overset{\oplus}{C}H}} \xrightarrow[\text{(an anion)}]{Z^-} Y\!\left[\!-CH_2-\underset{\displaystyle X}{\underset{|}{CH}}-\right]_n\!CH_2-\underset{\displaystyle X}{\underset{|}{CHZ}}$$

Polymer

Anionic Addition Polymerization

This involves the following steps:

Initiation: This involves the formation of negatively charged carbanions that act as chain carriers.

$$\underset{\text{(Anion)}}{\overset{\ominus}{Z}:} + CH_2=\underset{\displaystyle X}{\underset{|}{CH}} \longrightarrow \underset{\text{(Carbanion)}}{Z-CH_2-\underset{\displaystyle X}{\underset{|}{\overset{\ominus}{C}H}}:}$$

Propagation:

$$Z-CH_2-\underset{\displaystyle X}{\underset{|}{\overset{\ominus}{C}H}}: + CH_2=\underset{\displaystyle X}{\underset{|}{CH}} \longrightarrow Z-CH_2-\underset{\displaystyle X}{\underset{|}{CH}}-CH_2-\underset{\displaystyle X}{\underset{|}{\overset{\ominus}{C}H}}:$$

$$\xrightarrow[+\,nCH_2\,=\,CHX]{} Z\!\left[\!-CH_2-\underset{\displaystyle X}{\underset{|}{CH}}-\right]\!CH_2-\underset{\displaystyle X}{\underset{|}{\overset{\ominus}{C}H}}:$$

Termination:

$$Z\!\left[\!-CH_2-\underset{\displaystyle X}{\underset{|}{CH}}-\right]\!CH_2-\underset{\displaystyle X}{\underset{|}{\overset{\ominus}{C}H}}: \xrightarrow[\text{A cation}]{H^{\oplus}} Z\!\left[\!-CH_2-\underset{\displaystyle X}{\underset{|}{CH}}-\right]\!CH_2-\underset{\displaystyle X}{\underset{|}{CH_2}}$$

Polymer

12.8 THE MECHANISM OF CONDENSATION POLYMERIZATION

The chemistry of condensation polymerization deals generally with the chemistry of the functional groups. In condensation polymerization, also called step growth polymerization, the polymer molecules are built up through many separate reactions of functional groups. Polycondensation involves the condensation reaction of two or more reactive functional groups of monomers.

$$\text{O}\boxed{\text{H} + \text{HO}}\text{—}\underset{\underset{\text{O}}{\|}}{\text{C}}\text{—} \longrightarrow \text{—O—}\underset{\underset{\text{O}}{\|}}{\text{C}}\text{—} + H_2O$$

The following generalizations can be made regarding polymerization:

- Monomers should have two reactive functional groups for polymerization to occur.
- Polymerization proceeds by stepwise reaction between reactive functional groups.
- Only one type of reaction, that is a condensation reaction, between two functional groups is involved in a polymer formation.
- A polymer produced still contains both the reactive functional groups at its chain ends and hence is active and not dead as in chain polymerization.

Two types of polycondensation are possible:

AA—BB Type Polycondensation: When a pair of bifunctional monomers undergoes polycondensation.

$$n\text{A—A} + n\text{B—B} \longrightarrow \text{A—(—AB—)}_n\text{—B} + (2n-1) \text{ by product}$$

A—B Type Polycondensation: When a single bifunctional monomer undergoes self-condensation.

$$n\text{A—B} \longrightarrow \text{B—(—AB—)}_n\text{—A} + (n-1) \text{ by product}$$

If in the AA—BB type of the polycondensation reaction, one of the monomers is trifunctional, the polymer obtained is a 3D network polymer.

12.9 METHODS OF POLYMERIZATION

The polymerization reaction may be carried out in the solid phase, liquid phase, and gas phase. Most of the commercial polymers are prepared in liquid phase.

Solid Phase Polymerization

Solid phase polymerization is used for chain polymerization processes that are carried out at a low temperature.

These processes are very slow. An example of such a solid phase polymerization is the preparation of polyformaldehyde by the radiation polymerization of solid trioxane.

Gas Phase Polymerization

Gas phase polymerization occurs with a very few olefinic polymers. The methods are:

- Spraying the catalyst, generally the Zeigler-Natta Catalyst (a mixture of titanium chloride and alkyl aluminium in pentane medium), into the gaseous monomer.
- Feeding the gaseous monomer into a fluidized bed that is made up of the catalyst particles.

Examples of a gas phase polymerization are the polymerization of ethylene and *p*-xylene.

Liquid Phase Polymerization

The liquid phase polymerization may be further subdivided into four categories depending on the nature of the physical system:

Bulk Polymerization: This is the simplest process and is widely used for synthesis of condensation polymers. In this process, the monomer and initiator are kept in a reactor and heated to suitable temperature. The major drawback of this process is heat transfer. If the exothermic process control becomes more difficult, sometimes it may lead to explosions. The product obtained by bulk polymerization is of high purity. This technique is used in the free radical polymerization of methyl methacrylate.

Solution Polymerization: In solution polymerization, the reaction is carried out in the presence of a solvent. A monomer is dissolved in a suitable inert solvent along with the chain transfer agent. The free radical initiator is also dissolved in the solvent. The solvent facilitates the contact of the monomer and initiator. It also controls the viscosity increase.

It is advantageous when the polymer formed is insoluble in solvent and precipitates out as a slurry. The disadvantage of this process is the chain transfer due to the solvent, so it is difficult to get high molecular weight polymers. Evaporation of the solvent is used to separate the formed polymer. The product can also be isolated by precipitation in a nonsolvent. Block copolymers are exclusively prepared by this technique. Polyacrylonitrile is also prepared by this method.

Suspension Polymerization: By using this technique, only water-insoluble monomers can be polymerized. The monomer is suspended as droplets in a dilute aqueous solution containing protective colloids, such as polyvinylalcohol, surfactants, and so on. Suspension is brought about by the thorough agitation of the mixture. Protective colloids prevent coalescence of the droplets. A monomer soluble initiator is used. The product is obtained by filtration or spray drying. The product formed is obtained as spherical beads or pearls, so the process sometimes is known as bead or pearl polymerization.

On a commercial scale, this technique is used to obtain polyvinyl chloride, polyvinylacetate, and so on. Initiators are used in this case.

Emulsion Polymerization: This is superficially similar to suspension polymerization. But in this process a monomer dispersed in water in the presence of a surface active agent is polymerized to give a stable polymer latex. The surface active agent (surfactant) may be cationic, anionic, or neutral. The most commonly used surfactant is sodium lauryl sulphate. The role of the surfactant is to lower the surface tension at the monomer-water interface and also to facilitate emulsification of the monomer in water. Water-soluble initiators are used. The technique can provide high molecular weight polymers. This is the preferred method for obtaining synthetic rubbers that require a high molecular weight. The monomers—such as vinyl chloride, butadiene, chloroprene, acrylates, and methacrylates—are polymerized by this technique.

12.10 PLASTICS

The term plastics or plastic material generally is given to organic materials of a high molecular weight, that can be molded into any desired form when subjected to heat and pressure in the presence of a catalyst. The term *plastic* must be differentiated from the *resins*. Resins are the basic binding materials that form a major part of the plastics and that actually have undergone polymerization and condensation reactions during their preparation. However, the terms resin and plastic are now considered synonymous.

Classification of Plastics/Resins

Plastics or resins are of two types:

Thermoplastic resins: Thermoplastics soften when heated and subsequently melt. When cooled, they become hard and rigid once again. In this type of polymers, there is no crosslinking between the chains. Two chains of thermoplastics are capable of sliding over each other as the secondary van der Waals forces are weaker than the primary covalent forces. Under pressure, polymers deform especially at high temperature because the secondary van der Waals forces acting between different molecules become more and more weak and are easily overcome. The plasticity of such polymers is reversible, as the plasticity of these materials decreases as the temperature falls. The material can be molded and remolded without damage. If a relatively soft, flexible plastic is desired, thermoplastics are put to use. One disadvantage is that thermoplastics cannot be used where service conditions exceed the softening temperature. Needless to say, thermoplastics have some scrap value. Examples of thermoplastic resins include polyethylene, PVC, and nylon. Uses include plastic bags, mugs, toys, combs, buckets, and so on.

Thermosetting resins: Thermosetting resins do not become soft on heating, and they never melt once set. They are normally made from semifluid polymers with low molecular masses. In this case, the polymer chains are entangled with one another and cannot slide over each other. Deformation does not occur on heating, because only primary covalent bonds are present throughout the entire structure, forming a 3D network. They have no scrap value. Bakelite is an example and thermosetting resins are used in electric switches, telephone parts, cooker handles, and so on.

12.11 THE MOLDING CONSTITUENTS OF PLASTIC

Usually, the high polymer material is mixed with 4 to 10 ingredients, each of which imparts some useful property to the molded products. Some molding constituents are as follows:

- *Resin* is the binder that holds the different constituents together. Thermosetting resins are usually supplied as linear polymers of comparatively low molecular weights because they are fusible and easily moldable. The fusible form gets converted into the crosslinked infusible form during molding in the presence of a catalyst. Resins form 40–50 percent of the total molding mixture.
- *Plasticizers* are the substances that are added to resins to increase their plasticity and flexibility. The action of plasticizers is considered to be due to the neutralization of part of the intermolecular forces of attraction between the macromolecules of resins. Thus, they impart greater freedom of movement between the polymeric macromolecules of the resins, thereby increasing the flexibility and plasticity of the compounded material but decreasing the tensile strength and chemical resistance. A plasticizer is generally a high boiling liquid and is relatively nonvolatile at room temperature. The most commonly used plasticizers are castor oil, dibutylphthalate, alkyl, aryl phosphates, and so on.
- *Fillers* are added to impart valuable service properties such as high strength, heat resistance, finish, and workability, besides reducing the cost, shrinkage on setting, and brittleness. Fillers provide special characteristics to the products, for example fibrous fillers ensure high-strength properties. Graphite improves the antifriction properties. Asbestos and mica provide an increased heat resistance. The most widely used fillers are wood-flour, asbestos, china clay, talc, gypsum, sawdust, marble flour, paper pulp, mica, carbon black, powdered metals (Fe, Pb, Cu, and Al), metal oxides, and so on. Fillers form up to 60 percent of the mass of plastics.

- *Lubricants* are added to facilitate the molding operations (particularly in cold molding compounds). Lubricants also impart a flawless, glossy finish to the products and prevent the plastic material from sticking to the fabrication equipment. Some lubricants used are stearates, oleates, waxes, and soaps.
- *Catalysts/accelerators* are added only in the case of thermosetting resins with the object of accelerating the polymerization of the fusible resin during molding into a crosslinked infusible form. Catalysts used for compounding are hydrogen peroxide, benzoyl peroxide, acetyl sulphuric acid, metals (Ag, Cu, and Pb), metallic oxides (ZnO), ammonia, and its salts.
- *Stabilizers* offer protection against thermal degradation. Chlorine atoms present in PVC give rise to the formation of a small amount of hydrochloric acid within the plastic on storage. This leads to degradation of the PVC. Stabilizers react with the acid and neutralize it. Stabilizers commonly used are opaque molding compounds (white lead, lead chromate, litharge, red lead, lead silicate, lead naphthenate) and transparent molding compounds (stearates of lead, cadmium, and barium).
- *Coloring agents,* such as organic dye stuffs and inorganic pigments, are used to impart colors to plastics. Organic dyes soluble in that plastic are added to produce transparent articles, whereas inorganic pigments are used to impart colors to produce nontransparent articles.

12.12 MOLDING PLASTICS INTO ARTICLES

A number of methods of fabricating plastics into desired shapes have been employed depending upon whether the plastic or resin is thermosetting or thermoplastic. The methods are as follows:

Compression molding: The most popular processing technique for thermosetting plastics as well as thermoplastic resins. The predetermined quantity of plastic ingredients in proper proportions get filled between the two half pieces of the mold (Fig. 12.4). The lower half contains the cavity, whereas the upper half has a projection that fits into the cavity when the mold is closed. The gap between the projected part and the cavity determines the shape of the molded article. Two halves are closed very slowly. Finally, curing is done either by heating (in thermosetting) or cooling (in thermoplastics). After curing, the molded article is taken out by opening the mold parts. Plastic buttons, furniture handles, telephone parts, and so on are made by the compression molding technique.

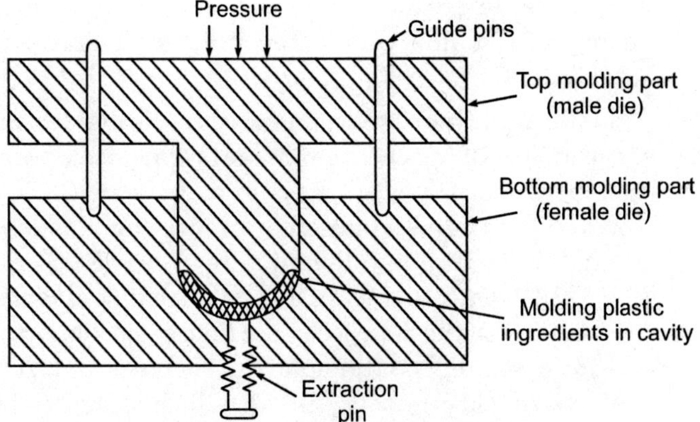

Fig.12.4 Compression molding of plastics.

Injection molding: This process is most suitable for the mass production of thermoplastic articles. The molding plastic powder is fed into a heated cylinder from where it is injected at a controlled rate into the tightly locked mold by means of a screw arrangement or by a piston plunger (Fig. 12.5). The mold is kept cold to allow the hot plastic to cure and become rigid. After sufficient curing, half of the mold is opened to allow the injection of the finished article without any deformation, and so on. Heating is done by oil or electricity. Almost all rigid thermoplastic articles, such as buckets, bowls, and furniture parts, are manufactured by this technique.

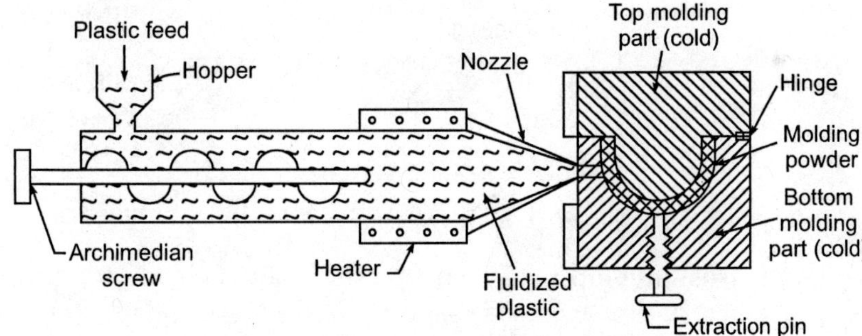

Fig. 12.5 Injection molding of plastics.

Blow molding: Hollow plastic materials such as bottles, jerry cans, and so on, are produced by the blow molding technique that originated from the glass industry. A hot softened tube of a thermoplastic is properly placed inside a two-piece hollow mold. When the two halves of the mold are closed, it pinches and closes one end of the soft tube and encloses a blowing pin at the other end. Compressed air is blown through the blowing pin while the other end of the tube remains pinched. The hot tube is inflated and assumes the shape of the hollow cavity. The mold is then cooled, opened, and the article is taken out.

Transfer molding: This method is applicable for thermosetting resins with the principle of injection molding. The molding powder is placed in a heated chamber, maintained at the minimum temperature at which the molding powder just begins to become plastic. The plastic material is then injected through an orifice into the mold by a plunger, working at a high pressure (Fig. 12.6). Due to the very great friction developed at the orifice, the temperature of the material at the time of ejection from the orifice rises to such an extent that the molding powder becomes almost liquid, and consequently it flows quickly into the mold, which is being heated up to the curing temperature required for setting. The molded article is then ejected mechanically.

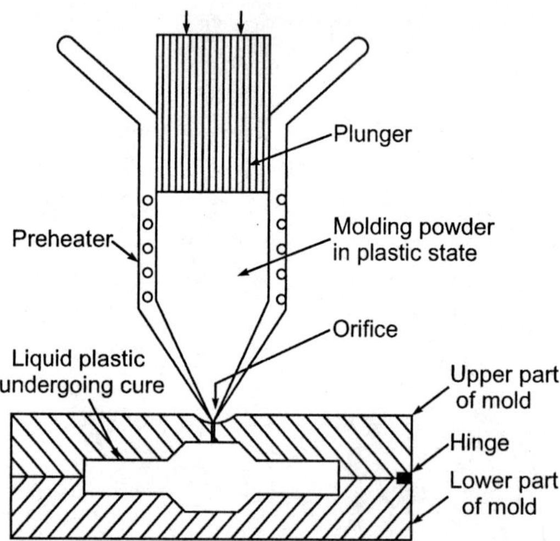

Fig. 12.6 Transfer molding of plastics.

Extrusion molding: When a plastic is softened and forced through a nozzle that forms the required shape, the technique is called *extrusion*. The technique has been used for the continuous molding of thermoplastic materials into articles of uniform cross section, such as tubes, rods, strips, and insulated electric cables. The thermoplastic ingredients are heated to a plastic condition and then pushed by means of a screw conveyor into a die that has the required outer shape of the article to be manufactured (Fig. 12.7). Here, the plastic mass gets cooled due to atmospheric exposure. A long conveyor continuously carries away the cooled product. Plastic bags can be manufactured by combining extrusion and blow molding.

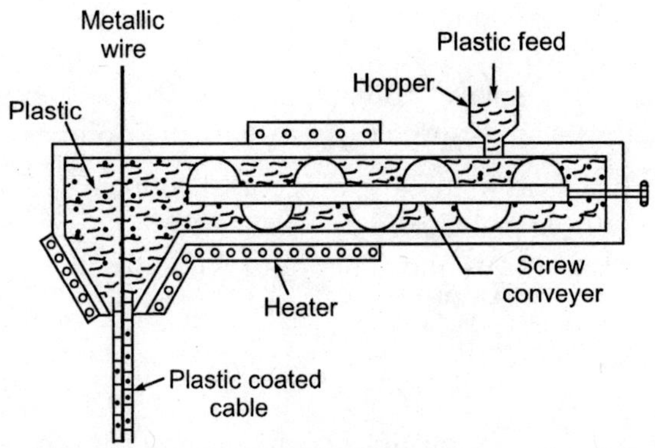

Fig. 12.7 Molding of insulated electric cable by vertical extrusion.

12.13 POLYMERIC MATERIALS

Polyethylene

More commonly known as polythene, this is the most widely used plastic. Polyethylene is obtained by the high-pressure polymerization of ethylene that uses oxygen as the initiator. The reaction takes place at a pressure as high as 1500 atmospheres and a temperature range of 180–250°C.

$$n \left[\begin{array}{c} H \quad H \\ | \quad\ | \\ C{=}C \\ | \quad\ | \\ H \quad H \end{array} \right] \longrightarrow \left[\begin{array}{c} H \quad H \\ | \quad\ | \\ -C{-}C- \\ | \quad\ | \\ H \quad H \end{array} \right]_n$$

Ethylene → Polyethylene

Polyethylene is a rigid, waxy, white, transluscent material that crystallizes very easily. Commercial polyethylene can be subdivided into three groups:

Low-density polyethylene (LDPE): Density (0.91–0.925 g/cm^3) M.P (110–125°C).

Polymerized under very high pressure (1000–5000 atmospheres and temperature range of 180–250°C in the presence of O_2 as initiator.

Medium-density polyethylene (MDPE): Density (0.925–0.940 g/cm^3) M.P. (130–140°C).

Polymerized under medium pressure.

High-density polyethylene (HDPE): Density (0.941–0.965 g/cm^3) M.P. (145–150°C).

Polymerized under atmospheric pressure (6–7 atmospheres) and temperature at 60°C in the presence of the Zeigler-Natta Catalyst [$TiCl_4$ + Al $(C_2H_5)_3$].

The high-density polyethylene, which is completely linear, has better chemical resistance and a higher softening point but is relatively brittle.

LDPE is used in making film and sheeting. Pipes made of LDPE are used for agricultural, irrigation, and domestic water line connections. HDPE is used in the manufacture of toys and other household articles.

Polyvinyl Chloride

Polyvinyl chloride is produced from vinylchloride by suspension polymerization. Vinyl chloride is obtained by the reaction of ethylene with HCl.

$$CH_2{=}CH_2 + HCl \longrightarrow \underset{\text{Vinyl chloride}}{CH_2{=}CHCl}$$

$$\underset{\text{VC}}{n\left[\begin{array}{cc} H & Cl \\ | & | \\ C & = C \\ | & | \\ H & H \end{array}\right]} \longrightarrow \underset{\text{PVC}}{\left[\begin{array}{c} H \quad Cl \\ | \quad\; | \\ -C-C- \\ | \quad\; | \\ H \quad H \end{array}\right]_n}$$

Polyvinyl chloride occurs as a colorless, rigid material, has a high density, and has a low softening point. It is resistant to light, atmospheric oxygen, inorganic acids, and alkalis. The material is the most widely used synthetic plastic mainly for cable insulation, leather cloth, packaging, and toys. It is also used for the manufacturing of film, sheet, and floor covering. PVC pipes are used for carrying corrosive chemicals in petrochemical factories.

Polystyrene

Polystyrene is one of the lightest plastic materials. Bulk and suspension polymerization are the most commonly used techniques. The monomer styrene is obtained from benzene and ethylene.

$$C_6H_6 + CH_2{=}CH_2 \xrightarrow{AlCl_3} C_6H_5{-}CH_2{-}CH_3 \xrightarrow{MgO} C_6H_5{-}CH{=}CH_3$$

(Styrene)

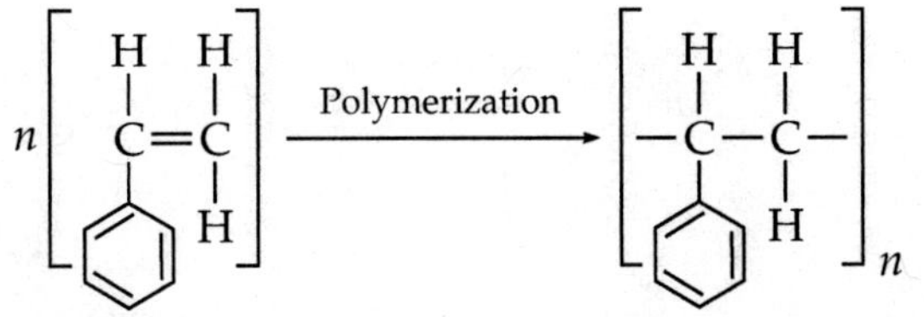

$$n\left[\begin{array}{c}\text{H}\quad\text{H}\\ |\quad\ |\\ \text{C}=\text{C}\\ |\quad\ |\\ \text{C}_6\text{H}_5\ \ \text{H}\end{array}\right] \xrightarrow{\text{Polymerization}} \left[\begin{array}{c}\text{H}\quad\text{H}\\ |\quad\ |\\ -\text{C}-\text{C}-\\ |\quad\ |\\ \text{C}_6\text{H}_5\ \ \text{H}\end{array}\right]_n$$

Styrene Polystyrene

Polystyrene is transparent, light, has excellent moisture resistance, has a low softening range, and is brittle. It has good electrical insulation characteristics.

Polystyrene is used for toys, combs, buttons, radio and television parts, food containers, food packaging, umbrella handles, and so on.

Polypropylene

Polypropylene is the lightest known industrial polymer. The monomer propylene is obtained as a byproduct in the gasoline refineries. Polypropylene is obtained by polymerizing propylene in the presence of the Ziegler-Natta Catalyst [Al $(C_2H_5)_3$ + $TiCl_4$].

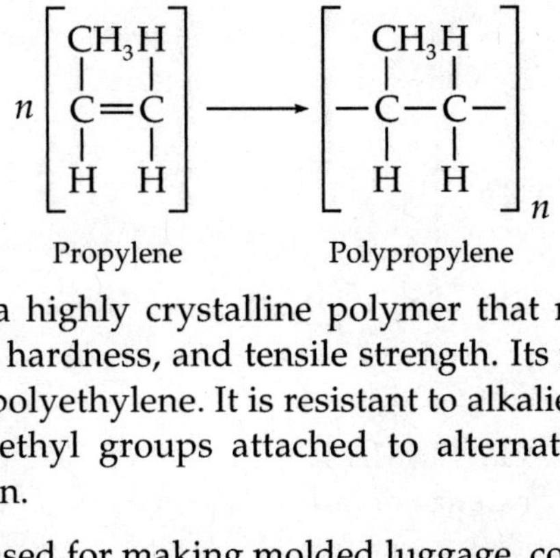

$$n\left[\begin{array}{c}\text{CH}_3\ \text{H}\\ |\quad\ |\\ \text{C}=\text{C}\\ |\quad\ |\\ \text{H}\quad\text{H}\end{array}\right] \longrightarrow \left[\begin{array}{c}\text{CH}_3\ \text{H}\\ |\quad\ |\\ -\text{C}-\text{C}-\\ |\quad\ |\\ \text{H}\quad\text{H}\end{array}\right]_n$$

Propylene Polypropylene

Polypropylene is a highly crystalline polymer that melts at 160–170°C. It exhibits high stiffness, hardness, and tensile strength. Its moisture resistance is comparable to that of polyethylene. It is resistant to alkalies, acids, and oils. Due to the presence of methyl groups attached to alternate carbon atoms, it is susceptible to oxidation.

Polypropylene is used for making molded luggage, colorful furniture, water buckets, and components of refrigerators, radios, and TVs. It is also used for producing packaging films and pipes. Its fibers are extremely strong and are used in making ropes. It is also used as an insulating material for electrical wires, and in the manufacture of drinking glasses, dinnerware, and so on.

Polyvinyl Acetate (PVA)

Polyvinyl acetate is a polymerized product produced by heating vinyl acetate in the presence of benzoyl peroxide as initiator.

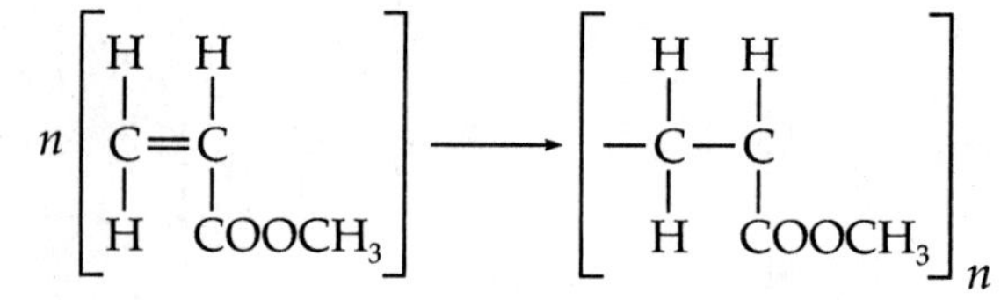

$$n\left[\begin{array}{cc} H & H \\ | & | \\ C = & C \\ | & | \\ H & COOCH_3 \end{array}\right] \longrightarrow \left[\begin{array}{cc} H & H \\ | & | \\ -C - & C \\ | & | \\ H & COOCH_3 \end{array}\right]_n$$

Vinyl acetate — Polyvinyl acetate

Polyvinyl acetate is a solid, soft, and sticky material that is resistant to atmospheric oxygen, water, and chemicals.

It is used as a basic material for chewing gums and surgical dressings. It is also used for the manufacture of lacquers, paints, and adhesives.

Polymethyl Methacrylate (Trade Name: Lucite or Plexiglass)

Polymethyl methacrylate (PMMA) is obtained by the polymerization of methyl methacrylate in the presence of H_2O_2. The monomer methyl methacrylate can be obtained from acetone as follows:

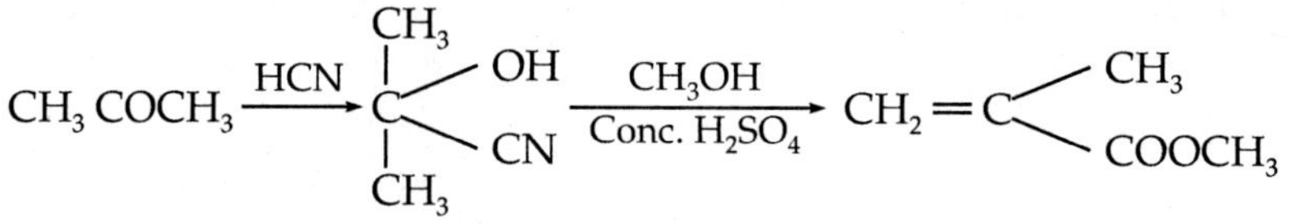

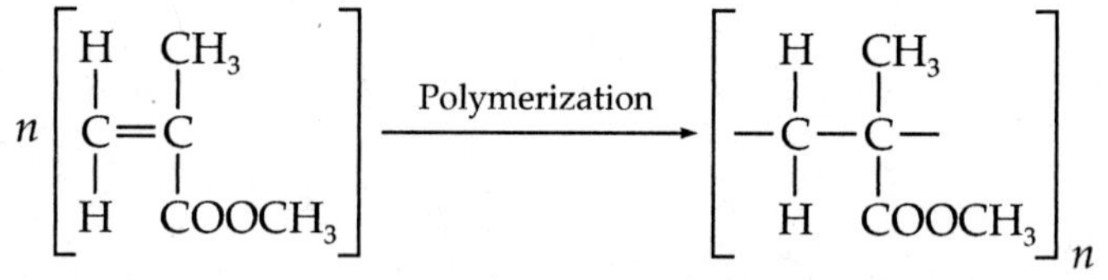

$$CH_3COCH_3 \xrightarrow{HCN} (CH_3)_2C\begin{cases} OH \\ CN \end{cases} \xrightarrow[\text{Conc. } H_2SO_4]{CH_3OH} CH_2 = C\begin{cases} CH_3 \\ COOCH_3 \end{cases}$$

$$n\left[\begin{array}{cc} H & CH_3 \\ | & | \\ C = & C \\ | & | \\ H & COOCH_3 \end{array}\right] \xrightarrow{\text{Polymerization}} \left[\begin{array}{cc} H & CH_3 \\ | & | \\ -C - & C- \\ | & | \\ H & COOCH_3 \end{array}\right]_n$$

PMMA

The commercial PMMA is a transparent thermoplastic, which is amorphous because of the presence of bulky side groups. The softening range is 130–140°C. It has low chemical resistance to hot acids and alkalis.

This material is an excellent substitute for glass. It has numerous applications, such as lenses, paints, TV screens, aircraft, and so on.

Polytetrafluoro Ethylene (PTFE) (Trade Name: Teflon)

This material was invented by the Dupont company.

Polytetrafluoro ethylene (PTFE) is a highly polymerized resin obtained by the polymerization of tetrafluoroethylene gas under pressure in the presence of benzoyl peroxide as initiator. The monomer is prepared from chloroform and hydrofluoric acid as follows:

$$CHCl_3 + HF \longrightarrow CHF_2Cl \longrightarrow CF_2{=}CF_2$$

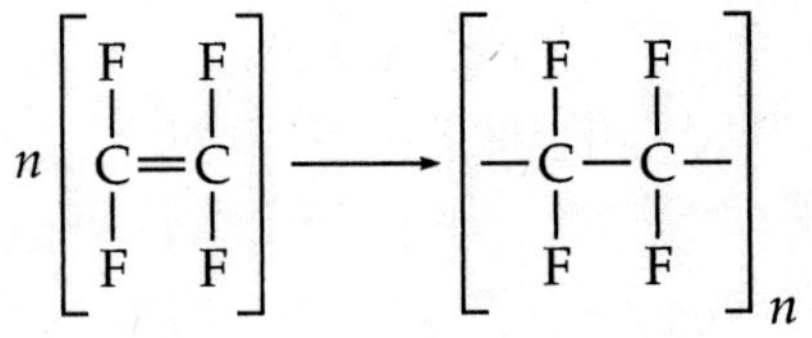

$$n\left[\begin{matrix} F & & F \\ | & & | \\ C & = & C \\ | & & | \\ F & & F \end{matrix}\right] \longrightarrow \left[\begin{matrix} & F & & F & \\ & | & & | & \\ - & C & - & C & - \\ & | & & | & \\ & F & & F & \end{matrix}\right]_n$$

TFE PTFE (Teflon)

This is marketed under the trade names Teflon or Fluon.

PTFE has good heat resistance, good moisture resistance, and exceptionally high chemical resistance. This material has a high softening point (327°C).

Its most important application is as a nonstick coating for utensils and cookingware. It is used for making packings, tank linings, tubings, pump parts, and so on. It has also been used for coating and impregnating glass fibers, asbestos fibers, and cloths.

Polycarbonate

Polycarbonate belongs to a group of plastics called engineering plastics; these plastics have better properties as construction material and are used in the engineering industry. These are the polyesters of phenol and carbonic acid and are obtained by the condensation polymerization of the diphenoxy methylene derivatives with diphenyl carbonate. Bisphenol-A [2, 2,-bis (4-hydroxy phenyl) propane] on interaction with diphenyl carbonate gives a polycarbonate.

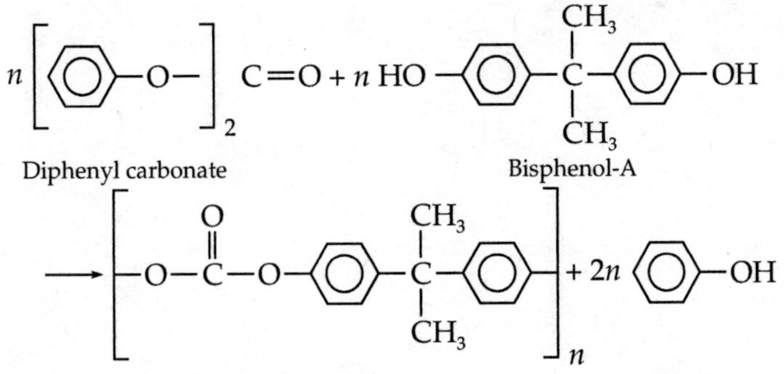

$$n\left[C_6H_5{-}O{-}\right]_2 C{=}O + n\,HO{-}C_6H_4{-}C(CH_3)_2{-}C_6H_4{-}OH$$

Diphenyl carbonate Bisphenol-A

$$\longrightarrow \left[{-}O{-}\overset{\overset{\displaystyle O}{\|}}{C}{-}O{-}C_6H_4{-}C(CH_3)_2{-}C_6H_4{-}\right]_n + 2n\,C_6H_5{-}OH$$

Polycarbonate Phenol

Polycarbonate is an amorphous powder that melts at around 265°C. It has a very high impact strength, is resistant to water, and has good electrical-insulation characteristics.

Many useful products, such as safety shields, baby bottles, telephone parts, and machinery housings are made from it.

Bakelite (Phenol-formaldehyde Resin)

These are produced by the polycondensation of a phenol with formaldehyde in the presence of acids or alkalis as catalysts.

$$CH_2{=}O + H^{\oplus} \longrightarrow \overset{\oplus}{C}H_2{-}OH$$

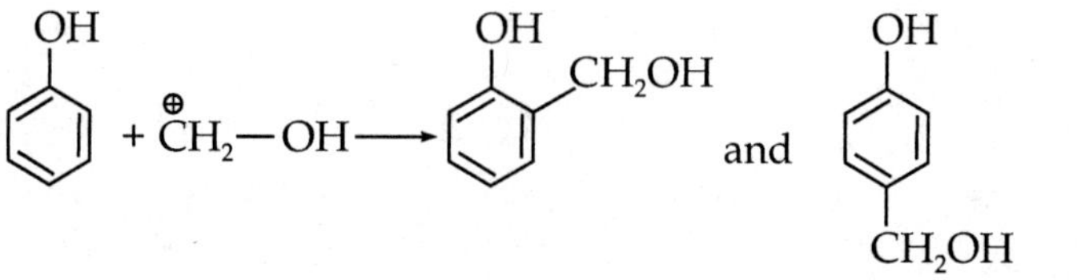

OH OH OH

CH_2OH

$+ \overset{\oplus}{C}H_2{-}OH \longrightarrow$ and

CH_2OH

o-hydroxy methyl phenol

p-hydroxy methyl phenol

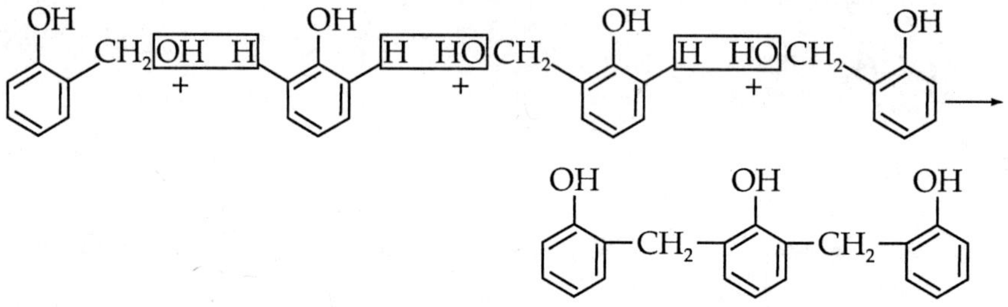

OH OH OH OH

CH_2 OH H + H HO CH_2 + H HO CH_2 + $\longrightarrow$

OH OH OH

CH_2 CH_2

Novolac

The initial reaction results in the formation of a linear polymer novolac. During molding, hexamethylene tetramine is added, which converts the fusible novolac into hard, infusible crosslinked structure bakelite.

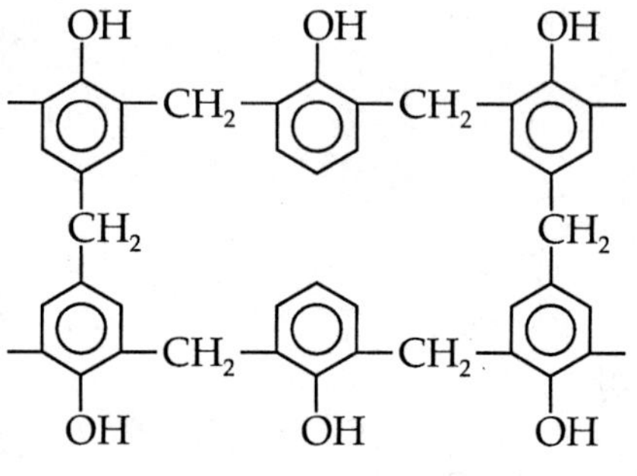

OH OH OH

CH_2 CH_2

CH_2 CH_2

CH_2 CH_2

OH OH OH

Bakelite

Bakelite is a rigid, hard, scratch-resistant infusible substance, which is attacked by alkalies. The substance possesses an excellent electrical insulating character.

Bakelite is used for good electrical, automotive, radio and TV parts; adhesives; bearings; paints; varnishes; and so on.

Nylon (Polyamides)

Nylon was invented by W.H. Carothers (1896–1937) at the Dupont company. The word *nylon* is now accepted as a generic term for the synthetic polyamides that are characterized by a repeating amide linkage (—NHCO—). The nylons have been named on the basis of the number of carbon atoms in the monomer chain.

Nylon 6: This is prepared from a polyamide called caprolactum, which contains six carbon atoms.

$$C_6H_{11}OH \xrightarrow{[O]} C_6H_{10}{=}O \xrightarrow{NH_2OH} C_6H_{10}{=}NOH \xrightarrow{[O]}$$

Cyclohexanol Cyclohexanone Cyclohexanone oxime

$$\text{NH}-CH_2-CH_2-CH_2-CH_2-CH_2-C{=}O \text{ (ring)} \xrightarrow{\text{Polymerization}} \left[-(CH_2)_5-\overset{O}{\overset{\|}{C}}-\overset{H}{\overset{|}{N}}-\right]_n$$

Caprolactum Nylon 6

Nylon 6 is a polycaprolactum that gets softened around 210°C. This is soluble in mineral acids and organic solvents. Its abrasion resistance is excellent. Nylon is used in the manufacture of tires and nonwoven fabrics.

Nylon 6, 6: Hexamethylene diamine and adipic acid are polymerized in a 1:1 ratio.

$$n\left[H-\overset{H}{\underset{H}{N}}-\left(\overset{H}{\underset{H}{C}}\right)_6-\overset{H}{\underset{H}{N}} + HO-\underset{O}{\overset{\|}{C}}-\left(\overset{H}{\underset{H}{C}}\right)_4-\underset{O}{\overset{\|}{C}}-OH\right] \xrightarrow{\text{Polymerization}}$$

Hexa methylene diamine Adipic acid

$$\left[-\underset{H}{N}-\left(\overset{H}{\underset{H}{C}}\right)_6-\underset{H}{N}-\underset{O}{\overset{\|}{C}}-\left(\overset{H}{\underset{H}{C}}\right)_4-\underset{O}{\overset{\|}{C}}-\right]_n + 2nH_2O$$

Polyamide (Nylon 6, 6)

This is a linear polymer that is not resistant to alkali and mineral acids. Oxidizing agents such as hydrogen peroxide, potassium permanganate, and so on are able to degrade the fiber.

This is mainly used in the manufacture of tire cords. Other uses include the manufacture of carpets, rope, fiber, cloth, and so on.

Nylon 6, 10: This is manufactured by the polymerization of hexamethylene diamine and sebacic acid.

$$n\left[H-\underset{H}{\overset{}{N}}-\left[\underset{H}{\overset{H}{C}}\right]_6-\underset{H}{N}-H + HO-\underset{O}{\overset{}{C}}-\left[\underset{H}{\overset{H}{C}}\right]_8-\underset{O}{\overset{}{C}}-OH\right]$$

Hexa methylene diamine   Sebacic acid

$$\longrightarrow \left[-\underset{H}{N}-\left[\underset{H}{\overset{H}{C}}\right]_6-\underset{H}{N}-\underset{O}{\overset{\|}{C}}-\left[\underset{H}{\overset{H}{C}}\right]_8-\underset{O}{\overset{\|}{C}}-\right]_n$$

Nylon 6, 10

It is mainly used in the production of monofilaments for brushes, sporting equipment, and so on.

Nylon 11: This is more expensive than other nylons. It is manufactured by self-condensation of the *w*-amino decanoic acid with the removal of water.

$$n[H_2N-(CH_2)_{10}-COOH] \longrightarrow \left[-(CH_2)_{10}-\overset{O}{\overset{\|}{C}}-\overset{H}{\overset{|}{N}}-\right]_n$$

Nylon-11

This is less water sensitive than other nylons because of the lengths of the hydrocarbon chain in the monomer. Due to its hydrophobic nature, it finds little commercial use. It has very good electrical insulating properties and melts at 185°C.

Nylon 11 is used mainly as a textile fiber.

Terylene (Polyester)

Polyester was discovered by the Imperial Chemical Company in the United Kingdom during World War II. Terylene is a polyester fiber made from ethylene glycol and terephthalic acid. The terephthalic acid required for the manufacture of terylene is produced by the catalytic atmospheric oxidation of *p*-xylene.

$$\underset{p\text{-xylene}}{CH_3-C_6H_4-CH_3} \xrightarrow[V_2O_5,\ 360°C]{O_2} HOOC-C_6H_4-COOH$$

$$\underset{\text{Ethylene glycol}}{\left[HO-\left[\begin{matrix}H\\|\\C\\|\\H\end{matrix}\right]_2-OH\right]} + \underset{\text{Terephthalic acid}}{HO-\overset{O}{\overset{\|}{C}}-C_6H_4-\overset{O}{\overset{\|}{C}}-OH} \xrightarrow[\text{Polymerization}]{\text{Condensation}} \left[\left[\begin{matrix}H\\|\\-C-\\|\\H\end{matrix}\right]_2-O-\underset{O}{\underset{\|}{C}}-C_6H_4-\underset{O}{\underset{\|}{C}}-\right]_n$$

(Polyethylene terephthalate)
PETP

This occurs as a colorless rigid substance. This is highly resistant to mineral and organic acids but is less resistant to alkalis. This is hydrophobic in nature. This has a high melting point due to the presence of aromatic ring.

It is mostly used for making synthetic fibers and can be blended with wool and cotton for better use and wrinkle resistance. Another application of the polyethylene terephthalate film is in electrical insulation.

Epoxy Resin

These are crosslinked polymers in which crosslinking is achieved by the reaction of epoxy groups. These are basically polyethers. The raw materials are bisphenol and epichlorohydrin, which are obtained as follows:

$$2\,C_6H_5OH + CH_3-CO-CH_3 \xrightarrow{H^+} \underset{\text{Bisphenol}}{HO-C_6H_4-\overset{CH_3}{\overset{|}{\underset{CH_3}{\underset{|}{C}}}}-C_6H_4-OH}$$

$$CH_3-CH=CH_2 + Cl_2 \xrightarrow[-\,HCl]{} CH_2=CH-CH_2Cl$$

$$\xrightarrow{HOCl} CH_2Cl-CHOH-CH_2Cl \xrightarrow{CaO} \underset{\text{Epichlorohydrin}}{CH_2Cl-\overset{\quad O \quad}{CH-CH_2}}$$

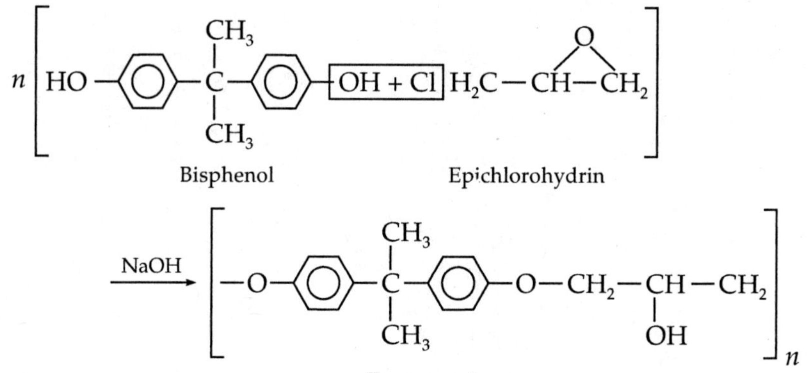

$$n\left[HO-C_6H_4-\underset{CH_3}{\overset{CH_3}{C}}-C_6H_4-\boxed{OH + Cl}\,H_2C-\overset{O}{CH-CH_2}\right]$$

Bisphenol Epichlorohydrin

$$\xrightarrow{NaOH}\left[-O-C_6H_4-\underset{CH_3}{\overset{CH_3}{C}}-C_6H_4-O-CH_2-\underset{OH}{CH}-CH_2\right]_n$$

Epoxy resin

They are resistant to chemical attack, are flexible, and are also resistant to water. They have good electrical properties and are heat resistant.

They find use in casting, potting, encapsulating, industrial flooring, adhesive, foams, and so on.

Silicones (Inorganic Polymer)

Most polymers (natural or synthetic) are organic, but there is one important class known as silicones in which silicon is present in the chain. Silicon is capable of forming long chains (such as carbon), and in these chains, silicon is always combined with oxygen.

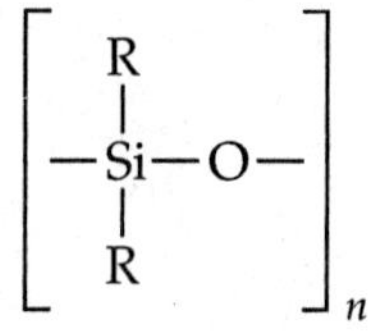

$$\left[-\underset{R}{\overset{R}{Si}}-O-\right]_n \quad R = \text{alkyl or phenyl radical}$$

The monomers for silicone polymer are produced by the hydrolysis of chlorosilanes. The unstable monomers condense to yield silicon polymer.

- Dimethyl silicon dichloride gives linear polymer:

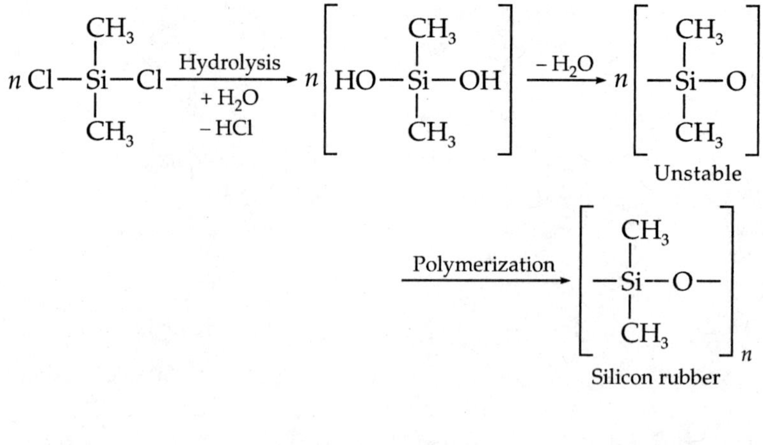

$$n\,Cl-\underset{CH_3}{\overset{CH_3}{Si}}-Cl \xrightarrow[+H_2O \; -HCl]{\text{Hydrolysis}} n\left[HO-\underset{CH_3}{\overset{CH_3}{Si}}-OH\right] \xrightarrow{-H_2O} n\left[-\underset{CH_3}{\overset{CH_3}{Si}}-O\right]$$

Unstable

$$\xrightarrow{\text{Polymerization}}\left[-\underset{CH_3}{\overset{CH_3}{Si}}-O-\right]_n$$

Silicon rubber

- Trimethyl silicon chloride gives polymer of a limited chain length:

$$2\,CH_3-\underset{CH_3}{\overset{CH_3}{\underset{|}{\overset{|}{Si}}}}-Cl \xrightarrow[-2HCl]{+2H_2O} 2\left[CH_3-\underset{CH_3}{\overset{CH_3}{\underset{|}{\overset{|}{Si}}}}-OH\right] \xrightarrow{-H_2O} CH_3-\underset{CH_3}{\overset{CH_3}{\underset{|}{\overset{|}{Si}}}}-O-\underset{CH_3}{\overset{CH_3}{\underset{|}{\overset{|}{Si}}}}-CH_3$$

Unstable

- Mono methyl silicon chloride is trifunctional and gives crosslinking to the product.

$$n\,CH_3-\underset{Cl}{\overset{Cl}{\underset{|}{\overset{|}{Si}}}}-Cl \xrightarrow[-3HCl]{+3H_2O} n\left[CH_3-\underset{OH}{\overset{OH}{\underset{|}{\overset{|}{Si}}}}-OH\right] \xrightarrow{\text{Polymerization}} \begin{matrix} & | & & | & \\ CH_3- & Si & -O- & Si & - \\ & | & & | & \\ & O & & O & \\ & | & & | & \\ - & Si & -O- & Si & - \\ & | & & | & \\ & CH_3 & & CH_3 & \end{matrix}$$

Unstable

Silicones may be liquid, viscous liquid, semisolid, rubber-like, and solids depending upon the proportion of various alkyl silicon halides used during the polymerization reactions. They exhibit excellent stability at high temperatures, good water resistance, good oxidation stability, and poor chemical resistance.

Silicones are used for the following:

Silicone fluids: The product gets separated as an oil. They are relatively low molecular weight silicones. They are used as high-temperature lubricants, an antifoaming agent for paper, and a textile treatment. Silicone fluids are modified and used as silicone greases and silicone compounds.

Silicone rubbers or elastomers: These are obtained by mixing high-molecular weight linear dimethyl silicone polymers with fillers (SiO_2) and curing agents. They are used as gaskets and seals, wires, and cable insulation. They also are used in surgical and prosthetic devices and as adhesive in the electronics industry.

Solid silicone resins: These are highly crosslinked silicone polymers obtained by condensing bifunctional and trifunctional silicon halides. They can withstand a temperature up to 200°C. They are used as insulating varnishes, encapsulating agents, and in industrial paints. Silicone resins give a nonstick surface to frying pans.

Synthetic Rubber

According to the American Society for Testing and Materials (ASTM), synthetic rubber or elastomer is defined as a "polymeric material which at room temperature can be stretched to at least twice its original length and upon immediate release of the stress will return quickly to approximately its original length." Following are the types of some elastomers:

Styrene Rubber (GR-S or Buna-S): This is produced by the copolymerization of butadiene and styrene:

$$nx\left[\mathrm{CH_2{=}CH{-}CH{=}CH_2}\right] + n\left[\mathrm{CH_2{=}CH(C_6H_5)}\right] \longrightarrow \left[\left[\mathrm{{-}CH_2{-}CH{=}CH{-}CH_2{-}}\right]_x \mathrm{{-}CH_2{-}CH(C_6H_5){-}}\right]_n$$

Butadiene Styrene (Styrene butadiene rubber, SBR)

It resembles natural rubber and has excellent abrasion resistance and resilience. Synthetic rubber is not resistant to oil, ozone, and weather, and it has good electrical properties.

This is used in the manufacture of motor tires. Other uses include floor tiles, shoe soles, gaskets, adhesives, tank lining, and so on.

Nitrile Rubber (GR-A or Buna-N): A copolymer of butadiene and acrylonitrile:

$$m\left[\mathrm{CH_2{=}CH{-}CH{=}CH_2}\right] + n\left[\mathrm{CH_2{=}CH(CN)}\right] \longrightarrow \left[\mathrm{{-}CH_2{-}CH{=}CH{-}CH_2{-}}\right]_m\left[\mathrm{{-}CH_2{-}CH(CN){-}}\right]_n$$

1, 3 Butadiene Acrylonitrile Poly butadiene co-acrylonitrile

Nitrile rubber has excellent resistance to heat, sunlight, oils, acids, and salts, but it is less resistant to alkalis because of the presence of cyano groups.

This is used for making conveyor belts, high-altitude aircraft components, tank linings, hoses, gaskets, adhesives, automobile parts, and so on.

Neoprene (GR-M): Made by polymerization of chloroprene:

$$n\left[\mathrm{CH_2{=}CH{-}CH{=}CH_2}\right] \xrightarrow{\text{Polymerization}} \left[\mathrm{{-}CH_2{-}CCl{=}CH{-}CH_2{-}}\right]_n$$

Chloroprene Polychloroprene (Neoprene)

Chemically, it has close resemblance to natural rubber. It has outstanding resistance to oils, chemicals, sunlight, temperature, and so on, and is highly soluble in polar solvents.

Neoprene is used for making gaskets, belts, hoses, adhesives, and so on.

Butyl Rubber (GR-I): This is made by the copolymerization of isobutene and isoprene (small amount):

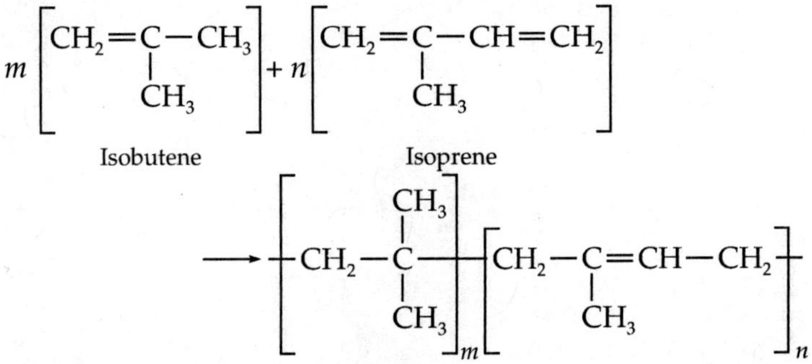

$$m\left[\begin{array}{c}CH_2{=}\underset{\displaystyle CH_3}{\underset{|}{C}}{-}CH_3\end{array}\right] + n\left[\begin{array}{c}CH_2{=}\underset{\displaystyle CH_3}{\underset{|}{C}}{-}CH{=}CH_2\end{array}\right]$$

Isobutene Isoprene

$$\longrightarrow \left[\left[-CH_2-\overset{\displaystyle CH_3}{\overset{|}{\underset{\displaystyle CH_3}{\underset{|}{C}}}}-\right]_m\left[-CH_2-\underset{\displaystyle CH_3}{\underset{|}{C}}{=}CH-CH_2-\right]\right]_n$$

It has excellent resistance to sunlight, weathering, heat, abrasion, acids, alkalies, and so on. It is not flame resistant and is nonoil resistant.

It is used in cable insulation, bicycle and automobile tubes, hoses, and so on.

Polysulphide Rubber (Thiocol or GR-P): This is made by condensation polymerization between sodium polysulphide and ethylene dichloride.

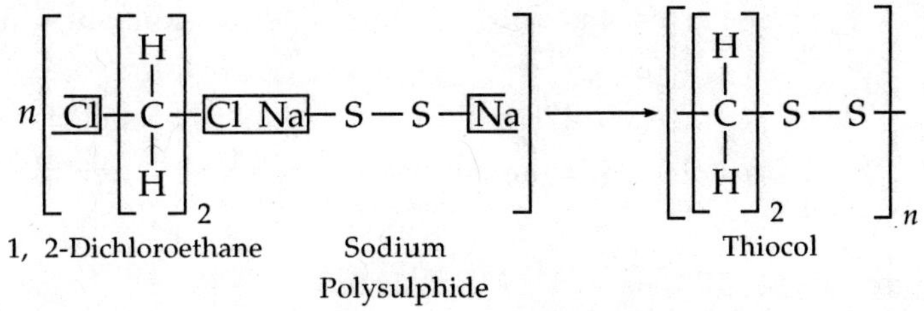

$$n\left[\boxed{Cl}-\left[\overset{\displaystyle H}{\overset{|}{\underset{\displaystyle H}{\underset{|}{C}}}}\right]_2-\boxed{Cl\ Na}-S-S-\boxed{Na}\right] \longrightarrow \left[-\left[\overset{\displaystyle H}{\overset{|}{\underset{\displaystyle H}{\underset{|}{C}}}}\right]_2-S-S-\right]_n$$

1, 2-Dichloroethane Sodium Polysulphide Thiocol

Polysulphide rubber has excellent resistance to mineral oils, fuels, solvents, oxygen, ozone, and sunlight. It cannot be vulcanized and hence does not form hard rubber. It possesses poor abrasion resistance.

This material is used for making oil and gasoline hoses, gaskets, washers, and so on.

Chlorosulphonyl Polyethylene (Hypalon): This is obtained by the reaction of polyethylene with chlorine and SO_2.

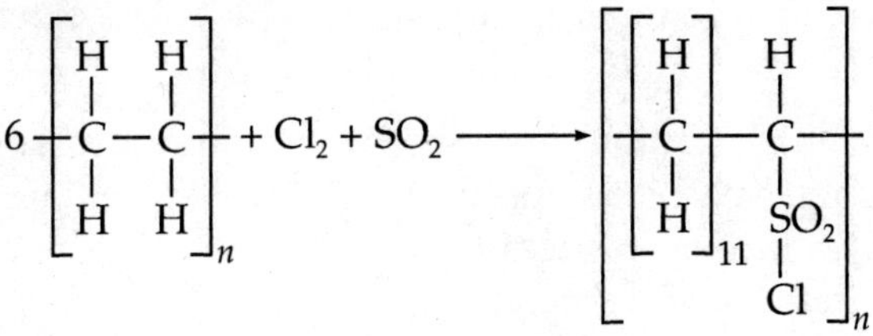

$$6\left[-\overset{\displaystyle H}{\overset{|}{\underset{\displaystyle H}{\underset{|}{C}}}}-\overset{\displaystyle H}{\overset{|}{\underset{\displaystyle H}{\underset{|}{C}}}}-\right]_n + Cl_2 + SO_2 \longrightarrow \left[\left[-\overset{\displaystyle H}{\overset{|}{\underset{\displaystyle H}{\underset{|}{C}}}}-\right]_{11}-\overset{\displaystyle H}{\overset{|}{\underset{\underset{\displaystyle Cl}{\underset{|}{SO_2}}}{\underset{|}{C}}}}-\right]_n$$

It has excellent resistance to oxidation, sunlight, ozone, and chemicals. It has moderate oil resistance and unlimited colorability. The abrasion resistance is excellent.

This material is used in making shoe soles, seals, gaskets, tank linings, and so on.

Polyurethane Rubber: This is produced by the reaction of polyalcohols with diisocyanates.

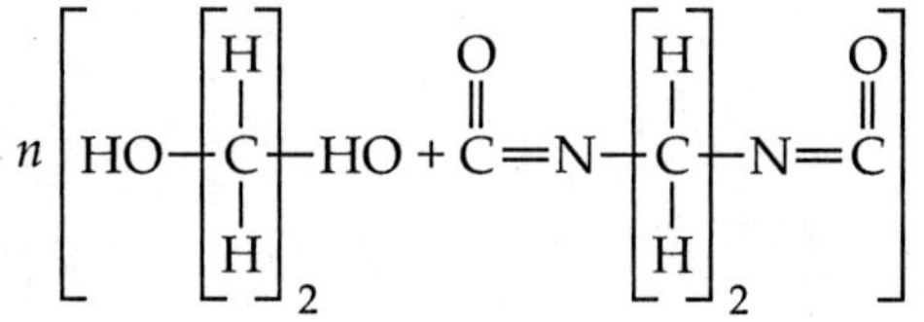

$$n\left[HO{-}\left(\begin{array}{c}H\\|\\C\\|\\H\end{array}\right)_2{-}HO + \overset{O}{\overset{\|}{C}}{=}N{-}\left(\begin{array}{c}H\\|\\C\\|\\H\end{array}\right)_2{-}N{=}\overset{O}{\overset{\|}{C}}\right]$$

Ethylene glycol Ethylene diisocyanate

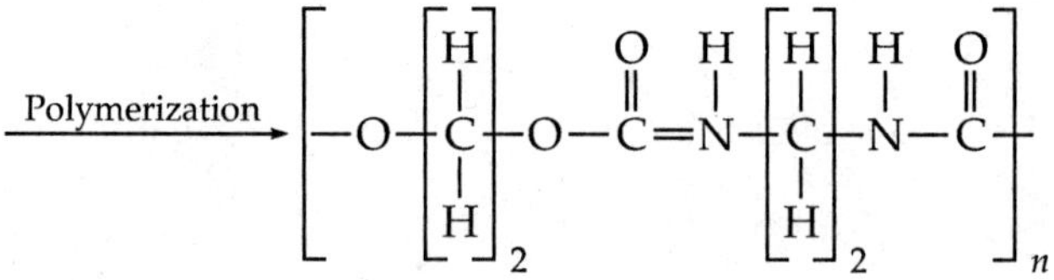

$$\xrightarrow{\text{Polymerization}}\left[{-}O{-}\left(\begin{array}{c}H\\|\\C\\|\\H\end{array}\right)_2{-}O{-}\overset{O}{\overset{\|}{C}}{=}\overset{H}{\overset{|}{N}}{-}\left(\begin{array}{c}H\\|\\C\\|\\H\end{array}\right)_2{-}\overset{H}{\overset{|}{N}}{-}\overset{O}{\overset{\|}{C}}{-}\right]_n$$

Polyurethane rubber

These are the hardest type and have extremely good abrasion resistance. They show good resistance to organic solvents but are attacked by acids and alkalis.

This material is used in making foams, shoe heels, surface coatings, and so on.

Polysiloxanes (Silicone Rubber): Discussed under silicones.

Vulcanization

The drawback that rubber becomes soft and sticky in warm weather and hard and stiff in cold weather remained a major problem, limiting its use. The difficulty was finally overcome through the discovery of the vulcanization of rubber using sulphur by Charles Goodyear in 1839. Vulcanization is a process in which the polymer chains of rubber are crosslinked through sulphur atoms. The process is carried out by heating rubber with sulphur or sulphur compounds at a temperature up to 150°C for a few hours. It converts rubber into a tougher product that has much better elasticity over a wide range of temperature and is no longer dissolved in organic solvents.

The vulcanization process converts the natural rubber into either soft rubber (containing 5 percent sulphur) or hard rubber (containing 30–35 percent sulphur). The vulcanization process can be enhanced in the presence of certain organic substances known as accelerators.

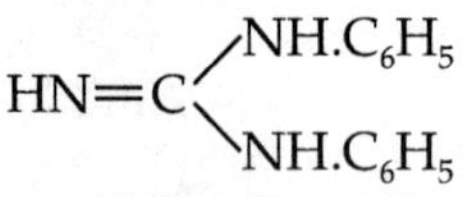

$$HN{=}C\begin{array}{l}\diagup NH.C_6H_5\\ \diagdown NH.C_6H_5\end{array}$$

Diphenylguanidine

$$C_6H_4\begin{array}{l}\diagup N \diagdown\\ \diagdown S \diagup\end{array}C.SH$$

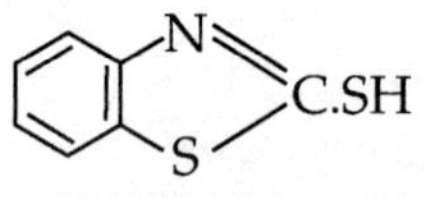

Mercaptobenzathiazole

Organic accelerators are further promoted by the inorganic compounds known as activators, for example, oxides or salts of zinc, lead, and calcium. The solubility of the activators is generally increased by the addition of a long chain of fatty acid or its salts. The chemistry of vulcanization is complex and is yet to be established.

EXERCISES

1. (*a*) Briefly discuss the principles underlying the synthesis of PVC.
   (*b*) Distinguish between addition and condensation polymerization, and provide examples.
2. (*a*) Discuss the properties and applications of polystyrene, PVC, and polyethylene.
   (*b*) Write notes on silicones.
3. (*a*) What are the methods of polymerization? Describe what you know about thermoplastic and thermosetting resins.
   (*b*) Write notes on synthetic rubbers.
4. (*a*) Discuss the mechanism of addition polymerization. Define the following terms: monomer, homopolymer, copolymer, and graft copolymer.
   (*b*) Write notes on silicones.
5. Discuss the mechanism of addition and condensation polymerization. Give examples.
6. Write brief notes on:
   (*a*) Epoxy resins (*b*) Synthetic rubbers.
7. Give a brief account of thermoplastic and thermosetting resins. What is a copolymer?
8. Write notes on synthetic rubber.
9. Write short notes on:
   (*a*) Bakelite (*b*) Silicones
10. Describe the differences between addition and condensation polymerizations, and give suitable examples.
11. (*a*) Differentiate between homopolymer and copolymer.
    (*b*) What are resins?
12. Explain the terms thermoplastic and thermosetting plastic. Give examples of each.
13. (*a*) Write the mechanism of the free-radical polymerization of styrene using benzoyl peroxide as the initiator.
    (*b*) What do thermoplastic and thermosetting resins mean? Give examples.

(c) What are the main differences between LDPE and HDPE?
(d) Write the equation for the synthesis of terylene.

14. Differentiate between thermoplastics and thermosetting plastics. Give one example of each type.
15. (a) What are the molding constituents of plastics?
    (b) What are the types of molding of plastics? Discuss with their applications.

MULTIPLE CHOICE QUESTIONS

1. High polymers are
   (a) liquids (b) gaseous
   (c) solids (d) colloids
2. Phenol-formaldehyde is commercially known as
   (a) PVC (b) elastomer
   (c) Bakelite (d) nylon
3. Raw materials used for the manufacture of polyester are
   (a) vinyl chloride (b) urea + formaldehyde
   (c) glycol + terephthalic acid (d) phenol + formaldehyde
4. Bakelite is an example of
   (a) thermoelastic (b) thermoplastic
   (c) thermosetting (d) thermite
5. Natural rubber is a polymer of
   (a) propylene (b) isoprene
   (c) ethylene (d) propane
6. Nylon is
   (a) a polythene derivative (b) a polyester fiber
   (c) a polyamide fiber (d) none of these
7. Tetra fluoroethylene is the monomer of
   (a) nylon-66 (b) teflon
   (c) polythene (d) PVC
8. Chloroprene is the monomer of
   (a) polystyrene (b) neoprene
   (c) PVC (d) polythene
9. Molecular mass of polymer is
   (a) small (b) very small
   (c) large (d) negligible
10. Which one is a macromolecule but not a polymer?
    (a) cellulose (b) ice
    (c) PVC (d) polystyrene

11. Terylene is a
   (*a*) polyamide (*b*) polyester
   (*c*) polyglycol (*d*) polycarbonate

12. Which one is a copolymer
   (*a*) buna-S (*b*) teflon
   (*c*) PVC (*d*) polystyrene

13. Nylon 6, 6 is
   (*a*) polyester (*b*) polyamide
   (*c*) polyglycol (*d*) polyacrylate

14. Which of the following polymers contains nitrogen?
   (*a*) nylon (*b*) teflon
   (*c*) PVC (*d*) terylene

15. Neoprene is a polymer of
   (*a*) isoprene (*b*) chloroprene
   (*c*) butadiene (*d*) acrylnitrate

Chapter 13

FUEL AND COMBUSTION

13.1 INTRODUCTION

Fuel is a combustible substance containing carbon as the major constituent, which on proper burning, gives large amount of heat that can be used economically for domestic and industrial purposes. In other words, any source of heat energy is fuel, for example wood, coal, charcoal, kerosene, petrol, producer gas, and so on.

During the process of combustion of a fuel such as coal, the atoms of carbon, hydrogen, and so on, combine with oxygen simultaneously liberating heat at a rapid rate. The energy is liberated due to the rearrangement of valence electrons in these atoms resulting in the formation of compounds (CO_2, H_2O, etc.). The products have comparatively less energy. So the energy released during the process of combustion is the difference of the energy of the reactants and that of the products.

$$\text{Fuel} + O_2 \longrightarrow \text{Products} + \text{Heat}$$

However, combustion is not necessary for a fuel to give out energy. According to the modern concept of fuel, any chemical or reactant that produces energy in a form that can be used for production of power is called fuel. A typical example of this is a fuel cell.

13.2 CLASSIFICATION

The classification of fuels is according to the occurrence and state of aggregation. According to occurrence, fuel is of the natural (primary) or derived (secondary) type, and according to the state of aggregation, the fuels are classified as solid, liquid, and gaseous fuels.

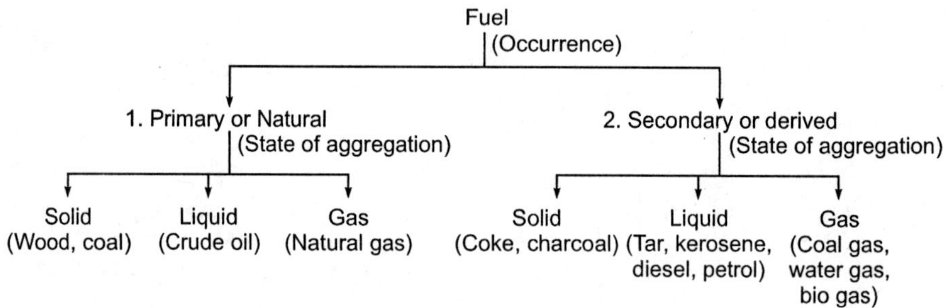

13.3 UNITS OF HEAT

Following are the units of heat:

C.G.S. System (Calorie): The unit of heat on the C.G.S. system is the calorie, which may be defined as the amount of heat required to raise the temperature of 1 gram of water through 1°C (15–16°C).

M.K.S. System (Kilocalorie): The unit of heat on the M.K.S. system is the kilocalorie, which may be defined as the amount of heat required to raise the temperature of 1 kilogram of water through 1°C.

$$1 \text{ kcal} = 1000 \text{ cal.}$$

British Thermal Unit (B.T.U.): The unit of heat on the British system is the British thermal unit (B.T.U.), which may be defined as the amount of heat required to raise the temperature of 1 pound of water through 1°F (60–61°F).

$$1 \text{ B.T.U.} = 252 \text{ cal} = 0.252 \text{ kcal}$$

$$1 \text{ kcal} = 3.968 \text{ B.T.U.}$$

Centigrade Heat Unit (C.H.U.): This is the amount of heat required to raise the temperature of 1 pound of water through 1°C.

$$1 \text{ kcal} = 3.968 \text{ B.T.U.} = 2.2 \text{ C.H.U.}$$

Calorific Value

The calorific value is defined as the total quantity of heat liberated by the complete combustion of one unit of fuel in oxygen.

Higher or Gross Calorific Value: Hydrogen is present in almost all fuels, and when the calorific value of fuel is determined experimentally, hydrogen is converted into steam. If the products of combustion are condensed to room temperature (25°C or 77°F), the latent heat of the condensation of the steam is also included in the measured heat. The total value calculated is known as the higher or gross calorific value and may be defined as the total amount of heat liberated when one unit of the fuel is burnt completely, and the combustion products are cooled to room temperature.

Lower or Net Calorific Value: In actual practice, during combustion of a fuel, the water vapors escape along with hot combustion gases and are not condensed. Hence, a lesser amount of heat is available. This is called the lower or net calorific value and may be defined as the amount of heat liberated when one unit of fuel is burnt completely, and the combustion products are allowed to escape. Thus,

$$\text{LCV} = \text{HCV} - \text{Latent heat of water vapor formed}$$

Now because 1 part by mass of hydrogen gives 9 parts by mass of water, the equation becomes:

$$\text{LCV} = \text{HCV} - \text{Mass of } H_2 \times 9 \times \text{Latent heat of steam}$$

Units of calorific value for solid, liquid, and gaseous fuels are shown in Table 13.1.

TABLE 13.1 Units of Calorific Value

Fuels	*Units (Calorific Value)*	
Solid and Liquid	Calories/gm k calories/kg B.T.U./lb	(C.G.S.) (M.K.S.) (F.P.S.)
Gas	kcal/Cubic meter B.T.U./Cubic feet	(kcal/m^3) (B.T.U./ft^3)

Determination of Calorific Value

A *bomb calorimeter* is used for determining the calorific value of solid and liquid fuels.

Construction

A simple sketch of a bomb calorimeter is shown in Fig. 13.1. It consists of a strong cylindrical stainless steel bomb in which the combustion of fuel is carried out. The bomb has a lid, which can be screwed to the body of the bomb to make a perfect gas tight seal. The lid is provided with two stainless steel electrodes and an oxygen inlet valve. A small ring is attached to one of the electrodes. In this ring, a nickel or stainless steel crucible can be supported. The bomb is placed in a copper calorimeter that is surrounded by an air jacket and water jacket to prevent heat loss due to radiation. The calorimeter is provided with an electrically operated stirrer and Beckmann's thermometer, which is sensitive enough to read upto 0.01°C.

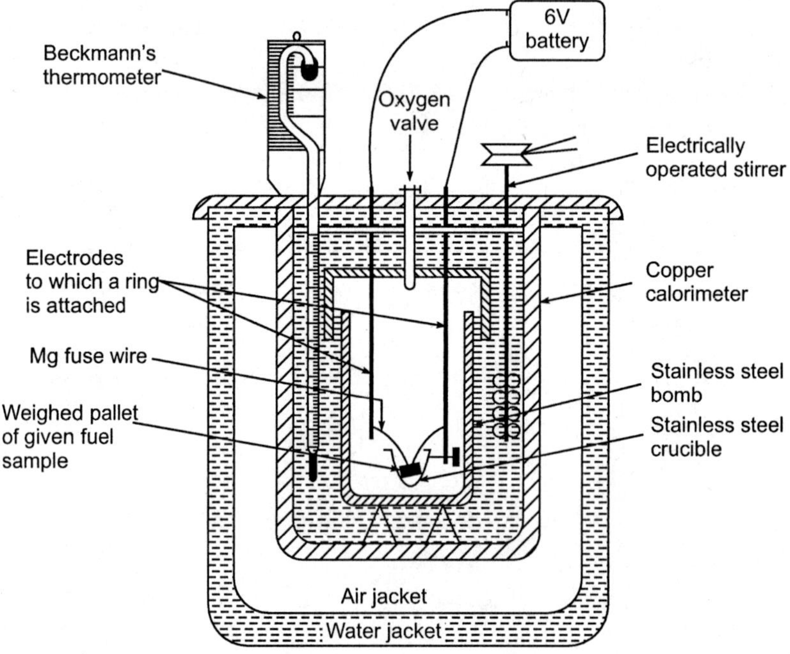

Fig.13.1 Bomb calorimeter.

Working

A weighed amount of the fuel is taken in the clean crucible. The crucible is supported over the ring. A fine magnesium wire touching the fuel sample is then stretched across the electrodes. The bomb lid is tightly screwed, and the bomb is filled with oxygen to 25 atmospheric pressure. The bomb is then lowered into the copper calorimeter, containing a known mass of water. The stirrer is worked and the initial temperature of water is noted. The electrodes are then connected to a

6-volt battery, and then circuit is completed. The sample burns, and heat is liberated. Uniform stirring of the water is continued, and the maximum temperature attained is recorded.

Observation

Weight of fuel taken = x gm

Weight of water in calorimeter = W gm

Water equivalent of the calorimeter, stirrer, thermometer, and bomb

= Wt. of apparatus × specific heat = W gm

Initial temperature of water in calorimeter = t_1°C

Final temperature of water in calorimeter = t_2°C

Let the higher calorific value of fuel = L cal/g

Calculation

Heat gained by water = $W \times (t_2 - t_1)$ cal

Heat gained by calorimeter = $w \times (t_2 - t_1)$ cal

Total heat gained = $W (t_2 - t_1) + w (t_2 - t_1)$ cal

= $(W + w)(t_2 - t_1)$ cal

Heat liberated by fuel = $x \times L$

Now, Heat liberated by fuel = Heat gained by water and calorimeter

$$x \times L = (W + w)(t_2 - t_1)$$

$$\therefore \quad HCV (L) = \frac{(W + w)(t_2 - t_1)}{x} \text{ cal/g}$$

To calculate lower (net) calorific value:

Let the percentage of hydrogen in fuel = H

Wt. of water produced from 1 gm of fuel = $\frac{9H}{100}$ gm = 0.09 H gm

∴ Heat taken by water in forming steam = 0.09 H × 587 cal

[Latent heat of steam = 587 cal/gm]

Hence, LCV = HCV – Latent heat of water formed

= (L – 0.09 H × 587) cal/gm

Corrections

For accuracy, the following corrections must be taken into consideration:

Cooling correction: Rate and time taken for cooling the water in the calorimeter from maximum temperature to room temperature must be considered. From the rate of cooling ($dt°$/minute) and the actual time taken for cooling (t minutes), the cooling correction of $dt \times t$ is added to the rise in temperature:

$$\therefore \quad L = \frac{(W + w)(t_2 - t_1 + \text{Cooling correction}) - [\text{Acid + fuse corrections}]}{\text{Mass of fuel } (x)}$$

Fuse wire correction: The heat liberated, as measured previously, includes the heat given out by the ignition of the fuse wire used.

Acid correction: During ignition, sulphur and nitrogen (if present) in the fuel are oxidized to the corresponding acids along with the evolution of heat:

$$S + 2H + 2O_2 \longrightarrow H_2SO_4 + \text{Heat}$$

$$2N + 2H + 3O_2 \longrightarrow 2HNO_3 + \text{Heat}$$

So this heat is also included in the measured heat and must be subtracted. The amount of these acids are analyzed from washings of the bomb by titration, whereas sulphuric acid alone is determined by precipitation as $BaSO_4$. The correction for 1 mg of S is 2.25 cal, whereas for 1 ml N/10 HNO_3 formed, it is 1.43 cal.

Boy's Gas Calorimeter

The Boy's gas calorimeter is the apparatus generally used for measuring the calorific values of gaseous fuels and liquid fuels that can be easily vaporized.

Construction

The apparatus consists of a suitable gas burner in which a known volume of gas can be burnt at a known pressure and at a uniform rate (3–4 liters per minute). Around the burner, there is a chimney or combustion chamber with a copper tubing coiled inside as well as outside. Water is passed at a constant rate at the top of the outer coil; it leaves the chamber at the top of the inner coil (Fig. 13.2). Two thermometers, T_1 and T_2, are fitted for recording the temperature of the incoming and outgoing water, respectively.

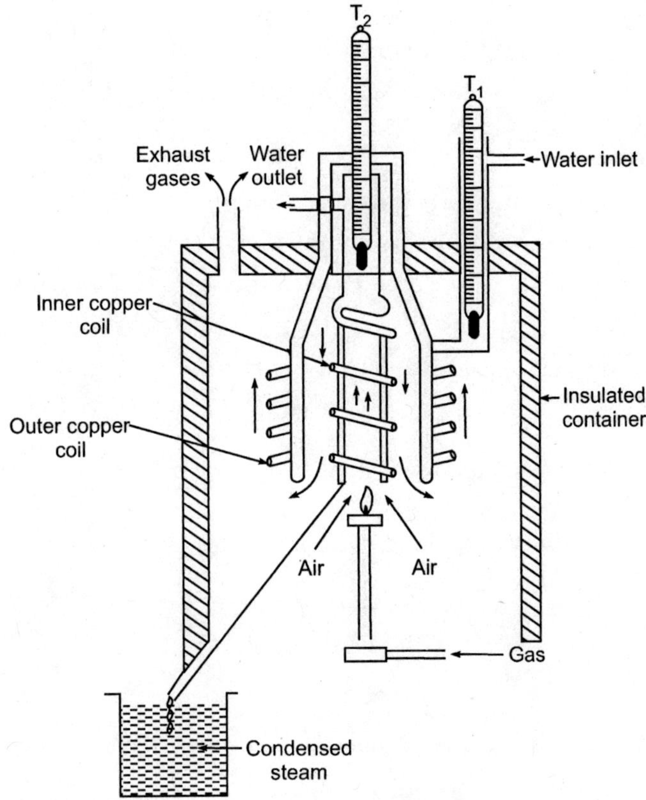

Fig. 13.2 Boy's gas calorimeter.

The whole assembly is enclosed in an insulated jacket.

Working

Circulation of water and burning gaseous fuel are continued at a constant rate for about 15 minutes for warming up the calorimeter. When the calorimeter gets warmed, the rates of the gaseous fuel and water are adjusted so that the outgoing water leaves the apparatus nearly at atmospheric pressure. Heat produced by the burning gaseous fuel is transferred to water in the copper coil, and the steam formed (if any) during combustion is condensed back into water, which is collected when the steady conditions are reached. The following readings are taken.

Observation

Volume of gas burnt at NTP in time T = V

Weight of water passed through coil in time T = W

Temperature of incoming water = t_1

Temperature of outgoing water = t_2
Weight of condensed water (steam) = W′
Higher (gross) calorific value of fuel = L

Calculation

$$\text{Heat absorbed by circulating water} = W(t_2 - t_1)$$
$$\text{Heat produced by combustion of fuel} = V \times L$$
$$\text{Heat produced} = \text{Heat absorbed}$$
$$V \times L = W(t_2 - t_1)$$
$$L = \frac{W(t_2 - t_1)}{V}$$

Now, to calculate lower (net) calorific value,

$$\text{Wt. of water condensed from 1 m}^3 \text{ of gas} = \frac{W'}{V}$$

$$\text{Latent heat of steam per m}^3 \text{ of gas} = \frac{W' \times 587}{V}$$

$$\text{LCV} = \text{HCV} - \text{Latent heat of steam per m}^3 \text{ of gas} = L - \frac{W' \times 587}{V}$$

Calorific Value of Fuel (Theoretical)

The calorific value of fuel can be approximately calculated by noting the amounts of constituents of the fuel. The higher calorific value of some combustible constituents of fuel are given in Table 13.2.

TABLE 13.2 Calorific Value of Fuel Constituents

Constituent	*Hydrogen*	*Carbon*	*Sulphur*
HCV (kcal/kg)	34,500	8,080	2,240

The oxygen (if present) in the fuel is assumed to be present in a combined form with hydrogen, that is, in the form of fixed hydrogen (H_2O). So the amount of hydrogen available for combustion

= Total mass of hydrogen in fuel – Fixed hydrogen

= Total mass of hydrogen in fuel – (1/8) mass of oxygen in the fuel

[$\because$ 8 parts of oxygen combine with one part of hydrogen to form H_2O]

Dulong's formula for calorific value from the chemical composition of fuel is

$$\text{HCV} = \frac{1}{100}\left[8{,}080\,\text{C} + 34{,}500\left(\text{H} - \frac{\text{O}}{8}\right) + 2{,}240\,\text{S}\right]\ \text{kcal/kg}$$

where, C, H, O and S are the percentages of carbon, hydrogen, oxygen, and sulphur in the fuel, respectively.

Oxygen is assumed to be present in combination with hydrogen as water.

$$\text{LCV} = \left[\text{HCV} - \frac{9}{100}\text{H} \times 587\right]\ \text{kcal/kg}$$

$$= [\text{HCV} - 0.09\,\text{H} \times 587]\ \text{kcal/kg}$$

As, 1 part of H by mass gives 9 parts of H_2O

$$2H_2 + O_2 \longrightarrow 2H_2O$$

(By Wt.) 4 36

Latent heat of steam = 587 kcal/kg.

13.4 CHARACTERISTICS OF GOOD FUEL

A good fuel should have the following characteristics:

High calorific value: A fuel should have high calorific value because the amount of heat liberated is independent of this value.

Moderate ignition temperature: Ignition temperature is the lowest temperature to which the fuel must be preheated so that it starts burning smoothly. Fuel with low ignition temperature is dangerous for storage and transport because it can be a fire hazard. High ignition temperature is not favorable for starting a fire, so an ideal fuel must have a moderate ignition temperature.

Low moisture content: Fuel should possess low moisture content because the moisture content of fuel reduces the heating value. It is also a loss of money because it is paid at the same rate as fuel.

Low noncombustible matter content: The noncombustible matter present in fuel remains as ash after combustion. The noncombustible matter also reduces the heating value. One percent of noncombustible matter in fuel means heat loss of about 1.5 percent. Moreover, storage, handling, and disposal of ash bears additional cost. A good fuel should have low contents of noncombustible matter.

Moderate velocity of combustion: Fuel must burn with a moderate velocity for a continuous supply of heat.

High pyrometric effect: The highest temperature obtained with the fuel is known as the *pyrometric effect*. A good fuel should have a high pyrometric effect, for example gaseous fuels.

Nonpollutant: The gaseous products of a combustion should not pollute the atmosphere, for example, CO, SO_2, NO_2, and so on, are the gaseous pollutants.

Easy availability: A good fuel should be available easily in bulk at a cheap rate.

Transportation: The transportation of fuel should be easy and low cost. It is easy to transport solid and liquid fuels, whereas for gaseous fuels, it is costly and hazardous.

Controlled combustion: Combustion of a fuel should be easy to start or stop according to requirement. It should be under control.

Low storage cost: Storage cost in bulk should be low.

Uniform particle size: In case of solid fuel, the particle size should be uniform so that combustion is regular.

13.5 SOLID FUELS

Wood: Wood consists of cellulose $(C_6H_{10}O_5)_n$, lignite, resins, inorganic substances, and nearly 60 percent of water. The calorific value varies from 3000–4000 kcal/kg of air-dried wood. Wood has a low ash content (0.3–0.6%). On burning, wood gives carbon dioxide and water.

$$(C_6H_{10}O_5)_n + 6O_2 \longrightarrow 6nCO_2 + 5nH_2O$$

Because it has a low calorific value, wood is very rarely used in industry. The principal use of wood is as a domestic fuel.

Coal: Coal is a primary fuel and a highly carbonaceous matter being formed as a result of the alteration of vegetable matter under favorable conditions. It is composed of C, H, N, and O, besides some noncombustible inorganic substance.

Various types of coal commonly recognized on the basis of rank or degree of alteration from the parent material wood are the following:

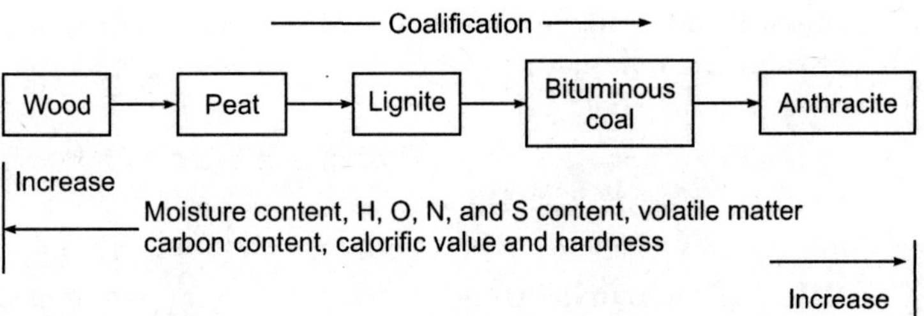

- *Peat* is a brown fibrous jelly-like mass. This is regarded as the first stage in the coalification of wood. It may contain 80–90 percent water. The average composition of air-dried peat is C = 57 %, H = 6 %, O = 35 %, and ash = 2.5–6 %. Its calorific value is about 5400 kcal/kg.
- *Lignite (brown coal)* is a brown-colored intermediate stage between the peat and bituminous coal. Lignite is compact in texture containing 20–60 % moisture; on air drying, it breaks up into small pieces. Air-dried lignite contains C = 60–70 %, O = 20 %. Lignite burns with a long smoky flame. Its calorific value is about 6500–7100 kcal/kg.
- *Bituminous coal (common coal)* is a compact black substance and is the most widely used of all the mineral coals. They are further classified as
  (*a*) *Sub-bituminous coals* are black in color and are more homogeneous and smooth. Their moisture and volatile matter contents are high. Carbon content varies from 75–83% and oxygen content varies from 10–20%. Their calorific value is about 7,000 kcal/kg.
  (*b*) *Bituminous coals* show a typically banded appearance and C-content ranges from 78–90 percent and volatile matter 20–45 percent. Their calorific value is about 8,000–8,500 kcal/kg. They are used in industries for making metallurgical coke, making coal gas, and for steam raising.
  (*c*) *Semi bituminous coals* are rich in carbon (90–95 percent) and have low volatile matter. These are used for coke manufacture. Their calorific value is about 8,500–8,600 kcal/kg.
- *Anthracite* is a class of highest rank coal containing the highest percentage of carbon (92–98 percent) and has the lowest volatile matter and moisture content. Their calorific value is about 8,650–8,700 kcal/kg. They are used for steam raising, for household purposes, and as a metallurgical fuel.

Selection of Coal

The following factors are taken into consideration when selecting of coal for different uses:

Calorific value: Should be high, which causes large quantities of heat from a small quantity of coal.

Moisture content: Should be low because moisture reduces the heating value of fuel. Also, moisture causes a loss of money because it is paid for at the same rate as coal.

Ash content: Should be low because it is noncombustible and reduces the heating value of coal. The composition of ash and the melting point are the important factors in metallurgical operations.

Calorific intensity: Defined as "the maximum temperature reached when the coal is completely burnt in the theoretical amount of air." It depends on quantity, nature, and the specific heat of the gaseous products of coal combustion. In most cases, the heat liberated due to combustion preheats the air and thus affects the calorific intensity. Calorific intensity can be calculated theoretically as follows:

$$\text{Flame temperature} = \frac{\text{Heat of combustion} + \text{Sensible heat of air}}{\sum(\text{Combustion products} \times \text{Specific heat})}$$

In general, coal should have a high calorific intensity.

Size of coal: Should be uniform to facilitate both its handling and the regulation of combustion process.

Coking quality: The coals, which on heating in the absence of air becomes soft, plastic, and fuse together into large coherent mass, are called *caking coals*. Such coals are difficult to oxidize. If the residue obtained after heating is porous, hard, strong, and usable for metallurgical purpose, the coal is known as *coking coal*.

All coking coals are caking, but all caking coals are not essentially coking coals. The coals that on heating undergo no fusing effect are called *free burning* or *noncoking coals*. Coking coals are suitable for metallurgical purposes because they are strong and not crushed under the weight of ore, flux, and so on in furnaces.

Sulphur and phosphorus contents: For coal used particularly for metallurgical purposes, this should be as low as possible because they negatively affect the properties of the metal produced. Also, the gaseous products during combustion cause corrosion of equipment and environmental pollution.

13.6 ANALYSIS OF COAL

To assess the quality of coal, the following two types of analysis are used.

Proximate Analysis

The *proximate analysis* gives information regarding the practical utility of coal. This involves the following steps:

Moisture: About 1 gm of a finely powdered air-dried coal sample is weighed in a crucible. The crucible is placed inside an electric oven maintained at 105—110°C for 1 hour. The crucible is then taken out, cooled in a desiccator, and weighed. The loss in weight is reported as moisture on a percentage basis.

$$\% \text{ moisture} = \frac{\text{Loss in weight}}{\text{Weight of coal taken}} \times 100$$

Volatile matter: Dried sample of coal left in a crucible (1) is then covered with a lid and placed in a muffle furnace maintained at 950°C for exactly 7 minutes. The crucible is then taken out, cooled in air, cooled in a desiccator, and weighed. The loss of weight is reported as volatile matter on a percentage basis.

$$\% \text{ volatile matter} = \frac{\text{Loss in weight due to removal of volatile matter}}{\text{Weight of coal taken}} \times 100$$

Ash: The residual coal in the crucible (2) is then heated in a muffle furnace at 700–750°C. The crucible is then taken out, cooled in air, cooled in a desiccator, and weighed. The experiment is repeated to get a constant weight of residue. The residue is reported as ash on a percentage basis.

$$\% \text{ ash} = \frac{\text{Weight of ash left}}{\text{Weight of coal taken}} \times 100$$

Fixed carbon: This is determined by deduction:

$$\% \text{C} = 100 - \% \text{ (Moisture + volatile matter + ash)}$$

A high percentage of fixed carbon is desirable.

Ultimate Analysis

Ultimate analysis is the elemental analysis of coal that deals with the determination of carbon, hydrogen, nitrogen, oxygen, sulphur, and ash contents. These determinations are made as given here:

Carbon and hydrogen: About 1–2 gm of an accurately weighed coal sample is burnt in a current of oxygen in a combustion apparatus. C and H of coal are converted into CO_2 and H_2O, respectively. The gaseous products are absorbed in KOH and $CaCl_2$ tubes, respectively, of known weights. The increase in weights of these are determined.

$$\underset{12}{C} + O_2 \longrightarrow \underset{44}{CO_2} \qquad \begin{bmatrix} \text{12 parts by mass of C gives} \\ \text{44 parts by mass of } CO_2 \end{bmatrix}$$

$$2KOH + CO_2 \longrightarrow K_2CO_3 + H_2O$$

$$\underset{2}{H_2} + \tfrac{1}{2}O_2 \longrightarrow \underset{18}{H_2O} \qquad \begin{bmatrix} \text{2 parts by mass of H gives} \\ \text{18 parts by mass of } H_2O \end{bmatrix}$$

$$7H_2O + CaCl_2 \longrightarrow CaCl_2 \, . \, 7H_2O$$

$$\%\,C = \frac{\text{Increase in weight of KOH tube} \times 12 \times 100}{\text{Weight of coal sample taken} \times 44}$$

$$\%\,H = \frac{\text{Increase in weight of } CaCl_2 \text{ tube} \times 2 \times 100}{\text{Weight of coal sample taken} \times 18}$$

Nitrogen (Kjeldahl method): About 1 gm of an accurately weighed powdered coal sample is heated with concentrated H_2SO_4 along with K_2SO_4 as the catalyst in a long-necked flask called a *Kjeldahl flask*. After the solution becomes clear, it is treated with excess KOH, and liberated ammonia is distilled over and absorbed in a known volume of a standard acid solution. The unused acid is determined by back titration with standard NaOH. From the volume of acid used by the ammonia liberated, the percentage N in coal is calculated.

$$\%\,N = \frac{\text{Vol. of acid used} \times \text{normality} \times 1.4}{\text{Weight of coal sample}}$$

Sulphur: This is determined from the washings obtained from the known mass of coal used in a bomb calorimeter, for determination of calorific value. During this determination of calorific value, S is converted into sulphate. The washings are treated with a barium chloride solution when barium sulphate is precipitated. This precipitate is filtered, washed, and heated to constant weight.

$$\%\,S = \frac{\text{Weight of } BaSO_4 \text{ obtained} \times 32 \times 100}{233 \times \text{Weight of coal sample taken in bomb}}$$

Ash: This is done just as in proximate analysis.

Oxygen: Oxygen is calculated by deduction.

$$\% O = 100 - \% (C + H + S + N + ash)$$

13.7 CARBONIZATION

When coal is heated strongly in the absence of air, it loses volatile matter and is converted into a white, lustrous, dense, strong, porous, and coherent mass known as coke. This process of converting coal into coke is called *carbonization*. The coal has a tendency to soften, swell, and stick together during combustion and can be converted into coke. Only bituminous coal can be coked. There are two types of carbonization of coal.

Low-temperature carbonization: In this process, the heating of coal is carried out at 500–700°C. The yield of coke is 75–80 percent containing 5–15 percent volatile matter. It is not mechanically strong and cannot be used as a metallurgical coke. However, it burns easily giving a smokeless, hot, and radiant fire and is an ideal fuel for domestic purposes. The gaseous product of this process has a high calorific value (6500–9500 kcal/m^3) and is therefore a valuable fuel.

By the disco process method, strongly coking coals are carbonized at a low temperature in a continuous process (Fig. 13.3).

Coal is fed to roasters, which are cast iron plates heated from below. Coal is stirred over these plates until it attains a temperature of 600°F. When coal is strongly coking, air is passed over it to destroy to some extent its agglutinating properties. However, if coal is weakly coking, no air is required for its oxidation.

After roasting, coal is allowed to fall into the raised end of a kiln type carbonizer. This carbonizer consists of two concentric cylinders. The inner cylinder rotates, whereas the outer is stationary. Heat is applied in the annular spaces between these two cylinders by hot combustion gases at 1040°F from a furnace. The byproducts in low temperature carbonization method are tar, ammonia, and so on.

High-temperature carbonization: In this process, the heating of coal is carried out at a high temperature (900–1200°C). The yield of coke is about 65–75 percent containing about 1–3 percent volatile matter. The coke obtained by this process is porous, hard, pure, compact, and strong and thus can be used in metallurgy. The gaseous byproduct of this process is high in volume, but its calorific value is low (5400–6000 kcal/m^3).

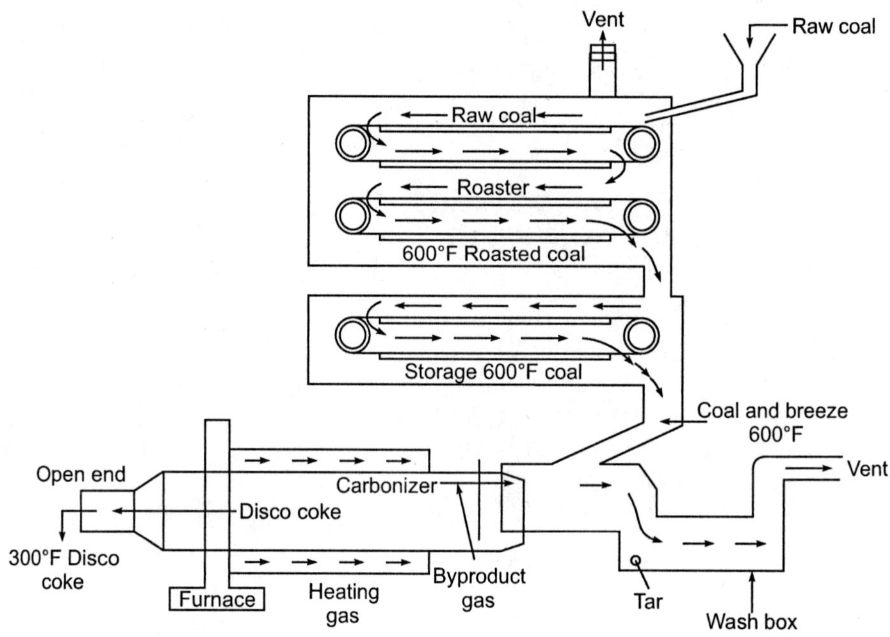

Fig. 13.3 The Disco process of coking coal.

High-temperature carbonization of coal is carried out in coking ovens in either of the following two types:

Beehive Oven

The earliest and cheapest process for the carbonization of coal is the Beehive oven (Fig. 13.4). The Beehive oven is a dome-shaped, brick structure about 4 m wide and 2.5 m high. The oven has two openings: one at the top for charging the coal and the other on the side for discharging the coke/supplying air. These openings can be opened or closed as desired.

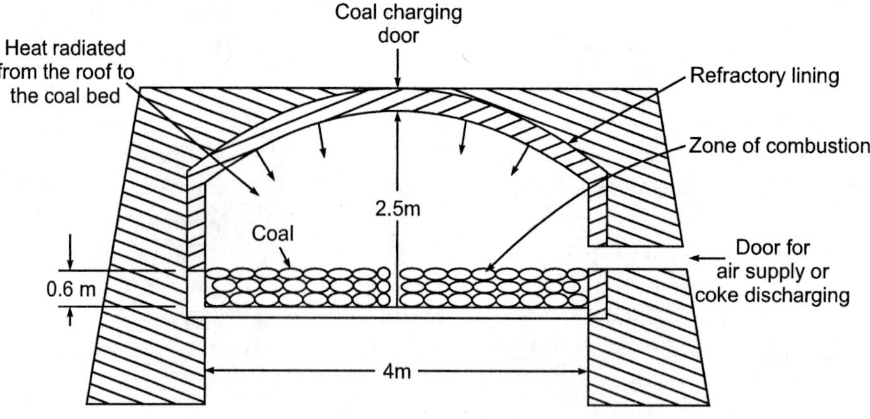

Fig. 13.4 Beehive coke oven.

Coal is charged through the top opening and levelled evenly to produce a layer about 0.6 m deep. Air is supplied through the side door, and coal is ignited. The volatile matter escapes or burns inside the side door. When the carbonization is complete, which generally takes three to four days, the hot coke is quenched with water and raked out through the side door. The yield of coke is 80 percent of the charged coal and averages about 5–6 tons of coke per firing.

Otto Hoffmann Byproduct Oven

Otto Hoffmann developed the modern byproduct coke oven, which has the advantage that the valuable byproducts are recovered. Here, heating is done on the principle of the "regenerative system" of heat economy, that is using the waste flue gases for heating the checker work of bricks.

Construction

The byproduct coke oven consists of narrow silica chambers separated by interspaces for burning gas. Each silica chamber is about 10–12 m long, 3–4 m high, and 0.40–0.45 m wide. Each chamber is provided with a charging hole at the top, a gas off-take, and a refractory-lined cast iron door at each end for discharging coke (Fig. 13.5).

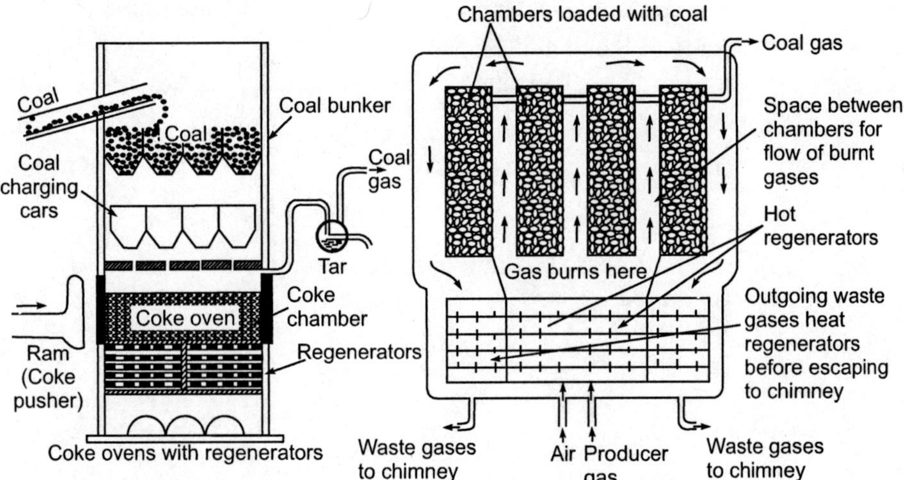

Fig. 13.5 Otto Hoffmann's byproduct coke oven with regenerators.

Working

Finely crushed coal is charged into chambers, and the chambers are closed. The ovens are heated to 1200°C by burning gaseous fuel (producer gas) on the regenerative principle to achieve economical heating. Flue gases produced during combustion are passed to the regenerators whose brick work takes up the heat and gets heated to about 1000°C. The waste gases then go to the chimney. The checker work serves to preheat the inlet gases. The heating is continued until the evolution of volatile matter ceases completely, which takes about 24 hours.

When coke is formed, a massive ram pushes the red hot coke outside. It is quenched by a water spray (wet quenching). In dry quenching, the red hot coke is cooled by an inert gas such as nitrogen. The coke obtained by dry quenching is cheaper, drier, cleaner, and stronger than wet-quenched coke. The yield of coke is about 75 percent of coal charged.

Recovery of Byproducts

The volatile matter coming in the form of a gas known as *coke oven gas* is composed of ammonia, hydrogen sulphide, napthalene, benzene, and so on. These products are separated from each other.

Recovery of tar: The coke oven gas is passed through a tower where liquor ammonia is sprayed. Here, dust and tar gets collected in a tank that is heated by steam coils to recover back ammonia. The ammonia is used again.

Recovery of NH_3: The gas free from tar is passed through a tower in which water is sprayed. Here, NH_3 goes into solution as NH_4OH.

Recovery of naphthalene: The gas is then passed through a tower in which water is sprayed at a very low temperature when naphthalene is condensed.

Recovery of benzene: The gas is passed through a tower in which petroleum is sprayed. Here, benzene and its higher homologues are removed.

Removal of H_2S: The gases are passed through purifiers containing ferric oxide.

$$Fe_2O_3 + 3H_2S \longrightarrow Fe_2S_3 + 3H_2O$$

When whole oxide is converted into sulphide, it is exposed to air when Fe_2O_3 is regenerated.

$$Fe_2S_3 + 4O_2 \longrightarrow 2FeO + 3SO_2 \uparrow$$

$$4FeO + O_2 \longrightarrow 2Fe_2O_3$$

Sulphur is removed as SO_2.

13.8 METALLURGICAL COKE

Among the solid fuels, coke is used in metallurgical process. A good coke used in metallurgical process should possess following requisites:

Purity: A metallurgical coke should have a minimum possible amount of moisture, ash, sulphur, and phosphorus. The presence of moisture in coke reduces its calorific value. An excess of ash hinders the heating and also helps in slag formation. The presence of S and P in coke on burning forms SO_2, P_2O_3, and P_2O_5, adversely affect the quality of metal to be produced. The presence of S in coke makes it brittle.

Strength: Coke used in metallurgical furnaces should be compact, hard, and strong so that it can withstand the load of overlying solids (ore + flux).

Porosity: Coke should be porous enough so that oxygen can easily come in contact with the carbon, so that complete combustion at a high rate takes place.

Size: Metallurgical coke should be of medium size. A very small size of coke causes choking, whereas a very big size does not give uniform heating.

Rate of combustion: A good metallurgical coke has a low rate of combustion because the lower the rate of combustion of coke, the higher will be its calorific intensity and hence the better the quality of fuel.

Combustibility: There should be no difficulty in the burning of coke; it should burn easily.

Cost: It should be cheaply available.

Calorific value: It should be high.

It should be noted that coal is not used in metallurgical processes particularly in a blast furnace because of the following facts:

- Coal has a lower strength and porosity than coke.
- Coal has a high sulphur content, which is very low in coke.
- Coal burns with a long flame because of the presence of volatile matter, which is suitable only for reverberatory furnace.

All these objectionable properties in coal are removed during its carbonization to form coke.

13.9 PULVERIZED COAL

In general, the combustion rate of solid fuels (coal) is slow because of the difficulty of thorough contact between the fuel and oxygen. The combustion rate can be increased by any of the following methods:

- By increasing the rate of oxygen supply; this method involves a wastage of heat carried by air itself.
- By finely powdering/pulverizing the coal, so that its free surface area is increased and the fuel comes in contact with air more easily. Generally, the

pulverization method is more satisfactory for volatile coals. The volatile matter present in the coal is liberated quickly and burns, thereby helping the burning of fixed carbon.

13.10 LIQUID FUELS

The liquid fuels are characterized by a low flash point, high calorific value, low viscosity at ordinary temperature, and low moisture and sulphur contents.

Petroleum

Because it is found in rocks, it is called petroleum, derived from the Latin *Petra* (rock) and *oleum* (oil). Thus, petroleum means rock oil. Petroleum is often called liquid gold because modern civilization has found it to be more valuable than precious metal gold.

Petroleum or crude oil is a dark greenish-brown viscous oil found deep in the earth's crust. The average composition of crude petroleum is

$$C = 79.5\text{–}87.1\%,\ H = 11.5\text{–}14.8\%,\ S = 0.1\text{–}3.5\%,\ N + O = 0.1\text{–}0.5\%$$

Classification of Petroleum

Petroleum is classified into three categories according to its composition.

Paraffin base petroleum: It contains hydrocarbons of the paraffin series. These paraffin hydrocarbons are branched or straight.

Asphaltic base petroleum: It contains hydrocarbons of the napthalene series.

Mixed base petroleum: It contains both paraffinic and asphaltic hydrocarbons and are generally rich in semi solid waxes.

Origin of Petroleum

Three widely accepted theories that explain the origin of petroleum are given here:

Carbide theory: According to this theory, hydrocarbons present in petroleum are formed by the action of water on inorganic carbides. Inorganic carbides are formed by the reaction of metal and carbon under high temperature and pressure inside the earth.

$$Ca + 2C \longrightarrow \underset{\text{Calcium carbide}}{CaC_2}$$

$$4Al + 3C \longrightarrow \underset{\text{Aluminium carbide}}{Al_4C_3}$$

$$\underset{}{CaC_2} + 2H_2O \longrightarrow Ca(OH)_2 + \underset{\text{Acetylene}}{C_2H_2}$$

$$Al_4C_3 + 12H_2O \longrightarrow 4Al(OH)_3 + \underset{\text{Methane}}{3CH_4}$$

These lower hydrocarbons then undergo hydrogenation and polymerization to give various types of hydrocarbons.

$$\underset{\text{Acetylene}}{C_2H_2} + H_2 \longrightarrow \underset{\text{Ethylene}}{C_2H_4} \xrightarrow{H_2} \underset{\text{Ethane}}{C_2H_6}$$

$$3C_2H_2 \longrightarrow \underset{\text{Benzene}}{C_6H_6}$$

$$3C_2H_4 \longrightarrow \underset{\text{Cyclohexane}}{C_6H_{12}}$$

$$2C_2H_4 \longrightarrow \underset{\text{Butene-2}}{CH_3.CH{=}CH.CH_3} \xrightarrow{H_2} \underset{\text{Butane}}{CH_3.CH_2.CH_2.CH_3}$$

However, this theory fails to explain the following facts:

- Presence of nitrogen and sulphur compounds
- Presence of chlorophyll and haemin
- Presence of optically active compounds

Engler theory: Engler (1900) suggested that petroleum is formed by the decay and decomposition of marine animals under high pressure and temperature. The sulphur dioxide given out by the volcanoes beside the seaside kills the fish and other sea animals that go on the piling beside the volcano. After hundreds of years, these animals start decomposing under the influence of high pressure and temperature to form petroleum. The theory fails to account for the presence of chlorophyll in petroleum.

Modern theory: According to modern theory, petroleum has resulted from the partial decomposition of marine animals and vegetable organisms of prehistoric forests. Changes in earth (volcanoes, etc.) had buried these materials underground. During the course of time under favorable conditions, biological matter decomposed into petroleum. This theory explains the presence of brine, coal, N and S compounds, chlorophyll, and haemin and also optically active compounds in petroleum.

Mining of Petroleum

Mining of petroleum is done by drilling holes in the earth's crust and sinking pipes up to the oil-bearing porous rocks. Usually, oil gushes out itself due to the hydrostatic pressure of natural gas, but as the pressure of gas subsides, air pressure is applied through pumps to force the oil out of the well.

Refining of Petroleum

Petroleum as comes from the ground is a viscous, dark colored, and foul-smelling liquid. It is always found mixed with a considerable amount of water and earth particles. Technically, it is called crude oil. Chemically, it is a complex mixture of hydrocarbons, so it becomes necessary to separate these hydrocarbons to make each fraction available for the appropriate use. It can be separated into a number of useful fractions by fractional distillation. This process of dividing petroleum into different useful fractions with different boiling ranges is called refining petroleum, and the plants set up for the purpose are called oil refineries. The process of refining involves the following steps.

Step I. Separation of Water (Cottrell's Process)

The crude oil is an emulsion of oil and salt water (brine). The process of separating water from oil consists of allowing the crude to flow between two highly charged electrodes. The colloidal water droplets coalesce to form large drops, which separate out from oil.

Step II. Removal of Sulphur Compounds

This involves the treatment of oil with copper oxide. A reaction occurs with sulphur compounds that results in the formation of copper sulphide (precipitate), which is then removed by filtration.

The test with the doctor solution, sodium plumbite, indicates whether many kinds of organic sulphides are present.

Step III. Fractional Distillation

The crude oil is then heated to about 400–430°C in an iron retort, whereby all volatile constituents are evaporated. The components, such as tar and asphalt, do not vaporize, and they settle at the bottom of the column. The hot vapors are then passed through a distillation column.

The distillation column is a cylindrical structure of steel, about 31 meters high and 3 meters in diameter. Inside the column, at short distances (25–50 cms.), horizontal plates called trays are fitted. In every tray, there are many holes, each having an upgoing short tube with a bubble-cap on each (Fig. 13.6). As the vapors go up, they begin to cool and condense in fractions at different heights of the column. A higher boiling fraction condenses first; whereas lower and lower boiling fractions condense one after the other. For this reason, aviation-petrol (a liquid with lowest b.p.) is collected in the uppermost tray, and fuel-oil (a liquid with highest b.p.) is collected in the lowest tray. Various principal fractionation products thus obtained are given in Table 13.3.

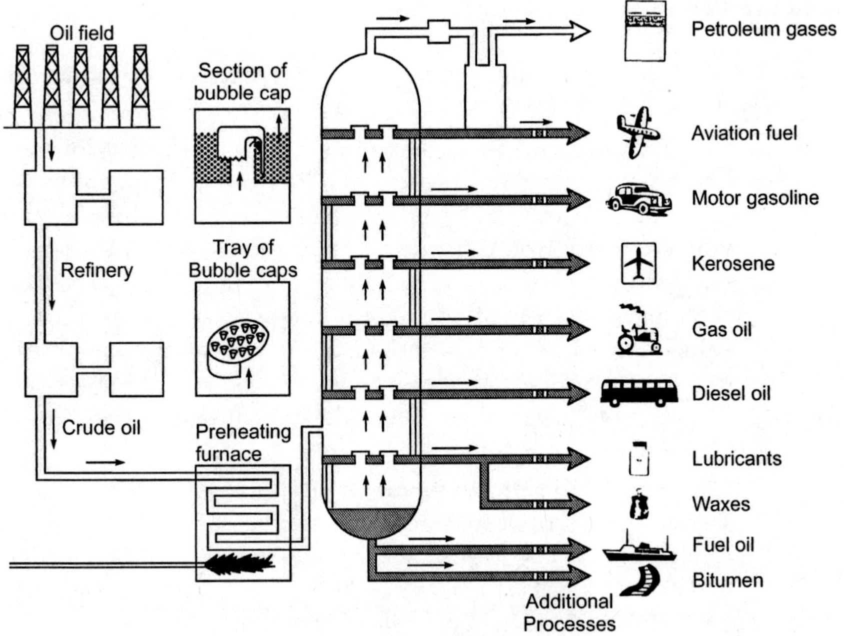

Fig. 13.6 Distillation column.

TABLE 13.3 Fractions by Distillation of Crude

Name of Fraction	*Boiling Range*	*Approx. Composition*	*Uses*
1. Uncondensed gases	Upto 30°C	C_1—C_5	Domestic fuel (LPG), synthesis of organic chemicals, production of carbon black
2. Petroleum ether	30–70°C	C_5—C_7	As a solvent for oil, varnish
3. Gasoline or petrol	40–120°C	C_5—C_{10}	Fuel for internal combustion engine of automobiles, airplanes, solvent, and dry cleaning
4. Naphtha	120–180°C	C_9—C_{10}	As solvent and dry cleaning
5. Kerosene	180–250°C	C_{10}—C_{16}	Illuminant, fuel for stoves
6. Diesel oil	250–320°C	C_{10}—C_{18}	Fuel for diesel engines
7. Heavy oil on refractionation gives	320–400°C	C_{17}—C_{30}	Lubrication
(*i*) Lubricating oil			Lubricant, cosmetics
(*ii*) Petroleum jelly			Medicines
(*iii*) Grease			Lubricant
(*iv*) Paraffin wax			Candles, boot polish, wax paper, tarpolin cloth, and so on
8. Residue pitch or petroleum coke	above 400°C	C_{30} and above	Water proofing of roofs, road making, fuel

Petroleum Products and Their Uses

We can now enumerate the most important petroleum products along with their uses:

LPG (Petroleum gases): Liquefied petroleum gas (LPG) is obtained as a byproduct during the cracking of heavy oils or from natural gas and sold as bottled gas. These gases consist mainly of ethane, propane, butane, and other volatile hydrocarbons. The main constituents of LPG are *n*-butane, isobutane, butylene, propane, and very little ethane. LPG is supplied under pressure in containers under the trade names Indane, Bharat gas, and so on. Its calorific value is about 27,800 kcal/m^3. LPG is dehydrated, desulphurized, and traces of odorous organic sulphides (mercaptans) are added to warn of gas leaks. LPG is widely used as a domestic and industrial fuel. Today, LPG is also accepted as motor fuel.

Gasoline or petrol: Aviation fuel and motor petrol come under this group. This is a mixture of hydrocarbons such as C_5H_{12} to C_8H_{18}. Its approximate composition is C = 84%, H = 15%, N + S + O = 1%. Its calorific value is about 11,250 kcal/kg. It is highly volatile, inflammable, and used as fuel for internal combustion engines of automobiles and airplanes.

Kerosene: This is the fraction obtained between 180 and 250°C and is a mixture of hydrocarbons such as $C_{10}H_{22}$ to $C_{16}H_{34}$. Its approximate composition is C = 84%, H = 16%, with less than 0.1% S. Its calorific value is 11,100 kcal/kg. It has served as fuel for cooking, especially where bottled gas is not obtainable. Kerosene is also used as jet engine fuel and for making oil gas.

Diesel oil: This is a fraction obtained between 250 and 320°C and is a mixture of $C_{15}H_{32}$ to $C_{18}H_{38}$ hydrocarbons. Its calorific value is about 11,000 kcal/kg. It is used as a diesel engine fuel.

13.11 CRACKING

With the increasing demand for some oil fractions, such as motor gasoline and kerosene, supply from mere distillations proved inadequate. So chemists made use of certain conversion reactions in this field. One of the major conversion processes used in an oil refinery is called *cracking*. Cracking is defined as "the decomposition of bigger hydrocarbon molecules into simpler, low boiling hydrocarbons of a lower molecular weight."

$$\underset{\text{(Decane)}}{C_{10}H_{22}} \xrightarrow{\text{Cracking}} \underset{(n\text{-pentane})}{C_5H_{12}} + \underset{\text{(Pentene)}}{C_5H_{10}}$$

B.P. = 174°C (decane); B.P. = 36°C (products)

The middle and heavy fractions are cracked to get gasoline, which has the largest demand as a motor fuel. There are two methods of cracking:

Thermal cracking: The heavy oils are subjected to a high temperature and pressure, when the bigger hydrocarbon molecules break down to give smaller molecules of the paraffins, olefines, with the same hydrogen. The process may be carried out either in liquid phase or vapor phase.

- Liquid phase cracking is carried out at a temperature of 470–540°C and a pressure of 40–60 atms. The cracked products are then separated in a fractionating column. The yield is 50–60 percent, and the octane rating of the petrol produced is 65–70.
- Vapor phase cracking is carried out at 550°C and higher and under 2–5 atms. Gasoline obtained from vapor phase cracking has better antiknock properties but poorer stability than gasoline from liquid phase cracking.

Catalytic cracking: When cracking is done in the presence of a catalyst, usually a mixture of silica and alumina or aluminium silicate, it is known as catalytic cracking. There are two methods of catalytic cracking in use:

- *In fixed-bed catalytic cracking,* the oil vapors are preheated to cracking temperatures (420–450°C) and then forced through a catalytic chamber (containing artificial clay mixed with Zirconium oxide) maintained at 425–450°C and 1.5 kg/cm^2 pressure. During their passage through the tower, about 40 percent of the charge is converted into gasoline and about 2–4 percent carbon is formed. The latter gets adsorbed on the catalyst bed. The vapors produced are then passed through a fractionating column, where heavy oil fractions condense. The vapors are then led through a cooler, where some of the gases are condensed along with gasoline, and uncondensed gases move on. The gasoline containing some dissolved gases is then sent to a stabilizer, where the dissolved gases are removed, and pure gasoline is obtained (Fig. 13.7).

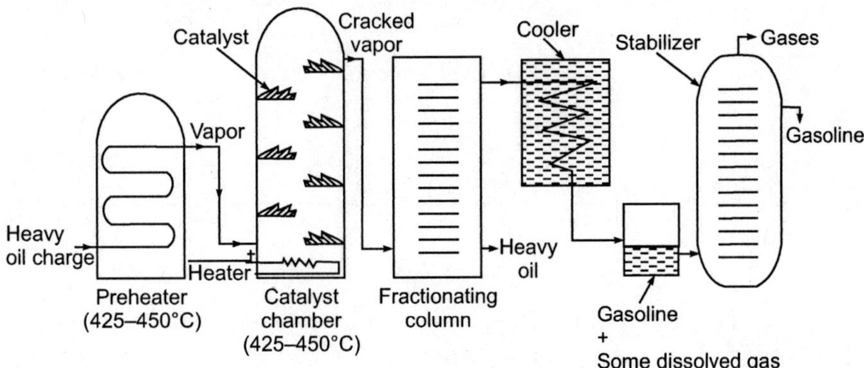

Fig. 13.7 Fixed-bed catalytic cracking.

The catalyst stops functioning after 8–10 hours due to the deposition of carbon. This is reactivated by burning off the deposited carbon.

- *In moving-bed catalytic cracking,* the solid catalyst is very finely powdered, so that it behaves as a fluid, which can be circulated in the gas stream. The vapors of the cracking stock (gas oil, heavy oil, etc.) mixed with a fluidized catalyst is forced up into a large reactor bed in which cracking occurs (Fig. 13.8). Near the top of the reactor, there is a centrifugal separator called a cyclone, which allows only the cracked oil vapors to pass on to the fractionating column but retains the catalyst powder in the reactor itself. The catalyst powder gradually becomes heavier, due to the coating with carbon, and settles to the bottom. From there, it is forced by an air blast to the regenerator maintained at 600°C.

  In the generator, carbon is burnt, and the regenerated catalyst then flows through a stand pipe for mixing with fresh batch of incoming cracking oil. At the top of the regenerator, there is a separator that permits only gases (CO_2, etc.) to pass out but holds back catalyst particles.

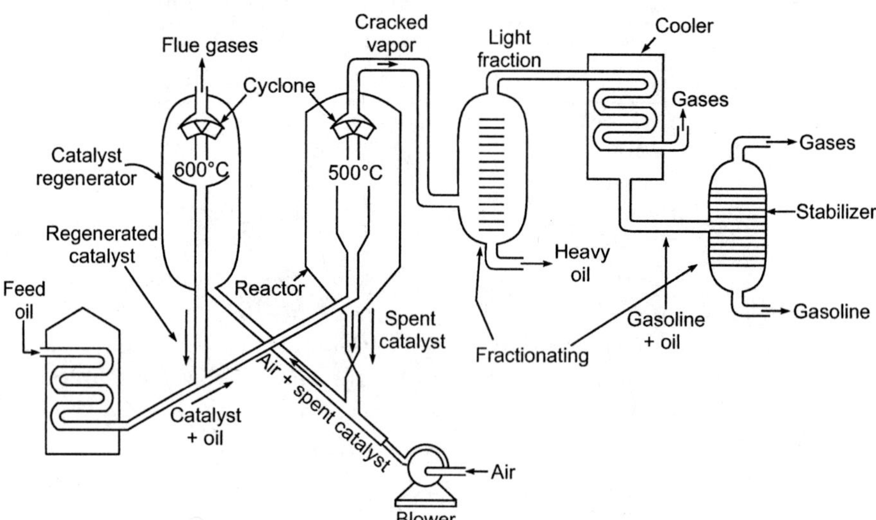

Fig. 13.8 Moving-bed type catalytic cracking.

13.12 SYNTHETIC GASOLINE

Gasoline is synthesized by any of the following three methods:

Polymerization: The gases obtained as a byproduct from cracking heavy oils and so on contain olefins and alkanes. When this gaseous mixture is subjected to a high pressure and a high temperature with or without the presence of a catalyst, it polymerizes to form higher hydrocarbons resembling gasoline, which is called polymer gasoline.

$$\underset{\text{Propene}}{CH_3.CH=CH_2} + \underset{\text{Butene-1}}{CH_3.CH_2.CH=CH_2}$$

$$\xrightarrow{\text{Polymerization}} \underset{\text{5-Methyl hexene-1}}{CH_2=CH.CH_2.CH_2.CH(CH_3)_2}$$

The polymerization of olefins provides a high octane number gasoline.

Fischer-Tropsch process: Fischer and Tropsch (1962) first developed this method. The water gas ($CO + H_2$) produced by passing steam over heated coke is mixed with hydrogen. The gas is purified by passing through Fe_2O_3 (to remove H_2S) and then into a mixture of $Fe_2O_3 . Na_2CO_3$ (to remove organic sulphur compounds). The purified gas is compressed to 5–25 atm. pressure and then led through a converter (containing a catalyst consisting of a mixture of 100 parts cobalt, 5 parts thoria, 8 parts magnesia, and 200 parts kieselguhr earth), maintained at about 200–300°C. A mixture of saturated and unsaturated hydrocarbons results:

$$nCO + 2nH_2 \longrightarrow C_nH_{2n} + nH_2O$$

$$nCO + (2n+1)H_2 \longrightarrow C_nH_{2n+2} + nH_2O$$

The reaction is exothermic, so the outcoming hot gaseous mixture is led to a cooler, where a liquid resembling crude oil is obtained. The crude oil obtained is subjected to fractionation to yield gasoline and diesel oil (Fig. 13.9).

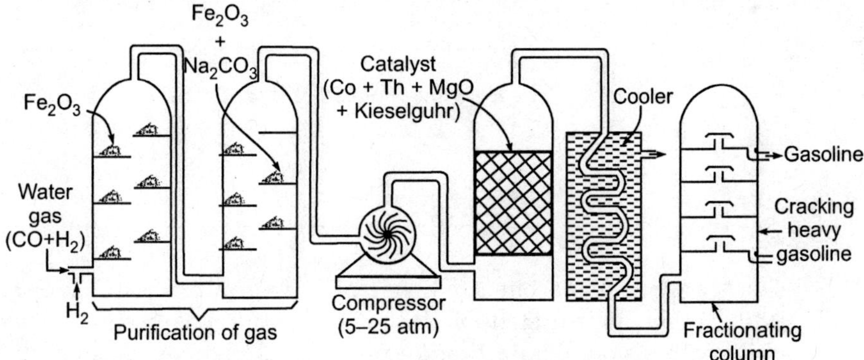

Fig. 13.9 Fischer-Tropsch method.

Bergius process: This method was first proposed by Bergius in Germany during the first world war. The low ash coal is finely powdered and made into a paste with heavy oil, and then a catalyst (organic compound of tin) is incorporated.

The paste is then pumped to the convertor where it is heated to 400–450°C under 200–250 atm for about 1.5 hours in the presence of hydrogen. Hydrogenation takes place to form higher saturated compounds, which undergo cracking and hydrogenation to yield a mixture of alkanes. Thus, the vapors leaving the convertor upon condensation give crude oil or synthetic petroleum. Crude oil is fractionally distilled to give gasoline, middle oil, and heavy oil. The middle oil is again hydrogenated in the vapor phase in the presence of a solid catalyst to give more gasoline. Usually, the processing of middle oil gives four times the gasoline obtained by the primary hydrogenation of coal. Heavy oil is again used for making a paste with fresh coal dust. One ton of soft coal is found to yield 140 gallons of gasoline on above treatment (near about 60 percent yield) (Fig. 13.10).

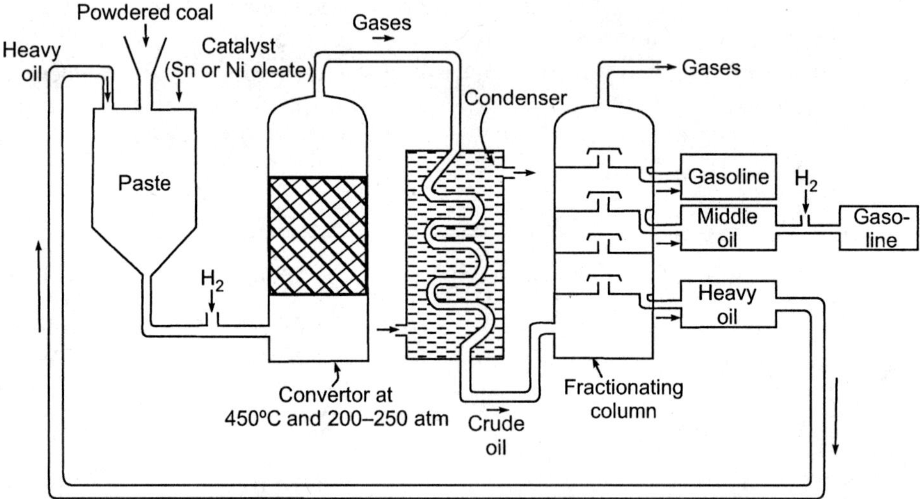

Fig. 13.10 The Bergius process of hydrogenation of coal to gasoline.

13.13 PURIFICATION OF GASOLINE

The straight-run gasoline (either from crude oil or synthetic) contains some undesirable unsaturated olefins and sulphur compounds. The unsaturated hydrocarbons are oxidized and polymerized, thereby causing gum and sludge formation when stored. On the other hand, sulphur compounds lead to corrosion of the internal combustion engine. The following processes are used for removal of such undesirable contents from gasoline:

- The gasoline fraction is agitated with sulphuric acid when aromatic compounds such as thiophene, and so on, are converted into their sulphonic acids. The contents are allowed to settle, and the upper layer is withdrawn and treated with alkali to remove excess acid. It is washed with water finally.

- In the sweetening process (doctors treatment), the sulphur compounds are generally removed by treating gasoline with an alkaline solution of sodium plumbite (Na_2PbO_2) with the controlled addition of S.

$$2RSH + Na_2PbO_2 + S \longrightarrow \underset{\text{(disulphide)}}{R.S.S.R.} + PbS + 2NaOH$$

- Olefins and the coloring matter of gasoline are removed by passing gasoline through kieselguhr, fuller's earth, clay, and so on.
- After purification of the gasoline, some inhibitors are added to retard the oxidation of olefins that cause the formation of gums on storage, for example, Common inhibitors are butylamino phenols, benzyl aminophenols and phenylene diamines.

13.14 KNOCKING AND OCTANE NUMBER

In an internal combustion engine, the mixture of fuel (gasoline vapor) and air is highly compressed before it is ignited to have maximum efficiency. During the process, there is a sharp metallic sound known as knocking. Knocking causes a loss of energy and increased engine wear. The tendency to knock is in the following order:

Straight chain paraffins > branched chain paraffins > olefins
> cycloparaffins (naphthalenes) > aromatics

The octane number is the rating of gasoline as to its antiknock property. *n*-heptane knocks very badly, and its antiknock value is given the arbitrary value 0; whereas 2, 2, 4 tri-methyl pentane (iso octane) gives very little knocking, and its antiknock value is given as 100.

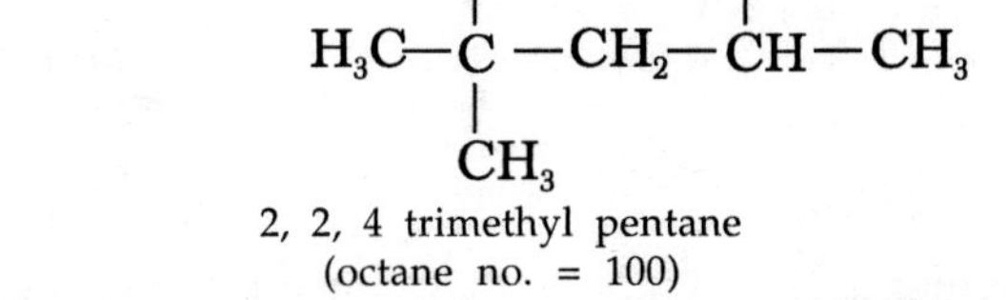

$$H-\underset{H}{\overset{H}{C}}-\underset{H}{\overset{H}{C}}-\underset{H}{\overset{H}{C}}-\underset{H}{\overset{H}{C}}-\underset{H}{\overset{H}{C}}-\underset{H}{\overset{H}{C}}-\underset{H}{\overset{H}{C}}-H$$

n-heptane (octane no. = 0)

$$H_3C-\underset{CH_3}{\overset{CH_3}{C}}-CH_2-\overset{CH_3}{CH}-CH_3$$

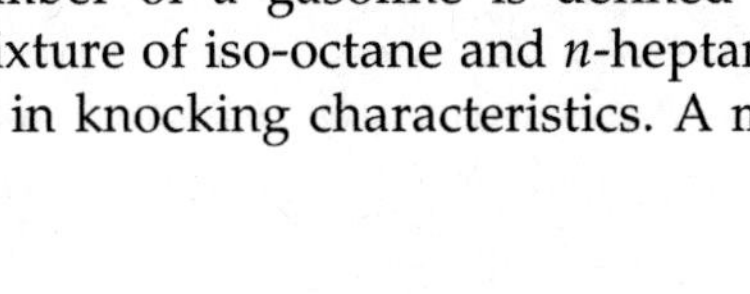

2, 2, 4 trimethyl pentane
(octane no. = 100)

Thus, the octane number of a gasoline is defined as the percentage of iso-octane present in a mixture of iso-octane and *n*-heptane, which matches the fuel (gasoline) under test in knocking characteristics. A modern good grade of gasoline has an octane number of 80 which is obtained by blending straight-run gasoline with cracked gasoline and hydrocarbons synthesized from refinery gases, and adding little amount of tetraethyl lead. Octane number 80 indicates that the fuel could give a same knocking as given by the mixture of 80 parts of iso-octane and 20 parts of *n*-heptane.

13.15 DIESEL FUEL AND CETANE NUMBER

Diesel fuel is used in engines where the fuel is directly ignited into cylinders and is atomized there before being burnt. In a diesel engine, the fuel is exploded by the application of heat and pressure and not by a spark. The main characteristic of a diesel engine fuel is that it should easily ignite below the compression temperature. Chemically, diesel consists of long straight hydrocarbons with a minimum amount of branched chain and aromatic hydrocarbons. The ignition quality of the diesel fuel is expressed in a cetane number.

Cetane (*n*-hexadecane), which ignites easily, is given the number 100; 2-methyl naphthalene, which ignites very slowly, is assigned a cetane number 0.

CH_3 (on naphthalene ring)

2-methyl naphthalene
(C.N. = 0)

$H_3C\!-\!CH(CH_3)\!-\![CH_2]_{10}\!-\!CH(CH_3)\!-\!CH_3$

n-hexadecane
(C.N. = 100)

Thus, the cetane number of diesel fuel is defined as the percentage of cetane present in a mixture of cetane and 2-methyl naphthalene, which matches the fuel under test in the ignition property.

The cetane number of a diesel fuel can be improved by adding small amount (1–5%) of certain compounds called additives (ethyl nitrate, isoamyl nitrate, acetone peroxide, etc.). Generally, diesel engines use a fuel with a cetane number of 45 or above.

13.16 GASEOUS FUELS

Gaseous fuels are classified into natural (natural gas) and artificial gases.

Natural Gas

Natural gas, which is a mixture of methane, ethane, propane, butane, pentane, carbon dioxide, nitrogen, and so on, is the most important fuel found mainly in the vicinity of coal mines or oil fields. It is also associated with petroleum deposits. The natural gas derived from oil wells may be either dry or wet. When natural gas occurs along with petroleum in oil wells, it is called *wet gas*. The wet gas is treated to remove propane, propene, butene, and butane, which are used as LPG. On the other hand, when the gas is associated with crude oil, it is called *dry gas*. The calorific value of wet natural gas is higher than that of the dry type because of a higher percentage of the heavier unsaturated molecules. The approximate composition of natural gas is

$$CH_4 = 70–90\%,\ C_2H_6 = 5–10\%,\ H_2 = 3\%,\ CO + CO_2 = \text{rest}$$

Following are the uses of natural gas:

- As an excellent domestic fuel that can be conveyed over very large distances in pipelines.
- For synthesis of a number of chemicals
- As a raw material for the manufacture of carbon black

Artificial Gases

There are several artificial gases as described in the following sections:

Water Gas

Water gas is also called blue gas because of the blue flame when it is burnt. Water gas is essentially a mixture of combustible gases (CO and H_2) with a little noncombustible gases (CO_2 and N_2). This is produced by passing alternatively steam and a little air through a bed of red hot coal or coke, maintained at about 900–1000°C in a reactor. The reactor consists of a steel vessel about 3 m wide and 4 m high. It is lined inside with refractory bricks. It has a cup and cone feeder at the top and an opening near the top for the exit of water gas. At the base, it is provided with inlet pipes for passing air and steam. Moreover, at the bottom, it has an arrangement for the removal of the ash formed (Fig. 13.11).

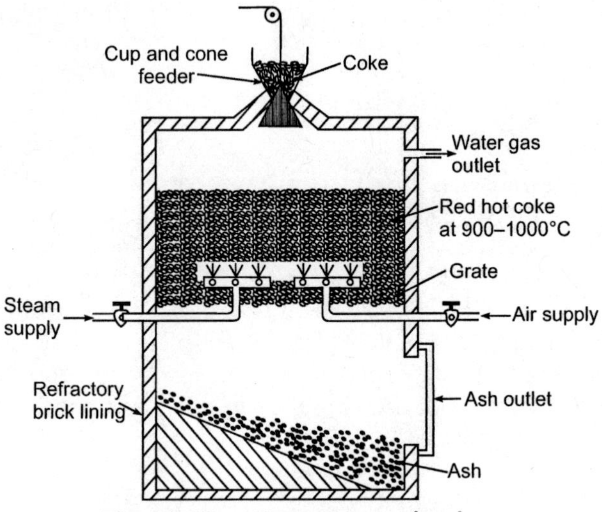

Fig. 13.11 Water gas production.

Reactions Involved

When steam reacts with red hot coke at 900–1000°C, the following reaction takes place:

$$C + H_2O \longrightarrow CO + H_2 - 29 \text{ kcal} \qquad \text{(Endothermic)}$$

At lower temperature $$C + 2H_2O \longrightarrow CO_2 + 2H_2 - 19 \text{ kcal} \qquad \text{(Endothermic)}$$

Both of these reactions are endothermic with the result that when steam is passed, the fuel bed cools down. To keep up the required temperature of the bed, for a short period air is blown into the fuel bed just before the blowing of steam. When air is blown into the fuel bed, the following reactions take place:

$$C + O_2 \longrightarrow CO_2 + 97 \text{ kcal} \qquad \text{(Exothermic)}$$

$$2C + O_2 \longrightarrow 2CO + 59 \text{ kcal} \qquad \text{(Exothermic)}$$

Both of these reactions are exothermic with the result that the temperature of the fuel bed rises. When steam is passed through the fuel bed, it is called the run period; when air is passed through the fuel bed, it is called the blow period. These are alternately repeated to maintain the proper temperature.

The average composition of water gas is

$$H_2 = 51\%, N_2 = 4\%$$
$$CO = 41\%, CO_2 = 4\%$$

Its calorific value is about 2,800 kcal/m^3.

Following are the uses of water gas:

- Source of hydrogen gas.
- An illuminating gas and a fuel gas. Its flame is short but very hot.

Water gas as it is has low calorific value. It is enriched with some gaseous hydrocarbons to make it of such a standard that it can be used for domestic purposes. This so-called carburetted water gas contains about H_2 (35 percent), CO (25 percent), hydrocarbons (35 percent), and $N_2 + CO_2$ (5 percent). Its calorific value is 4,500 kcal/m^3.

Producer Gas

Producer gas is essentially a mixture of combustible gases, that is carbon monoxide and hydrogen associated with a large percent of noncombustible gases such as N_2, CO_2, and so on. It is prepared by forcing air mixed with steam over a red hot coal or coke bed maintained at about 1,100°C in a special reactor called a gas producer (Fig. 13.12). The gas producer is a steel cylindrical vessel about 3 m in diameter and 4 m in height. The vessel is lined inside with refractory bricks. It is provided with a cup and cone feeder at the top and a side opening for producer gas exit. At the base, it has inlets for passing air and steam. Ash is removed at the base of the producer.

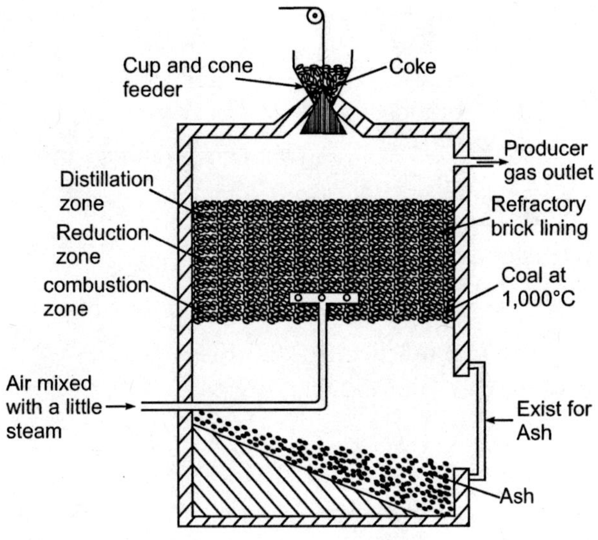

Fig. 13.12 Gas producer.

Reactions Involved

Following are the reactions involved:

Oxidation zone: When air is passed through the ignited fuel bed, the following reactions take place in the oxidation zone:

$$C + O_2 \longrightarrow CO_2 + 97.65 \text{ cal}$$

$$2C + O_2 \longrightarrow 2CO + 58.8 \text{ cal}$$

Both the reactions are exothermic reactions. The temperature of this zone is about 1,100°C.

Reduction zone: Carbon dioxide and steam pass on to the next layer of coal when the following reactions take place:

$$CO_2 + C \longrightarrow 2CO \quad - 36 \text{ kcal}$$

$$C + H_2O \longrightarrow CO + H_2 \quad - 29 \text{ kcal}$$

$$C + 2H_2O \longrightarrow CO_2 + 2H_2 \quad - 19 \text{ kcal}$$

All these reactions in the reduction zone are endothermic, and the heat is taken up from the fuel bed with the result that the temperature of the fuel bed falls. To maintain proper temperature for producing producer gas, the rate and quantity of steam supply are controlled.

Distillation zone (400–800°C): Mixture of carbon monoxide, hydrogen, and carbon dioxide pass on the third zone where the volatile matter is distilled from the coal. So CO, H_2, and volatile products of coal pass out from the coal. The average composition of producer gas is

$$CO = 22.30\%, N_2 = 52.55\%$$

$$H_2 = 8.12\%, CO_2 = 3\%$$

Its calorific value is about 1,300 kcal/m^3.

The uses of producer gas include the following:

- In furnaces (open-hearth and muffle)
- As a reducing agent in metallurgy

Coal Gas

Coal gas is produced during the carbonization of coal, that is when coal is heated in the absence of air. The coal is fed in closed silica retorts that are then heated to about 1,300°C. The heating is carried out by burning a producer gas and air mixture.

$$\text{Coal} \xrightarrow[\text{In absence of air}]{1300^\circ\text{C}} \text{Coke} + \text{Coal Gas} \uparrow$$

The outcoming gas from the retort is first scrubbed by passing through a hydraulic main. Much of the tar is then removed by cooling the gas in a huge water-cooled heat exchanger called a condenser.

Any remaining tar and ammonia present in the gas are removed by scrubbing with water in a scrubber. The cooled gas is then scrubbed with creosote oil, which dissolves benzol, naphthalene, and so on. The gas is then purified by passing it over moist ferric oxide when hydrogen sulphide gas, if present, is removed (Fig. 13.13).

$$2Fe(OH)_3(s) + 3H_2S(g) \longrightarrow Fe_2S_3(s) + 6H_2O(l)$$

When the iron oxide of the purifier is exhausted, it is taken out from the purifier and exposed to air when it gets oxidized to ferric oxide:

$$2Fe_2S_3(s) + 3O_2(g) \longrightarrow 2Fe_2O_3(s) + 6H_2S(g)$$

The purified coal gas is finally stored over water in gas holders.

Coal gas is a colorless gas with a characteristic odor. It is lighter than air and burns with a long smoky flame.

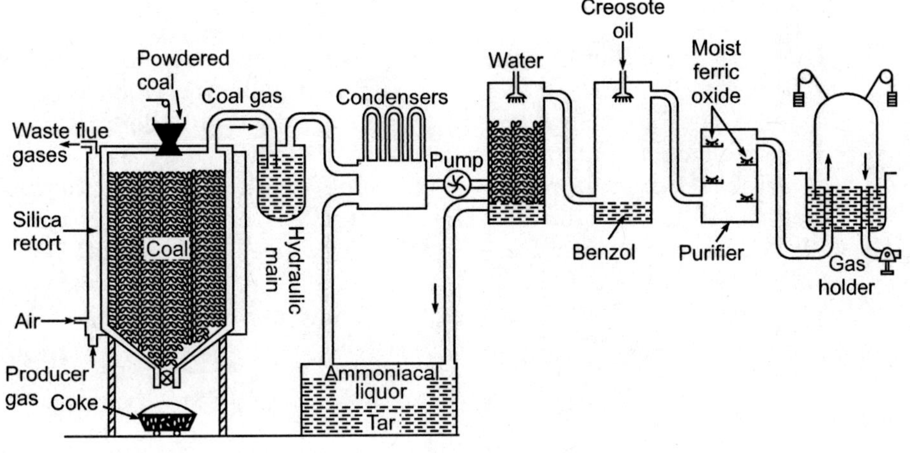

Fig. 13.13 Manufacture of coal gas.

The average composition of coal gas is

$H_2 = 40\%$	$N_2 = 4\%$
$CH_4 = 32\%$	$CO_2 = 1\%$
$CO = 7\%$	Rest = 4%
$C_2H_2 = 2\%$	
$C_2H_4 = 3\%$	

Its calorific value is about 4,900 kcal/m^3.

The following are the uses of coal gas:

- For illumination purposes
- As a fuel
- In metallurgy for providing reducing atmosphere

Biogas

Biogas is produced by the degradation of biological matter by the anaerobic bacterias in the absence of free oxygen. Natural gas is also a biogas, which results after a long time decay of animal and vegetable matters (buried under the earth), brought about by bacterias in the presence of high pressure, high temperature, radioactive rays, and so on. The cheapest and easily obtainable biogas is gobar gas, which is produced by the anaerobic fermentation of cattle dung.

Gobar Gas

The raw material for the gas is cow dung, which is subjected to the action of a kind of microorganism under anaerobic conditions. The process is carried out in a closed steel or concrete tank made above ground or under ground into which a slurry made by mixing equal parts of fresh cattle dung and water is poured. Anaerobic bacterias present in the dung digest this slurry forming mainly methane and carbon dioxide. The optimum temperature for this fermentation is 34–48°C. The gas generated is collected in a steel gas holder placed on the top of the digestion tank (Fig. 13.14). About 160 litres of gobar gas is produced per kg of cow dung. The average composition of gobar gas is

$$CH_4 = 55\% \qquad CO_2 = 35.0\%$$
$$H_2 = 7.4\%, \qquad N_2 = 2.6\% \quad \text{Rest} = H_2S$$

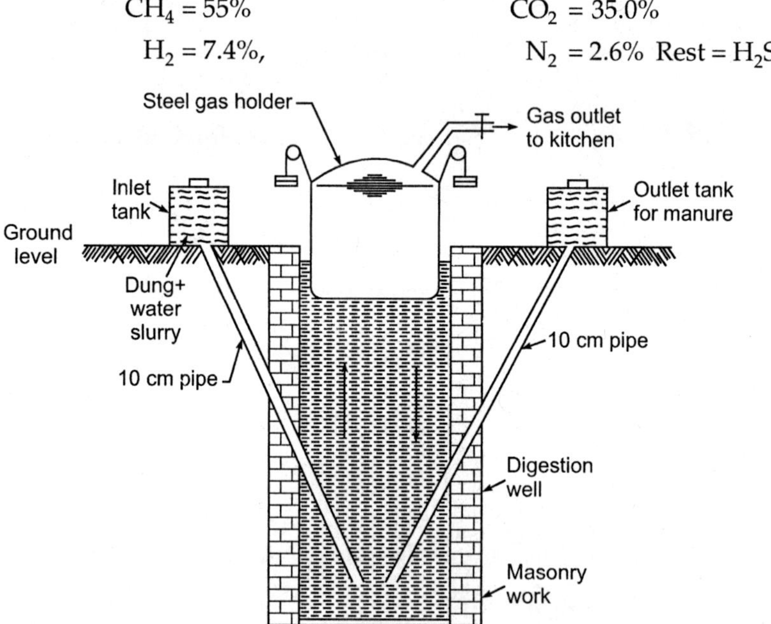

Fig. 13.14 Cross-section of a gobar gas plant.

Its calorific value is about 1,200 kcal/m^3.

Following are the uses of gobar gases:

- Nonpolluting as a domestic fuel
- Excellent yield of good manure

13.17 PETROCHEMICALS

Natural gas is a mixture of lower hydrocarbons (alkanes and alkenes). These hydrocarbons are raw materials from which a large variety of organic compounds termed as petrochemicals are obtained. The important compounds derived from petroleum and natural gas directly are listed here:

- Saturated hydrocarbons, that is methane, ethane, propane, butane, pentane, hexane and so on.
- Unsaturated hydrocarbons, that is ethylene, propylene, butylene, acetylene, and so on.
- Aromatic hydrocarbons, that is benzene, toluene, and so on obtained by aromatization of the *n*-hexane and *n*-heptane. They are further used for synthesis of a large number of aromatic compounds, dyes, drugs, and explosives.
- Iso-octane used in aviation fuel.
- Methyl alcohol, formaldehyde, acetaldehyde, and so on are obtained by the cracking of petroleum hydrocarbons.

These petroleum chemicals are used in industry for preparing solvents, explosives, resins, rubber, plastics, insecticides, pesticides, adhesives, fibers, detergents, dyes, and so on.

13.18 COMBUSTION

Combustion is an exothermic chemical reaction during which combustible elements or gases, when brought to their ignition temperature, burn in the presence of air instantly with the development of heat and a considerable increase in temperature.

Calculation of Air Required for Combustion

- Substances always combine in definite proportions, which are determined by their molecular weights. Thus, to know the amount of oxygen required for the combustion of an element or a gas, we must know its combustion reaction. The combustion of carbon in oxygen:

$$C + O_2 \longrightarrow CO_2$$

	C	O_2	CO_2
(By Wt.)	12	32	44
(By Vol.)	(1 Vol.)	(1 Vol.)	(1 Vol.)

12 parts by weight of C require 32 parts by weight of O_2, and 44 parts by weight of CO_2 is produced by 12 parts by weight of C.

1 Vol. of carbon requires 1 Vol. of oxygen.

1 Vol. of CO_2 is produced by 1 Vol. of carbon.

The following are the most common combustion reactions. From these, the amount of oxygen required by the substance may be calculated as shown previously.

$$2H_2 + O_2 \longrightarrow 2H_2O$$
$$S + O_2 \longrightarrow SO_2$$
$$2CO + O_2 \longrightarrow 2CO_2$$
$$CH_4 + 2O_2 \longrightarrow CO_2 + 2H_2O$$
$$2C_2H_6 + 7O_2 \longrightarrow 4CO_2 + 6H_2O$$
$$C_2H_4 + 3O_2 \longrightarrow 2CO_2 + 2H_2O$$

Nitrogen, ash, and carbon dioxide (if any) present in the fuel or the air are incombustible, so they do not take any oxygen during combustion. The total amount of oxygen consumed by the fuel will be given by the sum of the amounts of oxygen required by the individual combustible constituent present in the fuel.

- 22.4 L (22,400 ml.) of any gas at STP (0°C and 760 mm pressure) weighs equal to its 1 gram molecule, that is its molecular weight in grams. For example: 22.4 L of CO_2 at STP will weigh 44 gm. (Mol. wt. of CO_2 = 44)
- The molecular weight of air is taken as 28.94.
- Air contains 21 percent oxygen by volume and 23 percent oxygen by weight. Hence, from the amount of oxygen required by the fuel, the amount of air can be calculated.

$$1 \text{ kg of oxygen is supplied by } \frac{1 \times 100}{23} = 4.35 \text{ kg of air.}$$

$$1 \text{ m}^3 \text{ of oxygen is supplied by } \frac{1 \times 100}{21} = 4.76 \text{ m}^3 \text{ of air.}$$

- Weight can be converted into volume at a certain temperature and pressure by using the gas equation

$$PV = nRT$$

- It may be noted that of the total amount of hydrogen, some of it is present in combined form (*i.e.*, as H_2O), which is noncombustible. So, it does not take part in combustion. The rest of hydrogen (called available hydrogen) only takes part in the combustion reaction.

$$4H + O_2 \longrightarrow 2H_2O + \text{Heat}$$

(by wt.) 4 32

Now 1 part of hydrogen combines chemically with 8 parts by mass of oxygen

$$= \text{Mass of hydrogen} - \left(\frac{\text{Mass of oxygen}}{8}\right)$$

So the quantity of hydrogen available for the combustion reaction will be $\left(H - \frac{O}{8}\right)$ where H represents the total quantity of hydrogen, and O represents the total quantity of oxygen in fuel.

∴ The theoretical amount of oxygen required for the complete combustion of 1 kg of solid or liquid fuel is

$$= \left[\frac{32}{12} \times C + 8\left(H - \frac{O}{8}\right) + S\right] \text{kg}$$

C, H, S, and O represent the weight of carbon, hydrogen, sulphur, and oxygen, respectively, per kg of fuel. Because the percentage of oxygen in air by weight is 23, the amount of air required theoretically for combustion of 1 kg of fuel is

$$= \frac{100}{23}\left[\frac{32}{12} \times C + 8\left(H - \frac{O}{8}\right) + S\right] \text{kg}$$

13.19 ANALYSIS OF FLUE GAS

The analysis of flue gas is carried out to have a correct idea about the complete or incomplete combustion of fuel. If there is a high percentage of carbon monoxide in the flue gases, it shows incomplete combustion and that the oxygen supply is less than required. If there is a high percentage of oxygen in the flue gases, the combustion may be complete, but it indicates that the oxygen supply is in excess.

The analysis of flue gases is carried out by Orsat's apparatus (Fig. 13.15).

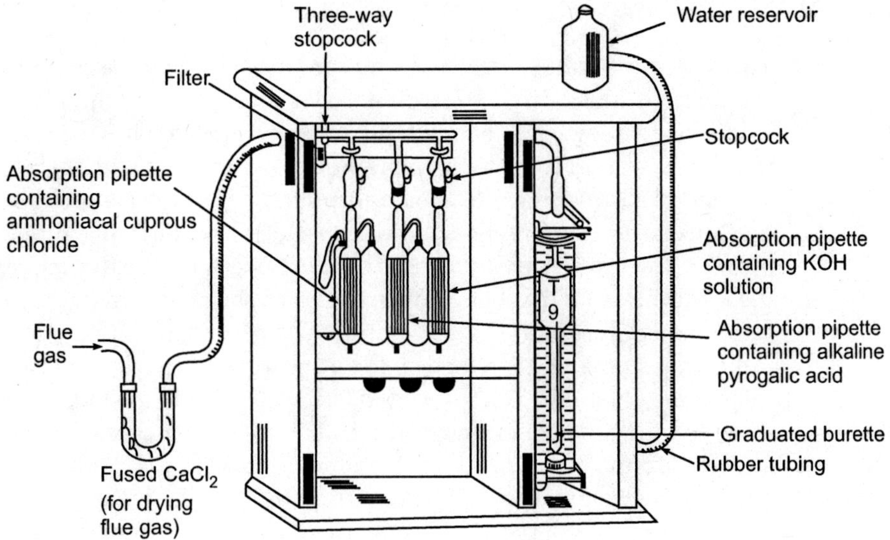

Fig. 13.15 Orsat's gas analysis apparatus.

Description of Orsat's Apparatus

It consists of a water-jacketed measuring burette connected in series to a set of three absorption bulbs each through a stopcock, the free end of which is further connected to a U-tube packed with glass wool (to avoid the incoming smoke particles, etc.). The graduated burette is surrounded by a water jacket to keep the temperature of gas constant during the experiment. The lower end of the burette is connected to a water reservoir by means of long rubber tubing. The absorption bulbs are usually filled with glass tubes so that the surface area of contact between the gas and the solution is increased.

The absorption bulbs have solutions for the absorption of CO_2, O_2, and CO, respectively. The first bulb has a potassium hydroxide solution (250 g KOH in 500 ml of boiled distilled water) and absorbs only CO_2. The second bulb has a solution of alkaline pyrogallic acid (25 g pyrogallic acid + 200 g KOH in 500 ml. distilled water) and can absorb CO_2 and O_2. The third bulb contains ammoniacal cuprous chloride (100 g cuprous chloride + 125 ml. liquor ammonia + 375 ml. water) and can absorb CO_2, O_2, and CO. Hence, the flue gas must be passed first through the potassium hydroxide bulb where CO_2 is absorbed, then through the alkaline pyrogallic acid bulb when only O_2 will be absorbed (because CO_2 has already been removed), and finally through the ammoniacal cuprous chloride bulb, where only CO will be absorbed.

Working

- The whole apparatus is cleaned, and the stoppers are greased and then tested for air-tightness. The absorption bulbs are filled with their respective solutions to a level just below their rubber connections. Their stopcocks are then closed. The jacket and leveling reservoir are filled with water. The three-way stopcock is opened to the atmosphere, and the reservoir is raised until the burette is completely filled with water, and air is excluded from the burette. The three-way stopcock is now connected to the flue gas supply, and the reservoir is lowered to draw in the gas to be analyzed in the burette. The sample gas present in the apparatus may be mixed with some air. So, the three-way stopcock is opened to the atmosphere, and the gas is expelled by raising the reservoir. This process of sucking and exhausting gas is repeated three to four times to exclude all the air in the apparatus. Finally, the gas is sucked into the burette, and the volume of the flue gas is adjusted to 100 ml. at atmospheric pressure. This is ensured by equalizing the level of water in the burette and reservoir. The three-way stopcock is now closed.
- The stopper of the absorption bulb containing the caustic potash solution is opened, and the gas is passed into this bulb by raising the water reservoir. The gas is again sent to the burette. This process is repeated several times to ensure the complete absorption of CO_2. The unabsorbed gas is finally taken

back to the burette until the level of solution in the CO_2 absorption bulb stands at the constant mark, and then its stopcock is closed. The levels of water in the burette and reservoir are equalized, and the volume of residual gas is noted. The decrease in volume gives the volume of CO_2 in 100 ml. of the flue gas sample.

- The volumes of O_2 and CO are similarly determined by passing the remaining gas through the alkaline pyrogallic acid bulb and the ammoniacal cuprous chloride bulb, respectively. The gas remaining in the burette after the absorption of CO_2, O_2, and CO is taken as nitrogen.

SOLVED NUMERICAL PROBLEMS

1. *The ultimate analysis of coal gives carbon = 84%, sulphur = 1.5%, nitrogen = 0.6%, hydrogen = 5.5%, and oxygen = 8.4%. Calculate the gross and net calorific values using Dulong's formula.*

Sol. Gross calorific value (GCV) or HCV is

$$\text{GCV} = \frac{1}{100}\left[8080 \times \text{C} + 34500 \times \left(\text{H} - \frac{\text{O}}{8}\right) + 2240 \times \text{S}\right]$$

$$= \frac{1}{100}\left[8080 \times 84 + 34500 \times \left(5.5 - \frac{8.4}{8}\right) + 2240 \times 1.5\right]$$

$$= 8356 \text{ kcal/kg}$$

Net calorific value (LCV) is

$$\text{LCV} = \left(\text{HCV} - \frac{9\text{H}}{100} \times 587\right) \text{kcal/kg}$$

$$= \left(8356 - \frac{9 \times 5.5}{100} \times 587\right) \text{kcal/kg}$$

$$= 8065.44 \text{ kcal/kg}$$

2. *A coal has the following composition by weight: C = 90%, O = 3.0%, S = 0.5%, N = 0.5%, and ash = 2.5%. The net calorific value of coal was found to be 8,490.5 kcal/kg. Calculate the percentage of hydrogen and the higher calorific value of the coal.*

Sol. $\text{HCV} = \left(\text{LCV} + \frac{9\text{H}}{100} \times 587\right) \text{kcal/kg}$

$$= (8490.5 + 52.8 \text{ H}) \text{ kcal/kg} \qquad (13.1)$$

Also, $$\text{HCV} = \frac{1}{100}\left[8080 \times 90 + 34{,}500\left(\text{H} - \frac{3.0}{8}\right) + 2240 \times 0.5\right]\text{kcal/kg}$$

$$= (7754.8 + 345\ \text{H})\ \text{kcal/kg} \qquad (13.2)$$

Now from Equation (13.1) and Equation (13.2)

$$8490.5 + 52.8\ \text{H} = 7754.8 + 345\ \text{H}$$

or $$\text{H} = 4.575$$

So, the percentage of hydrogen is 4.575%

$$\therefore \qquad \text{HCV} = (8{,}490.5 + 52.8 \times 4.575)\ \text{kcal/kg}$$

$$= 8{,}731.8\ \text{kcal/kg}$$

3. *0.72 g of a fuel containing 80% carbon, when burnt in a bomb calorimeter, increased the temperature of water from 27.3°C to 29.1°C. If the calorimeter contains 250 g of water, and its water equivalent is 150 g, calculate the HCV of fuel in kJ/kg.*

Sol. x = 0.72 g, W = 250 g, w = 150 g, t_1 = 27.3 °C, t_2 = 29.1°C

$$\text{HCV(L)} = \frac{(W + w)(t_2 - t_1)}{x}\ \text{kcal/kg}$$

$$= \frac{(250 + 150)(29.1 - 27.3)}{0.72}\ \text{kcal/kg} = 1{,}000\ \text{kcal/kg}$$

4. *When 0.84 g of a sample of fuel was completely burnt in excess of oxygen, the increase in the temperature of the water in a calorimeter containing 1060 g of water was 2.5°C. Calculate the HCV of fuel, if the water equivalent of the calorimeter, and so on is 135 g.*

Sol. x = 0.84 g, W = 1060 g, w = 135 g, $t_2 - t_1$ = 2.5°C

$$\text{HCV(L)} = \frac{(W + w)(t_2 - t_1)}{x}\ \text{kcal/kg}$$

$$= \frac{(1060 + 135)(2.5)}{0.84} = 3556.55\ \text{kcal/kg}$$

5. *The following data were obtained in a Boy's gas calorimeter experiment:*

Volume of gas use = 0.1 m^3 at STP

Wt. of water heated = 25 kg

Temperature of inlet water = 20°C

Temperature of outlet water = 33°C

Wt. of steam condensed = 0.025 kg

Calculate HCV and LCV per m^3 at STP. Take the heat liberated in condensing water vapor and cooling the condensate as 580 kcal/kg.

Sol. $V = 0.1\ m^3$, $W = 25\ kg$, $t_1 = 20°C$, $t_2 = 33°C$, $m = 0.025\ kg$

$$HCV(L) = \frac{W(t_2 - t_1)}{V} = \frac{25(33-20)}{0.1}\ kcal/m^3 = 3{,}250\ kcal/m^3$$

$$LCV = HCV - (m/V) \times 580$$

$$= 3{,}250\ kcal/m^3 - (0.025\ kg/0.1\ m^3) \times 580\ kcal/kg$$

$$= 3{,}105\ kcal/m^3$$

6. *A sample of coal was analyzed as follows: Exactly 2.500 g was weighed into a silica crucible. After heating for 1 hour at 110°C, the residue weighed 2.415 g. The crucible next was covered with a vented lid and strongly heated for exactly 7 minutes at 950 ± 20°C. The residue weighed 1.528 g. The crucible was then heated without the cover, until a constant weight was obtained. The last residue was found to weigh 0.245 g. Calculate the percentage results of the preceding analysis.*

   Sol. Mass of moisture in coal sample = 2.500 – 2.415 = 0.085 g

$$\text{Mass of volatile matter} = 2.415 - 1.528 = 0.887\ g$$

$$\text{Mass of ash} = 0.245\ g$$

$$\text{Percent of moisture} = \frac{0.085 \times 100}{2.500} = 3.40\%$$

$$\text{Percent volatile matter} = \frac{0.887 \times 100}{2.500} = 35.48\%$$

$$\text{Percent ash} = \frac{0.245 \times 100}{2.500} = 9.80\%$$

$$\text{Percent fixed carbon} = 100 - (3.40 + 35.48 + 9.80) = 51.32\%$$

7. *Calculate the amount of theoretical air by weight and volume for the complete combustion of 2 kg of coke. (Air contains 21% by volume of oxygen and 23.2% by weight of oxygen.)*

   Sol.

$$\underset{12}{C} + \underset{32}{O_2} \longrightarrow \underset{44}{CO_2}$$

The weight of oxygen required for the combustion of 2 kg of coke

$$= 2 \times \frac{32}{12} = 5.333\ kg = 5333\ g$$

Hence, the amount of air required for the complete combustion of 2 kg coke

$$= 5.333 \times \frac{100}{23.2}\ kg = 22.987\ kg$$

The volume of air needed for the complete combustion of 2 kg coke

$$= \text{No. of moles of air} \times 22.4 = \frac{22.987}{28.94} \times 22.4$$

$$= 17.792 \text{ liters at NTP } (\because \text{ Molecular mass of air} = 28.94)$$

8. *A sample of a fuel on analysis gave the following results:*

$$C = 86\%, H = 4\%, N = 1.3\%$$

$$S = 3\%, O = 4\%, Ash = 1.7\%$$

Calculate the minimum weight of air (containing 23% of oxygen by weight) required for the complete combustion of 500 g of this fuel.

Sol. Wt. of C = 0.86 kg   Wt. of S = 0.03 kg

Wt. of H = 0.04 kg   Wt. of O = 0.04 kg

Nitrogen and ash are incombustible, so they do not require oxygen.

$$\text{Wt. of } O_2 \text{ required by carbon} = \frac{32}{12} \times 0.86 = 2.29 \text{ kg}$$

$$\text{Wt. of } O_2 \text{ required by sulphur} = \frac{32}{32} \times 0.03 = 0.03 \text{ kg}$$

$$\text{Wt. of } O_2 \text{ required by hydrogen} = \frac{32}{4} \times 0.04 = 0.32 \text{ kg}$$

$$\therefore \text{ Total wt. of } O_2 \text{ required} = 2.64 \text{ kg}$$

$$\text{Wt. of } O_2 \text{ already present} = 0.04 \text{ kg}$$

$$\therefore \text{ Wt. of } O_2 \text{ needed} = 2.64 - 0.04 = 2.60 \text{ kg}$$

$$\text{Wt. of air (containing 23\% } O_2\text{) required} = \frac{100}{23} \times 2.60 = 11.30 \text{ kg/kg of fuel}$$

$$\therefore \text{ Wt. of air required by 500 g of fuel} = \frac{11.30 \times 500}{1000} = 5.65 \text{ kg}$$

9. *Calculate the weight and volume of air required for the combustion of 3 kg of carbon.*

Sol. Wt. of air required for combustion of 3 kg C

$$= \text{Wt. of C} \times \frac{32}{12} \times \frac{100}{23} = 3 \text{ kg} \times \frac{32}{12} \times \frac{100}{23} = 34.783 \text{ kg}$$

Vol. of air required for combustion of 3 kg C

$$= \text{Wt. of air in g} \times (22.4 \text{ L/mol}) \times (\text{mol/mol. wt.})$$

$$= 34{,}783 \text{ g} \times (22.4 \text{ L/mol}) \times (\text{mol}/28.94 \text{ g})$$

$$= 2.692 \times 10^4 \text{ L or } 26.92 \text{ m}^3$$

10. *Calculate the volume of air required for the complete combustion of 1 cubic meter of producer gas, the composition of which (by volume) is H_2 = 30%, CO = 10%, CH_4 = 4%, and N_2 = 56%.*

Sol. 1 m^3 of producer gas has the composition:

$$H_2 = 0.30\ m^3,\ CO = 0.10\ m^3,\ CH_4 = 0.04\ m^3,\ \text{and}\ N_2 = 0.56\ m^3.$$

The combustion reactions are

$$H_2(g) + \tfrac{1}{2}O_2(g) \longrightarrow H_2O(l)$$

(by Vol.) 1 0.5 1

$$CO(g) + \tfrac{1}{2}O_2(g) \longrightarrow CO_2(g)$$

(by Vol.) 1 0.5 1

$$CH_4(g) + 2O_2(g) \longrightarrow CO_2(g) + 2H_2O(l)$$

(by Vol.) 1 2 1

Volume of oxygen required for complete combustion of 1 m^3 of producer gas:

$$0.30 \times 0.5 + 0.10 \times 0.5 + 0.04 \times 2\ m^3 = 0.28\ m^3$$

Volume of air required for complete combustion of 1 m^3 of producer gas:

$$= \frac{0.28 \times 100}{21} = 1.333\ m^3$$

11. *A sample of gasoline was found to contain 15.4% hydrogen and 84.6% carbon by weight. Calculate the volume of oxygen at N.T.P. required for the complete combustion of 1 kg of this fuel.*

Sol. Wt. of gasoline = 1 kg

Wt. of carbon = 0.846 kg

Wt. of hydrogen = 0.154 kg

The combustion reactions are

$$C + O_2 \longrightarrow CO_2$$

(by Wt.) 12 32

$$2H_2 + O_2 \longrightarrow 2H_2O$$

(by Wt.) 4 32

$$\therefore \quad \text{Wt. of } O_2 \text{ required by 0.846 kg of C} = \frac{32}{12} \times 0.846 = 2.256\ \text{kg}$$

Wt. of O_2 required by 0.154 kg of H_2 $= \frac{32}{4} \times 0.154 = 1.232$ kg

$\therefore$ Total weight of O_2 required $= 2.256 + 1.232 = 3.488$ kg

Because 32 gm of O_2 occupies a volume of 22.4 liters at N.T.P.

3.488×1000 gm of O_2 will occupy $= \frac{22.4}{32} \times 3.488 \times 1000 = 2441.6$ liters.

12. *A sample of coal was found to have the following composition: C = 75%, H = 5.2%, O = 12.1%, N = 3.2%, and ash = 4.5%. Calculate the minimum air required for the complete combustion of 1 kg of coal. Also calculate the higher calorific value and lower calorific value of the coal sample.*
[Gross calorific value in kcal/kg : C = 8,080 ; H = 34,500 ; S = 2,240]

Sol. 1 kg of coal contains C = 750 g, H = 52 g, O = 121 g, N = 32 g, ash = 45 g
Minimum weight of air required for complete combustion of 1 kg of coal

$$= \text{Wt. of}\left[C \times \left(\frac{32}{12}\right) + H \times \left(\frac{16}{2}\right) - O\right] \times \left(\frac{100}{23}\right)$$

$$= \left[750\,\text{g} \times \left(\frac{32}{12}\right) + 52\,\text{g} \times \left(\frac{16}{2}\right) - 121\,\text{g}\right] \times \left(\frac{100}{23}\right)$$

$$= [2{,}000\,\text{g} + 416\,\text{g} - 121\,\text{g}] \times \left(\frac{100}{23}\right)$$

$$= 2{,}295\,\text{g} \times \left(\frac{100}{23}\right) = 9{,}978\,\text{g or } 9.978\,\text{kg}$$

$$\text{HCV} = \frac{1}{100}\left[8{,}080 \times C + 34{,}500\left(H - \frac{O}{8}\right) + 2{,}240 \times S\right] \text{kcal/kg}$$

$$= \frac{1}{100}\left[8{,}080 \times 75 + 34{,}500\left(5.2 - \frac{12.1}{8}\right) + 2{,}240 \times O\right] \text{kcal/kg}$$

$$= \frac{1}{100}[6{,}06{,}000 + 1{,}27{,}219 + 0] \text{ kcal/kg}$$

$$= [6{,}060 + 1{,}272] \text{ kcal/kg} = 7{,}332 \text{ kcal/kg}$$

$$\text{LCV} = (\text{HCV} - 0.09\,H \times 587) \text{ kcal/kg}$$

$$= (7{,}332 - 0.09 \times 5.2 \times 587) \text{ kcal/kg} = 7{,}057 \text{ kcal/kg}$$

13. *Find the minimum amount of air required for the complete combustion of 10 kg coal having the following composition by weight:*

C = 81%, H = 8%, O = 5%, N = 2%, and remaining is ash.

Also calculate the higher calorific value and lower calorific value of the coal sample given the gross calorific value in kcal/kg of C = 8,080 and H = 34,500.

Sol. In 1 kg coal:

$$C = 810 \text{ g}, H = 80 \text{ g}, O = 50 \text{ g}, N = 2 \text{ g}, \text{ash} = 40 \text{ g}$$

Minimum weight of air required for the complete combustion of

$$1 \text{ kg of coal} = \text{wt. of} \left(C \times \frac{32}{12} + H \times \frac{16}{2} - O \right) \times \frac{100}{23}$$

$$= \left(810\,\text{g} \times \frac{32}{12} + 80\,\text{g} \times \frac{16}{2} - 50\,\text{g} \right) \times \frac{100}{23}$$

$$= 11{,}956.5 \text{ g} = 11{,}957 \text{ kg}$$

$$\text{HCV} = \frac{1}{100}\left[8{,}080 \times C + 34{,}500\left(H - \frac{O}{8} \right) \right] \text{kcal/kg}$$

$$= \frac{1}{100}\left[8{,}080 \times 81 + 34{,}500\left(8 - \frac{5}{8} \right) \right] \text{kcal/kg}$$

$$= 9.089 \text{ kcal/kg}$$

$$\text{LCV} = (\text{HCV} - 0.09\text{ H} \times 587) \text{ kcal/kg}$$

$$= (9{,}089 - 0.09 \times 8 \times 587) \text{ kcal/kg}$$

$$= 8{,}666.4 \text{ kcal/kg}$$

EXERCISES

1. What is a fuel? List the different types of fuels. What are the requirements of a fuel to be used in an industry?
2. What are the gross and net calorific values of a fuel? How would you express them in the case of gaseous fuel?
3. Write a note on production, composition, and applications of LPG.
4. Define an octane number. What is LPG? Discuss the large-scale manufacture of water gas, and mention a few of its uses.
5. What are the compositions of producer gas and water gas? Discuss the large-scale preparation of producer gas, and mention some of its uses.
6. How do you define a fuel? Classify it, and write a short note on the refining of petroleum.

7. Explain the terms *knocking, octane number*, and *cetane number.*
8. (*a*) Write a short note on Biogas.
   (*b*) With a neat diagram, describe the production of coal gas. Mention its applications.
9. Explain the term *proximate analysis*, and describe its usefulness.
10. What is the ultimate analysis of coal? Write briefly.
11. (*a*) Write briefly on metallurgical coke.
    (*b*) What is the difference between low-temperature and high-temperature carbonization?
12. (*a*) Define the calorific value of a fuel.
    (*b*) What are the constituents of bio gas?
    (*c*) What products are obtained by the refining of petroleum?
13. (*a*) Give the units of calorific value of solid, liquid and gaseous fuels.
    (*b*) How is the calorific value of a fuel determined?
14. What does the term *refining petroleum* mean? Explain with a neat diagram how petroleum is refined.
15. Write a note on the source, composition, and uses of natural gas.
16. Name the parameters to be determined under the proximate and ultimate analysis of coal. Why are these two analysis required? Out of many coal samples, which coal samples can be considered the best and why?
17. (*a*) What do the terms *knocking* and *octane number* mean?
    (*b*) Write short notes on octane number and cetane number.
18. What is coke? Describe the manufacture of coke by the beehive-coke oven method.
19. (*a*) Describe the fixed-bed catalytic cracking process for obtaining gasoline.
    (*b*) Give a brief account of the refining of petroleum.
20. (*a*) What is the difference between caking coal and coking coal?
    (*b*) Describe the manufacture of producer gas. Mention its application.
21. (*a*) What is cracking? Mention the catalysts used for catalytic cracking.
    (*b*) What do you mean by a 75-octane fuel?

MULTIPLE CHOICE QUESTIONS

1. Coke is a pure variety of carbon that contains
   (*a*) 60% C (*b*) 50% C
   (*c*) 80–95% C (*d*) 70–80% C
2. The calorific value of coal decreases because of
   (*a*) carbon (*b*) hydrogen
   (*c*) oxygen (*d*) sulphur

3. Anthracite contains
   (*a*) 60% C (*b*) 78% C
   (*c*) 70% C (*d*) 90% C
4. A bomb calorimeter is used for determining the calorific value of
   (*a*) a solid fuel (*b*) a liquid fuel
   (*c*) a gaseous fuel (*d*) both (*a*) and (*b*)
5. Analysis of flue gases is done by
   (*a*) an orsat apparatus (*b*) a bomb calorimeter
   (*c*) a boy's gas calorimeter (*d*) a retort
6. The purest form of coal is
   (*a*) peat (*b*) lignite
   (*c*) bituminous (*d*) anthracite
7. Isooctane has an octane rating of
   (*a*) 50 (*b*) 100
   (*c*) 0 (*d*) 120
8. The cetane rating of hexadecane is
   (*a*) 100 (*b*) 0
   (*c*) 50 (*d*) 120
9. Gas with the least calorific value is
   (*a*) coal gas (*b*) water gas
   (*c*) producer gas (*d*) natural gas
10. The moisture is determined in proximate analysis of coal at
    (*a*) 75°C (*b*) 105°C
    (*c*) 35°C (*d*) 55°C
11. An example of a primary fuel is
    (*a*) natural gas (*b*) petrol
    (*c*) wood charcoal (*d*) coke
12. An example of a secondary fuel is
    (*a*) wood (*b*) coal
    (*c*) natural gas (*d*) gobar gas
13. Petrol is a mixture mainly of
    (*a*) alkenes (*b*) alkanes
    (*c*) alkynes (*d*) aromatic hydrocarbons
14. Water gas is prepared by passing steam over red hot
    (*a*) Fe (*b*) Cu
    (*c*) Ni (*d*) C
15. A fuel with a high ignition temperature is
    (*a*) petrol (*b*) wood
    (*c*) kerosene (*d*) LPG

16. The calorific value of a fuel is expressed as

(*a*) kcal cm (*b*) kcal cm^{-3}

(*c*) kcal m^{-3} (*d*) cal m^{-3}

17. The composition of a producer gas is

(*a*) $CO + N_2$ (*b*) $CO + H_2$

(*c*) $CO + CH_4$ (*d*) $CH_4 + H_2$

18. Water gas is composed of

(*a*) $N_2 + CO$ (*b*) $CH_4 + H_2$

(*c*) $CO + H_2$ (*d*) $N_2 + H_2$

19. Gobar gas contains

(*a*) CH_4 (*b*) CO_2

(*c*) H_2 (*d*) all of these

20. Fuel suitable for use in motor engines is

(*a*) kerosene (*b*) gasoline

(*c*) diesel (*d*) heavy oil

Chapter 14

WATER TREATMENT

14.1 INTRODUCTION

Water is used for industrial purposes and for municipal supply. Water treatment is used to ensure the right quality and quantity of water for these purposes. Water is available in nature in abundance via the following prime sources:

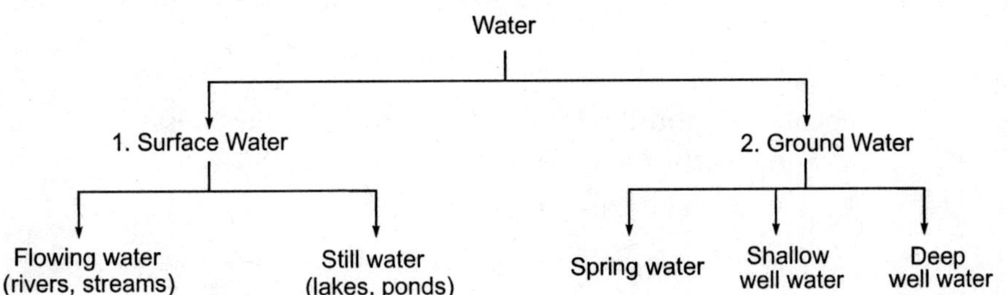

The other sources of limited use are sea water and rain water. Rain water is the purest form of water.

The surface water (river and lake water) contains dissolved CO_2; upon contact with basic materials, it dissolves them as bicarbonates.

$$\underset{\text{(Insoluble)}}{CaCO_3} + CO_2 + H_2O \longrightarrow \underset{\text{(Soluble)}}{Ca(HCO_3)_2}$$

$$\underset{\text{(Insoluble)}}{MgCO_3} + CO_2 + H_2O \longrightarrow \underset{\text{(Soluble)}}{Mg(HCO_3)_2}$$

In a similar manner, chlorides and sulphates present in limestone ($CaCO_3$) are also dissolved.

14.2 WATER QUALITY PARAMETERS AND STANDARD

The various types of impurities present in water impart certain properties to water. Color, decaying aquatic organisms, and certain chemicals may impart odor; suspended matter may impart turbidity to the water; the presence of dissolved salts makes the water hard, and excess quantities of metals and dissolved gases makes it corrosive. Bacteriological impurities due to pathogenic bacteria make water unfit for drinking or domestic purposes.

Specifications for Industrial Water

The quality of water required for industries depends upon the requirements of a particular industry. Each industry has its own specifications for water. The water required for boiler-feed purposes must be of high quality, that is free from dissolved salts and dissolved gases.

Specifications for Drinking Water

Drinking water standards fall into two categories:

Primary standards: Primary standards specify the Maximum Contaminant Levels (MCL) of various dissolved minerals based on their effect on human health.

Secondary standards: These standards may vary from place to place depending upon the taste, odor, color, and hardness of water, and do not have any anticipated effect on health.

The common standards for drinking water are:

- Free from objectionable minerals that have a detrimental effect on health
- pH value ranging from 6.5–8.5
- Total dissolved solids less than 500 ppm
- Free from pathogenic bacteria
- Free of iron (permissible limit 0.3 mg/L) and manganese (permissible limit 0.05 mg/L)
- Noncorrosive
- Free from dissolved gases such as H_2S
- Odorless with a pleasant taste

14.3 HARDNESS OF WATER

The water is called hard because it is hard to get lather with soap. This is due to the presence of certain salts of calcium, magnesium, and other heavy metals dissolved in it. A sample of hard water when treated with soap (which is sodium or potassium salts of higher fatty acids such as stearic, oleic, and palmitic acids) does not produce lather, rather soap is precipitated in the form of insoluble salts of calcium and magnesium.

No lather is formed until all these ions are completely removed, so a large amount of soap is wasted.

$$\underset{\text{Sodium stearate}}{2C_{17}H_{35}COONa} + CaCl_2 \longrightarrow \underset{\text{Calcium stearate}}{(C_{17}H_{35}COO)_2Ca\downarrow} + 2NaCl$$

Thus, water that does not produce lather with soap solution readily but forms a white curd is called *hard water*. On the other hand, water that lathers easily on shaking with soap solution is called *soft water*. There are no dissolved calcium and magnesium salts in soft water.

Types of Hardness

The two types of water hardness are the following:

Temporary or carbonate hardness: This is caused due to the bicarbonates of calcium and magnesium in water and can be removed by boiling. The bicarbonates are decomposed on boiling yielding insoluble carbonates or hydroxides, which are deposited as a crust at the bottom of the vessel.

$$Ca(HCO_3)_2 \xrightarrow{\Delta} CaCO_3\downarrow + H_2O + CO_2\uparrow$$

$$Mg(HCO_3)_2 \xrightarrow{\Delta} Mg(OH)_2\downarrow + 2CO_2\uparrow$$

Permanent or noncarbonate hardness: This is because of the presence of sulphates and chlorides of calcium and magnesium and cannot be removed by boiling. It can be removed by the use of chemical agents.

14.4 UNITS OF HARDNESS

The following are the units of hardness:

Parts per million (ppm): The parts of the calcium carbonate equivalent to the hardness per 10^6 parts of water, that is 1 ppm = 1 part of $CaCO_3$ eq. hardness in 10^6 parts of water.

Milligrams per liter (mg/L): The number of milligrams of $CaCO_3$ equivalent to the hardness present per liter of water.

$$1\ mg/L = 1\ mg\ of\ CaCO_3\ eq.\ hardness\ per\ L\ of\ water$$

But 1 L of water weighs $= 1\ kg = 1{,}000\ g = 1{,}000 \times 1{,}000\ mg = 10^6\ mg$

$$\therefore \quad 1\ mg/L = 1\ mg\ of\ CaCO_3\ eq.\ per\ 10^6\ mg\ of\ water$$

$$= 1\ part\ of\ CaCO_3\ eq.\ per\ 10^6\ parts\ of\ water$$

$$= 1\ ppm$$

Degree French (°Fr): The parts of $CaCO_3$ equivalent to the hardness per 10^5 parts of water.

$$1°Fr = 1 part\ of\ CaCO_3\ hardness\ eq.\ per\ 10^5\ parts\ of\ water$$

Degree Clark (°Cl): The hardness is also measured on Clark's scale.

$$1°Clark = 1\ grain\ of\ CaCO_3\ eq.\ hardness\ per\ gallon\ of\ water$$

or

$$1°Cl = 1\ part\ of\ CaCO_3\ eq.\ hardness\ per\ 70{,}000\ parts\ of\ water$$

Soft water has been found to have less than 10° hardness, whereas hard water has been found to have a hardness between 20° and 30°.

The relationship between various units of hardness follows:

$$1\ ppm = 1\ mg/L = 0.1°Fr = 0.07°Cl$$

$$1\ mg/L = 1\ ppm = 0.1°Fr = 0.07°Cl$$

$$1°Fr = 10\ ppm = 10\ mg/L = 0.7°Cl$$

$$1°Cl = 1.43°F = 14.3\ ppm = 14.3\ mg/L$$

14.5 DISADVANTAGES OF HARD WATER

The disadvantages of hard water in domestic uses, industrial uses, and steam generation in boilers are described in the following sections.

Domestic Use

Following are the disadvantages:

Washing, bathing: Hard water does not form lather with soap easily. As a result, the cleansing quality of soap is decreased, and a lot of soap is wasted.

Cooking: The boiling point of water is increased because of the presence of salts. Hence, more fuel and more time are required for cooking.

Drinking: Hard water causes negative effects on our digestive system.

Industrial Use

Following are the disadvantages in various industries:

Textile industry: Hard water wastes soap. Precipitates of calcium and magnesium soaps adhere to fabrics and cause problems.

Sugar industry: Water containing sulphates, nitrates, carbonates, and so on causes difficulties in the crystallization of sugar.

Dyeing industry: The dissolved salts in hard water may react with costly dyes to form precipitates.

Paper industry: Calcium, magnesium, and iron salts in water may affect the quality of paper.

Pharmaceutical industry: Hard water may cause some undesirable products while preparating pharmaceutical products.

Steam Generation in Boilers

Boilers are used for steam generation. The presence of hard water may cause the following problems:

- Scale and sludge formation
- Priming and foaming
- Corrosion
- Caustic embrittlement

Scale and Sludge Formation in Boilers

Because of the continuous evaporation of water in boilers, the concentration of salts increases progressively. After the saturation point is reached, precipitates form on the inner walls of the boiler (Fig. 14.1).

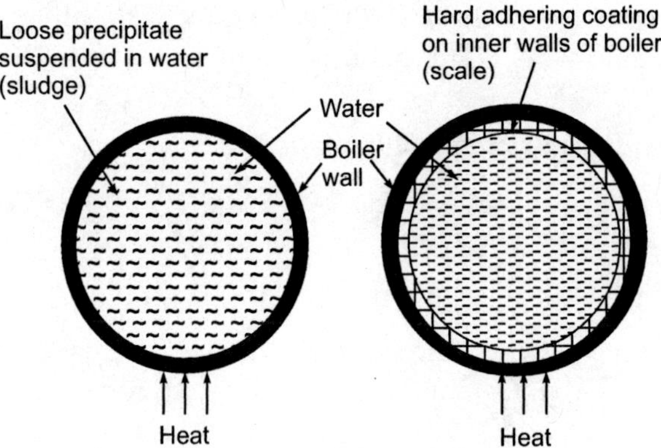

Fig. 14.1 Scale and sludge in boilers.

Sludge

Sludge is a soft, loose, and slimy precipitate formed within the boiler. Sludge is formed at comparatively colder portions of the boiler and collects in the area where the flow rate is slow. These are formed by substances that have greater solubilities in hot water than in cold water, for example $MgCO_3$, $MgCl_2$, $CaCl_2$, $MgSO_4$, and so on.

The disadvantages of sludge formation follow:

- The sludges are poor conductors of heat, so they cause a loss of heat.
- The working of the boiler is disturbed because of choking of the pipes by the sludge.

Sludge can be prevented

- By using well-softened water
- By drawing off a portion of concentrated water frequently

Scale

Scale refers to hard deposits that stick very firmly to the inner surface of the boiler. Scales are difficult to remove even by hammer and chisel.

The following are the causes of scale formation:

Decomposition of calcium bicarbonate:

$$Ca(HCO_3)_2 \longrightarrow \underset{\text{(Scale)}}{CaCO_3\downarrow} + H_2O + CO_2\uparrow$$

In low-pressure boilers, $CaCO_3$ causes scale formation.

In high-pressure boilers, $CaCO_3$ becomes soluble.

$$CaCO_3 + H_2O \longrightarrow Ca(OH)_2 + CO_2\uparrow$$

Deposition of calcium sulphate: The solubility of calcium sulphate in water decreases with the rise of temperature. In super-heated water, $CaSO_4$ is insoluble. This is the main cause of scales in high-pressure boilers.

Hydrolysis of magnesium salts:

$$MgCl_2 + 2H_2O \longrightarrow \underset{\text{(Scale)}}{Mg(OH)_2\downarrow} + 2HCl\uparrow$$

$Mg(OH)_2$ so formed by hydrolysis of magnesium salts is a soft type of scale.

Presence silica: Silica present in small quantities deposits as silicates ($CaSiO_3$, $MgSiO_3$) and is very difficult to remove.

The disadvantages of scale formation follow:

Wasting fuel: The scale formation causes a decrease of heat transfer. As a result, overheating is required, which causes the consumption of fuel.

Danger of explosion: The hot scale cracks because of expansion, and water suddenly comes in contact with overheated iron plates. This causes the formation of a large amount of steam suddenly. This results in high pressure that causes the boiler to burst.

To prevent scale formation:

External treatment: Efficient softening of water.

Internal treatment: Suitable chemicals are added to the boiler water either to precipitate or to convert the scale into compounds.

Important treatment methods are as follows:

Colloidal conditioning: In low-pressure boilers, organic substances such as kerosene, agar and so on are added in water to avoid scale formation.

Phosphate conditioning: Sodium phosphate is added in high-pressure boilers, which forms soft sludge of calcium and magnesium phosphates:

$$3CaCl_2 + 2Na_3PO_4 \longrightarrow \underset{\text{(Sludge)}}{Ca_3(PO_4)_2} \downarrow + 6NaCl$$

Carbonate conditioning: Sodium carbonate is added to boiler water when $CaSO_4$ is converted into $CaCO_3$:

$$CaSO_4 + Na_2CO_3 \longrightarrow \underset{\text{(Sludge)}}{CaCO_3} + Na_2SO_4$$

$CaCO_3$ is loose sludge and can be removed by a blow-down operation.

Calgon conditioning: Calgon (sodium hexa meta phosphate) is added to boiler water. It forms a soluble complex compound with $CaSO_4$:

$$Na_2[Na_4(PO_3)_6] \rightleftharpoons 2Na^+ + [Na_4P_6O_{18}]^{2-}$$

$$2CaSO_4 + [Na_4P_6O_{18}]^{2-} \longrightarrow \underset{\text{Soluble complex ion}}{[Ca_2P_6O_{18}]^{2-}} + 2Na_2SO_4$$

Treatment with Sodium Aluminate: Sodium aluminate is hydrolyzed into aluminium hydroxide (precipitate):

$$NaAlO_2 + 2H_2O \longrightarrow NaOH + Al(OH)_3 \downarrow$$

$$MgCl_2 + 2NaOH \longrightarrow Mg(OH)_2 \downarrow + 2NaCl$$

The flocculent precipitate of $Mg(OH)_2$ and $Al(OH)_3$ produced inside the boiler entraps the suspended and colloidal impurities along with silica. The precipitate now can be removed by a blow-down operation. Sodium aluminate is very cheaply available from bauxite refining units.

Electrical conditioning: Sealed glass bulbs, containing mercury, are connected to a battery and set rotating in the boiler. Mercury bulbs emit electrical discharges when water boils to prevent scale formation.

Radioactive conditioning: Energy radiations emitted by radioactive salts placed inside the boiler prevent scale formation.

Boiler Corrosion

Boiler corrosion is the decay of boiler material by a chemical or electrochemical attack by its environment. The reasons are as follows.

Dissolved Oxygen

The dissolved oxygen content of water is nearly 8 ppm at room temperature. The dissolved oxygen in water attacks the boiler material.

$$2Fe + 2H_2O + O_2 \longrightarrow 2Fe(OH)_2 \downarrow$$

$$4Fe(OH)_2 + O_2 \longrightarrow \underset{\text{Rust}}{2[Fe_2O_3 . 2H_2O]} \downarrow$$

The following are ways to prevent dissolved oxygen:

- By adding sodium sulphite, hydrazine, or sodium sulphide, dissolved oxygen from the boiler water can be removed completely.

$$2Na_2SO_3 + O_2 \longrightarrow 2Na_2SO_4$$

$$N_2H_4 + O_2 \longrightarrow N_2 + 2H_2O$$

$$Na_2S + 2O_2 \longrightarrow Na_2SO_4$$

Hydrazine is mostly used for the removal of dissolved oxygen because it does not increase dissolved salts.

- The process of mechanical deaeration (Fig. 14.2) involves spraying water in a perforated, plate-fitted tower heated from the sides and connected to a vacuum pump. In the presence of a high temperature, low pressure, and large exposed surface, the dissolved oxygen level is reduced in water.

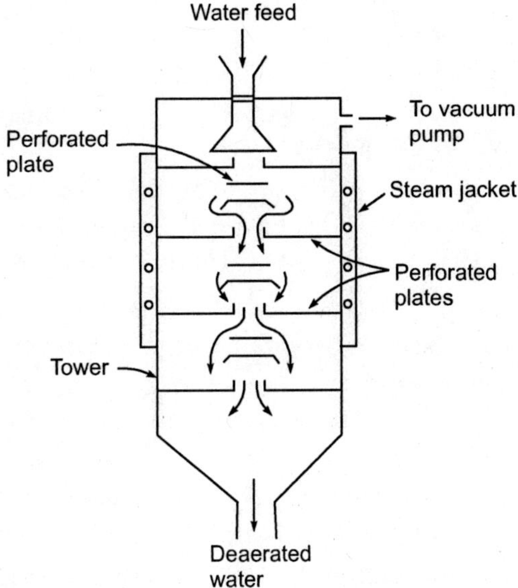

Fig. 14.2 Mechanical deaeration of water.

Dissolved Carbon Dioxide

The dissolved CO_2 content of boiler water causes boiler corrosion:

$$CO_2 + H_2O \longrightarrow H_2CO_3$$

Carbonic acid (H_2CO_3) has a corrosive effect on boiler material. If water used in the boiler contains bicarbonate, CO_2 is released.

$$Mg(HCO_3)_2 \xrightarrow{\Delta} MgCO_3 + H_2O + CO_2 \uparrow$$

Dissolved carbon dioxide can be prevented:

- By adding ammonia quantitatively, dissolved carbon dioxide from the boiler water can be removed.

$$2NH_4OH + CO_2 \longrightarrow (NH_4)_2CO_3 + H_2O$$

- By mechanical deaeration.

Acids from Dissolved Salts

Water containing dissolved magnesium salts liberate acids on hydrolysis:

$$MgCl_2 + 2H_2O \longrightarrow Mg(OH)_2 \downarrow + 2HCl$$

The liberated HCl reacts with the iron of the boiler producing HCl again and again in a chain reaction

$$2HCl + Fe \longrightarrow FeCl_2 + H_2 \uparrow$$

$$FeCl_2 + 2H_2O \longrightarrow Fe(OH)_2 \downarrow + 2HCl$$

A small amount of $MgCl_2$ can cause corrosion of iron to a large extent.

Priming and Foaming

Foaming is the production of foam or bubbles in boilers due to the presence of oils. Some particles of liquid water are carried out along with steam, during the production of steam from a boiler. This process of "wet steam" formation is called *priming*, which depends on the following:

- Presence of dissolved solids in large excess
- High steam velocities
- Sudden boiling
- Improper boiler design
- Instant increase in steam production rate

The disadvantages of priming and foaming:

- Dissolved salts in boiler water are carried by wet steam to turbine blades where they get deposited as the water evaporates and reduces the efficiency.
- Dissolved salts may enter the parts of the machinery, thereby decreasing the life of machinery.
- The actual height of the water column cannot be assessed.

To prevent priming:

- Fit mechanical steam purifiers.
- Provide efficient softening of boiler-feed water.
- Maintain low water levels in boilers.

To prevent foaming:

- Add antifoaming chemicals such as castor oil.
- Add sodium aluminate to remove oil from the boiler water.

Caustic Embrittlement

This is a type of boiler corrosion caused by using highly alkaline water in the boiler. Na_2CO_3 is usually present in water softened by the lime soda process. In high-pressure boilers, Na_2CO_3 decomposes as

$$Na_2CO_3 + H_2O \longrightarrow 2NaOH + CO_2$$

Another source of formation of sodium hydroxide in boiler water is the presence of natural alkalinity in water. Some sodium salts that may also be present in water form sodium hydroxide. This causes the boiler water to become caustic. Caustic soda attacks the surrounding area dissolving the iron of the boiler as sodium ferroate (Na_2FeO_2). This embrittlement of boiler parts causes failure of the boiler. Such a phenomenon is called *caustic embrittlement*.

To prevent caustic embrittlement:

- Use sodium phosphate as a softening agent replacing sodium carbonate.
- Add tannin or lignin to the boiler water, which blocks the hair-cracks and prevents the caustic soda solution to come in contact.
- Add Na_2SO_4 into the boiler water which also blocks hair-cracks.

14.6 SOFTENING WATER

The removal of hardness-causing constituents such as Ca^{2+} and Mg^{2+} is known as water softening. The following three methods are adopted for softening water.

Lime-Soda process: In this process, water is treated with lime [$Ca(OH)_2$] followed by the addition of soda ash (Na_2CO_3) to convert the soluble calcium and magnesium salts in water into insoluble compounds. This process can be accomplished either in cold or hot water.

- In the *cold lime-soda process,* the hard water is first analyzed and then fed into the inner vertical circular chamber fitted with a stirrer. Calculated quantities of lime and soda ash are added into the tank. It is essential to add small amounts of coagulant [$Al_2(SO_4)_3$], which entraps the fine precipitate of aluminium hydroxide, thereby causing easy filteration. It also removes silica and oil if present in water.

$$Al_2(SO_4)_3 + 3Ca(HCO_3)_2 \longrightarrow 2Al(OH)_3 \downarrow + 3CaSO_4 + 6CO_2 \uparrow$$

  Vigorous stirring and mixing at room temperature causes the softening of water. The heavy sludge settles down in the outer chamber, and the softened water reaches up. The softened water is then passed through the filtration unit, and finally filtered soft water flows out through the outlet at the top (Fig. 14.3). The cold lime-soda process provides water containing a residual hardness of 50–60 ppm.
- In the *hot lime-soda process,* a mixture of raw water and softening chemicals is heated near the boiling point of water. The reactions are much faster, and the precipitation is almost complete. The precipitation occurs more quickly and also settles down rapidly, so no coagulants are needed. Much of the dissolved gases (O_2, CO_2) are also removed (Fig. 14.4).

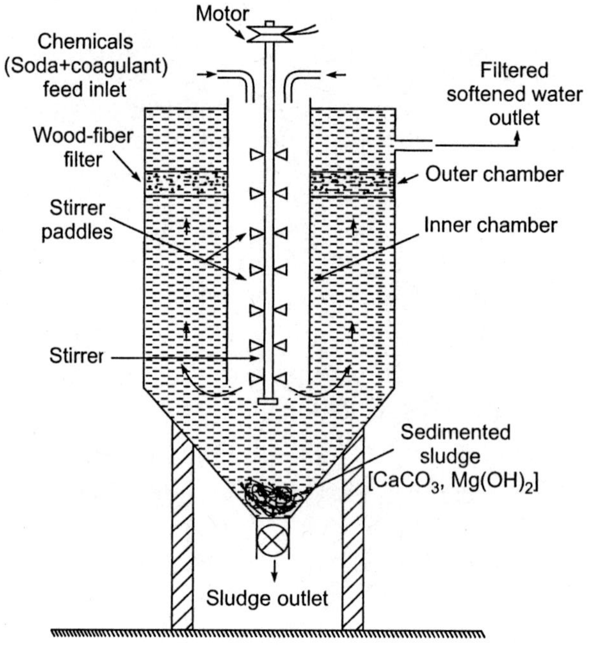

Fig. 14.3 Continuous cold lime-soda softener.

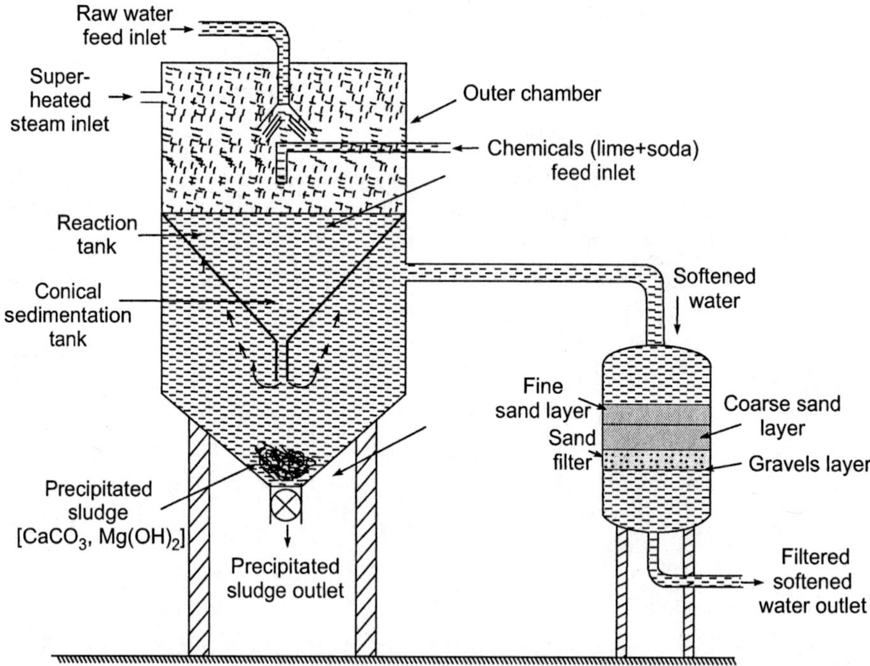

Fig. 14.4 Continuous hot lime-soda softener.

The hot lime-soda process produces water with a residual hardness of 15–30 ppm.

Zeolite or permutit process: In this process, water is softened with the help of a natural or artificial zeolite. Permutit is an artificial zeolite known as hydrate of sodium aluminium orthosilicate. Zeolites, known as green sand, are used for water softening. Artificial zeolite, known as permutit, is more common and has the general formula

$$Na_2O \,.\, R_2O_3 \,.\, xSiO_2 \,.\, nH_2O \; (x = 5 - 13, n = 3 - 4)$$

R_2O_3 stands for the oxide of an amphoteric metal such as aluminium or iron.

The zeolite or permutit is put in a suitable column as shown in Fig. 14.5. Hard water is percolated through it. The permutit can be represented as Na_2Z; after the base exchange, it is converted into CaZ and MgZ. Sodium present in the zeolite is replaced by divalent Ca^{2+} and Mg^{2+} ions present in hard water. Water is thus softened because sodium salts do not cause hardness.

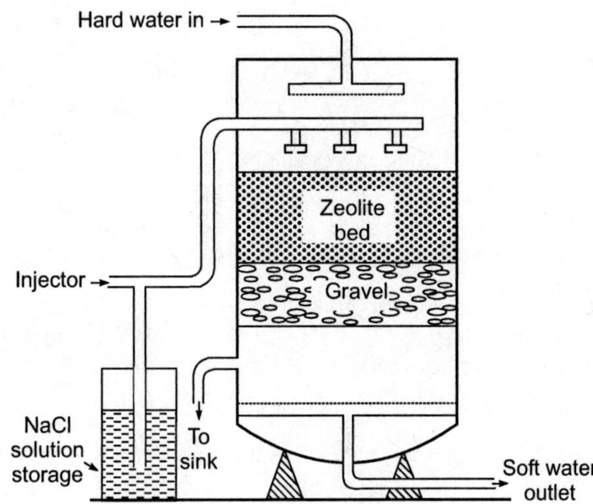

Fig. 14.5 Zeolite softener.

$$Na_2Z + Ca(HCO_3)_2 \longrightarrow 2NaHCO_3 + CaZ$$

$$Na_2Z + Mg(HCO_3)_2 \longrightarrow 2NaHCO_3 + MgZ$$

$$Na_2Z + CaSO_4 \longrightarrow Na_2SO_4 + CaZ$$

$$Na_2Z + MgCl_2 \longrightarrow 2NaCl + MgZ$$

The zeolite can be recovered by a passing a 10 percent NaCl (brine) solution into the bed.

$$CaZ + 2NaCl \longrightarrow Na_2Z + CaCl_2$$

$$MgZ + 2NaCl \longrightarrow Na_2Z + MgCl_2$$

The soft water obtained by this method is used mostly for laundry purposes. Water softened by this method cannot be used for boiler purposes because it may cause caustic embrittlement.

Ion exchange process: Ion exchange resins are insoluble, crosslinked, and long-chain organic polymers; the functional groups attached to the chains can exchange the cation and anion present in the water. This process can remove all soluble minerals of water without subjecting it to the costly distillation process.

- *Cation exchanger resins* are styrene-divinyl benzene copolymers, which on sulphonation or carboxylation, are capable of exchanging their hydrogen ions with the cations in water.

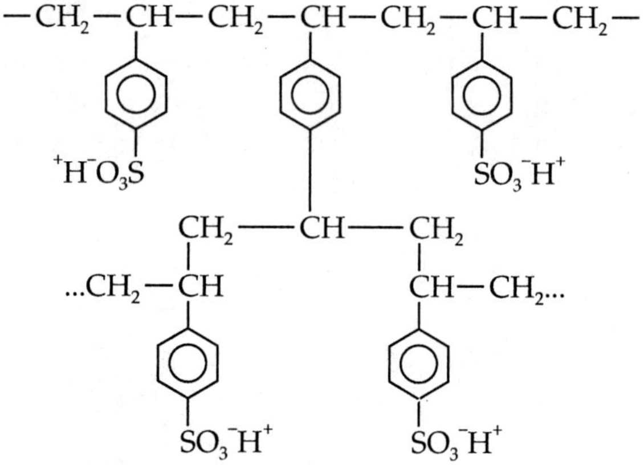

$-CH_2-CH-CH_2-CH-CH_2-CH-CH_2-$

$^+H^-O_3S$ $SO_3^-H^+$

$CH_2-CH-CH_2$

$...CH_2-CH$ $CH-CH_2...$

$SO_3^-H^+$ $SO_3^-H^+$

The resin may be represented as $(RSO_3^-)H^+$ where R represents the resin network.

- *Anion exchanger resins* are styrene-divinyl benzene copolymers containing amine or quarternary ammonium groups as an integral part of the resin and equivalent amount of anions such as Cl^-, SO_4^{2-}, OH^- ions, and so on.

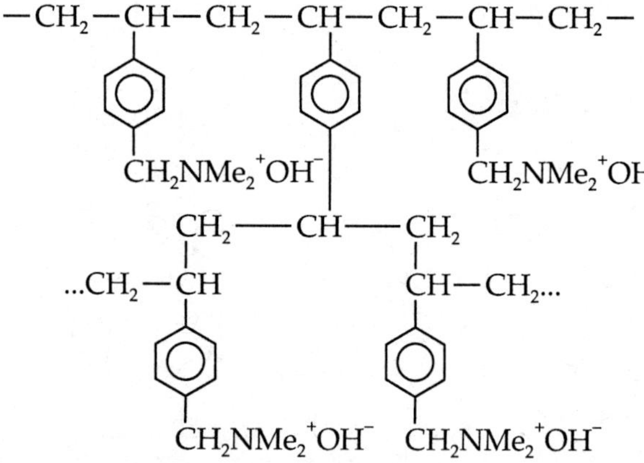

$-CH_2-CH-CH_2-CH-CH_2-CH-CH_2-$

$CH_2NMe_2^+OH^-$ $CH_2NMe_2^+OH^-$

$CH_2-CH-CH_2$

$...CH_2-CH$ $CH-CH_2...$

$CH_2NMe_2^+OH^-$ $CH_2NMe_2^+OH^-$

The resin may be represented as $R'OH^-$.

Working

The softener essentially consists of two steel tanks interconnected with a pipe (Fig. 14.6). In one of these, the cation exchange resin is put on the top of a glass wool plug; in the second tank the anion exchange resin is put on the top of a glass wool plug. The hard water is first passed through the cation exchange column, where all the cations, such as Ca^{2+}, Mg^{2+}, and so on, are removed. An equivalent amount of H^+ ions are released into the water.

$$2RH^+ + Ca^{2+} \longrightarrow R_2Ca^{2+} + 2H^+$$

$$2RH^+ + Mg^{2+} \longrightarrow R_2Mg^{2+} + 2H^+$$

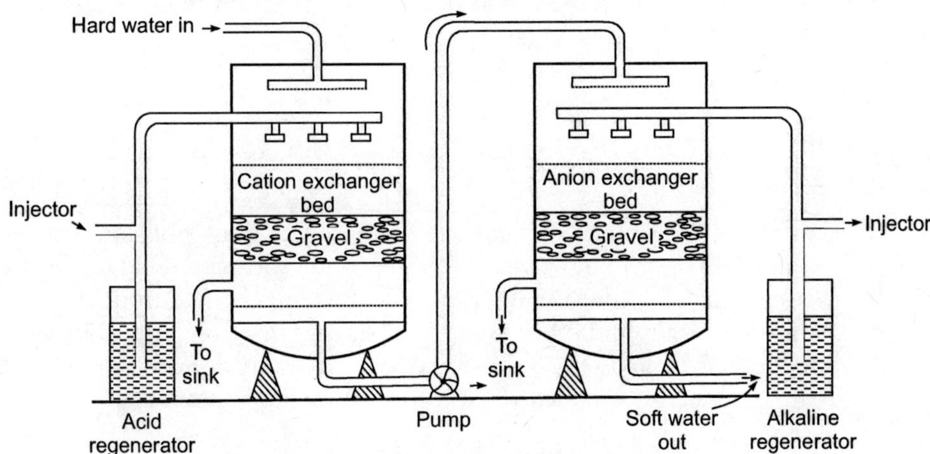

Fig. 14.6 Demineralization of water.

Then, the hard water is passed through the anion exchange column, where anions such as SO_4^{2-}, Cl^-, and so on are removed, and an equivalent amount of OH^- ions are released into the water.

$$R'OH^- + Cl^- \longrightarrow R'Cl^- + OH^-$$

$$2R'OH^- + SO_4^{2-} \longrightarrow R'_2SO_4^{2-} + 2OH^-$$

H^+ and OH^- ions so released get combined to produce a water molecule:

$$H^+ + OH^- \longrightarrow H_2O$$

The water obtained is free from cations and anions and is known as deionized or demineralized water.

Regeneration

The column gets exhausted when used for a long period and can be regenerated by passing a solution of an appropriate ion. The cation exchanger is regenerated by passing suitable acid (dil. HCl or dil. H_2SO_4), and the anion exchanger is regenerated by passing an alkali (dil. NaOH):

$$R_2Ca^{2+} + 2H^+ \longrightarrow 2RH^+ + \underset{\text{(Washing)}}{Ca^{2+}}$$

$$R'_2SO_4^{2-} + 2OH^- \longrightarrow 2R'OH^- + \underset{\text{(Washing)}}{SO_4^{2-}}$$

The column is washed with deionized water, and the washing (Ca^{2+}, SO_4^{2-}) is passed into the sink or drain.

14.7 PURIFICATION OF WATER FOR DOMESTIC USE

Natural water from rivers, canals, and so on does not conform to all the specification of drinking water. Water for domestic use should be potable, that is safe and palatable for drinking purposes. It should be free from, disease-causing germs, chemicals, and other foreign substances that are harmful to humans. The water is thus treated suitably to remove undesirable impurities present in it. The treatment methods are described in the following sections.

Removal of Suspended Impurities

Suspended impurities can be removed by the following:

Screening: The raw water is passed through screens to screen the floating matter.

Sedimentation: This is the process of allowing water to stand undisturbed in tanks (5 m deep) when suspended particles settle down at the bottom due to the force of gravity. The supernatant water is then drawn from the tank by pumps. The retention period in a sedimentation tank ranges from two to six hrs. When there are fine clay particles, coagulant is added to the water before sedimentation. The most commonly used coagulants are alum [$K_2SO_4 \cdot Al_2(SO_4)_3 \cdot 24H_2O$], sodium aluminate ($NaAlO_2$), and so on.

Filtration: This is the process of removing colloidal matter and most bacteria, and microorganisms by passing water through a bed of fine sand.

Removal of Microorganisms

Water contains some disease-causing bacteria (Pathogenic). The process of destroying/killing the disease-causing bacteria and microorganisms from the water and making it safe for use is called *disinfection*. The chemicals used for the purpose are called *disinfectants*. The following methods are used for disinfection:

Boiling: Water becomes safe for use after boiling for 10–15 minutes, when all the disease-causing bacteria are killed.

By adding bleaching powder: Bleaching powder is soluble in water (1 part in 20 parts of water). Generally, 1 kg of bleaching powder per 1000 kiloliters of water is mixed. The hypochlorous acid produced acts as the germicide.

$$CaOCl_2 + H_2O \longrightarrow Ca(OH)_2 + Cl_2$$

$$Cl_2 + H_2O \longrightarrow HCl + \underset{\text{(Germicide)}}{HOCl}$$

The disinfecting action of bleaching powder is due to chlorine produced by it.

Chlorination: Chlorine (gas/solution) on reaction with water produces hypochlorous acid, which acts as a germicide.

$$Cl_2 + H_2O \longrightarrow \underset{\text{(Germicide)}}{HOCl} + HCl$$

This is the most widely used disinfectant worldwide. The apparatus used for it is known as the chlorinator. The excess of chlorine is removed by sulphites known as antichlors.

Chloramine process: When chlorine and ammonia are mixed, a compound chloramine is formed. The sterilizing action of chloramine is due to the reaction with water that produces hypochlorous acid, which acts as a germicide.

$$Cl_2 + NH_3 \longrightarrow \underset{\text{Chloramine}}{ClNH_2} + HCl$$

$$ClNH_2 + H_2O \longrightarrow \underset{\text{(Germicide)}}{HOCl} + NH_3$$

The advantage of this process is that it does not impart a chlorinous taste to water.

Ozonization: Ozone is an excellent disinfectant. This is produced by passing silent electric discharge through oxygen.

$$3O_2 \longrightarrow 2O_3$$

Ozone is unstable and liberates nascent oxygen

$$O_3 \longrightarrow O_2 + O$$

The nascent oxygen is very powerful oxidizing agent that kills all the bacteria. The ozone and water are allowed to pass through separate inlets in a sterilizing tank. Ozone improves the taste of water. Very palatable water thus results from sterilization with ozone. The only disadvantage is the high cost involved in this process.

Potassium permanganate method: The water is sterilized by adding a calculated amount of $KMnO_4$. This method is not very popular because it is costly. It is mostly used for sterilizing well water in rural areas.

By using ultraviolet rays: Water is exposed to ultraviolet rays from a electric mercury vapor lamp immersed in water.

The flow sheet diagram for a municipal water treatment plant is shown in Fig. 14.7.

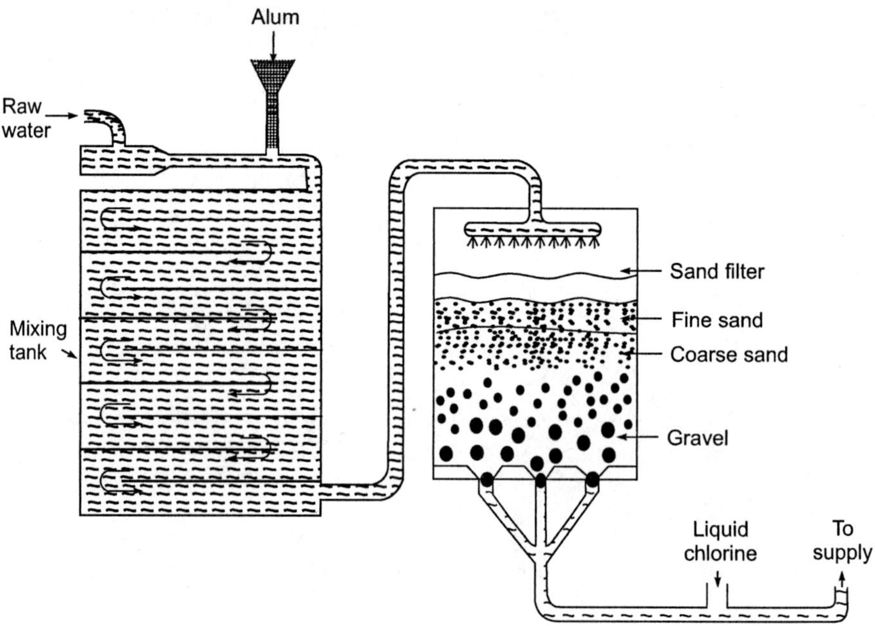

Fig. 14.7 Flow sheet of a drinking water treatment plant.

14.8 WATER ANALYSIS

Both physical and chemical analysis of water is essential.

Physical Analysis

Physical analysis includes the following:

Color: Color in water may be due to the presence of fine particles in suspension or due to certain mineral matter in solution. Color is measured with an instrument known as a *tintometer*. The unit of color is measured on the platinum cobalt scale. The standard color can be produced by dissolving 1 mg of platinum-cobalt in one liter of double distilled water. The number on the scale should be preferably less than 10 and should not exceed 20.

Odor: The extent of odor depends on the pH of water. A lower pH produces higher hydrogen sulphide. The presence of microscopic organisms, sewage, or wastes are the major causes of odor in polluted water. The intensity of odor is expressed in terms of the threshold odor number.

pH (Hydrogen ion concentration): pH— a measure of hydrogen ion activity— is used to express acidity or alkalinity of a sample of solution. Pure water (double distilled) has a pH equal to 7. pH can be determined electrometrically or colorimetrically.

Turbidity: Turbidity is a measure of the penetration of light into the medium of water and depends on either suspended silt and clay particles or dispersed planktonic organisms. Turbidity can be measured by Jackson's turbidimeter. For municipal water supply, turbidity should not exceed 10 ppm.

Chemical Analysis

Chemical analysis includes the following:

Estimation of free chlorine: Free chlorine in excess makes the water unfit for drinking purposes. The estimation of free chlorine in water is based on the oxidation of potassium iodide by free chlorine. Thus, when the water sample is treated with an excess of potassium iodide solution, the free chlorine present in water oxidizes potassium iodide and liberates an equivalent amount of iodine, which dissolves in excess KI to form a deepviolet colored complex.

$$Cl_2 + 2KI \longrightarrow 2KCl + I_2$$

$$KI + I_2 \longrightarrow \underset{\text{(complex)}}{KI_3}$$

The amount of liberated iodine can be estimated by titrating the resulting solution against a standard sodium thiosulphate solution using starch as the indicator.

$$I_2 + 2Na_2S_2O_3 \longrightarrow 2NaI + \underset{\substack{\text{Sodium} \\ \text{tetrathionate}}}{Na_2S_4O_6}$$

$$I_2 + \text{Starch} \longrightarrow \text{Deep blue complex}$$

Alkalinity: Alkalinity of water is due to the presence of carbonates and bicarbonates of calcium and magnesium. So the carbonates and bicarbonates in water can be determined by titrating the solution with standard sulphuric acid N/10 using first phenolphthalein and later methyl red as the indicators. In the first step of titration, using phenolphthalein (pH 8.2–10.5) as the indicator, conversion of carbonates into bicarbonates occurs.

$$2MgCO_3 + H_2SO_4 \longrightarrow Mg(HCO_3)_2 + MgSO_4$$

In the second step to that solution, using methyl red as indicator (pH 4.2–5.4), bicarbonates present in the water and bicarbonates formed from carbonates get neutralized.

$$Mg(HCO_3)_2 + H_2SO_4 \longrightarrow MgSO_4 + H_2O + CO_2$$

Take the case of ordinary water:

Titration with Indicator	*End Point*	*Inference*
Phenolphthalein	Zero	Alkalinity is due to HCO_3^-
Methyl orange (after phenolphthalein end point)	Zero	OH^-
Phenolphthalein		
Phenolphthalein end point >	Half the total titration	CO_3^-
Half of total titration		$CO_3^- + OH^-$
Phenolphthalein end point < Half of the total titration		$CO_3^{--} + HCO_3^-$

Hardness: This method is based on the following facts:

- Eriochrome black-T forms a wine-red colored unstable complex with calcium and magnesium ions (present in water).
- EDTA (ethylene diamine tetra-acetic acid) takes calcium and magnesium ions from the unstable complex and forms a colorless stable complex with the regeneration of blue-colored dye (eriochrome black-T).

$$\underset{\text{(of water)}}{\begin{bmatrix} Ca^{2+} \\ Mg^{2+} \end{bmatrix}} + \text{Eriochrome black - T} \longrightarrow \underset{\text{Unstable complex (wine red)}}{\begin{bmatrix} Ca \\ Mg \end{bmatrix} \text{Eriochrome black - T}}$$

$$\xrightarrow{\text{EDTA}} \underset{\text{Stable complex}}{\begin{bmatrix} Ca^{2+} \\ Mg^{2+} \end{bmatrix} \text{EDTA}} + \underset{\text{(blue)}}{\text{Eriochrome black-T}}$$

Thus, the change of wine-red color to distinct blue marks the end point of titration.

Dissolved oxygen: Determination of dissolved oxygen in a sample of water is based upon the oxidation of KI by dissolved oxygen present in the water sample. The liberated iodine is titrated against a standard 0.025 N $Na_2S_2O_3$ using starch as the indicator. Dissolved molecular oxygen in water is not capable of reacting with KI (*aq*). So an oxygen carrier such as $Mn(OH)_2$ is used to carry out the reaction between KI and O_2. $Mn(OH)_2$ is produced by the action of KOH on $MnSO_4$.

$$2KOH + MnSO_4 \longrightarrow Mn(OH)_2 + K_2SO_4$$

$$2Mn(OH)_2 + O_2 \longrightarrow \underset{\text{Basic manganic oxide}}{2MnO(OH)_2}$$

$$MnO(OH)_2 + H_2SO_4 \longrightarrow MnSO_4 + 2H_2O + \underset{\text{Nascent oxygen}}{[O]}$$

$$2KI + H_2SO_4 + [O] \longrightarrow K_2SO_4 + H_2O + I_2$$

$Mn(OH)_2$ produced in the first reaction reacts with dissolved molecular oxygen to form basic manganic oxide, which reacts with concentrated H_2SO_4 to liberate nascent oxygen. Nascent oxygen results in the oxidation of KI to I_2. The brown solution containing dissolved I_2 is then titrated with 0.025 N $Na_2S_2O_3$ using starch as the indicator.

$$2S_2O_3^{-} + 2e^{-} \xrightarrow{\text{Reduction}} S_4O_6^{--}$$

$$2I^{-} \xrightarrow{\text{Oxidation}} I_2 + 2e^{-}$$

$$2S_2O_3^{--} + 2I^{-} \longrightarrow I_2 + S_4O_6^{--}$$

Starch forms a deep blue starch iodide complex when the solution turns blue. At the end point, the blue color disappears, and the solution becomes colorless.

SOLVED NUMERICAL PROBLEMS

1. *A water sample contains 160 mg of $CaSO_4$ per liter. Calculate the hardness in terms of the $CaCO_3$ equivalent.*

Sol. $$CaSO_4 \equiv CaCO_3$$

$$\text{Mol. mass } 136 \text{ gm/mol} \equiv 100 \text{ mg/mol}$$

Hence, 136 mg/L of $CaSO_4$ = 100 mg/L of $CaCO_3$ eqv.

$$160 \text{ mg/L of } CaSO_4 = \frac{100 \text{ mg/L of } CaCO_3 \text{ eqv.} \times 160}{136}$$

$$= 117.64 \text{ mg/L of } CaCO_3 \text{ eqv.}$$

The hardness is 117.64 mg/L or ppm.

2. *Calculate temporary hardness and total hardness of a sample of water containing Mg $(HCO_3)_2$ = 9.3 mg/L, Ca $(HCO_3)_2$ = 17.4 mg/L, $MgCl_2$ = 8.7 mg/L, and $CaSO_4$ = 12.6 mg/L.*

Sol. Temporary hardness $= \left[9.3 \times \frac{100}{46} + 17.4 \times \frac{100}{162}\right]$ mg/L = 3.9 mg/L

Permanent hardness $= \left[8.7 \times \frac{100}{95} + 12.6 \times \frac{100}{136}\right]$ mg/L = 18.3 mg/L

Total hardness = (3.9 + 18.3) mg/L = 22.2 mg/L

3. *A sample of water contains Ca $(HCO_3)_2$ = 32.4 mg/L, Mg $(HCO_3)_2$ = 29.2 mg/L, and $CaSO_4$ = 13.5 mg/L. Find its temporary hardness and permanent hardness.*

Sol. Temporary hardness $= \left[32.4 \times \frac{100}{162} + 29.2 \times \frac{100}{146}\right]$ mg/L

= 40 mg/L or 40 ppm

Permanent hardness $= 13.5 \times \frac{100}{136}$ mg/L

= 9.9 mg/L or 9.9 ppm

4. *Calculate the quantity of lime of purity 70% and soda of purity 80% required for softening of 10,000 liter of water, which contains the following impurities: $CaCl_2$ = 180 mg/L, Ca $(NO_3)_2$ = 210 mg/L, $MgSO_4$ = 125 mg/L and Mg $(HCO_3)_2$ = 90 mg/L.*

Sol.

Constituent	*Amount*	*Multiplication Factor*	*$CaCO_3$ Equivalent*
$CaCl_2$	180 mg/L	100/111	$180 \times \frac{100}{111}$ = 162.2 mg/L
Ca $(NO_3)_2$	210 mg/L	100/164	$210 \times \frac{100}{164}$ = 128 mg/L
$MgSO_4$	125 mg/L	100/120	$125 \times \frac{100}{120}$ = 104.2 mg/L
Mg $(HCO_3)_2$	90 mg/L	100/146	$90 \times \frac{100}{146}$ = 61.6 mg/L

Lime required

$$= \frac{74}{100} [2 \times Mg(HCO_3)_2 + MgSO_4 \text{ as } CaCO_3 \text{ eq.}] \times (100/\% \text{ purity}) \times \text{Vol. of water}$$

$$= \frac{74}{100} [(2 \times 61.6 + 104.2) \text{ mg/L}] \times \frac{100}{70} \times 10{,}000 \text{ L}$$

$= 2.4 \times 10^6$ mg = 2.4 kg

Soda required

$$= \frac{106}{100} [CaCl_2 + MgSO_4 + Ca(NO_3)_2 \text{ as } CaCO_3 \text{ eq.}] \times (100/\% \text{ purity}) \times \text{Vol. of water}$$

$$= \frac{106}{100} [(162.2 + 104.2 + 128) \text{ mg/L}] \times \frac{100}{80} \times 10{,}000 \text{ L}$$

$= 5.2 \times 10^6$ mg = 5.2 kg

5. *How many grams of $FeSO_4$ dissolved per liter gives 215 ppm of hardness? (Fe = 56, S = 32, O = 16, Ca = 40, C = 12)*

Sol. $FeSO_4 \equiv CaCO_3$

$$56 + 16 + 64 = 136 \text{ gm} \equiv 100 \text{ gm}$$

∴ 100 ppm of hardness = 136 ppm of $FeSO_4$

$$215 \text{ ppm of hardness} = \frac{136 \times 215}{100}$$

$= 292.4$ mg/L of $FeSO_4$

$= 292.4$ ppm of $FeSO_4$

6. *If 50 ml. of a sample of hard water consumed 15 ml of 0.01 M EDTA. What is the hardness of water?*

Sol. $50 \text{ ml} \times N_{hardness} = 15 \text{ ml} \times 0.01 \text{ M EDTA}$

Molarity of hardness,

$$M_{hardness} = \frac{15 \times 0.01}{50} = 0.003 \text{ M}$$

$= 0.003 \times 2N$

$= 0.006$ N

∴ Normality of hardness, $N_{hardness} = 0.006$

∴ Strength of hardness = 0.006 × 50 g/L

= 0.3 g/L = 300 mg/L or ppm

7. *A sample of water has a CO_3^{2-} concentration of 16.5 ppm. What is the molarity of CO_3^{2-} in the sample?*

Sol. 16.5 ppm = 16.5×10^{-6} g/L of CO_3^{2-}

$$\therefore \qquad \text{Molarity} = \frac{16.5 \times 10^{-6}\ \text{g/L}}{60\ \text{g/mol}} = 2.7 \times 10^{-7}\ \text{mol/L}$$

EXERCISES

1. What are scales? What are the ill effects of scales?
2. Describe the hot lime-soda process for softening water. Mention its advantages over the cold lime-soda process.
3. What is hard water? How can it be removed? Briefly mention the disadvantages of hard water.
4. What disadvantages are expected if hard water is used for steam generation in boilers? What processes may be used to overcome these problems?
5. Describe a water treatment method for a municipal water supply.
6. (*a*) What do temporary and permanent hardness of water mean?
   (*b*) Write a short note on caustic embrittlement.
7. What factors are responsible for the hardness of water? Describe how water is purified for a city water supply.
8. How is the hardness of water removed by the ion exchange method?
9. Describe any two methods for softening hard water.
10. How is hardness of water determined by the EDTA method? Write the necessary calculations.
11. Describe the causes and harmful effects of scale formation.
12. What are the disadvantages of using hard water in a boiler operation? Discuss any four methods of disinfecting the water.
13. What are ion-exchange resins? How will you purify water by using the resins? What are the advantages of this method over other methods?
14. Write a short note on:
    (*i*) Scales and sludges
    (*ii*) Caustic embrittlement
15. (*a*) Distinguish between the internal and external treatment of water.
    (*b*) Write a short note on priming and foaming.
16. Describe the lime-soda process for water softening. Give the chemistry involved during the softening.
17. (*a*) Describe the ion-exchange process.
    (*b*) Describe the Zeolite process and its disadvantages.

18. Write a short note on:
    (*a*) Phosphate conditioning of boiler
    (*b*) Priming and foaming
19. What are the different units in which the hardness of water is expressed? How are they related to each other?
20. During the titration of hard water with EDTA, which type of buffer solution is added and why?
21. (*a*) Which indicator is used during the determination of hard water by the EDTA method, and what is the color change at the end point?
    (*b*) Explain the principle of water disinfection by bleaching powder and chloramine.

MULTIPLE CHOICE QUESTIONS

1. Purest form of natural water is
   (*a*) river water (*b*) sea water
   (*c*) rain water (*d*) lake water
2. Temporary hardness of water can be removed by
   (*a*) boiling (*b*) filtration
   (*c*) screening (*d*) sedimentation
3. Hardness in water is caused by
   (*a*) sodium chloride (*b*) sodium carbonate
   (*c*) calcium chloride (*d*) potassium nitrate
4. Which of the following can be used as coagulant?
   (*a*) $FeSO_4 \cdot 7H_2O$ (*b*) $Al_2(SO_4)_3 \cdot 18H_2O$
   (*c*) $Na_2Al_2O_4$ (*d*) all of these
5. Calgon is a trade name given to
   (*a*) sodium silicate (*b*) calcium phosphate
   (*c*) sodium hexa metaphosphate (*d*) sodium zeolite
6. A blow-down operation causes the removal of
   (*a*) scales (*b*) sludges
   (*c*) both scale and sludge (*d*) hot water only
7. Distilled water can be obtained by
   (*a*) boiling (*b*) the zeolite process
   (*c*) the lime-soda process (*d*) the ion-exchange process
8. Brackish water mostly contains dissolved
   (*a*) Ca salts (*b*) Mg salts
   (*c*) NaCl (*d*) suspended impurities

9. Hard water contains

(*a*) Ca^{2+}, Mg^{2+} or Fe^{2+}
(*b*) NO_3^- and PO_4^{3-}
(*c*) Na^+, K^+
(*d*) dissolved gases

10. Water can be sterilized by using

(*a*) O_2
(*b*) O_3
(*c*) H_2O_2
(*d*) KOH

11. Colloidal conditioning of a boiler is done by using

(*a*) calgon
(*b*) EDTA
(*c*) ion-exchangers
(*d*) lignin

12. Ultraviolet rays are used in water treatment for

(*a*) illumination
(*b*) sterilization
(*c*) coagulation
(*d*) sedimentation

Chapter 15

Environmental Pollution and Control

15.1 INTRODUCTION

The *environment* may be defined as the surroundings or control conditions influencing the development or growth of people, animals, plants, and microorganisms. The environment consists of four segments: atmosphere (the air envelope), hydrosphere (the water bodies), lithosphere (the land masses), and biosphere (life forms). Progress in science and technology is leading to the pollution of the environment and a serious ecological imbalance that in the long run may prove disastrous for mankind. Man has tried to domesticate the environment, which caused changes in the environmental conditions. Environmental pollution may be defined as "the excessive discharge or addition of undesirable foreign substances into the environment, thereby altering the natural quality of the environment and as such damaging human, plant, and animal life." A *pollutant* may also be defined as any solid, liquid, or gaseous substance present in such a concentration as may be or tend to be injurious to environment.

Pathway of a Pollutant

The pathway of a pollutant is the mechanism by which the pollutant gets distributed from its source into the environment segments. For example, tetra ethyl lead (TEL) present in gasoline/automobile gets distributed in nature by the pathway as follows:

$$\underset{\text{in gasoline}}{Pb(C_2H_5)_4} \xrightarrow[\text{Autoexhausts}]{} PbCl_2 + PbBr_2$$

$$\downarrow (\text{Air})$$

$$\text{To food crops and food chain} \longleftarrow \underset{(\text{Soil})}{PbCl_2 + PbBr_2}$$

Causes of Environmental Pollution

The following are the causes of environmental pollution:

- Population explosion
- Rapid industrialization
- Rapid urbanization
- Exploitation of nature
- Natural phenomena such as radioactivity, volcanic eruptions, strong winds, forest fires, and so on.

15.2 AIR POLLUTION

The atmosphere is about 150 km thick. The zone nearest to the earth is the troposphere, which is just up to 6.4 km above the earth. The upper zone is the stratosphere, which includes the ozone layer. The mesosphere on top of the stratosphere extends to 80 km. The region beyond the mesosphere is the thermosphere. The air pollution concerns mainly the state of the troposphere. The average percentage composition of the clean dry air near sea level is $N_2 = 78.08$, $O_2 = 20.95$, Ar = 0.9, $CO_2 = 0.0314$, and Ne = 0.018; other trace components are CH_4, O_3, CO, SO_2, NO_2, Xe, Kr, and so on.

Air pollution may be defined as "the excessive discharge of undesirable foreign substances into the atmospheric air, thereby altering the natural quality of air and as such damaging human, plant, and animal life." Air pollution is considered to be one of the most dangerous and common kinds of environmental pollution. Air pollution is most crucial from the public health point of view because every individual person breathes approximately 22,000 times a day, inhaling 15–22 kg of air.

Air pollutants are classified into two categories on the basis of their origin:

- *Primary pollutants* are directly emitted from the source into the atmosphere and are found as such, for example, CO, NO_2, SO_2, and so on.
- *Secondary pollutants* are derived from the primary pollutants due to the chemical or photochemical reaction in the atmosphere. Secondary pollutant examples include peroxy acetyl nitrate (PAN), photochemical smog, and so on.

15.3 TYPES OF AIR POLLUTANTS

There are four major groups of air pollutants: gases and vapors, particulates, deforestation, and internal combustion engines.

Gases and Vapors

Gas and vapor pollutants include the following:

Oxides of Sulphur (SO_x, *i.e.*, SO_2 and SO_3)

Sources of SO_2 are thermal power plants, the petroleum industry, sulphuric acid plants, paper manufacturing plants, and so on. The major source of SO_2 is the combustion of fuels, such as coal (the sulphur content in coal ranges between 1 and 4 percent). The primary effect of SO_2 is irritation of the respiratory tract. Even at low concentrations, it can have cumulative chronic effects. Lung cancer is known to result from raised levels of SO_2 in the atmosphere.

Effect on Plants

SO_2 is injurious to plants. A mild dose of SO_2 may cause interveinal chlorotic bleaching of leaves resulting in a change to the color leaves. SO_2 causes acid rain, which also damages plants. Acute exposure to high levels of SO_2 kills leaf tissues causing leaf necrosis. SO_3 is formed by the oxidation of SO_2 under the influence of sunlight. Oxidation can take place only in the presence of a catalyst such as NO_x, metal oxides, and dust. In SO_x, SO_2 is about 97–99 percent and SO_3 is 1–3 percent. 1 ppm of SO_3 in the air causes irritation to the respiratory tract.

Oxides of Nitrogen (NO_x, *i.e.*, NO and NO_2)

Sources of nitrogen oxide are combustion of fuels, acid manufacture, the explosive industry, transportation, and so on. Nitric oxide is less toxic than NO_2. NO_2 is more harmful to human health, causing irritation in the eyes, throat, and lungs. It also causes respiratory diseases, bronchitis, and oedema of the lungs. Higher levels of NO_x cause gum inflammation, internal bleeding, pneumonia, oxygen deficiency, and lung cancer. The biochemical mechanism of NO_2 toxicity is not clear. It may affect some cellular enzyme systems. The possible antidote is vitamin E (antioxidant).

Effect on Plants

A mild dose of NO_x causes bleaching and the suppressed growth of the leaf. NO_2 is highly injurious to plants.

Carbon Monoxide (CO)

Sources of CO are partial combustion of fuel in automobiles, industries, and oil refineries; cigarette smoke; and domestic heat appliances. The chief source of CO is automobile exhaust, which can cause headache, paralysis, and death. CO displaces oxygen from the haemoglobin to produce carboxyhaemoglobin:

$$O_2Hb + CO \longrightarrow COHb + O_2$$

The result is a reduction in the oxygen carrying capacity of blood. People exposed to 80 ppm of CO have their blood-oxygen carrying capacity reduced by 15 percent. The initial effect of CO poisoning is loss of awareness and judgement, which results in automobile accidents. CO is also responsible for heart attacks, and it affects the central nervous system. CO poisoning can be cured by exposing the affected person to fresh O_2.

Effect on Plants

CO has some detrimental effects on plants when exposed for a long period. A CO concentration from 100–10,000 ppm causes leaf drop, leaf curling, reduction in leaf size, and so on.

Hydrogen Sulphide (H_2S)

Hydrogen sulphide is well known for its rotten egg smell. The sources are the paper industry, petroleum refineries, coke oven plants, and the decomposition of organic wastes. Exposure to H_2S for short periods can result in fatigue. H_2S causes headache, nausea, collapse, coma, and finally death even at 1–3 ppm. H_2S at 5 ppm affects the digestive system and destroys appetite.

Carbon Dioxide (CO_2)

This is released into the atmosphere in the form of smoke, which is produced by the combustion of fuels such as coal, wood, petroleum, and so on. Plants and animals also release CO_2 by respiration. Deforestation and population explosion are also causes for the increase in the concentration of CO_2 in the atmosphere. Carbon dioxide is nontoxic, therefore it is not harmful to human health. The excess of CO_2 in the atmosphere causes respiratory problems and suffocation. The increased amount of CO_2 in the air is responsible for global warming.

Ozone (O_3)

Ozone is produced in the upper layer of the atmosphere from oxygen gas by the absorption of ultraviolet light.

$$3O_2 \xrightleftharpoons{UV} 2O_3$$

Therefore air in the upper layer is rich in ozone. Ozone is destructive to fabrics, rubber goods, crops, and so on. The ozone layer in the stratosphere is very important for the existence of life on earth. It absorbs most of the harmful UV radiations of the sun; otherwise, the UV radiation would damage plants and cause diseases such as skin cancer in animals and humans. The depletion of the ozone layer is due to excessive use of chemicals, called aerosol spray propellants (CFC).

Effect on Plants

A mild dose of ozone results in flecks on the upper surfaces of leaves, premature aging, and suppressed growth, whereas a severe dose of ozone may result in the collapse of the leaf and bleaching.

Depletion of the Ozone Layer

Chlorofluoro carbons (CFC) are the exhausts of supersonic aircraft and jumbo jets flying in the upper atmosphere. CFCs also find applications in refrigeration, air-conditioning, aerosols, sterilization, cleaning foams, and so on. CFCs accumulate at high altitudes and undergo decomposition under the influence of UV radiation. One of the main decomposition products is chlorine. Each atom of chlorine so released reacts with more than 10^5 molecules of ozone converting it into oxygen. This results in the gradual depletion of the ozone layer.

$$CFCl_3\,(g) \xrightarrow{h\nu} CFCl_2\,(g) + Cl_2\,(g)$$

$$CF_2Cl_2\,(g) \longrightarrow CF_2Cl\,(g) + Cl\,(g)$$

$$O_3\,(g) \xrightarrow{h\nu} O_2\,(g) + O\,(g)$$

$$Cl\,(g) + O_3\,(g) \longrightarrow ClO\,(g) + O_2\,(g)$$

$$O\,(g) + ClO\,(g) \longrightarrow Cl\,(g) + O_2\,(g)$$

$$Cl\,(g) + O_3\,(g) \longrightarrow ClO\,(g) + O_2\,(g)$$

$$O\,(g) + O_3\,(g) \longrightarrow 2O_2\,(g)$$

Emission of nitric oxide by high-flying supersonic aircrafts is responsible for the ozone depletion.

$$NO\,(g) + ClO\,(g) \longrightarrow Cl\,(g) + NO_2\,(g)$$
$$Cl\,(g) + O_3\,(g) \longrightarrow ClO\,(g) + O_2\,(g)$$
$$\overline{NO\,(g) + O_3\,(g) \longrightarrow NO_2\,(g) + O_2\,(g)}$$

Hydrochlorofluoro carbons (HCFCs) and hydrofluoro carbons (HFCs) are now considered short term ozone-friendly substitutes for CFCs in the air-conditioning and refrigeration industry.

Hydrogen Fluoride (HF)

The major sources of hydrogen fluoride are the manufacture of phosphate fertilizers, the aluminium industry, brick plants, and pottery and ferro enamel work. This is an important air pollutant even in a very low concentration. It causes irritation, bone disorders, and tooth disorders. This also causes fluorosis in cattle breeding affected plants.

Biochemical Effect of Fluorides

Fluorine does not concentrate in any tissue, only in the bones and teeth. Soluble fluorides are rapidly and almost completely absorbed from the gastrointestinal tract, even at high intakes and are rapidly distributed throughout the body as fluoride ion. It readily breaks the erythrogates. The disappearance of fluoride from the blood is quite fast. A fluoride concentration of more than 1 ppm causes erosion of the teeth. The fully formed enamel of an adult tooth is unaffected by fluorine, and deciduous teeth are only affected by a high fluoride intake. Fluorosis leads to the deformation of bones resulting in knock-knees, bowlegs, and stiffening of joints, joint pains, back pain, and so on. The uptake of fluoride by the bone depends on its vascularity and growth activity. Not only the old but also the youth and children are increasingly victims of this disease and have to use the support of a cane.

Aerosols

These are the chemicals released into the atmosphere with force in the form of vapor. The major source of aerosols is jet and aeroplane exhaust. The aerosols contain fluorocarbons used as propellants in jet engines. Aerosols containing fluorocarbons, NO_2, or SO_2 reach and deplete the ozone layer of the stratosphere.

Hydrocarbons

The major source is transportation. Methane (CH_4) is the main naturally occurring hydrocarbon emitted in the atmosphere. It is produced by bacteria during the anaerobic decomposition of organic matter in soil, water, and sediments.

$$2CH_2O \xrightarrow{\text{Bacteria}} CH_4 + CO_2$$

The hydrocarbons in the air are of concern only because they undergo chemical reactions in the presence of sunlight and nitrogen oxides to form peroxy acetyl nitrates (PAN). They cause irritation of the eyes and throat, and asthma in humans. They also cause photochemical smog. Hydrocarbons are carcinogenic in nature.

Effect on Plants

Hydrocarbons are injurious to plants. They inhibit plant growth and damage leaf tissues.

Particulates

Small, fine solid particles and liquid droplets are collectively called *particulates*. These are present in the air in excess and pose serious air pollution problems. The most important physical property is size. Particulates range in size from a diameter of 0.0002 μ to a diameter of 500 μ. The life period of particulates varies from a few seconds to several months, which depends on their setting rate, size, density of particles, and the turbulence of the air. Following are types of particulate:

Dust: Dust is produced by crushing or grinding organic and inorganic material. Major sources of dust are mines, vehicles, ceramics factories, combustion, natural phenomena, (forest fire, wind), and so on.

The effects of atmospheric dust are respiratory diseases. Dust causes silicosis if it contains silica.

Smoke: Smoke is composed of tiny particles of carbon, ash, oil, and so on. Smoke results from an incomplete combustion of fuel. The major sources are railroads, automobile engines, furnaces, and so on. Smoke is carcinogenic in nature.

Smog: The term comes from the words *smoke* and *fog*. Smog can be of two types:

- *London-type smog* from burning coal covers urban areas at night on cold days. This smog consists of smoke, sulphur compounds, and fly ash. This is also called classical smog, and is reducing in nature. Prolonged exposure to smog is very dangerous for elderly people and those with lung and heart diseases.

$$CH_3-\overset{\overset{\displaystyle O}{\|}}{C}-O-O-NO_2$$

PAN

- The *Los Angeles-type smog* is also called photochemical smog because it is formed through chemical reactions involving sunlight. The most notorious organic oxidant is peroxy acetyl nitrate (PAN). Nitrogen oxides, and hydrocarbons CO and O_3 are the constituents of this type of smog. This is oxidizing in nature. It reduces visibility, causes irritation of the eye, and damages vegetation.

Asbestos dust: The sources are mining, the processing and manufacture of asbestos gaskets, and building materials. The effect is the disease called asbestosis.

Lead dust: The sources are lead mining, auto exhaust, lead batteries, lead paints, and so on. Lead dust settles down on plants and food stuffs, which causes lead poisoning when consumed by humans.

Fly ash: The bulk of the mineral particulate matter in a polluted atmosphere exists as oxides and other compounds resulting from the combustion of high ash fossil fuel. Smaller ash particles called fly ash enter furnace flues and emerge from the stack, in the absence of collector devices, and enter the atmosphere. The fly ash emission from a variety of coal combustion units show a wide range of composition.

Deforestation

Green plants use CO_2 to manufacture food by photosynthesis, give out O_2, and purify atmospheric air. Plants absorb harmful gases from the atmosphere and help maintain the temperature balance. Therefore, deforestation causes air pollution.

Internal Combustion Engines

The major manmade air pollutant is the internal combustion engine used for running vehicles. In cities, vehicles cause air pollution to a great extent. The fuels burnt in the internal combustion engine are all a mixture of hydrocarbons of different molecular masses. The following reaction occurs when a mixture of gasoline vapor and air burns in ICE.

$$2C_8H_{18} + 25O_2 \longrightarrow 16CO_2 + 18H_2O + \text{Heat}$$

Besides, CO_2, CO, unburnt hydrocarbons, SO_2, NO_2, and so on enter the atmosphere.

Air pollution due to ICE can be controlled by:

- Using an engine with a better design
- Using a suitable catalyst in fuel so that complete oxidation can take place
- Mixing the exhaust gas with more air and burning them so that the concentration of toxic pollutants will be minimized
- Adding Tetra Ethyl Lead (TEL) to the gasoline, which emits less smoke

15.4 ACID RAIN

The term *acid rain* was first used by Robert Angus (1872). The presence of excessive acids in rain water is acid rain. This is one of the effects of air pollution. Every source of energy, that is coal, wood, or petroleum products has sulphur and nitrogen. These two elements when burnt in the presence of atmospheric oxygen are converted into their oxides (sulphur dioxide, nitrogen dioxide, and so on). These oxides dissolve in the moisture present in the atmosphere to form corresponding acids that then return to the earth's surface with rain water as acid rain. Pure rain water has a pH around 5.5–5.7.

$$SO_2 + H_2O \longrightarrow H_2SO_3$$

$$2SO_2 + 2H_2O + O_2 \longrightarrow 2H_2SO_4$$

$$4NO_2 + 2H_2O + O_2 \longrightarrow 4HNO_3$$

$$HCl(g) + H_2O \longrightarrow HCl\,(aq)$$

The adverse effects of acid rain include the following:

- Increases the acidity of rain water.
- Causes damage to aquatic life.
- Increases the acidity of soil, that is the pH of soil is decreased, which affects vegetation.
- Accelerates the rate of the corrosion of metals.
- Causes irritation to the eyes.
- Corrodes buildings, monuments, bridges, and so on.
- Contaminates the potable water and enters our bodies, thereby causing neurological disorders.

15.5 THE GREENHOUSE EFFECT

The *greenhouse effect* was first suggested by a Swedish chemist Arthenius. CO_2 is exclusively present in the troposphere. In excess, it can act as a serious pollutant.

The sun's rays consist of UV, IR, visible radiations, and so on. The ozone layer protects the UV and allows visible and IR radiations toward the earth. Because IR radiations are of a short wavelength, they easily pass through the CO_2 layer causing a heating effect. The heated earth reradiates this absorbed energy as radiations of a longer wavelength. This does not pass through the CO_2 layer but is absorbed by the CO_2 layer. Thus, the earth's atmosphere heats up, which is called the greenhouse effect (Fig 15.1). The thick CO_2 layer acts like the glass panels of a greenhouse. Every year, the worldwide CO_2 concentration is increasing at the rate of 0.75 ppm, and the temperature is rising at a rate of 0.05°C. The consequence is global warming, which will result in the melting of glaciers and ice caps and ultimately flood. The gases responsible for the greenhouse effect are known as greenhouse gases, including CO_2, CH_4, chlorofluorocarbons, O_3, N_2O, and water vapor.

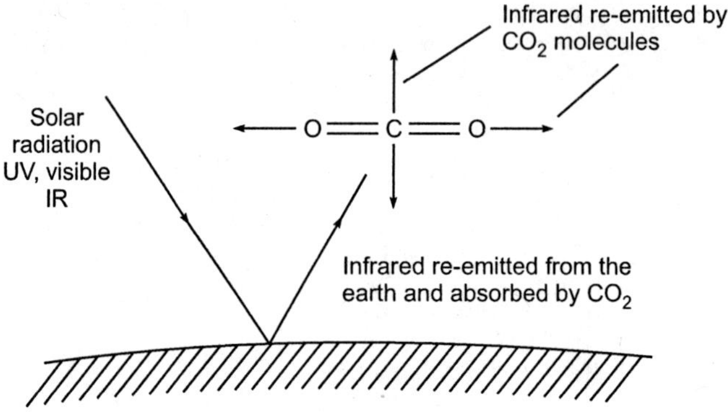

Fig. 15.1 The greenhouse effect.

15.6 MONITORING AIR QUALITY

Monitoring air quality is essential at strategic points. There are two types of monitoring:

- Biological monitoring
- Physio-chemical monitoring

Monitoring essentially consists of methods of air quality assessment. Biological monitoring directly provides *qualitative* information of the presence of pollutants in the natural stream of air.

Physio-chemical monitoring directly provides *quantitative* information about the presence of pollutants in natural air. Physio-chemical monitoring is based on actual measurements of physical parameters such as pH value, temperature, and turbidity, and the chemical parameters such as the presence of harmful gases (SO_2, NO_2, CO, and so on) and the presence of pesticides, insecticides, phenol, and heavy metals, for example Cu, Zn, Hg, Pb, Co, and so on. Physio-chemical monitoring provides suitable information regarding undesirable pollutants, which indicates the possible source of the effluents.

15.7 CONTROLLING AIR POLLUTION

The most effective method of controlling air pollution is the reduction of the pollutant discharge at the source by the application of control equipment and process control.

- Automobile exhausts may be minimized by using catalysts.
- The source discharge can be diluted by using tall stacks to reduce the concentration of air pollutants at the ground level.
- Smoke can be reduced or prevented during combustion of fuel by the following:
  - A correct method of firing.
  - The correct quantities of air should be admitted. Little supply of air results in smoke, whereas excess air wastes heat carried by high-temperature flue gases.
  - Maintain a high temperature because at low temperature, the combustion is incomplete and smoky.
  - Feed fuel continuously.

The equipment to control smoke is the Cottrell electrostatic precipitator. Smoke is a colloidal solution of negatively charged carbon particles in the air. Before passing the smoke to the chimney, it is passed through a chamber provided with a knob maintained to a high potential of + 30,000 volts or more. Under the influence of the electrical field, the smoke particles precipitate out and settle at the bottom of the chamber so that only hot, dust-free gases go out through the chimney. The efficiency of the electrostatic precipitator is 99.9 percent and it is capable of purifying 1,50,000 liters of gas per minute at a temperature of 600°C. The systematic diagram of the Cottrell electrostatic precipitator is shown in Fig. 15.2.

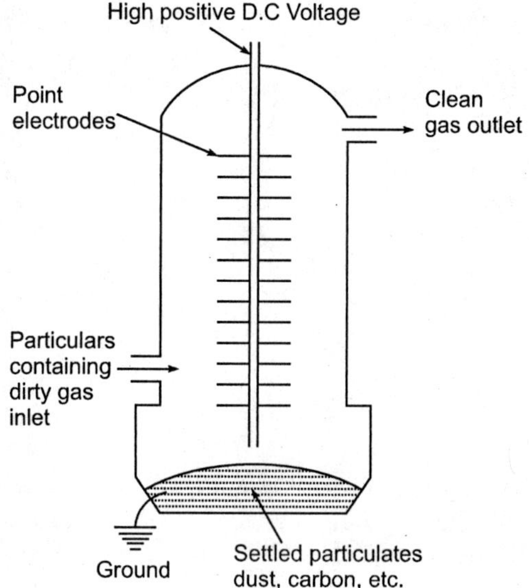

Fig. 15.2 The Cottrell electrostatic precipitator.

- For dust removal, extraction ventilation is used. The air stream with dust particles is maintained at a high velocity to keep the dust particles in suspension. Then, the rate is reduced suddenly to cause the particles to settle out under the action of gravity. Other methods are separators, cyclones, scrubbers, and so on.
- SO_2 pollution can be controlled either by removing sulphur after combustion or by the desulphurization of the fuel. The methods to reduce SO_2 in fuel gases are the following:

  By liquid ammonia: By passing SO_2 containing gases through ammonia, ammonium sulphite is obtained as a byproduct:

  $$2NH_3 + SO_2 + H_2O \longrightarrow (NH_4)_2SO_3$$

  Cairox method: SO_2 is oxidized when SO_2 containing gas is mixed with alkaline $KMnO_4$:

  $$KMnO_4 + SO_2 + KOH \longrightarrow MnO_2 + K_2SO_4 + H_2O$$

  Lime water wash: The SO_2 containing gas is passed through lime water when calcium sulphite is formed:

  $$Ca(OH)_2 + SO_2 \longrightarrow CaSO_3 + H_2O$$

- H_2S and SO_2 can be reduced by the Claus process when elemental sulphur is obtained. The process is a two-stage process that occurs in the following two steps:

  $$2H_2S + 3O_2 \longrightarrow 2H_2O + 2SO_2$$

$$2H_2S + SO_2 \longrightarrow 2H_2O + 3S$$

- Acids and chemical fumes can be removed by passing through a tower filled with coke. A counter current of water is applied at the top to recover volatile matters.
- Use modern energy resources, such as thermal power, electricity, nuclear power, solar energy, and so on, instead of using wood and coal.
- There should be a buffer zone between an industrial area and a residential area, so that the concentration of pollutants can be minimized. Proper zoning results in considerable improvement of health of the community as a whole.
- Plants use CO_2 during photosynthesis and reduce the concentration of CO_2 in the atmospheric air. There are also specific plants that can absorb specific pollutants. Planting trees is very helpful in reducing air pollution due to fly ash and coal dust from coal-handling systems. Air passing through the vegetation kingdom slows down and helps the deposition of suspended particulate matter on the leaves which are either washed by rain water or by a periodic shedding of leaves. Therefore, a plantation can control air pollution to a great extent.

15.8 WATER POLLUTION

Of the earth's water supply, 97 percent lies in oceans, although the high salt content does not allow its use for human consumption; 2 percent of water resource are locked in polar ice caps, and glaciers, and only 1 percent is found as fresh water for human consumption and other uses.

Water pollution may be defined as "any alteration in the physical (temperature, turbidity), chemical (toxic chemicals), and biological (useful microorganisms) property of water as well as contamination with any foreign substance that would constitute a health hazard or otherwise decreases the utility of water."

Any substance that disallows the normal use of water must be regarded as a water pollutant.

15.9 SOURCES OF WATER POLLUTION

There are numerous sources of water pollution as follows:

Domestic sewage: This sewage includes waste water from homes and commercial establishments. The sewage contains human excreta, kitchen wastes, street wastes, and organic substances that provide nutrition for bacteria and fungi. Municipal waste is the principal contributor of water pollution. Most municipal sewages are discharged in water without treatment. Sewage treatment deposits the suspended material, called sludge, whereas the liquid waste consists of ions such as Ca^{2+}, Mg^{2+}, Na^{+}, K^{+}, Cl^{-}, NO_2^{-}, SO_4^{2-}, and so on in a dissolved state.

The adverse effects of domestic sewage is that it is an excellent medium for the growth of pathogenic bacteria, viruses, and so on. Disposal of sewage into rivers, lakes, and canals causes several water-borne diseases such as dysentery, cholera, typhoid, hookworm infections, and so on.

Industrial effluents: Water gets polluted by acids, alkalis, detergents, soaps, phenols, aldehydes, ketones, amines, cyanides, heavy metals (Cu, Zn, Pb, Hg, Cd, Co, etc.), pesticides, insecticides, fungicides, and so on, which are released from chemical industries. Toxic substances enter the water resource from industries such as sugar, paper, textile, pharmaceuticals, steel mills, soap, oil refineries, tanneries, fertilizers, and so on, damaging the biological activity and killing useful organisms.

Agricultural waste: Pesticides, fungicides, insecticides, fertilizers, and so on enter waterways as runoff from agricultural lands causing heavy pollution to water resources.

Agricultural waste leads to the accumulation of nitrates in the water. When this water is used by humans the nitrates cause serious ill effects, including methemoglobinaemia, in children. Nitrates also affect the oxygen-carrying capacity of haemoglobin and cause suffocation and irritation to the respiratory and vascular system. An increased discharge of nitrates and phosphates into the water causes eutrophication, which results in a depletion of oxygen and an increase in the BOD (biochemical oxygen demand) (mg. of O_2 required to decompose the organic matter in 1 L waste water) of water. This causes death in fish and aquatic life.

Petroleum: The production, distribution, and use of large quantities of petroleum oil results in oil contamination; in some cases, it may be accidental. Oil slicks are responsible for the deaths of many birds. Oil reduces the thermal insulation and resistance to cold, irritates the digestive system, and produces toxic effects.

Radio nuclide effects: Radioactive pollutants enter into water streams from various sources such as nuclear tests in nuclear power plants, atomic explosions, nuclear installations, and processing of radioactive materials (*e.g.*, uranium, thorium, etc.). The solid nuclear waste filled in containers is dumped into the sea bed. The corrosive action of sea water may cause leakage of radioactive waste in water and may pose health hazards.

Radionuclides in water cause diseases such as cancer, leukemia, and cataracts in humans. They affect plant metabolism and aquatic life as well.

Thermal pollution: Thermal pollution takes place because power stations use water in the process of cooling their generators. Cooling water is discharged at a raised temperature causing a warming trend of the surface water. Some rivers may have their temperature so high (even upto 40°C) that aquatic life is eliminated.

At higher temperatures, blue-green algae and sewage fungus will grow more, which will result in plant death. As the temperature of water rises, the amount of dissolved oxygen in water decreases. So at a higher temperature, less dissolved oxygen will be present in the water, which may be fatal for aquatic life.

Siltation: This is the most widespread and damaging water pollution. In this type of pollution, the soil particles create a high turbidity in the water and may hinder the free movement of aquatic organisms, growth of fish, and fish productivity.

15.10 PREVENTION AND CONTROL OF WATER POLLUTION

Monitoring water pollution is essential at strategic points in natural streams and in waste water discharge. Water pollution can be measured by biological monitoring and physiochemical monitoring. Water pollution can be prevented and controlled by recycling wastes. Coconut and agricultural wastes can be converted into paper and board, and jute wastes can be converted into hard boards. The solidified wastes must be disposed of in landfills. Nonbiodegradable wastes must be dumped underground with the utmost care. Migration of these wastes into the soil should not be allowed. The sewage (domestic/industrial) must be pretreated to minimize the concentration of contaminants before being released into rivers or canals. Excess use of pesticides, insecticides, and fertilizers should be discouraged. The public should be made aware of water pollution issues.

15.11 SEWAGE

Sewage is the liquid waste that contains human and household discharged water, industrial wastes, street washings, and storm water. Generally, sewage contains about 99.5 percent water and the rest is organic and inorganic matter that is dissolved in water either in the colloidal state or as suspension. Fresh domestic sewage is usually grey-green to grey-yellow in color, but it darkens as time passes due to decomposition. Fresh domestic sewage is usually sweet smelling, but when stale, it has an offensive odor due to the evolution of hydrogen sulphide, ammonium sulphide, phosphine, and so on.

Aerobic and Anaerobic Oxidations

Generally, sewage consists of both aerobic and anaerobic bacteria that can oxidize organic compounds present in the sewage. With a sufficient amount of dissolved oxygen (about 8 ml. per liter), oxidation of the organic compound is carried out by aerobic bacteria. The oxidation products are inoffensive smelling, nonpurified nitrites, nitrates, sulphates, phosphates, and so on. This is called *aerobic oxidation.*

On the other hand, when free oxygen in water is below a certain limited value, the sewage is known as stale. In such a condition, anaerobic bacteria brings about putrefaction and the production of CH_4, H_2S, $(NH_4)_2S$, and PH_3, which have an offensive odor. This is known as *anaerobic oxidation.* When the anaerobic decomposition is continuing, the sewage is known as *septic sewage.*

Biological oxygen demand (BOD) is the amount of dissolved or free oxygen needed for the biological oxidation of the organic matter under aerobic condition at 20°C and for a period of 5 days. The unit of BOD is mg/liter or ppm.

Chemical oxygen demand (COD) is a measure of the oxidizable impurities present in the sewage. This is a measure of both the biologically oxidizable and biologically inert organic matter such as cellulose, and so on. Consequently, the value of COD is more than BOD. Its determination takes only 3 hours and is more scientific.

BOD Determination

The test is based upon the determination of the dissolved oxygen prior to and following a 5-day period at 20°C. A known volume of the sample of sewage is diluted with a known volume of the dilution water (water containing nutrients for a bacterial growth), whose dissolved oxygen content is predetermined. The entire solution is incubated in a closed bottle at 20°C for 5 days. After this, the unused oxygen is determined. The difference between the original oxygen content in the diluted water and the unused oxygen of the solution after 5 days provides BOD.

COD Determination

A known volume (say 250 ml.) of the sample is refluxed with a known excess of standard $K_2Cr_2O_7$ (say 1 N) and dil. H_2SO_4 in the presence of a little Ag_2SO_4 catalyst for 1½ hours. The unreacted $K_2Cr_2O_7$ is then titrated against the standard Mohr's salt solution. The oxygen equivalent of $K_2Cr_2O_7$ consumed is taken as a measure of the COD.

$$1 \text{ ml. of } 1\text{ N } K_2Cr_2O_7 \equiv 0.008 \text{ g oxygen}$$

Sewage Treatment

From a public health point of view, sewage treatment is an important process in which harmful constituents of sewage are transformed into either harmless or less harmful contents. Sewage treatment is generally carried out using the artificial treatment method, which is called *sewarage*, and involves the following steps:

Primary treatment: This removes the larger suspended particles from the sewage and produces a liquid that contains only colloidal and dissolved matter.

Secondary treatment: This removes the bulk of the organic matter present in the fluid, using biochemical oxidation processes, and provides an effluent that can be discharged into the sea and rivers to make use of natural biochemical processes to complete the purification.

Tertiary treatment: This produces water that is of drinking water standard. Most sewage plants only go as far as the secondary treatment standard.

Filters used during the process are trickling filters as shown in Fig. 15.3. These are either rectangular or circular and about 2 m deep. They are filled with either coarse, crushed rock; large anthracite coal; or graded clinkers. The underdrain system is provided at the bottom of the bed to collect and remove effluents. Sewage is delivered to the filters by means of a rotating distributor.

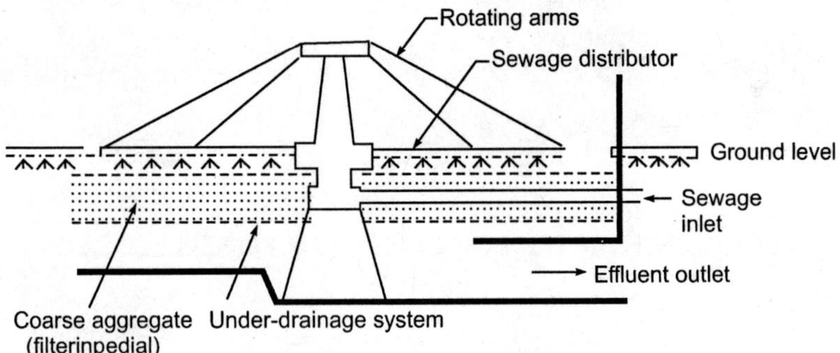

Fig. 15.3 Trickling filter.

The sewage is allowed to trickle over the filtering medium by means of a rotating distributor arm. As the trickled sewage starts percolating downward through the filtering media, microorganims (present in sewage) grow on the surface of the aggregates, using the organic material of the sewage as food. Aerobic conditions always prevail, and purified sewage is removed from the bottom. Normally, trickling filters remove about 90 percent BOD.

The sewage treatment process can be illustrated schematically in Fig. 15.4.

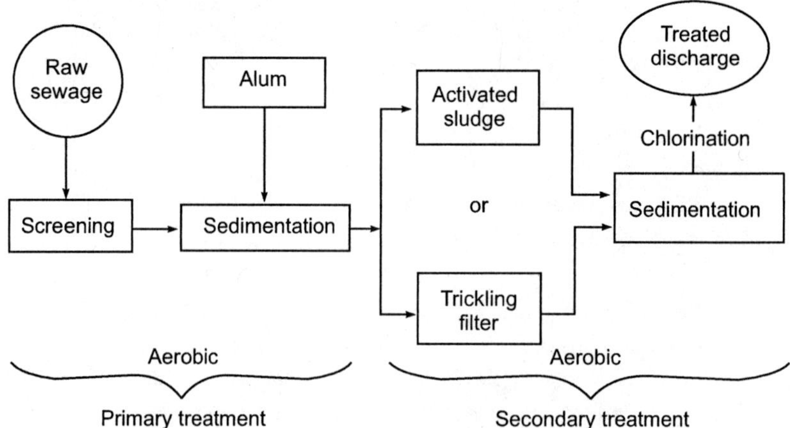

Fig. 15.4 Flow diagram of sewage treatment.

The Activated Sludge Process

The oxidation process is quickened if carried out in the presence of a part of sludge from the previous oxidation process. This sludge from the previous process is known as *activated sludge*. After the aeration, the effluent is sent to a settling or sedimentation tank, where sludge is deposited, and clean, nonputrefying liquid is drawn off. A part of the settled sludge is sent back for seeding a fresh batch of sewage, whereas the remainder is disposed of in landfills, or a sea burial.

15.12 THE BIOCHEMICAL EFFECTS OF TOXIC METALS

Metals that accumulate in the bodies of organisms remain for a long time and behave as cumulative poisons.

Lead: Lead is the most serious toxic metal. The major sources are lead batteries, transportation, lead pipes, lead paints and so on. Lead is a cumulative poison. Divalent lead easily replaces Ca^{2+} in bone and becomes fixed. Lead binds strongly to a large number of molecules such as amino acids, haemoglobins, enzymes, RNA, and DNA. It thus disrupts many metabolic pathways. The effects of lead poisoning are impaired blood synthesis, hypertension, and brain damage. Blood-lead levels are commonly used to monitor the status of lead poisoning in human beings. The major biochemical effect of Pb is its interference with haem synthesis, which gives rise to haematological damage. Lead may cause abnormalities in fertility and pregnancy.

Arsenic: Arsenic is well known for its toxicity because of its war time use. Arsenic is used primarily in agricultural chemicals, ceramics, glass, and pharmaceuticals. The symptom of acute arsenic poisoning is severe gastroenteritis; in chronic poisoning, this symptom is loss of weight and hair. The toxic action of As(III) is to attack SH groups of an enzyme to inhibit the enzyme action. The biochemical actions of As include coagulation of proteins and complexation with coenzymes.

Cadmium: Cadmium is toxic, that is lower levels have an adverse effect on human beings. The sources are batteries, cadmium-plated utensils, tobacco, and rubber tires. Cadmium accumulates in the liver and kidneys, resulting in kidney damage. Cadmium poisoning causes the disease called Itai-Itai or ouch ouch. At high levels, Cd causes kidney problems, anaemia, and bone marrow disorders. The major portion of Cd ingested into our body is trapped in the kidneys and eliminated. Excessive amounts of Cd^{2+} when ingested replaces Zn^{2+} at key enzymatic sites, bringing about metabolic disorders.

Mercury: Mercury is the most toxic heavy metal. The sources are the chlor-alkali industry, fungicides, electrical apparatuses, and paints. Metallic mercury vapor when inhaled goes through the lungs into the blood and then into the brain to damage the central nervous system. The most dangerous and toxic of the mercury compounds is methyl mercury CH_3Hg^+ or $(CH_3)_2Hg$. These compounds are soluble in fat. Hg or its salts are converted to methyl mercury by bacteria in water. Methyl mercury brings about brain damage and paralysis followed by death. Mercury as a toxic metal came to light after the incidence of minamata disease in Japan. Methyl mercury enters the food chain through plankton and gets concentrated by fish.

15.13 SOIL POLLUTION

The word *soil* is derived from a Latin word *solum*, which means earthy material in which plants grow. Soil pollution is the addition of chemical substances in an indefinite proportion to the soil system, whereby the fertility of the soil changes. Any substance capable of changing the productive capacity of the soil is termed a soil pollutant. Large quantities of waste products are deposited in soil each year. The major sources of soil pollution are the following:

- Improper disposal of solid and liquid wastes
- Dumping of domestic and industrial wastes on land
- Chemicals such as fertilizers, pesticides, and so on applied to plants and soils

- Radioactive wastes discharged from industrial and research laboratories and hospitals
- Soil erosion due to deforestation
- Heavy metals discharged from electroplating industries

The following are ways to prevent and control soil pollution:

- Treat and dispose of solid waste in landfills
- Determine the wastes needed for further processing, if the industry is to meet the new requirements for cost effectiveness
- Encourage recycling of wastes
- Perform extensive plantation
- Ban toxic chemicals.

15.14 ENVIRONMENTAL IMPACT ASSESSMENT (EIA)

The Environmental Impact Assessment (EIA) is a process of identifying possible consequences for the natural environment and for human health and welfare after implementing a particular activity of economic development. Unbalanced development has created environmental problems. It becomes necessary to determine the impact of development on the environment. EIA is the tool of environmental protection to test the compatibility of the environment before implementing of any private or government projects and industries that could damage the natural resources and cause environmental pollution. EIA essentially involves the following steps:

Screening: Before a project is launched, a number of sites are selected. The screening of these sites is done in relation to environmentally sensitive areas. A site away from the environmentally sensitive area is preferred, which will not adversely affect the local environment.

Scoping: Significant issues regarding environmental quality should be identified. These may include quality of water, air, land, flora, and fauna, that will be affected. The frequency and duration of the observations for checking the quality should be decided.

Baseline data collection: Baseline data (status of the environment and the natural resources) should be compiled. The data regarding water, air, wildlife, vegetations, human habitat, mineral deposits, and so on should be collected prior to the commencement of the project.

Identification of impacts: The impact may be for the short term or the long term. The long-term impact will alter the quality of the environment and have detrimental effects. The effects of the short-term impact will be minimized afterwards. The impact may be reversible, that is its effect will be neutralized by adopting certain corrective measures. It may be irreversible, if its harmful effects cannot be contained.

Prediction of impacts: The prediction of future environmental states is technically difficult. Forecasting future environmental conditions in the absence of a project is dependent on the quality of the economic and population forecasts for an area. Mathematical models can be extensively used in this step.

Evaluation of impacts: The evaluation criteria for comparing alternative plans/projects are often inexactly specified and create an element of uncertainty at the plan evaluation stage.

The EIA involves determining the likely economic, population, and environmental consequences that would result from implementation. The EIA process is designed so that developments cause minimum environmental damage.

EXERCISES

1. Mention the various air pollutants and describe the effects of the following substances on the environment.
   (*i*) Hydrogen sulphide (*ii*) Nitrogen oxide
   (*iii*) Sulphur dioxide
2. What are the common air pollutants and their effect on the environment?
3. What is meant by the greenhouse effect? Describe the two monitoring techniques used for the greenhouse effect.
4. Name the two toxic metals and discuss their biochemical effects. How are these metals contaminated with water?
5. Write short notes on:
   (*a*) Poisoning by mercury (*b*) Poisoning by lead
   (*c*) NO_x as a pollutant
6. Write short notes on:
   (*a*) Biological effect of As (*b*) Pb as a poison
   (*c*) Effect of CO on haemoglobin
7. What are the important sources of water pollution?
8. Write a short note on the greenhouse effect.
9. What is meant by the Environmental Impact Assessment (EIA)? How is it measured?
10. Write notes on:
    (*a*) Photochemical smog
    (*b*) Primary and secondary pollutants

11. Discuss the sources and harmful effects of the following pollutants:
    (*a*) Carbon monoxide (*b*) Particulate matter
    (*c*) Photochemical oxidants
12. Write a brief note on soil pollution.
13. (*a*) What are the various factors that cause air pollution?
    (*b*) Explain: (*i*) Ozone depletion (*ii*) Polluting effects of CO.
14. (*a*) Discuss the various methods and equipments used to control gaseous pollutants.
    (*b*) Enumerate the biochemical effects of
    (*i*) Arsenic (*ii*) Mercury (*iii*) Lead
15. (*a*) How does automobile exhaust contribute to air pollution?
    (*b*) Describe the sources and constituents of air pollution.
16. Write an account of the toxic chemicals present in the environment. Indicate their sources and toxic actions.
17. (*a*) Define pollution.
    (*b*) "The auto exhaust gases are believed to play an important role in air pollution." Explain.
18. Name important air pollutants and mention their effects on human life.
19. Write a note on acid rain.
20. What is global warming, what gases are responsible for this, and what are their sources?
21. (*a*) What does BOD refer to?
    (*b*) What is the reason behind the toxicity of lead?
    (*c*) Write the mechanism of depletion of the ozone layer by chlorofluoro carbon, $CFCl_3$.

MULTIPLE CHOICE QUESTIONS

1. World Environment Day is
   (*a*) 5 July (*b*) 5 June
   (*c*) 15 July (*d*) 15 June
2. Fossil fuel is
   (*a*) coal (*b*) petroleum
   (*c*) both of these (*d*) none of these
3. The presence of which of the following gases in air checks the ultraviolet light from sunlight?
   (*a*) SO_2 (*b*) CO_2
   (*c*) NO (*d*) O_3

4. Ozone depletion in the stratosphere is caused by
(*a*) SO_2 (*b*) NO_2
(*c*) CO_2 (*d*) chlorofluoro carbons

5. Damage to blood is caused by
(*a*) Ca (*b*) Mg
(*c*) As (*d*) Pb

6. Lead in water can cause
(*a*) eye disease (*b*) arthritis
(*c*) kidney damage (*d*) hair falling out

7. Which of the following is not a major air pollutant in automobile exhausts?
(*a*) CO (*b*) NO_x
(c) unburned petrol (*d*) SO_2

8. Haemoglobin of the blood forms carboxy haemoglobin with
(*a*) CO_2 (*b*) CO
(*c*) SO_2 (*d*) NO_2

9. Which is a greenhouse gas?
(*a*) CH_4 (*b*) CO_2
(*c*) chlorofluoro carbons (*d*) all of these

10. The Bhopal gas tragedy in 1984 was caused by
(*a*) carbonyl chloride (*b*) carbon monoxide
(*c*) methyl cyanide (*d*) methyl iso cyanate

11. Which of the following can cause photochemical smog?
(*a*) HCN (*b*) SO_2
(*c*) HCl (*d*) O_3

12. Incomplete combustion of fuel in internal combustion engines releases which poisonous gas into the atmosphere?
(*a*) CO (*b*) CO_2
(*c*) SO_2 (*d*) SO_3

Part 3

EXPERIMENT 1

Aim of the experiment:

Determination of sodium hydroxide and sodium carbonate in a mixture.

Apparatus required:

Burette
Pipette
Conical flask
Glazed tile

Chemicals required:

Given alkali mixture solution
$\frac{N}{10}$ HCl solution
Phenolphthalein indicator
Methyl orange indicator

Theory:

To determine NaOH and Na_2CO_3 when present together, the double indicator method is generally employed. The principle involved in this method is that when a sodium carbonate solution is titrated with hydrochloric acid, the neutralization occurs in two stages. The first one correspond to the hydrogen carbonate (HCO_3^-) stage, when the net reaction is

$$CO_3^{2-} + H^+ \longrightarrow HCO_3^-$$

The equivalence point for the primary stage of ionization of carbonic acid is at pH 8.3, which may be detected by employing a phenolphthalein indicator. The second stage of the titration corresponds to the displacement of all the carbonic acid, when the net reaction is

$$CO_3^{2-} + 2H^+ \longrightarrow H_2CO_3$$

$$HCO_3^- + H^+ \longrightarrow H_2CO_3$$

$$H_2CO_3 \longrightarrow H_2O + CO_2$$

The equivalence point for the secondary stage of ionization is at pH range 3.1 to 4.4, which may be detected by employing the methyl orange indicator. Thus, when a mixture of Na_2CO_3 and NaOH is titrated with standard HCl, the phenolphthalein end point corresponds to the neutralization of all the NaOH and the partial neutralization of all the Na_2CO_3 to the bicarbonate stage.

$$NaOH + HCl \longrightarrow NaCl + H_2O$$

$$Na_2CO_3 + HCl \longrightarrow NaCl + NaHCO_3$$

Now if the titration is continued using methyl orange as the indicator, the neutralization of the bicarbonate to the carbonic acid occurs.

$$NaHCO_3 + HCl \longrightarrow NaCl + H_2CO_3$$

$$H_2CO_3 \longrightarrow H_2O + CO_2$$

So, the methyl orange end point means the *total volume of the acid* run down from the beginning of the experiment.

Thus, if [P] and [M] correspond to the phenolphthalein and methyl orange end points, then

$$[P] \equiv NaOH + \frac{1}{2}\, Na_2CO_3$$

and

$$[M] \equiv NaOH + Na_2CO_3$$

Hence,

$$[M] - [P] \equiv \frac{1}{2}\, Na_2CO_3$$

$$2\{[M] - [P]\} \equiv Na_2CO_3$$

and

$$[M] - 2\{[M] - [P]\} \equiv NaOH$$

Procedure:

Pipette out 10 ml of the given alkali mixture solution into a conical flask and add 2 to 3 drops of the phenolphthalein indicator. The solution assumes a pink color. Titrate the solution with the $\frac{N}{10}$ HCl from the burette until the solution becomes colorless. Note down this titre value as phenolphthalein end point [P]. Then, add 2 to 3 drops of the methyl orange indicator to the same solution in the conical flask and continue the titration until a sharp color change from yellow to red occurs at the end point. Note down the total titre value from the beginning of the experiment as methyl orange end point [M].

Tabulation:

No. of Observations	Volume of Alkali Mixture Solution (ml)	I.B.R. (ml)	Titration with Phenolphthalein		Titration with Methyl Orange	
			F.B.R. (ml)	Vol. of HCl Used [P] (F.B.R. – I.B.R.)	F.B.R. (ml)	Vol. of HCl Used [M] (F.B.R. – I.B.R.)
1	10					
2	10					
3	10					

Calculation:

NaOH Content

We know that the equivalent weight of NaOH = 40

Hence, 1000 ml of 1N HCl ≡ 40 g of NaOH

$$1 \text{ ml of } 1\text{N HCl} \equiv \frac{40}{1000} \text{ g of NaOH}$$

$$\therefore \quad M-2(M-P) \text{ ml of } \frac{N}{10} \text{ HCl} \equiv \frac{40}{1000} \times \frac{1}{10} \times [M-2(M-P)] \text{ g of NaOH}$$

The strength of NaOH in the given solution

$$= \frac{40}{1000 \times 10} \times [M-2(M-P)] \times \frac{1000}{10} = x \text{ g/liter}$$

Na_2CO_3 Content

We know that the equivalent weight of Na_2CO_3 = 53

Hence, 1000 ml of 1N HCl ≡ 53 g of Na_2CO_3

$$1 \text{ ml of } 1\text{N HCl} \equiv \frac{53}{1000} \text{ g of } Na_2CO_3$$

$$2(M-P) \text{ ml of } \frac{N}{10} \text{ HCl} \equiv \frac{53}{1000} \times \frac{1}{10} \times 2(M-P) \text{ g of } Na_2CO_3$$

The strength of Na_2CO_3 in the given solution

$$= \frac{53}{1000} \times \frac{1}{10} \times 2(M-P) \times \frac{1000}{10} = y \text{ g/liter}$$

Conclusion:

The given alkali mixture solution contains x g/liter of NaOH and y g/liter of Na_2CO_3.

Preparation of reagents:

1. *Hydrochloric acid* $\left(\frac{N}{10}\right)$:
   Dilute 8.4 ml. of a concentrated hydrochloric acid to 1 liter with distilled water and standardize with $\left(\frac{N}{10}\right)$ Na_2CO_3.
2. *Phenolphthalien indicator:*
   Dissolve 1 g of phenolphthalein in 100 ml of distilled ethyl alcohol, and add 100 ml of distilled water with constant stirring. Filter if necessary.
3. *Methyl orange indicator:*
   Dissolve 50 mg of the free-acid form of the methyl orange solid in distilled water, and make the volume to 100 ml with distilled water. Filter the solution if necessary.

EXPERIMENT QUESTIONS

1. Define volumetric analysis.
2. What is the difference between end point and equivalence point?
   Ans. The point at which the indicator gives a clear visual change gives the end point of the titration. The point at which the completion of reaction occurs is called the equivalence point. The equivalence point determined by using an indicator is called the end point.
3. What is an indicator?
4. What do you mean by the alkalinity of water?
   Ans. The alkalinity of water is due to the presence of OH^-, CO_3^{2-}, and HCO_3^- ions, which causes an increased concentration of hydroxide ions due to hydrolysis.
5. What is the pH range of the phenolphthalein and methyl orange indicator?
6. What is caustic soda?
7. What is washing soda?
8. What is the color of the methyl orange in an acidic and basic medium?
9. Write the action of the methyl orange and phenolphthalein indicators.
   Ans. Phenolphthalein (pH range 8.0–9.6) has a benzenoid structure in an acidic medium (colorless) and a quinonoid structure in an alkaline medium (pink).

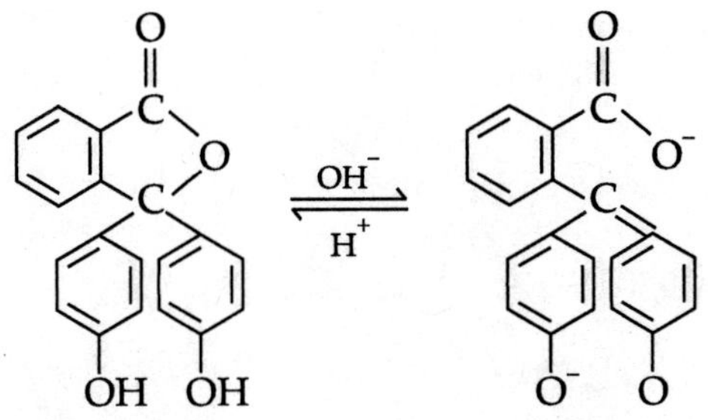

(Colorless) $\underset{H^+}{\overset{OH^-}{\rightleftharpoons}}$ (Pink)

Methyl orange has a benzenoid form in an alkaline medium (yellow) and a quinonoid form in an acidic medium (red). It has a pH range of 3.1–4.4.

EXPERIMENT 2

Aim of the experiment:

Determination of the total hardness of water by the EDTA method.

Apparatus required:

Burette
Pipette
Beaker
Conical flask
Volumetric flask
Funnel

Chemicals required:

Standard hard water
0.01 M EDTA solution
Buffer solution
Eriochrome Black-T indicator (EBT)

Theory:

The hardness of water is the property that prevents the lathering of soap. Hardness occurs due to the presence of certain salts of Ca and Mg ions in the water. These react with the sodium salts of long-chain fatty acids present in the soap and are precipitated out in the form of calcium and magnesium stearate, palmitate, or oleates. No lather is formed until all these ions are completely removed, and thus a large amount of soap is wasted.

$$2C_{17}H_{35}COONa + CaCl_2 \longrightarrow (C_{17}H_{35}COO)_2Ca \downarrow + 2NaCl$$

Sodium stearate — Calcium stearate

Hardness is of two types:

Temporary or carbonate hardness: Caused by the presence of carbonate and bicarbonate of calcium and magnesium and is removed on boiling. This is also called alkaline hardness.

$$Ca(HCO_3)_2 \xrightarrow{\Delta} CaCO_3 + H_2O + CO_2$$

Permanent or noncarbonate hardness: Caused by the presence of chlorides and sulphates of calcium and magnesium, which cannot be removed by boiling. This is also called nonalkaline hardness.

The hardness of water may be determined by complexometric titration. In an aqueous solution, EDTA ionizes to give two Na^+ ions and a strong chelating agent.

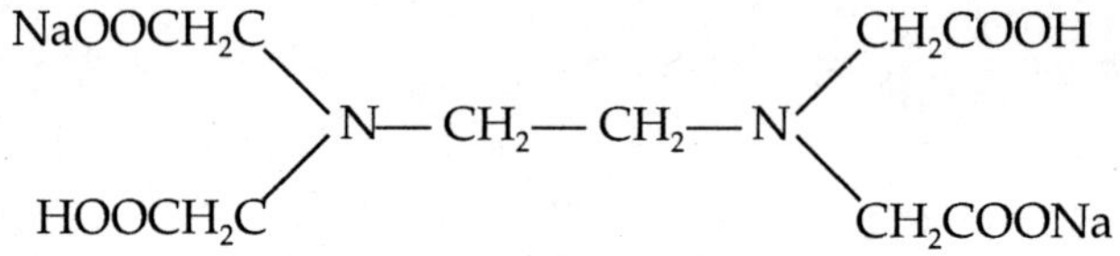

$$\begin{matrix} NaOOCH_2C \\ HOOCH_2C \end{matrix} \!\!>\! N-CH_2-CH_2-N \!<\!\! \begin{matrix} CH_2COOH \\ CH_2COONa \end{matrix}$$

Disodium salt of ethylene diamine tetra acetic acid (EDTA)

The titration is carried out in the presence of the indicator Eriochrome Black-T, which is a complex organic compound.

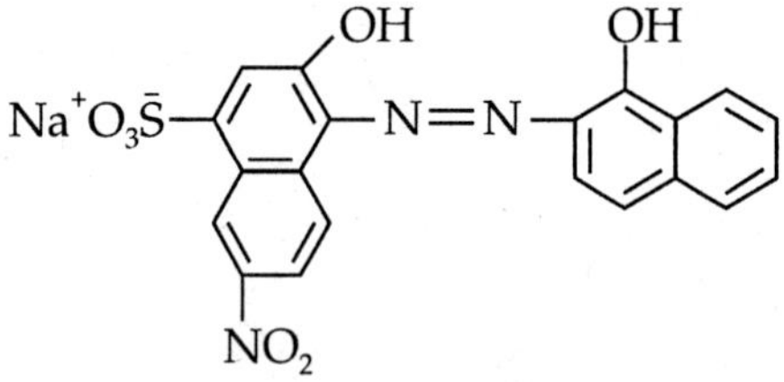

$Na^+O_3\bar{S}$–(naphthalene ring with OH, NO_2)–N=N–(naphthalene ring with OH)

Eriochrome Black-T

Sodium-1 (1 hydroxy 2-naphthyl azo)-6 nitro-2 naphthol-4-sulphonate

The indicator, when added in small amounts to hard water, buffered to a pH of about 10 combines with a few of the Ca^{2+} and Mg^{2+} ions to form a weak complex of wine-red color, which changes to blue when an excess drop of EDTA is added.

$$\begin{bmatrix} Ca^{2+} \\ Mg^{2+} \end{bmatrix} + EBT \longrightarrow \begin{bmatrix} Ca^{2+} \\ Mg^{2+} \end{bmatrix} EBT \xrightarrow{EDTA}$$

Unstable complex (Wine red)

$$\begin{bmatrix} Ca^{2+} \\ Mg^{2+} \end{bmatrix} EDTA + EBT$$

Complex (Blue)

Thus, the change of the wine-red color to distinct blue marks the end point of the titration. Hardness of water is expressed as a calcium carbonate equivalent.

Procedure:

Standardization of EDTA solution: Take 25 ml of standard hard water in a conical flask. Add 10 ml of buffer solution to it followed by 2–3 drops of the EBT indicator. The color of the solution becomes wine red. Titrate the solution against the standard EDTA solution from the burette until the color changes from wine red to blue at the end point. Take at least two concordant readings, and note the titre value (V_1 ml).

Determination of total hardness: Take 25 ml. of the supplied hard water in a conical flask. Add 10 ml of buffer solution into it followed by 2–3 drops of the EBT indicator. Titrate the solution slowly against the EDTA solution from the burette until the end point is reached when the color changes from wine red to blue. Take two concordant readings, and note the titre value (V_2 ml).

Determination of permanent hardness: Take 25 ml of the supplied hard water in a 50 ml beaker, boil gently for about 10 minutes, cool, filter into a 25 ml of volumetric flask, and make the volume up to mark with distilled water. Take this solution in a conical flask. Add 10 ml of buffer solution into it followed by 2–3 drops of the EBT indicator. Titrate the solution slowly against the EDTA solution from the burette until the color changes from wine red to blue. Take two concordant readings, and note the titre value (V_3 ml).

Tabulation:

TABLE E2.1 Standardization of the EDTA Solution

No. of Observations	*Volume of Standard Hard Water (ml)*	*I.B.R. (ml)*	*F.B.R. (ml)*	*Volume of EDTA Solution Used (Difference V_1 ml)*
1	25			
2	25			
3	25			

TABLE E2.2 Total Hardness

No. of Observations	*Volume of Supplied Hard Water (ml)*	*I.B.R. (ml)*	*F.B.R. (ml)*	*Volume of EDTA Solution Used (Difference V_2 ml)*
1	25			
2	25			
3	25			

TABLE E2.3 Permanent Hardness

No. of Observations	*Volume of Supplied Hard Water After Treatment for Removal of Temporary Hardness (ml)*	*I.B.R. (ml)*	*F.B.R. (ml)*	*Volume of EDTA Solution Used (Difference V_3 ml)*
1	25			
2	25			
3	25			

Calculation:

Standardization of EDTA solution

$$1 \text{ ml standard hard water} = 1 \text{ mg } CaCO_3$$

$$V_1 \text{ ml EDTA} = 25 \text{ ml standard hard water} = 25 \text{ mg } CaCO_3$$

$$1 \text{ ml EDTA} = \frac{25}{V_1} \text{ mg } CaCO_3$$

Total hardness

$$25 \text{ ml hard water} = V_2 \text{ ml EDTA}$$

$$= V_2 \times \frac{25}{V_1} \text{ mg } CaCO_3$$

$$1 \text{ ml hard water} = V_2 \times \frac{25}{V_1 \times 25} \text{ mg } CaCO_3$$

$$1000 \text{ ml hard water} = V_2 \times \frac{25}{V_1} \times \frac{1000}{25} \text{ mg } CaCO_3$$

$$= \frac{V_2}{V_1} \times 1000 \text{ mg } CaCO_3$$

$$\text{Total hardness} = \frac{V_2}{V_1} \times 1000 \text{ ppm} = x \text{ ppm}$$

Permanent hardness

$$25 \text{ ml hard water} = V_3 \text{ ml EDTA}$$

$$= V_3 \times \frac{25}{V_1} \text{ mg } CaCO_3$$

$$1000 \text{ ml hard water} = V_3 \times \frac{25}{V_1} \times \frac{1000}{25} \text{ mg } CaCO_3$$

$$= \frac{V_3}{V_1} \times 1000 \text{ mg } CaCO_3$$

$$\text{Permanent hardness} = \frac{V_3}{V_1} \times 1000 \text{ ppm}$$

$$\text{Temporary hardness} = \text{Total hardness} - \text{Permanent hardness}$$

$$= \left(\frac{V_2}{V_1} - \frac{V_3}{V_1}\right) \times 1000 \text{ ppm}$$

Conclusion:

Total hardness of supplied hard water = X ppm.

Preparation of reagents:

Standard hard water: Dissolve 1.0 g of pure dry $CaCO_3$ in a small amount of dilute hydrochloric acid, and then evaporate the solution to dryness on a water bath. Dissolve the residue in distilled water to make a 1L solution (1 ml ≡ 1 mg $CaCO_3$).

EDTA solution (0.01 M): Weigh 3.722 g of disodium dihydrogen ethylene diamine tetra acetic acid (Mol. wt. 372.25) after getting dried it in an oven at 75°C for 90 minutes and cooled. Dissolve it in distilled water in a volumetric flask and make up the volume to 1L with distilled water.

Buffer solution (pH 10): Dissolve 70 g of NH_4Cl and add 568 ml of conc. ammonia solution and dilute with distilled water to 1L.

Eriochrome Black-T indicator: Dissolve 0.5 g of the dye stuff in 100 ml rectified spirit or methanol. The solution is stable for the one month.

EXPERIMENT QUESTIONS

1. What do you mean by temporary and permanent hardness?
2. What is the difference between hard water and heavy water?
3. What is the principle of EDTA titration?
4. What is the trade name of EDTA?
5. How can temporary hardness of water be removed?
6. What are the disadvantages of hard water?
7. What are the buffer solutions?

8. Why is hardness generally expressed as a calcium carbonate equivalent?
   Ans. Hardness of water is generally expressed as a calcium carbonate equivalent because it is mostly insoluble salt, and its molecular weight is 100, which makes the calculation easier.
9. Why is the pH of the solution maintained at 10 when determining hardness by the EDTA method?
   Ans. At higher pH values, $CaCO_3$ or Mg $(OH)_2$ may get precipitated, and the indicator may change its color. At lower pH values, the Mg-indicator complex becomes unstable, and a sharp end point can be obtained.
10. What is a basic buffer? Give an example.
11. Write the structure of the Eriochrome Black-T indicator.

EXPERIMENT 3

Aim of the experiment:

Estimation of calcium in limestone.

Apparatus required:

Burette
Pipette
Conical flask
Volumetric flask
Funnel
Beaker

Chemicals required:

$\frac{N}{10}$ $KMnO_4$ solution
Ammonium oxalate (8%)
Hydrochloric acid (1 : 1)
Ammonia
5N H_2SO_4
Ammonium chloride
Given sample of ore

Theory:

Limestone essentially consists of calcium carbonate but is generally associated with small quantities of magnesium carbonate. The limestone powder is dissolved in HCl, and the calcium present in the solution is precipitated as calcium oxalate using ammonium oxalate in the presence of ammonia. The precipitate calcium oxalate, after washing, is treated with 5N H_2SO_4, and the oxalic acid that is liberated is treated with a standard $KMnO_4$ solution. From the volume of $KMnO_4$ required for the titration, the amount of Ca in the ore can be calculated.

$$1 \text{ ml of } 1\text{N } KMnO_4 \equiv 0.020 \text{ g of Ca}$$

$$\equiv 0.028 \text{ g of CaO}$$

$$\equiv 0.050 \text{ g of } CaCO_3$$

$$CaCO_3 + 2HCl \longrightarrow CaCl_2 + H_2O + CO_2$$

$$CaCl_2 + (NH_4)_2C_2O_4 \longrightarrow CaC_2O_4 + 2NH_4Cl$$

$$CaC_2O_4 + H_2SO_4 \longrightarrow CaSO_4 + H_2C_2O_4$$

$$2KMnO_4 + 3H_2SO_4 \longrightarrow K_2SO_4 + 2MnSO_4 + 5\,[O] + 3H_2O$$

$$5H_2C_2O_4 + 5\,[O] \longrightarrow 5H_2O + 10CO_2$$

Procedure:

Weigh accurately 1 g of the limestone into a beaker, and treat with a little amount of 1 : 1 HCl to dissolve the ore. Take care that there is no loss of the solution due to the quick evolution of CO_2 by covering the beaker with a watch glass. After all the ore is dissolved, wash the watch glass with distilled water to the same beaker. The solution is to be made alkaline by adding ammonia. Add 1 g of NH_4Cl, continue the stirring, boil, and then add 50 ml of 8 percent ammonium oxalate solution. Boil the solution, and allow the calcium oxalate precipitate to settle down. Decant the supernatant liquid through a filter paper. Wash the precipitate several times with hot distilled water containing a little ammonium hydroxide until the filtrate is free from the chloride and oxalate ions. Transfer the precipitate from the cone of the filter paper to a beaker with the help of a jet of water. Add about 50 ml of hot 5N H_2SO_4 and 40 ml of hot distilled water. Warm it with constant stirring. If the precipitate does not dissolve, add a little excess 5N H_2SO_4. Cool it, and transfer the entire contents of the beaker into a 250 ml volumetric flask. Make up the volume with distilled water, and shake well. Pipette out 25 ml of this solution into a conical flask. Add 20 ml 5N H_2SO_4. Heat to about 60–80°C. Titrate the hot solution against the standard $KMnO_4$ solution until a faint permanent pink color is obtained. Repeat the process to get concordant readings, and note the titre value.

Tabulation:

No. of Observations	*Volume of the Solution Taken in Conical Flask (ml)*	*I.B.R. (ml)*	*F.B.R. (ml)*	*Volume of $KMnO_4$ Used (Difference V ml)*
1	25			
2	25			
3	25			

Calculation:

Weight of the ore sample taken = W g

Volume of calcium ions/oxalate solution prepared = 250 ml
Volume of solution taken in conical flask for titration = 25 ml

$$\text{Normality of } KMnO_4 \text{ solution} = \frac{N}{10}$$

$$\text{Concordant volume of } \frac{N}{10} \; KMnO_4 \text{ used} = V \text{ ml}$$

We know that,

$$1 \text{ ml of } 1N \; KMnO_4 = 0.020 \text{ of Ca}$$

$$V \text{ ml of } \frac{N}{10} \; KMnO_4 = \frac{0.020 \times V}{10} \text{ g of Ca}$$

$$25 \text{ ml of the solution contains } \frac{0.020 \times V}{10} \text{ g of Ca}$$

$$250 \text{ ml of the solution contains } \frac{0.020 \times V}{10} \times 10 \text{ g of Ca} = Y \text{ g of Ca}$$

Now,
1 g of the ore contains Y g of Ca
100 g of the ore contains Y × 100 g of Ca = M%

Conclusion:

The given limestone ore sample contains M% of Ca.

Preparation of reagent:

Potassium permanganate solution $\left(\frac{N}{10}\right)$**:** Dissolve 3.1607 g of $KMnO_4$ (eq. wt. 31.607) in distilled water. Boil for 1 hour, cool, filter, and make up the volume to 1L with distilled water. Standardize it with oxalic acid to know the exact strength.

Sulphuric acid (5N): Add slowly 139 ml of conc. sulphuric acid (36N, 98%) to distilled water. After cooling, make up the volume to 1 L with distilled water.

Ammonium oxalate (8%): Dissolve 8 g of ammonium oxalate in 100 ml of distilled water.

Hydrochloric acid (1 : 1): Mix equal volumes of concentrated HCl and boiled out distilled water.

EXPERIMENT QUESTIONS

1. What are the different ores of calcium?
   Ans. Limestone, Dolomite etc.
2. What is difference between ore and mineral?
3. What is the procedure to estimate calcium in dolomite?
   Ans. If the ore is limestone, single precipitation is sufficient. In case of dolomite, where a large quantity of magnesium is present, there is every possibility that the precipitate of calcium oxalate may be contaminated with the magnesium oxalate giving higher values of Ca. In order to get the correct values of Ca, dissolve the calcium oxalate precipitate in about 5 ml 6N HCl. Make the solution slightly ammoniacal, dilute with 50 ml of boiled distilled water and treat again with ammonium oxalate solution (double precipitation is necessary).
4. Why is it necessary to wash the precipitate of calcium oxalate free of chloride and oxalate ions?
5. Why do we heat the solution to 70°C in case of oxalic acid vs. $KMnO_4$ titrations?
6. What is the significance of estimation of calcium in limestone or dolomite?

EXPERIMENT 4

Aim of the experiment:

Determination of the percentage of available chlorine in a given sample of bleaching powder.

Apparatus required:

Burette
Pipette
Volumetric flask
Conical flask
Funnel
Pestle and mortar

Chemicals required:

Given sample of bleaching powder
Potassium iodide
Glacial acetic acid
$\frac{N}{10}$ sodium thiosulphate solution (Hypo)
Starch indicator

Theory:

The major constituents of bleaching powder are calcium hypochlorite [$Ca(OCl)_2 . 4H_2O$], basic calcium chloride [$CaCl_2 . Ca(OH)_2 . H_2O$], and some free calcium hydroxide [$Ca(OH)_2$]. The active constituent of bleaching powder is hypochlorite ($CaOCl_2$), which is responsible for the disinfection and bleaching action of bleaching powder. When $Ca(OCl)_2$ is treated with HCl, free chlorine is liberated according to the equation:

$$Ca(OCl)_2 + 4HCl \longrightarrow CaCl_2 + 2H_2O + 2Cl_2$$

$$OCl^- + Cl^- + 2H^+ \longrightarrow Cl_2 + H_2O$$

$$Cl_2 + H_2O \longrightarrow HCl + \underset{\substack{\text{Hypochlorous acid} \\ \downarrow \\ \text{Kills germs}}}{HOCl}$$

$$Cl_2 + KI \longrightarrow KCl + I_2$$

$$I_2 + 2Na_2S_2O_3 \longrightarrow \underset{\text{(Sodium tetrathionate)}}{Na_2S_4O_6} + 2NaI$$

The available chlorine refers to the chlorine liberated by the action of the dilute acids and is expressed in a percentage weight of bleaching powder. The liberated chlorine from the bleaching powder sets free from KI an equivalent amount of iodine, which is titrated against a standard solution of hypo.

Procedure:

Weigh out accurately 3 g of air-dried bleaching powder in a clean, dried, and preweighed, weighing tube. Transfer the sample in a mortar, and make a paste with a little distilled water. Transfer the entire mass quantitatively to a 250 ml volumetric flask, and make up to the mark with distilled water. Pipette out 25 ml of the homogeneous solution from the volumetric flask to a conical flask. Add about 1 g of solid KI and about 20 ml of glacial acetic acid. Titrate the liberated iodine against the hypo solution until a light-yellow color appears. Then add 10 drops of freshly prepared starch solution, and continue the titration until the color disappears. Repeat the titration to get at least two concordant readings. Note the volume of the hypo solution used (V_2 ml).

Tabulation:

No. of Observations	*Volume of $CaOCl_2$ Solution (V_1 ml)*	*I.B.R. (ml)*	*F.B.R. (ml)*	*Volume of Hypo Solution Used (Difference V_2 ml)*
1	25			
2	25			
3	25			

Calculation:

Weight of the given powder sample = W g

N_1 = Normality of available chlorine = ?

V_1 = Volume of pipette out in conical flask = 25 ml

N_2 = Normality of the hypo solution = $\frac{N}{10}$

V_2 = Volume of the hypo solution used = x ml (say)

$$N_1V_1 = N_2V_2$$

or
$$N_1 \times 25 = \frac{N}{10} \times x$$

or
$$N_1 = \frac{x}{250} N$$

Therefore, the amount of chlorine per liter of the solution

$$= \frac{x \times 35.5}{250} \text{ g}$$

$$1000 \text{ ml } Ca(OCl)_2 \equiv \frac{x \times 35.5}{250}$$

$$1 \text{ ml } Ca(OCl)_2 \equiv \frac{x \times 35.5}{250 \times 1000}$$

$$250 \text{ ml } Ca(OCl)_2 \equiv \frac{x \times 35.5}{250 \times 1000} \times 250 \equiv \frac{x \times 35.5}{1000}$$

If W g of sample contains $\frac{x \times 35.5}{1000}$ g of Cl_2

100 g of sample contains $\frac{x \times 35.5}{1000} \times \frac{100}{W} = \frac{x \times 35.5}{W \times 10}$ g of Cl_2

Conclusion:

The percentage of available chlorine in the supplied sample of bleaching powder

$$= \frac{x \times 35.5}{W \times 10}.$$

Preparation of reagents:

Sodium thiosulphate solution $\left(\frac{N}{10}\right)$: Dissolve 24.82 g of $Na_2S_2O_3$, $5H_2O$ (eq. wt. 248.21) in 1 liter of boiled out distilled water. Standardize it with $K_2Cr_2O_7$ to know the exact strength.

Starch indicator (1%): Prepare a paste of 1 g of powdered starch with distilled water, and add to this paste 100 ml of boiled water with constant stirring. Boil for about 5 minutes and then cool.

EXPERIMENT QUESTIONS

1. What is the formula of bleaching powder?
2. What are the prime constituents of bleaching powder?
3. What do you mean by the available chlorine in bleaching powder?
4. What is the other name for bleaching powder?
   Ans. The other name of bleaching powder is chloride of lime or bleach.
5. Explain the mechanism of sterilizing water by bleaching powder.
6. What is iodometric titration?
   Ans. Iodometric titration is defined as the iodine titration in which some oxidizing agent liberates iodine from iodide. Then, the liberated iodine is titrated against a standard solution of reducing agent added from a burette. Iodometry is used to estimate the strength of the oxidizing agents.
7. What is starch?
   Ans. Starch is a polysaccharide with the general formula $(C_6H_{10}O_5)_n$.
8. What is the oxidation state of chlorine in bleaching powder?

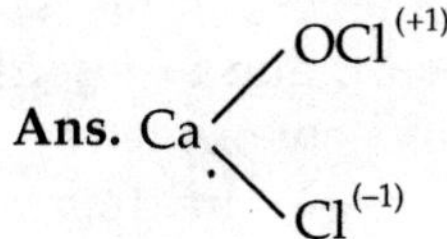

9. What is the percentage of available chlorine in commercial bleaching powder?
   Ans. 36–38%.

EXPERIMENT 5

Aim of the experiment:

Preparation of phenolphthalein.

Apparatus required:

Round-bottomed flask (250 ml)
Beaker (2 L)
Buchner funnel
Oil bath
Water bath

Chemical required:

Phenol
Phthalic anhydride
Conc. H_2SO_4
Dil. NaOH
Absolute alcohol
Dil. HCl
Dil. CH_3COOH

Theory:

Phenol condenses with phthalic anhydride in the presence of conc. H_2SO_4 or anhydrous $ZnCl_2$ to yield the colorless phenolphthalein as the main product. When dilute caustic alkali is added to an alcoholic solution of phenolphthalein, an intense red coloration appears. The alkali opens the lactone ring in phenolphthalein and forms a salt at one phenolic group. The reaction may be represented in steps, with the formation of a hypothetical unstable intermediate that changes to a colored ion. The color is probably due to resonance, which places the negative charge on either of the two equivalent oxygen atoms. With an excess of conc. caustic alkali, the first red color disappears, this is due to the production of the carbinol and attendant salt formation rendering resonance impossible. The various reaction may be represented as follows:

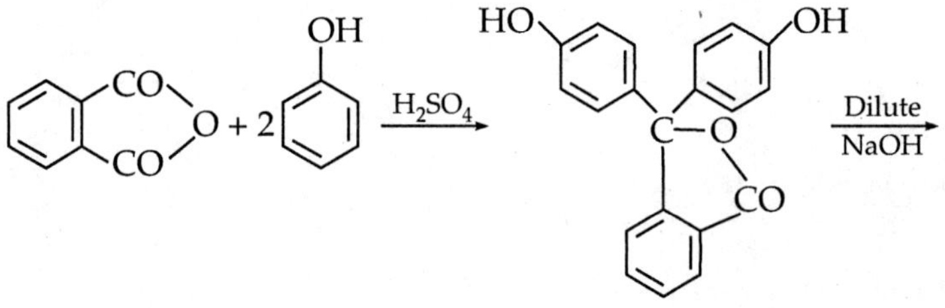

CO, O, CO + 2 OH —H_2SO_4→ HO, OH, C—O, CO —Dilute NaOH→

Phthalic anhydride Phenol Phenol phthalein (colorless)

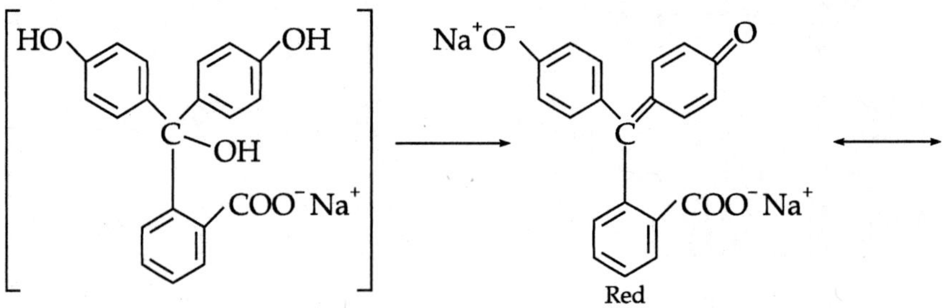

[HO, OH, C—OH, COO^-Na^+] ⟶ Na^+O^-, O, C, COO^-Na^+ ⟷

Red

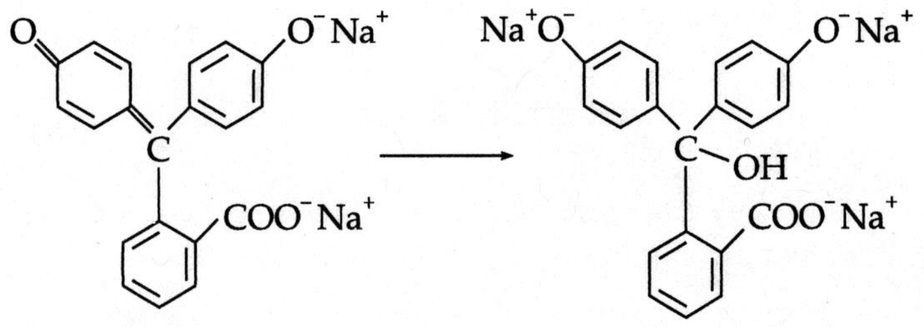

(Red) (Colorless)

Procedure:

To a mixture of pure phenol and 25 g of phthalic anhydride contained in a 250 ml round-bottomed flask, add 20 g (11 ml) of conc. sulphuric acid. Heat the flask in an oil bath at 115–120°C for 9 hours. Then, pour the reaction mixture while still hot into 1 liter of hot water contained in a 2 liter beaker, and boil until the odor of phenol disappears; add water to replace that lost by evaporation. When cold, filter the yellow, granular precipitate at the pump, and wash it with water. Dissolve the solid in a dilute sodium hydroxide solution, and filter from the undissolved residue (the byproducts of the reaction). Acidify the filtrate with dilute acetic acid and a few drops of dilute hydrochloric acid, and allow to stand overnight. The crude phenolphthalein separates as a pale yellow, sandy powder. Filter and dry this substance. Purify the crude product by dissolving it in six times its weight of absolute alcohol, and decoloring carbon and reflux on a water bath for 1 hour. Filter the hot solution through a preheated buchner funnel, wash the residue with 2 parts by weight of boiling absolute alcohol, and concentrate the combined filtrate and washing to two-thirds of its bulk on a water bath. Dilute the cooled solution with eight times the weight of cold water (it will become turbid), stir the mixture well, and after standing for a few seconds, filter through a wet filter to remove the resinous oil that separates. Heat the filtrate on a water bath to evaporate most of the alcohol; the turbidity disappears, and the phenolphthalein separates out in the form of a white powder. Filter this off and dry.

Conclusion:

Yield = 18 g

The melting point of phenolphthalein was found to be 256–258°C (pure).

Preparation of reagents:

Dilute acetic acid (5 N): Dilute 287 ml of glacial acetic acid to 1 L of distilled water.

Dilute hydrochloric acid (5 N): Dilute 430 ml of concentrated HCl to 1 L of distilled water.

EXPERIMENT QUESTIONS

1. What is carbolic acid?
2. Write the structure of phthalic acid.
3. Write the structure of phenolphthalein.
4. Why it is not advisable to take the melting point of dye?
   Ans. The dye decomposes before it melts.

EXPERIMENT 6

Aim of the experiment:

Preparation of aspirin (acetyl salicylic acid).

Apparatus required:

Conical flask
Beaker (100 ml)
Buchner funnel

Chemicals required:

Salicylic acid
Acetic anhydride
Conc. H_2SO_4

Theory:

Salicylic acid upon acetylation yields acetyl salicylic acid or aspirin. Acetylation proceeds rapidly with acetic anhydride in the presence of a little conc. H_2SO_4 as catalyst.

$$\text{C}_6\text{H}_4(\text{OH})(\text{COOH}) \xrightarrow[\text{H}_3\text{SO}_4]{(\text{CH}_3\text{CO})_2\text{O}} \text{C}_6\text{H}_4(\text{OCOCH}_3)(\text{COOH}) + \text{CH}_3\text{COOH}$$

Salicylic acid Acetyl salicylic acid

Procedure:

Take 5 g of salicylic acid and 7 ml of acetic anhydride in a 100 ml conical flask. Add 2–3 drops of conc. H_2SO_4, shake, and then heat the flask on a water bath keeping the temperature between 60° and 70°C for 20 minutes. Cool the mixture in an ice bath with stirring; add about 20 ml of ice cold water to decompose the excess of acetic anhydride. Pour the contents of the conical flask into a 250 ml beaker containing 50 g of crushed ice, and stir until a white solid appears. Scratch the side of the flask with a glass rod or stainless steel spatula. If a white solid does not appear, filter through the buchner funnel, and dry by pressing into the folds of filter paper. Recrystalize the crude acetyl salicylic acid from a mixture of equal volumes of acetic acid and water.

Conclusion:

Yield = 5 g

M.P. = 136–137°C

Acetyl salicylic acid decomposes when heated and does not possess a true clearly defined melting point.

EXPERIMENT QUESTIONS

1. What is acetylation?
2. What are the roles of acetic anhydride in this experiment?
   Ans. Acetic anhydride serves as a solvent for the reaction and converts salicylic acid to aspirin.
3. What are the acetylating agents?
   Ans. Acetyl chloride and acetic anhydride.
4. Why is the melting point of aspirin not a reliable indicator of its purity?
   Ans. Because aspirin decomposes at high temperatures.

EXPERIMENT 7

Aim of the experiment:

Preparation of a buffer solution and determination of the pH of a buffer solution.

Apparatus required:

pH meter with glass electrode and standard calomel electrode
Beaker (100 ml)

Chemicals required:

Buffer solution (pH 4 and 9)
Ammonium chloride
Ammonium hydroxide
Ammonium acetate
Glacial acetic acid

$\frac{M}{5}$ acetic acid

$\frac{M}{5}$ sodium acetate

$\frac{N}{5}$ hydrochloric acid

$\frac{N}{5}$ potassium chloride

Theory:

The modern pH meter is an electronic digital voltmeter that is scaled to read pH directly. All pH meters have a provision for standardizing the glass electrode in a buffer solution of a known pH. This is necessary because different electrodes have different asymmetry potentials. After the adjustment has been made so that the meter registers correctly the known pH of the buffer solution, the instrument gives the pH of other solutions without calculation.

Procedure:

1. Prepare the buffer solutions having pH = 4 and pH = 9 (buffer tablets BDH and E-Merck are available). These can also be prepared in the laboratory.

0.05 M solution of potassium hydrogen phthalate gives a buffer solution of pH 4.008 at 25°C; 0.05 M solution of borax gives a buffer solution of pH 9.18 at 25°C.

Ammonium chloride-Ammonium hydroxide buffer solution (pH = 10): Dissolve 70 g NH_4Cl, and add 568 ml of conc. ammonia solution having a specific gravity 0.88—0.90. Stir, and make up the volume to 1 L with distilled water.

Ammonium acetate buffer solution (pH = 5.0): Dissolve 250 g of ammonium acetate in 150 ml of distilled water. Add 700 ml of glacial acetic acid.

41.0 ml $\frac{M}{5}$ acetic acid + 9.0 ml $\frac{M}{5}$ sodium acetate (pH = 4.0)

5.3 ml of $\frac{N}{5}$ HCl + 25 ml $\frac{N}{5}$ KCl. Dilute the mixture to 100 ml with distilled water (pH = 2.0)

2. To determine pH:
   - Switch on the instrument, and allow the instrument to warm up for 10–15 minutes.
   - Wash the calomel and glass electrodes (or combined electrode) with distilled water.
   - Take the buffer solution (pH = 4) in a clean glass beaker. Lower the electrodes so that they are immersed in the solution to a depth of about 1 inch.
   - Measure the temperature of the solution by a thermometer, and set the temperature compensate control to this value.
   - Insert the electrode assembly into the same beaker, and set the selector switch (if available) of the instrument to read pH range (0–7).
   - Adjust the "set buffer" control until the meter reading agrees with the known pH of buffer solution (pH = 4).
   - Put the selector back to the zero position. Remove the electrode assembly, rinse in distilled water, dry with tissue paper, and place into a small beaker containing a second buffer solution (pH = 9). The same procedure is to be adopted, except now, the selector switch is put to a pH range of 7–14.
   - Remove the electrode assembly, rinse in distilled water followed by drying with tissue paper. Place in the first buffer solution, and confirm that the correct pH reading is shown on the meter. If not, repeat the calibration procedure.
   - If the calibration is satisfactory, rinse the electrodes, and so on with distilled water, and introduce into the test solution (buffer solutions) contained in a small beaker. Read off the pH of the solution.
   - Remove the electrodes, and so on rinse in distilled water, and leave standing in distilled water.

Conclusion:

The pH of the buffer solutions prepared is recorded as

1. ....
2. ....
3. ....
4. ....

Preparation of reagents:

Acetic acid $\left(\frac{\text{M}}{5}\right)$**:** Dilute 11.5 ml of 17.4 N and 99.5 percent glacial acetic acid to 1 L with distilled water in a volumetric flask.

Hydrochloric acid $\left(\frac{\text{N}}{5}\right)$**:** Dilute 18 ml of conc. hydrochloric acid to 1 L with distilled water, and standardize with Na_2CO_3 $\left(\frac{\text{N}}{5}\right)$.

Potassium chloride $\left(\frac{\text{N}}{5}\right)$**:** Dissolve 14.9 g of KCl in distilled water, and dilute to 1 liter.

Sodium acetate $\left(\frac{\text{M}}{5}\right)$**:** Dissolve 16.4 g of sodium acetate in distilled water, and dilute to one liter.

EXPERIMENT QUESTIONS

1. Define the term pH.
2. Which electrode is used in a pH meter universally?
   Ans. Glass electrode.
3. What is the effect of temperature on pH?
4. What are buffer solutions?
5. What is the pH scale?
6. What is the significance of pH titrations?
7. Mention the industrial applications of pH metric measurements.
8. Why is the hydrogen electrode generally not used in a pH measurement?
   Ans. Because it is difficult to set up and cannot be used in the presence of As, S, Fe^{3+}, $Cr_2O_3^-$, and so on.

EXPERIMENT 8

Aim of the experiment:

Standardization of $KMnO_4$ using sodium oxalate.

Apparatus required:

Burette
Pipette
Conical flask

Chemicals required:

Sodium oxalate (0.1 N)
Potassium permanganate solution (approx 0.1 N)
5 N sulphuric acid

Theory:

Potassium permanganate is not available in pure form and is associated with manganese dioxide, which catalyzes the auto decomposition of a permanganate solution on standing. Because of this, the concentration of $KMnO_4$ solution falls somewhat after preparation. Hence, the standard $KMnO_4$ solution cannot be prepared by exact weighing. It has to be standardized by a sodium oxalate of a known strength.

$$2MnO_4^- + 5C_2O_4^{2-} + 16H^+ \xrightarrow{60\text{–}70°C} 2Mn^{2+} + 10CO_2 + 8H_2O$$

MnO_4^- (having Mn^{7+}) is reduced to Mn^{2+} and oxalate ions are oxidized to CO_2.

$KMnO_4$ acts as a self indicator, and the appearance of a light-pink color that remains for 1–2 minutes is the end point.

Procedure:

Pipette out 10 ml of a standard sodium oxalate solution into a conical flask. Add 10 ml of 5 N H_2SO_4. Warm the solution over a water bath up to 50–60°C temperature. Now, titrate it against the $KMnO_4$ solution filled in a burette until a pale-pink color persists for 1 minute. Note the volume of $KMnO_4$ used (V_2 ml). Repeat the titration to get at least two concordant readings.

Tabulation:

No. of Observations	*Volume of Sodium Oxalate Solution (V_1)*	*I.B.R. (ml)*	*F.B.R. (ml)*	*Volume of $KMnO_4$ Used (V_2) (Difference in ml)*
1	10			
2	10			
3	10			

Calculation:

$$N_1 = \text{Normality of oxalic acid} = \frac{N}{10}$$

$$V_1 = \text{Volume of oxalic acid} = 10 \text{ ml}$$

$$N_2 = \text{Normality of } KMnO_4 = ?$$

$$V_2 = \text{Volume of } KMnO_4 = x \text{ ml (say)}$$

Applying normality equation:

$$N_1V_1 = N_2V_2$$

or

$$N_2 = \frac{N_1V_1}{V_2} = \frac{\frac{N}{10} \times 10}{x} = \frac{N}{x}$$

Conclusion:

The normality of the $KMnO_4$ solution is $\frac{N}{x}$.

Preparation of reagents:

Sodium oxalate $\left(\frac{N}{10}\right)$: Use oxalic acid $\left(\frac{N}{10}\right)$ to get better results. Dissolve 6.3 g of oxalic acid (eq. wt. 63 g) in distilled water, and dilute to 1 liter.

Sulphuric acid (5 N): Add slowly 139 ml of conc. sulphuric acid (36 N, 98%) to distilled water. After cooling, make up the volume to 1 liter with distilled water.

EXPERIMENT QUESTIONS

1. What is a standard solution?
2. What do you mean by standardization?
3. What is meant by an internal indicator, external indicator, and self indicator?

   Ans. Internal indicator: Any substance that is added to the reaction mixture to indicate the end point of the titration is called the internal indicator. Examples are phenolphthalein, methyl orange, and so on.

   External indicator: The indicator used externally is called the external indicator. A drop of the reaction mixture is removed and mixed with a drop of the indicator on a glazed tile. An example is potassium ferricyanide, $K_3[Fe(CN)_6]$, used in the titration of $K_2Cr_2O_7$ with ferrous ammonium sulphate.

   Self indicator: When one of the reactants itself acts as indicator that is called a self indicator. An example is the titrant $KMnO_4$ in the titration of oxalic acid with $KMnO_4$.
4. What do you mean by primary and secondary standards?
5. Why is $KMnO_4$ acidified with only sulphuric acid and not hydrochloric or nitric acid?

   Ans. Nitric acid is a strong oxidizing agent, and hydrochloric acid is a strong reducing agent; both react with $KMnO_4$.
6. Why is the oxalic acid solution heated before titration with $KMnO_4$?

   Ans. The reaction of oxalic acid and $KMnO_4$ is very slow. So heating to about 60–70°C is required.
7. Can we heat oxalic acid up to 100°C?

   Ans. No because oxalic acid may decompose to form CO_2 and CO.
8. What is the auto catalyst in this titration reaction?

   Ans. Mn^{2+} produced in the reaction.
9. What types of titrations are permanganate titrations?

   Ans. Redox titration.

EXPERIMENT 9

Aim of the experiment:

Determination of ferrous iron in Mohr's salt by potassium permanganate.

Apparatus required:

Burette
Pipette
Conical flask
Glazed tile

Chemicals required:

$\frac{N}{10}$ $KMnO_4$ solution
5 N sulphuric acid solution
Given Mohr's salt solution

Theory:

Potassium permanganate oxidizes ferrous iron into a ferric state in an acid medium in the cold and is itself reduced to colorless manganous ions. The estimation is based on the oxidation-reduction reaction as given here:

$$MnO_4^- + 8H^+ + 5e^- \longrightarrow Mn^{2+} + 4H_2O$$

$$[Fe^{2+} \longrightarrow Fe^{3+} + e^-] \times 5$$

$$MnO_4^- + 5Fe^{2+} + 8H^+ \longrightarrow Mn^{2+} + 5Fe^{3+} + 4H_2O$$

$$2KMnO_4 + 10FeSO_4 + 8H_2SO_4 \longrightarrow 2MnSO_4 + 5Fe_2(SO_4)_3 + K_2SO_4 + 8H_2O$$

Procedure:

Pipette out 10 ml of the Mohr's salt solution into the conical flask. Add 10 ml of a 5 N H_2SO_4 solution into this. Titrate it against a $\frac{N}{10}$ $KMnO_4$ solution filled in the burette until a pale-pink color persists for at least 1–2 minutes. Note the volume of $KMnO_4$ used (V_2 ml). Repeat the titration to get at least two concordant readings.

Tabulation:

No. of Observations	*Volume of Mohr's Salt Solution (V_1 ml)*	*I.B.R. (ml)*	*F.B.R. (ml)*	*Volume of $KMnO_4$ Solution Used (Difference V_2 ml)*
1	10			
2	10			
3	10			

Calculation:

$$N_1 = \text{Normality of Mohr's salt solution} = ?$$
$$V_1 = \text{Volume of Mohr's salt solution} = 10 \text{ ml}$$
$$N_2 = \text{Normality of } KMnO_4 = \frac{N}{10}$$
$$V_2 = \text{Volume of } KMnO_4 = x \text{ ml (say)}$$

Applying the normality equation:

$$N_1V_1 = N_2V_2$$

or $$N_1 = \frac{N_2V_2}{V_1} = \frac{\frac{N}{10} \times x}{10} = \frac{x}{100} N$$

$$\text{Strength of } Fe^{2+} \text{ ions in gm/liter} = \text{Eq. wt.} \times \text{normality}$$
$$= 56 \times \frac{x}{100} = 0.56 \times x \text{ gm/liter}$$

Conclusion:

The given Mohr's salt solution contains $0.56 \times x$ gm/liter of Fe^{2+} ion.

Preparation of reagents:

Potassium permanganate $\left(\frac{N}{10}\right)$: Dissolve 3.1607 g of $KMnO_4$ (eq. wt. 31.607 g) in distilled water. Boil for 1 hour, cool, filter, and make up the volume to 1 liter with distilled water. Standardize it with oxalic acid to know the exact strength.

Sulphuric acid (5 N): Add slowly 139 ml of conc. sulphuric acid (36 N, 98%) to distilled water. After cooling, make up the volume to 1 liter with distilled water.

EXPERIMENT QUESTIONS

1. What is redox titration? What is its significance?
2. What indicator is used in this titration?
3. What is Mohr's salt?
4. Give one example of a double salt.
5. What is the oxidation state of iron in Mohr's salt?
6. Can Fe^{2+} ions be determined with $KMnO_4$ in the presence of HCl? Explain.
   Ans. When HCl is present, some $KMnO_4$ may be consumed in the oxidation of Cl^- ions.

7. Why is Mohr's salt solution prepared in dil. H_2SO_4?
   Ans. The solution of ferrous ammonium salt (Mohr's salt) is prepared in dil. H_2SO_4 to avoid oxidation and hydrolysis.
8. Write the titration reaction of estimation of iron.
9. What is the equivalent mass of iron?

EXPERIMENT 10

Aim of the experiment:

Determination of the partition coefficient of iodine between benzene and water.

Apparatus required:

Four glass-stoppered reagent bottles
Separating funnel
Burette
Pipette (5 and 10 ml)
Conical flask

Chemicals required:

Saturated solution of iodine in benzene
$\frac{N}{10}$ and $\frac{N}{100}$ hypo solution
Potassium iodide
Starch indicator

Theory:

Iodine distributes itself in two immiscible solvents as water and benzene and has the same molecular state of iodine in both of these solvents. If C_1 and C_2 are concentrations of iodine in water and benzene layers in contact with each other and forming the heterogeneous equilibrium, then the ratio of these concentrations $\frac{C_1}{C_2}$, would be constant at a fixed temperature.

Mathematically,

$$K = \text{Partition coefficient} = \frac{C_1}{C_2} = \frac{\text{Conc. of iodine in water layer}}{\text{Conc. of iodine in benzene layer}}$$

Procedure:

1. Take four glass-stoppered bottles of about 250 ml capacity. Wash these bottles thoroughly first with a suitable reagent and then distilled water. Dry these bottles, and label them as 1, 2, 3, and 4.
2. Using a measuring cylinder, take out 10, 20, 30, and 40 ml of iodine solution in different glass bottles.
3. Add 40, 30, 20, and 10 ml of benzene in each bottle, respectively. Then, add 50 ml of distilled water in each bottle.
4. Stopper these bottles, and shake each bottle vigorously for about 20 minutes. Sufficient shaking of each bottle is necessary for accurate results.
5. Keep these bottles undisturbed for sometime so that two separated layers are formed distinctly. If droplets are found sticking to the walls of any bottle, swirl the bottle gently to bring down these droplets.
6. Pour the contents of each bottle separately into a separating funnel, and collect the benzene and water layers of each bottle in separate beakers.
7. Pipette out 10 ml of the organic layer from the separated layers of bottle no. 1 into a conical flask.
8. Add about 1 g of KI, add 5 drops of starch solution, and titrate against the standard $\frac{N}{10}$ hypo solution. The disappearance of the blue color marks the end point. Repeat this process with the organic layer of the other bottles in the same way.
9. Pipette out 10 ml of the aqueous layer of bottle no. 1 into a titration flask. Add about 1 g of KI, add 5 drops of starch solution, and titrate against the $\frac{N}{100}$ hypo solution. The disappearance of the blue color marks the end point. Repeat this process with the aqueous layer of the other bottles in the same way.

Tabulation:

No. of Observations	*% of Iodine (V/V)*	*Titration with Water Layer*			*Titration with Benzene Layer*		
		I.B.R. (ml)	*F.B.R. (ml)*	*Vol. of* $\frac{N}{100}$ *Hypo Used (Difference* V_1 *ml)*	*I.B.R. (ml)*	*F.B.R. (ml)*	*Vol. of* $\frac{N}{10}$ *Hypo Used (Difference* V_2 *ml)*
1	10						
2	20						
3	30						
4	40						

Calculation:

$$K = \frac{\text{Concentration in } H_2O \text{ layer}}{\text{Concentration in } C_6H_6 \text{ layer}}$$

$$\text{Conc. in water layer} = \frac{\text{Normality of hypo used with water layer} \times V_1}{\text{Vol. of water layer used}} \text{ g equi/L}$$

$$\text{Conc. in benzene layer} = \frac{\text{Normality of hypo used with benzene layer} \times V_2}{\text{Vol. of benzene layer}} \text{ g equi/L}$$

Values of K for each set

1. 10% solution, K = ...
2. 20% solution, K = ...
3. 30% solution, K = ...
4. 40% solution, K = ...

Conclusion:

The value of K in different sets come nearly constant. Hence, the molecular state of iodine in water and benzene is the same.

Preparation of reagents:

Sodium thiosulphate $\left(\frac{N}{10}\right)$: Dissolve 24.82 g of $Na_2S_2O_3 \cdot 5\,H_2O$ (eq. wt. 248.21) in 1 L of boiled out distilled water. Standardize it with $K_2Cr_2O_7$ to know the exact strength.

Sodium thiosulphate $\left(\frac{N}{100}\right)$: Dissolve 2.482 g of $Na_2S_2O_3 \cdot 5H_2O$ (eq. wt. 248.21) in 1 L of boiled out distilled water. Standardize it with $K_2Cr_2O_7$ to know the exact strength.

Starch indicator (1%): Prepare a paste of 1 g of powdered starch with distilled water, and add to this paste 100 ml boiled out distilled water with constant stirring. Boil for about 5 minutes and then cool.

EXPERIMENT QUESTIONS

1. Define the distribution law.
2. What is a partition coefficient?
3. What are the applications of the distribution law?

   Ans. (*a*) The molecular state of a solute in a given solvent can be ascertained.
   (*b*) Extraction of a solute from a solution by another immiscible solvent.
   (*c*) Study of complex ions.
   (*d*) Desilverization of lead.
4. Name the indicator, and describe the end point in hypo titration.
5. Why do we use different concentrations of the hypo solution in estimating I_2 in water and benzene layers?
6. What is hypo?

EXPERIMENT 11

Aim of the experiment:

Determination of the value of rate constant (K) for the hydrolysis of ethyl acetate catalyzed by hydrochloric acid.

Apparatus required:

Burette
Pipette (5 ml, 10 ml)
Conical flask (six 50 ml and one 250 ml)
Stopwatch
Thermostat

Chemicals required:

Ethyl acetate
$\frac{N}{2}$ HCl solution
$\frac{N}{10}$ NaOH solution
Phenolphthalein indicator

Theory:

The hydrolysis of ester (ethyl acetate) in the presence of dilute acid is an example of a first-order reaction as the concentration of water is in large excess and does not change significantly during the course of the reaction. This is, therefore, a pseudo-unimolecular reaction.

$$\underset{\text{Ethyl acetate}}{CH_3COOC_2H_5} + H_2O \xrightarrow{H^+} \underset{\text{Acetic acid}}{CH_3COOH} + \underset{\text{Ethyl alcohol}}{C_2H_5OH}$$

The velocity constant is calculated from the equation of a first-order reaction.

$$K = \frac{2.303}{t} \log \frac{a}{a - x}$$

where, a = Initial concentration

$(a - x)$ = Concentration after time 't'

The reaction is acid-catalyzed in the presence of HCl. Because free acetic acid is produced during hydrolysis, the reaction is investigated by titrating the acetic acid against a standard alkali at a different time 't'.

If V_o = Initial concentration, when $t = 0$

V_t = Titration after different timings

V_∞ = Final titration *i.e.*, when hydrolysis is complete ($t = \infty$)

Then, $a \propto (V_\infty - V_o)$

$(a - x) \propto (V_\infty - V_t)$

$$K = \frac{2.303}{t} \log \frac{(V_\infty - V_o)}{(V_\infty - V_t)}$$

Procedure:

Take 100 ml of $\frac{N}{2}$ HCl in a 250 ml conical flask, stopper it, and keep it in a thermostat or water bath at room temperature. Take 20 ml of ethyl acetate in a 50 ml conical flask, stopper it, and keep the flask in the same water bath or thermostat. Allow the solutions to stand for 20 minutes to attain bath temperature. In the meantime, fit the burette properly, and fill it with a 0.1 N NaOH solution. Also add 15 g of pounded ice in each of the six 50 ml conical flasks.

Pipette out 5 ml of the ester from the conical flask, and add it to the flask containing 100 ml of 0.5 N HCl.

Shake the contents, pipette out 10 ml of the reaction mixture, and transfer it immediatey to the first conical flask containing pounded ice. Titrate this against a 0.1 N NaOH solution taken in the burette after adding a drop of the phenolphthalein indicator. The appearance of a pale-pink color marks the end point. Note the titre value as V_0.

After 10 minutes of the initial withdrawal, again pipette out 10 ml of the reaction mixture, and add to the second conical flask containing ice; shake well, and titrate against 0.1 N NaOH as described previously. Note the titre value V_t after 10 minutes. Repeat the same procedure after every 10 minutes to get a total of six readings.

Finally, remove the flask from the thermostat or water bath, and place it in a separate water bath at a temperature of 50–60°C for about 1 hour. Pipette out 10 ml of the reaction mixture, and titrate against 0.1 N NaOH with phenolphthalein as the indicator. Note the titre value as V_∞.

Tabulation:

Temperature of the water bath = t °C.

No. of Observations	*Time in Minutes*	*I.B.R. (ml)*	*F.B.R. (ml)*	*Vol. of NaOH Solution Used Difference in (ml)*	$V_\infty - V_t$
1	0			V_0	$V_\infty - V_0$
2	10			V_{10}	$V_\infty - V_{10}$
3	20			V_{20}	$V_\infty - V_{20}$
4	30			V_{30}	$V_\infty - V_{30}$
5	40			V_{40}	$V_\infty - V_{40}$
6	50			V_{50}	$V_\infty - V_{50}$
7	∞			V_∞	

Calculation:

$$K = \frac{2.303}{t} \log \frac{a}{a-x} = \frac{2.303}{t} \log \frac{V_\infty - V_o}{V_\infty - V_t} \text{ minute}^{-1}$$

Calculate the value of rate constant (K) at different time intervals.

Graph:

The rate equation for the first-order reactions:

$$t = \frac{2.303}{K}\log\frac{a}{a-x} = \frac{2.303}{K}\log a - \frac{2.303}{K}\log(a-x)$$

$$= -\frac{2.303}{K}\log(a-x)$$

It indicates that a plot of t against log $(a - x)$ or log $(V_\infty - V_t)$ will be a straight line whose slope is given by $-\dfrac{2.303}{K}$.

Thus, $$-\frac{2.303}{K} = \tan\theta = -\frac{OA}{OB}$$

or $$K = 2.303 \times \frac{OB}{OA}$$

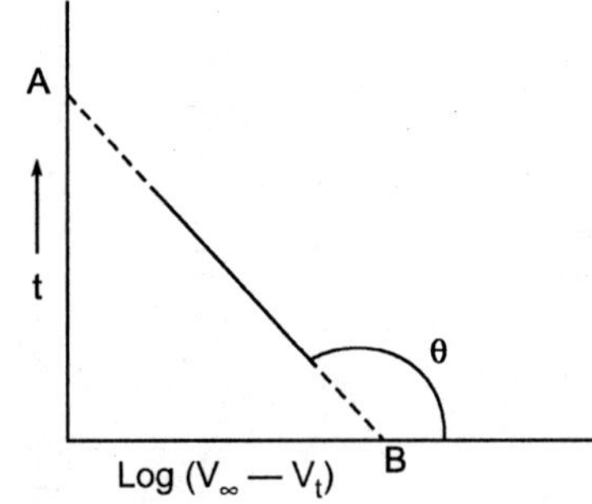

Fig. E11.1 Plot of t vs. log (V_∞ - V_t).

Conclusion:

The hydrolysis of ester (ethyl acetate) catalyzed by dilute acid is a first-order reaction because the value of K is constant, and the graph is a straight line.

Preparation of reagents:

Hydrochloric acid solution $\left(\frac{N}{2}\right)$: Dilute 45 ml of conc. HCl to 1 L with distilled water, and standardize with a $\frac{N}{2}$ sodium carbonate solution using a methyl orange indicator.

Sodium hydroxide solution $\left(\frac{N}{10}\right)$: Dissolve 4 g of NaOH in 1 L of distilled water.

Phenolphthalein indicator: Dissolve 1 g of phenolphthalein in 100 ml of distilled ethyl alcohol, and add 100 ml of distilled water with constant stirring. Filter if necessary.

EXPERIMENT QUESTIONS

1. Define the term order and molecularity of a reaction.
2. What is the significance of chemical kinetics?
3. What is the velocity of reaction?
4. Give examples of first- and second-order reactions.
5. Give the units of rate constant for first- and second-order reactions.
6. Why are third- or higher-order reactions very rare?
   Ans. Third- or higher-order reactions are very rare because in such reactions more reacting molecules should meet simultaneously, which is improbable.
7. What is the difference between rate constant and reaction rate?
   Ans. The rate constant is numerically equal to the reaction rate when the reactants are at their unit concentration.
8. What factors does the reaction rate depend on?
   Ans. Concentration and temperature.
9. How does the reaction rate vary with temperature?
   Ans. The rate of reaction increases with the increase in temperature.
10. Outline the usefulness of the study of chemical kinetics.

EXPERIMENT 12

Aim of the experiment:

Determination of the concentration of a potassium permanganate solution (given) by spectrophotometer.

Apparatus required:

UV-visible spectrophotometer
Beaker
Tissue paper

Chemicals required:

Standard solution of $KMnO_4$
Distilled water

Theory:

When an electromagnetic radiation is passed through a sample, certain characteristic wavelengths are absorbed by the sample. As a result, the intensity of the transmitted light is decreased. The measurement of the decrease in intensity of radiation is the basis of spectrophotometry. A spectrophotometer is a device that detects the percentage transmittance of light radiation when light of a certain intensity and frequency range is passed through the sample. Thus, the instrument compares the intensity of the transmitted light with that of incident light. The modern ultraviolet-visible spectrometer consists of a light source, a monochromator, a detector, an amplifier, and the recording devices (Fig. E 12.1).

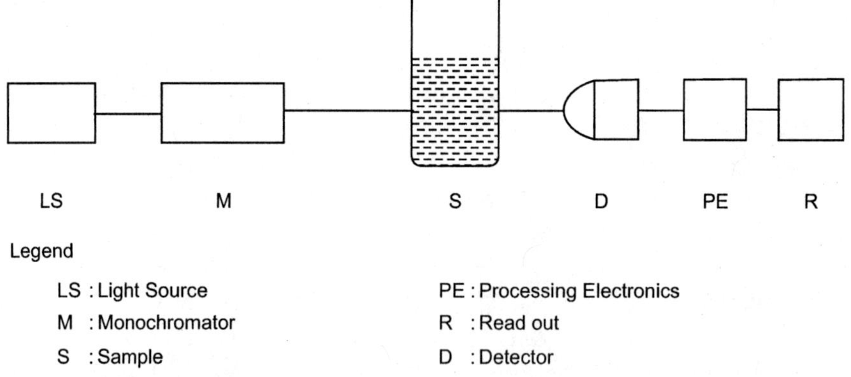

Fig. E12.1 UV-VIS spectrophotometer.

When monochromatic light falls on a homogeneous medium, a portion of the incident light is reflected, a portion is absorbed within the medium, and the remainder is transmitted.

$$I_o = I_a + I_t + I_r$$

In aqueous solutions, I_r is negligible as compared to I_a and I_t.

$$I_o = I_a + I_t$$

According to Beer Lambert's law, the intensity of the incident light is proportional to the length of the thickness of the absorbing medium and the concentration of the solution

$$I_t = I_o \,.\, 10^{-\varepsilon cl} \qquad \text{(E12.1)}$$

where 'c' is the concentration of the solute expressed in mole/L, 'l' is the length of the cell, and 'ε' is a constant characteristic of the solute called the molar extinction coefficient or molar absorptivity. Further, $\log \frac{I_o}{I_t}$ is also called optical density (OD) or absorbance (A). Because absorbance 'A' of the medium is given by

$$A = \varepsilon c l$$

Transmittance 'T' of a solution is the ratio of $\frac{I_t}{I_o}$, that is the fraction of incident light transmitted by solution.

$$T = \frac{I_t}{I_o}$$

$$-\log T = \log \frac{I_o}{I_t} = A$$

$$A = -\log T = \varepsilon c l \qquad \text{(E12.2)}$$

Procedure:

To set the instrument:

1. Ensure that the meter initially reads 0 on the transmittance scale (T). If 0 is not adjusted, set the mechanical 0 with the adjusting knob.
2. Connect the instrument to the power supply, and turn on the power switch.
3. Adjust the wavelength knob to the required wavelength region on the scale (approx.)
4. Choose the position of the wavelength switch correspondingly either to 340–400 nm or 400–960 nm.
5. Adjust the "set zero" knob so that meter needle reads 0 on the T-scale and 100 on the OD scale.

To work with the instrument:

1. Open the lid of the cell compartment, and insert a cuvette containing a blank solution, that is distilled water. Close the lid properly.
2. Adjust the control knob (set 100) to 100% transmittance or 0 optical density (OD).
3. Remove the cuvette, and close the lid tightly again. Empty the cuvette, and rinse it with the standard solution of $KMnO_4$. Fill it with the standard solution.
4. Now place the cuvette containing the standard solution in the cell compartment. Note the OD and the transmittance.

To determime λ_{max}:

1. Now change the wavelength by 20 nm, and note the absorbance (OD) and transmittance for each wavelength.
2. Plot a graph between the wavelength on the X-axis and the absorbance (OD) on the Y-axis.

To verify Beer's Law:

1. Fix the wavelength of the λ_{max} position.
2. Prepare $KMnO_4$ solutions with concentrations 0.2%, 0.5%, 1.0%, 1.5%, 2.0%, 2.5%, 3.0%, and so on (20 ml each).
3. Note the absorbance (OD) of the series of solutions of $KMnO_4$ prepared in the preceding step by the method described previously.
4. Plot a graph between the OD against the concentration. (A straight line shows verification of Beer's law.)
5. Now find the OD of the unknown solution of the $KMnO_4$. Find the concentration of this solution from the graph.

Tabulation:

TABLE E12.1 Determination of λ_{max}

No. of Observations	*Wavelength (nm)*	*Absorbance (OD)*

TABLE E12.2 Verification of Beer's Law

No. of Observations	*Concentration (C) Mole/L*	*Absorbance (OD)*

Calculation:

1. A curve is plotted between the wavelength and absorbance (OD). From the graph (Fig. E 12.2), the maximum value of A gives λ_{max}.

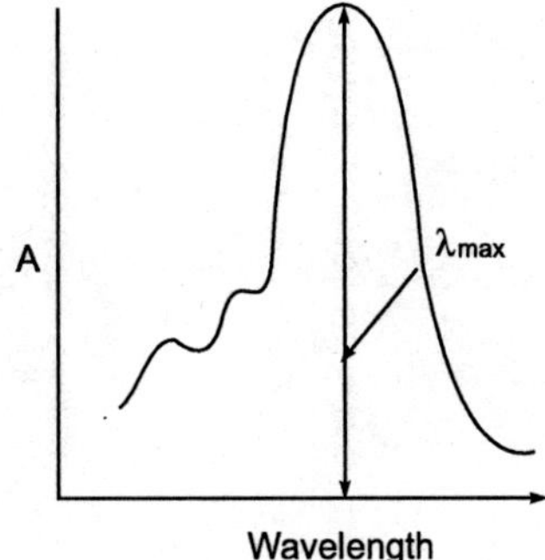

Fig. E12.2

2. A curve is plotted between the OD and concentration, and if a straight line is obtained as shown by Equation E12.2. Beer's law is verified (Fig. E 12.3).

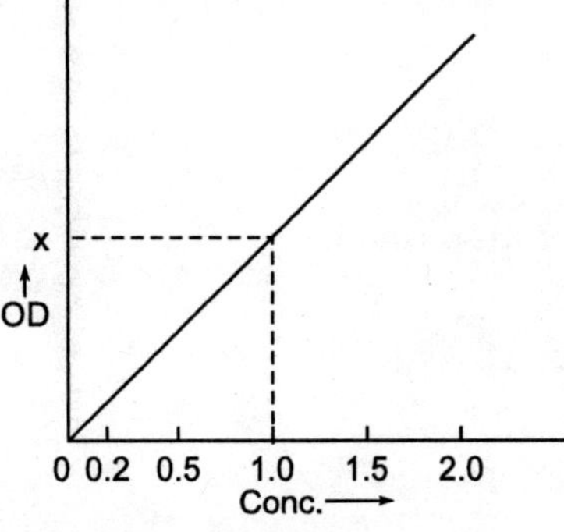

Fig. E12.3

3. From the graph of OD versus concentration, the concentration of the unknown solution can be found. For example, in Fig. E12.3, if X is the OD of unknown solution, then its concentration will be 1.0%.

Conclusion:

Concentration of the unknown solution = ... mg/L

EXPERIMENT QUESTIONS

1. What is spectrophotometry?
2. What is absorbance?
3. How you can verify Beer's law?
4. What is transmittance?
5. Discuss Lambert and Beer's law.

EXPERIMENT 13

Aim of the experiment:

Determination of dissolved oxygen in a given water sample (Winkler's method).

Apparatus required:

Glass-stoppered bottle
Burette
Pipette
Conical flask
Measuring cylinder
Glazed tile

Chemicals required:

$\frac{N}{40}$ $Na_2S_2O_3$ solution (hypo)
$MnSO_4$ (48%)
Alkaline potassium iodide
Conc. H_2SO_4
Starch

Theory:

The deficiency of oxygen in water is a kind of water pollution. Dissolved oxygen (DO) is needed for living organisms to maintain their biological processes. In the absence of a sufficient amount of dissolved oxygen in water, the anaerobic degradation of the pollutants make the water foul smelling. DO in boiler water is responsible for boiler corrosion. The DO test helps to assess the raw water quality for different kinds of water uses.

The determination of dissolved oxygen in a sample of water is based upon the oxidation of KI by the dissolved oxygen present in the water sample. The liberated iodine is titrated against a standard hypo solution using starch as the indicator. Dissolved molecular oxygen in water is not capable of reacting with KI, therefore an oxygen carrier such as manganese hydroxide is used. $Mn(OH)_2$ is produced by the action of KOH on $MnSO_4$. The $Mn(OH)_2$ so obtained reacting with dissolved molecular oxygen to form a brown precipitate of a basic manganic oxide $MnO(OH)_2$, which reacts with concentrated H_2SO_4 to liberate nascent oxygen. Nascent oxygen results in oxidation of KI to I_2. This liberated I_2 is then titrated against a standard $Na_2S_2O_3$ solution using starch as the indicator. The reactions involved are

$$MnSO_4 + 2KOH \longrightarrow Mn(OH)_2 + K_2SO_4$$

$$2Mn(OH)_2 + O_2 \longrightarrow \underset{\text{Basic manganic oxide}}{2MnO(OH)_2}$$

$$MnO(OH)_2 + H_2SO_4 \longrightarrow MnSO_4 + 2H_2O + \underset{\text{Nascent oxygen}}{[O]}$$

$$2KI + H_2SO_4 + [O] \longrightarrow K_2SO_4 + 2H_2O + I_2$$

$$2Na_2S_2O_3 + I_2 \longrightarrow Na_2S_4O_6 + 2NaI$$

Starch forms a deep-blue colored starch iodide complex when the solution turns blue. At the end point, the blue color disappears, and the solution becomes colorless. The nitrites present in water interfere with the titration because they can also liberate I_2 from KI.

$$2HNO_2 + H_2SO_4 + 2KI \longrightarrow 2NO + K_2SO_4 + 2H_2O + I_2$$

Sodium azide is used, which decomposes nitrites into nitrogen.

$$2NaN_3 + H_2SO_4 \longrightarrow \underset{\text{Hydrazoic acid}}{2HN_3} + Na_2SO_4$$

$$HNO_2 + HN_3 \longrightarrow N_2O + N_2 + H_2O$$

Procedure:

Collect 300 ml of water sample in a glass-stoppered bottle, avoiding contact with air. Add 2 ml of 48 percent $MnSO_4$ followed by 3 ml of alkaline potassium iodide reagent by means of a pipette. Stopper the bottle, and shake vigorously. Allow the brown precipitates to settle down. Now add 1 ml of conc. H_2SO_4, and mix until the precipitate is completely dissolved. Measure 100 ml of this solution by a measuring cylinder into a conical flask, and titrate slowly against a $\frac{N}{40}$ hypo solution, until the solution becomes pale yellow. Add about 2 ml of starch solution (freshly prepared), and continue the titration until the blue color disappears. Note the titre value (V ml). Repeat the titration to get at least two concordant readings.

Tabulation:

No. of Observations	*Volume of the Solution Taken in the Conical Flask (ml)*	*I.B.R. (ml)*	*F.B.R. (ml)*	*Volume of Hypo Solution Used (Difference V ml)*
1	100			
2	100			
3	100			

Calculation:

$$1000 \text{ ml of } 1 \text{ N } Na_2S_2O_3 \equiv 8000 \text{ mg of } O_2$$

$$1 \text{ ml of } 1 \text{ N } Na_2S_2O_3 \equiv 8 \text{ mg of } O_2$$

$$V \text{ ml of } \frac{N}{40} Na_2S_2O_3 \equiv 8 \times V \times \frac{1}{40} \text{ mg of } O_2$$

$$100 \text{ ml of the solution contain} = 8 \times V \times \frac{1}{40} \text{ mg of } O_2$$

$$1000 \text{ ml. of the solution contain} = 8 \times V \times \frac{1}{40} \times \frac{1000}{100} = X \text{ mg of } O_2$$

Conclusion:

The amount of dissolved oxygen in the given water sample = X ppm.

Preparation of reagents:

Sodium thiosulphate solution $\left(\frac{N}{40}\right)$: Dissolve 6.20 g of $Na_2S_2O_3$ solution in distilled water, and make up this volume to 1 L. Standardize it with a $\left(\frac{N}{40}\right)$ $K_2Cr_2O_7$ solution to know the exact strength.

Manganese sulphate solution (48%): Dissolve 480 g of $MnSO_4 \cdot 4H_2O$ in distilled water, and dilute to 1 L.

Alkaline potassium iodide solution: Dissolve 700 g of potassium hydroxide and 150 g of potassium iodide in distilled water. Cool, and dilute to 1 L.

Starch indicator (1%): Prepare a paste of 1 g of powdered starch with distilled water, and add to this paste 100 ml of boiled water with constant stirring. Boil for about 5 minutes and then cool.

EXPERIMENT QUESTIONS

1. What is the significance of determining the dissolved oxygen in water?
2. What is hypo?
3. What is starch?
4. What is an oxygen carrier?
5. What is the principle of this titration reaction?
6. What is the formula of sodium tetrathionate?
7. Why is azide added to the reaction mixture?

EXPERIMENT 14

Aim of the experiment:

Determination of the viscosity of a lubricating oil by Redwood viscometer.

Apparatus required:

Redwood viscometer
Thermometer
Stopwatch
Kohlrausch flask

Materials:

Given sample of lubricating oil

Theory:

Viscosity is one of the most important properties of any lubricating oil. Viscosity describes the suitability of the oil for lubricating purposes. A lubricant must reduce friction between the sliding parts of any machine. This avoids the direct metal to metal contact. The main criteria of a lubricant is that it should be sufficiently viscous under a high temperature and pressure exerted by the machine to adhere to the surface. If the viscosity is low, then a thin film of lubricant cannot adhere to the sliding surfaces. If viscosity is high, there will be excessive friction. The absolute viscosity of a fluid oil can be determined by measuring the rate of flow of the oil through a capillary tube kept at a uniform temperature. The specific viscosity is generally determined by measuring the time taken for a given quantity of oil to flow through an orifice or jet of standard dimension under standard conditions.

The measurement of lubricating oil viscosity is made with the help of an apparatus called a Redwood viscometer. The Redwood viscometer is commonly used for determining viscosities of thin lubricating oils. There are two types of Redwood viscometers: Redwood viscometer No. 1 and Redwood viscometer No. 2. Redwood viscometer No. 1 will correctly indicate the viscosity for a liquid having a time flow between 30 seconds and 2000 seconds. If the time flow measured with this apparatus for any oil exceeds 2000 seconds, the test should be repeated with Redwood viscometer No. 2, which will give the correct value of viscosity for such highly viscous oil.

Description of the apparatus:

Redwood viscometer consists of the following parts (as shown in Fig. E14.1):

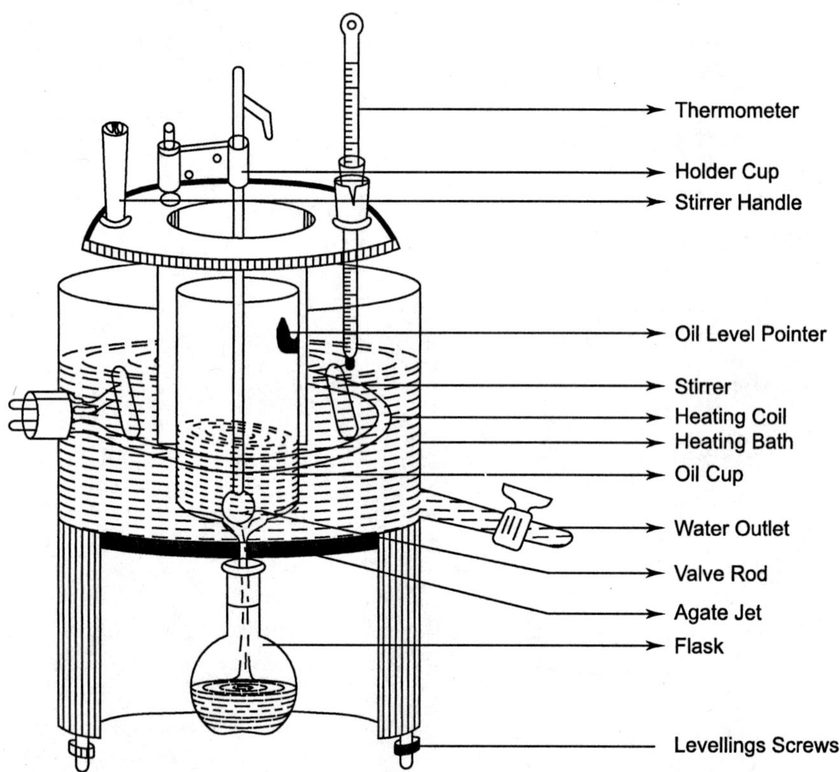

Fig. E14.1 Coss-section of the Redwood viscometer.

Oil cup: The cup is a 90 mm in height and 46.5 mm in diameter silver-plated brass cylinder. Its upper end is open, and its lower end is fitted with an *agate jet* with a bore of diameter 1.62 mm and length 10 mm. The jet can be opened and closed by a valve rod. The *valve rod* is a small silver-plated brass ball fixed to a stout wire. There is a pointer to indicate the level to which the cylinder is to be filled with oil. The *oil level pointer* is fixed on the inner side of the cylinder. The cover of the cup is filled with a *thermometer* to indicate the temperature of oil.

Heating bath: There is a cylindrical copper bath that surrounds the oil cup. This copper bath contains water. It is provided with an outlet tap to let out water and a long side tube projecting outward. This is needed to heat the bath water by means of a burner. There is a thermometer to indicate the temperature of the water.

Stirrer: The heating bath is provided with a stirrer that stirs the water in the heating bath to maintain a uniform desired temperature. The stirrer is sealed at the top to prevent water from rushing into the oil cylinder.

Spirit level: The cover of the cap is provided with a spirit level for vertical levelling of the jet.

Levelling screws: The entire apparatus rests on three legs provided at the bottom with levelling screws.

Kohlarausch flask: This flask receives the oil from the jet outlet. Its capacity is 50 ml up to the mark in its neck.

Procedure:

Accurately level the entire apparatus. Clean the oil cup with the help of a suitable solvent, and properly dry to remove any trace of the solvent. Fill the bath with water to determine the viscosity at 80°C and below. For a higher temperature, fill the bath with an oil of suitable viscosity at the test temperature. Keep the brass ball in position to seal the orifice. Now pour the supplied oil sample under test carefully into the oil cup up to the mark. Keep the Kohlarausch flask in the position below the jet. Insert a thermometer and a stirrer, and cover the lid. Keep stirring the water in the bath and oil in the oil cup, and adjust the bath temperature until the oil attains the desired constant temperature. When the temperature of the oil has become steady in the oil cup and shows a constant reading, lift the ball valves, and simultaneously start the stopwatch. Allow the oil to fill in the flask up to the 50 ml mark. Stop the stopwatch, and note the time in seconds. Replace the ball valve in position to seal the cup to prevent overflow of the oil. Again, refill the oil cup up to the mark, and repeat the experiment to get nearly reproducible results. Repeat the experiment at five elevated temperatures, say 40°C, 50°C, 60°C, 70°C, and 80°C, and note the respective time of flow as described in the following table.

Tabulation:

No. of Observations	*Temperature (°C)*	*Time of Flow (sec.) Redwood Viscosity*	*Kinematic Viscosity (Centistokes)*
1	Room temperature		
2	40		
3	50		
4	60		
5	70		
6	80		

Calculation:

The ratio of absolute viscosity to density for any fluid is known as its *absolute kinematic viscosity*. Because the instruments used are of standard dimensions, the kinematic viscosity of the oil in centistokes can be calculated from the time taken by the oil to flow through the standard orifice of the instrument with the help of the following equation.

The viscosity of the given oil sample with the help of the Redwood viscometer at t °C = ... Redwood seconds

$$V = At - \frac{B}{t} \text{ s}$$

Where, V = Kinematic viscosity of the oil in centistokes

t = Time of flow in seconds

A and B are instrument constants. The value of

A = 0.264 and B = 190, when t = 40 to 85 seconds

B = 0.247 and B = 65, when t = 85 to 2000 seconds

Conclusion:

The given sample of lubricating oil has

Redwood viscosity at t_1°C = ... Redwood seconds

Kinematic viscosity at t_1°C = ... centistokes

EXPERIMENT QUESTIONS

1. What are lubricants?
2. What are the units of viscosity?
3. What is the effect of temperature on the viscosity of liquid and gas?
4. What is kinematic viscosity?
   Ans. The coefficient of viscosity by density is called the kinematic viscosity.
5. What is the unit of kinematic viscosity?
6. Mention the names of other viscometers.
   Ans. Ostwald viscometer and Saybolt viscometer.
7. What is viscosity? Discuss its significance for a lubricant.

EXPERIMENT 15

Aim of the experiment:

Determination of the flash point and fire point of a given oil sample by the Pensky-Marten's closed-cup apparatus.

Apparatus required:

Pensky-Marten's closed-cup apparatus

Materials:

Given sample of lubricating oil

Theory:

The *flash point* is the lowest temperature at which it gives off just enough vapors to produce a combustible mixture, which will flash over the entire surfaces of the sample. After the flash point has been obtained, if the temperature is further raised, a point is reached at which the emission of inflammable vapors occurs rapidly enough to support a flame. This is called the *fire point*. The fire point temperature is generally higher by 5 to 40°C than that of the flash point. A good lubricating oil has a high flash and fire point and thus has a high working temperature. This ensures safety against fire hazards during the storage, transport, and use of the lubricating oil. If a liquid has a flash point less than 140°F, it is called a flammable liquid; those with flash points above 140°F are called combustible liquids. The Pensky-Marten's closed-cup apparatus is used for oils with flash points above 120°F.

Description of the apparatus:

The Pensky-Marten's closed-cup apparatus consists of the following parts (Fig. E 15.1):

Oil cup: The cup is about 5 cm in diameter and 5.5 cm deep. The level to which oil is to be filled is marked inside the cup. The cup lid has 4 standard size openings. Through one passes a thermometer, through the second the test flame is introduced, through the third the stirrer carrying 2 brass blades are introduced into the apparatus, and the fourth is meant for the admission of air.

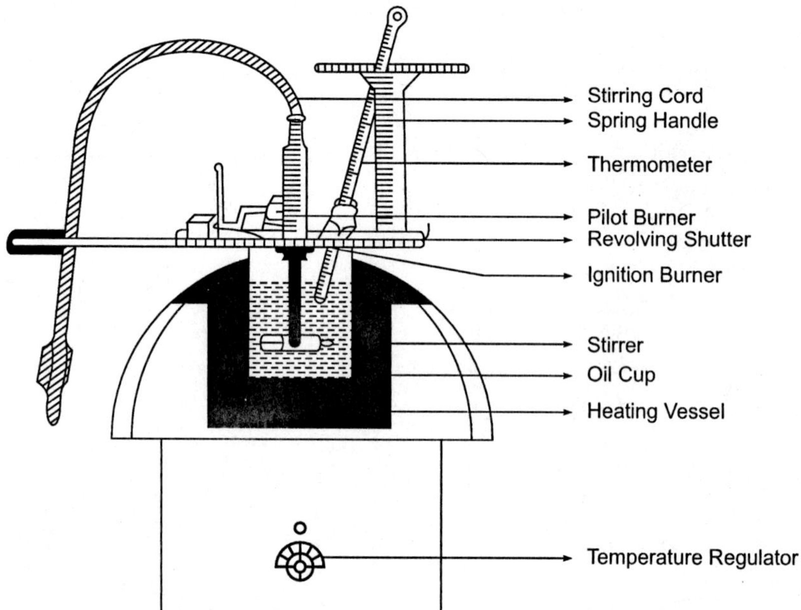

Fig. E15.1 Cross-section of the Pensky-Marten's closed-cup apparatus.

Shutter: This level mechanism is provided at the top. By moving the shutter, the opening in the lid opens, and the flame carried by the flame exposure device is introduced into the opening (Fig. E15.2). Thus, the flame can be brought over the oil surface.

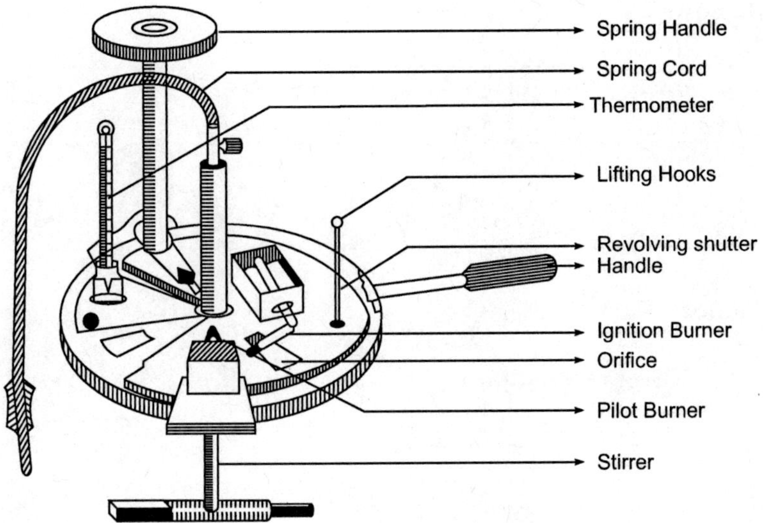

Fig. E15.2 Revolving shutter and accessories.

Flame exposure device: A small flame is connected to the shutter by a level mechanism.

Air bath: The oil cup is supported by its oil flange on an air bath heated by a gas burner.

Pilot burner: As the test flame is introduced in the opening, it gets extinguished. But when the test flame is returned to the original position, it is automatically lighted by the pilot burner.

Procedure:

Thoroughly clean and dry all parts of the cup and its accessories. Take the lubricating oil free from the suspended impurities, fill up to the mark in the oil cup, and then heat the air bath with a bunsen burner. Rotate the stirrer in the oil cup at the rate of 2 revolutions per second. Heat the oil bath to raise the temperature by 5°C in 1 minute. At every 1°C rise in temperature, introduce the test flame for a moment by moving the shutter. The temperature at which a distinct flash appears inside the cup gives the flash point of the oil. At the flash point, there is a combination of a weak sound and light. The heating is continued when the oil ignites and continues to burn for 5 seconds. The recorded temperature gives the fire point of the oil.

Tabulation:

No. of Observations	*Flash Point* (°C)	*Fire Point* (°C)
1		
2		
3		

Conclusion:

The given sample of lubricating oil has

Flash point = ... °F

Fire point = ... °F

EXPERIMENT QUESTIONS

1. Define the flash point and fire point of a lubricating oil.
2. What should be the flash point of a good lubricant?
   Ans. A flash point must be at least above the temperature at which the lubricant is to be used to avoid the risk of a fire hazard.
3. What are the factors that affect the flash and fire points?
   Ans. Moisture, vapor pressure, apparatus used, frequency of application of test flame, rate of heating the test oil, and so on.
4. What is the significance of a flash point and fire point measurement?
5. What happens to the flash point of an oil if it is contaminated with moisture?
   Ans. If moisture is present in the lubricating oil, it increases the flash point because steam prevents vapor from igniting.

Appendix

ANSWERS

CHAPTER 1

1. (*a*) 3. (*b*) 5. (*d*) 7. (*b*) 9. (*c*)
11. (*b*) 13. (*a*) 15. (*e*) 17. (*a*) 19. (*c*)
21. (*a*) 23. (*d*) 25. (*b*)

CHAPTER 2

1. (*a*) 3. (*c*) 5. (*a*) 7. (*c*)

CHAPTER 3

1. (*d*) 3. (*b*) 5. (*c*) 7. (*a*) 9. (*d*)
11. (*d*) 13. (*a*)

CHAPTER 4

1. (*d*) 3. (*b*) 5. (*a*) 7. (*a*)

CHAPTER 5

1. (*a*) 3. (*c*) 5. (*d*) 7. (*c*) 9. (*b*)
11. (*d*) 13. (*b*) 15. (*c*)

CHAPTER 6

1. (*b*)	**3.** (*b*)	**5.** (*d*)	**7.** (*b*)	**9.** (*b*)
11. (*b*)	**13.** (*c*)	**15.** (*a*)		

CHAPTER 7

1. (*d*)	**3.** (*c*)	**5.** (*b*)	**7.** (*c*)	**9.** (*c*)
11. (*c*)	**13.** (*c*)			

CHAPTER 8

1. (*b*)	**3.** (*c*)	**5.** (*a*)	**7.** (*b*)	**9.** (*d*)
11. (*d*)	**13.** (*a*)			

CHAPTER 9

1. (*b*), (*c*)	**3.** (*c*)	**5.** (*b*)	**7.** (*a*)	**9.** (*a*)
11. (*c*)	**13.** (*b*)	**15.** (*c*)		

CHAPTER 10

1. (*a*)	**3.** (*b*)	**5.** (*d*)	**7.** (*c*)	**9.** (*b*)
11. (*b*)				

CHAPTER 11

1. (*a*)	**3.** (*c*)	**5.** (*b*)	**7.** (*b*)	**9.** (*b*)

CHAPTER 12

1. (*d*)	**3.** (*c*)	**5.** (*b*)	**7.** (*b*)	**9.** (*c*)
11. (*b*)	**13.** (*b*)	**15.** (*b*)		

CHAPTER 13

1. (*c*)	**3.** (*d*)	**5.** (*a*)	**7.** (*b*)	**9.** (*c*)
11. (*a*)	**13.** (*b*)	**15.** (*b*)	**17.** (*a*)	**19.** (*d*)

CHAPTER 14

1. (*c*)	**3.** (*c*)	**5.** (*c*)	**7.** (*d*)	**9.** (*a*)
11. (*d*)				

CHAPTER 15

1. (*b*)	**3.** (*d*)	**5.** (*d*)	**7.** (*d*)	**9.** (*d*)
11. (*d*)				

INDEX

D

E

F

G

H

I

K

L

M

N

S

T

U

RECEIVED
MAR 3 1 2008